Plasma Physics

Alexander Piel

Plasma Physics

An Introduction to Laboratory, Space, and Fusion Plasmas

Prof. Dr. Alexander Piel
Christian-Albrechts-Universität Kiel
Institut für Experimentelle und
Angewandte Physik
Olshausenstrasse 40
24098 Kiel
Germany
piel@physik.uni-kiel.de

ISBN 978-3-642-43631-4 ISBN 978-3-642-10491-6 (eBook)
DOI 10.1007/978-3-642-10491-6
Springer Heidelberg Dordrecht London New York

Softcover re-print of the Hardcover 1st edition 2010

Cover design: eStudio Calamar, Girona/Spain

Printed on acid-free paper

Springer is part of Springer Science+Business Media (www.springer.com)

To Hannemarie,
Christoph and Johannes

Preface

This book is an outgrowth of courses in plasma physics which I have taught at Kiel University for many years. During this time I have tried to convince my students that plasmas as different as gas dicharges, fusion plasmas and space plasmas can be described in a unified way by simple models.

The challenge in teaching plasma physics is its apparent complexity. The wealth of plasma phenomena found in so diverse fields makes it quite different from atomic physics, where atomic structure, spectral lines and chemical binding can all be derived from a single equation—the Schrödinger equation. I positively accept the variety of plasmas and refrain from subdividing plasma physics into the traditional, but artificially separated fields, of hot, cold and space plasmas. This is why I like to confront my students, and the readers of this book, with examples from so many fields. By this approach, I believe, they will be able to become discoverers who can see the commonality between a falling apple and planetary motion.

As an experimentalist, I am convinced that plasma physics can be best understood from a bottom-up approach with many illustrating examples that give the students confidence in their understanding of plasma processes. The theoretical framework of plasma physics can then be introduced in several steps of refinement. In the end, the student (or reader) will see that there is something like the Schrödinger equation, namely the Vlasov-Maxwell model of plasmas, from which nearly all phenomena in collisionless plasmas can be derived.

My second credo as experimentalist is that there is a lack of plasma diagnostics in many textbooks. We humans have only an indirect experience of plasmas, we cannot touch, hear, smell or taste plasma. Even the visual impression of a plasma is only the radiation from embedded atoms. Therefore, we must use indirect evidence to deduce plasma properties, like density, temperature and motion. Each time my students have grasped the principle of a plasma process, I ask what we can learn about the plasma by studying this process.

In preparing this book, I have been supported by many colleagues. My special thanks go to John Goree, Thomas Klinger and André Melzer for many fruitful discussions which led to the concept of this book and for critically reading selected chapters. Holger Kersten commented on Chap. 11 and permitted photographing some of his gas discharges. Many examples in this book were taken from papers published together with my PhD students and Post-Docs, which I gratefully acknowledge (in alphabetical order): Günther Adler, Oliver Arp, Dietmar Block, Rainer Flohr, Franko Greiner, Knut Hansen, Axel Homann, Markus Klindworth, Gerd Oelerich-Hill, Markus Hirt, Iris Pilch, Volker Rohde, Christian Steigies, Thomas Trottenberg and Ciprian Zafiu. Special thanks go to John Goree and Vladimir Nosenko for the fruitful cooperation at The University of Iowa during my sabbatical leaves in 2001 and 2005. Many recent results were obtained from collaborations within the Transregional Collaborative Research Centre TR-24 *Fundamentals of Complex Plasmas*. My special thanks go to Michael Bonitz and his group.

Several colleagues made their original data available: I thank Tom Woods and Rodney Viereck for their efforts in providing the WHI Solar Irradiance Reference Spectrum, and Stephan Bosch who made his fit functions for the fusion cross sections and fusion rates accessible. Horst Wobig provided historic data from the stellarators WIIa and W7-AS. Matthias Born informed me about the mercury problem in high-pressure lamps. Permission to reproduce figures were given by André Bouchoule, John R. Brophy, David Criswell, Fabrice Doveil, John Goree, Greg Hebner, Noah Hershkowitz, Rolf Jaenicke, John Lindl, Jo Lister, Salvatore Mancuso, Richard Marsden, Bob Merlino, Gregor Morfill, Jef Ongena, and Steven Spangler. Our librarian, Frank Hohmann, was indispensible in retrieving rare literature.

The following institutions gave permission to use informations from their websites: NASA Hubble Heritage Team, NASA/JPL-Caltech, NASA/SOHO, NASA/TRACE, EFDA-JET, ITER Organization and NIF/LLNL. IPP/MPG kindly granted permissions to use figures of the Wendelstein 7-A and 7-X stellarators.

Kiel, Germany
November 2009

Alexander Piel

Contents

Acronyms

ac	alternating current
ADC	analog to digital converter
AGM	anode glow mode
CCD	charge coupled device
CME	coronal mass ejection
CRI	color rendering index
DAC	digital to analog converter
DAW	dust acoustic wave
dc	direct current
DDW	dust density wave
DIAW	dust ion-acoustic wave
DL	double layer
DS1	Deep Space-1
D–T	deuterium–tritium
EEDF	electron energy distribution function
EEPF	electron energy probability function
ES-1	electrostatic code 1
FCC	Federal Communications Committee
ICF	inertial confinement fusion
ICP	inductively coupled plasma
IRC	internal reflective coating
IRI	International Reference Ionosphere
ITER	International Thermonuclear Experimental Reactor
JET	Joint European Torus
LASCO	Large-Angle Spectrometric COronograph
MHD	magnetohydrodynamics
NASA	National Aeronautics and Space Administration
NIF	National Ignition Facility
NSTAR	NASA solar electric propulsion technology application readiness
PDP1	plasma device planar code
PIC	particle-in-cell
OML	orbital motion limit

rf	radio frequency
RIE	reactive ion etching
SMART-1	Small Missions for Advanced Research in Technology
STEREO	Solar TErrestrial RElations Observatory
SWOOPS	Solar Wind Observations Over the Poles of the Sun
TEXTOR	Tokamak Experiment for Technology Oriented Research
TFTR	Tokamak Fusion Test Reactor
TLM	temperature limited mode
TNT	trinitrotoluene
UV	ultraviolet
VCR	video cassette recorder
WIIa	Wendelstein IIa stellarator
W7-A	Wendelstein 7-A stellarator
W7-AS	Wendelstein 7 advanced stellarator
W7-X	Wendelstein 7-X stellarator

Chapter 1
Introduction

"Begin at the beginning", the King said gravely, "and go on till you come to the end; then stop."

Lewis Carroll, Alice in Wonderland

In this chapter we take a short tour through the history of plasma physics and make the reader acquainted with natural plasmas on the grand scale of the solar system, cold plasmas on the small scale of discharges, and with the hottest plasmas produced by man in experiments on controlled nuclear fusion.

In physics, the word *plasma*[1] designates a fully or partially ionized gas consisting of electrons and ions. The term plasma was introduced 80 years ago by Irving Langmuir (1881–1957) [1] to describe the charge-neutral part of a gas discharge. As his co-worker Harold M. Mott-Smith recollected later [2], "[Langmuir] pointed out that the 'equilibrium' part of the discharge acted as a kind of sub-stratum carrying particles of special kinds, like high-velocity electrons from thermionic filaments, molecules and ions of gas impurities. This reminds him of the way blood plasma carries around red and white corpuscles and germs." This shows that the relationship of Langmuir's choice of name with blood plasma was intentional.

David A. Frank-Kamenezki identified plasma as the *fourth state of matter* [3]. This view, on the one hand, alludes to the four elements of pre-Socratic Greek philosophy, Earth (solid), Water (liquid), Air (gaseous) and Fire. On the other hand, the ideas on a fourth state of matter go back to Michael Faraday (1791–1867), who, in 1809, speculated about a *radiant* state of matter he associated with the luminous phenomena produced by electric currents flowing in gases.

From a phenomenological point of view, the identification of plasma as a new state of matter can be justified because the splitting at high temperature of neutral atoms into electrons and ions is associated with a new energy barrier, the ionisation energy. Today we know that plasma is not only the hot, disordered state of matter described above. Rather, we have learned during the last 20 years that plasma systems can attain gaseous, liquid and even solid phases.

[1] The Greek verb $\pi\lambda\acute{\alpha}\sigma\sigma\varepsilon\iota\nu$ means: to form, to mould, to shape. The noun $\pi\lambda\acute{\alpha}\sigma\mu\alpha$ means figure, shape, effigy.

A. Piel, *Plasma Physics*, DOI 10.1007/978-3-642-10491-6_1,

The plasma state, as an electrically conductive medium, possesses a number of new properties that distinguish it from neutral gases and liquids. Here, one can think of the ragged shape of a lightning discharge or the magnetically confined plasma in a solar prominence. Most of the visible matter in space is in the plasma state. This is certainly true when we compare the mass of the stars with that of planets and dust regions. To be honest, dark matter (if it exists!) may take the lead in the comparison with plasmas. However, it is our human experience with the *cold* conditions on planet Earth that gives us the impression of the first three states of matter being the *natural* ones.

Our technical age is unthinkable without plasma. Plasma arc switches are used in the distribution of electric energy; high-pressure lamps illuminate our streets and serve as light sources in modern data projectors; fluorescent tubes light our offices and homes; computer chips are etched with plasma technologies; plasma-assisted deposition processes result in flat computer screens and large-area solar cells. The future energy supply may benefit from electricity produced by controlled nuclear fusion. These different phenomena can be described in a unified way by fundamental concepts.

1.1 The Roots of Plasma Physics

Surprisingly, plasma science is an old discipline of physics, although it was only named so in 1928. The roots of plasma physics are intimately related to the history of electricity [4]. Modern electricity was born in about 1600, when William Gilbert (1544–1603) described triboelectricity. One generation later, Otto von Guericke (1602–1686) invented the vacuum pump (in 1635), generated electricity with a rotating sulphur sphere (in 1663), and discovered the corona discharge at sharp-pointed tips. Another century later, in 1745, Ewald Georg von Kleist (1700–1748) and independently, in 1746, Pieter van Musschenbroek (1692–1686) invented the Leyden jar, a high-voltage capacitor. When such a Leyden jar produced a spark in air, it sounded like a gun shot, from which the terminology *gas discharge* arose. In the age of enlightenment systematic experiments were performed to study nature. In the 1770s, the famous physicist Georg Christoph Lichtenberg (1742–1799), see Portrait in Fig. 1.1, built the largest high-voltage generator of his time, an *electrophor*, that could produce more than 200,000 V. The traces of discharges on the surface of his electrophor, known as Lichtenberg figures, established the link to lightning discharges.

When high-current electric batteries became available, the electric arc was discovered, in 1803, by Vasily V. Petrov (1761–1834) and independently by Humphrey Davy (1778–1829). Such an electric arc forms when the contact between two carbon electrode tips is opened while a strong current flows. In 1831, Michael Faraday, see Portrait in Fig. 1.1, discovered electric glow discharges in rarefied gases and made systematic investigations during the next 4 years. This field was further explored by Julius Plücker (1801–1868), Johann Wilhelm Hittorf (1824–1914), and William Crookes (1832–1919), who made experiments with such low-pressure discharges.

Fig. 1.1 Georg Christoph Lichtenberg and Michael Faraday—early pioneers in gas discharge physics

Several discoveries resulted from this research on gas discharges, such as cathode rays by Hittorf, in 1869; X-rays by Wilhelm Conrad Röntgen (1845–1923), in 1895; and finally the electron by Joseph John Thomson (1856–1940), in 1897. Nicola Tesla (1856–1943), in 1891, started investigating electric discharges driven by high-frequency electric fields. In this pre-historic era of plasma physics, it was found that gas discharges involved the motion of electrons and positive ions, which represents the electric current flowing in a gas.

The discovery of collective phenomena in gas discharges, which define the modern concept of a plasma, and their proper explanation by mathematical models was left to the 20th century. The systematic investigation of the plasma state and the formulation of general laws was founded in the work of Irving Langmuir and his co-workers, during the 1920s, on gas-filled diodes [5], as well as by investigations on gas discharges with cold and hot cathodes by Walter Schottky (1886–1976) [6].

In the 1930s, many groups started with systematic studies of the plasma state. The textbooks by Alfred Hans von Engel (1897–1990) and Max Steenbeck (1904–1981) [7] and by Rudolf Seeliger (1886–1965) [8], became early classics and were extended by the review article by Mari Johan Druyvesteyn (1901–1995) and Frans Michel Penning (1894–1953) on the mechanisms in low-pressure discharges [9].

A second pillar, on which today's plasma physics rests, is *radio science*, which deals with the propagation of electromagnetic waves in the ionosphere. This field was pioneered by Edward V. Appleton (1892–1965) [10] who won a Nobel prize in 1947, Sydney Chapman (1888–1970) [11], and Appleton's colleagues at the Cavendish Laboratory in Cambridge, John Ashworth Ratcliffe (1902–1987) [12] and K. G. Budden (1915–2005) [13]. The art of predicting the conditions for short-wave radio communications was developed by Karl Rawer (1913–2003) [14] and others.

Since the mid-1950s, research on controlled nuclear fusion established the field of hot-plasma physics. Scientific questions like confinement of hot plasmas by magnetic fields and plasma instabilities became important. Lyman Spitzer (1914–1997) and Igor V. Kurchatov (1903–1960) laid the foundations of magnetically confined fusion plasmas. Progress in this field also cross-fertilized similar problems in solar and magnetospheric physics. With the availability of high-power lasers, controlled nuclear fusion was also attacked with the concept of inertial confinement. This field was shaped by many researchers which cannot be individually listed here. The quest for solving the energy problem of the 21st century remains the driving force behind fusion research.

1.2 The Plasma Environment of Our Earth

We start our grand tour through natural plasmas in the solar system. The physics of the Sun–Earth system is governed by many plasma processes and comprises nuclear reactions in the Sun's interior, plasma eruptions from the Sun's surface, a steady-state solar wind, and the interaction of the solar wind with the Earth's magnetosphere and the formation of an ionosphere.

1.2.1 The Energy Source of Stars

The most important plasma object in our space vicinity is the Sun, which provides the thermal radiation that makes the Earth habitable. Because the Sun is our nearest star, it is a well studied object and its inner mechanisms are well understood. The Sun, and the stars in general, are examples for working steady-state fusion reactors that convert protons to heavier elements and radiate the produced energy away. In stars with about one solar mass, the proton-proton cycle burns hydrogen into helium according to the main nuclear reaction chain

$$\begin{aligned} \mathrm{p}+\mathrm{p} &\rightarrow {}^{2}\mathrm{D}+\mathrm{e}^{+}+\nu_{\mathrm{e}} \\ {}^{2}\mathrm{D}+\mathrm{p} &\rightarrow {}^{3}\mathrm{He} \\ {}^{3}\mathrm{He}+{}^{3}\mathrm{He} &\rightarrow {}^{4}\mathrm{He}+2\mathrm{p} \end{aligned}$$

In each cycle there is a resulting energy of 26.21 MeV, which is available as heat while 0.51 MeV escape with the neutrino. The key features of the Sun are compiled in Table 1.1.

Between 1920 and 1950, the understanding of the inner structure of stars has grown in parallel with the development of plasma and nuclear physics. In the center of a star, the high densities and temperatures are sufficient to ignite nuclear fusion reactions. On the other hand, the produced energy keeps the interior hot to provide the pressure that balances the weight of the outer layers and prevents the collapse of the star. The transport of energy to the surface involves radiation and convection. Star spectra give us information about the surface temperature and the chemical composition of a stellar atmosphere, which are linked to the state of evolution of this star. The plasma physics of stellar atmospheres can be found in classical astrophysical textbooks, e.g. [15].

Table 1.1 Characteristics of the Sun

Mass, $m_\odot$		1.989×10^{30} kg
Radius		6.955×10^{8} m
Pressure	(center)	1.3×10^{9} bar
Temperature	(center)	15×10^{6} K
	(surface)	15,000 K
	(corona)	$(1–2)\times 10^{6}$ K
	(prominences)	(5,000–10,000) K
Luminosity		3.90×10^{26} W
Magnetic field	(polar)	$\approx 10^{-4}$ T
	(prominences)	$\approx 10^{-3}–10^{-2}$ T
	(sun spots)	≈ 0.3 T
Plasma density	(corona)	1.7×10^{14} m^{-3}
	(prominences)	$(10^{16}–10^{17})$ m^{-3}

1.2.2 The Active Sun

Already in 1616, Galileo Galilei (1564–1642) had detected dark spots on the Sun. Today, we know that these spots are the footpoints of strong magnetic fields. On a smaller scale, magnetic dipolar structures appear where a plasma-filled magnetic flux tube rises above the solar surface and forms so-called *coronal loops*. Figure 1.2 shows the light emission from coronal loops in the soft X-ray regime as observed by the Transition Region and Coronal Explorer (TRACE) satellite. TRACE is a mission of the Stanford-Lockheed Institute of Space Research and part of the NASA Small Explorer program. The magnetic fields are produced by a dynamo mechanism that takes its geometry and energy from the Sun's differential rotation.

Fig. 1.2 Coronal loops filled with hot plasma that emits in the soft X-ray regime. Observed at 17.1 nm wavelength by the Transition Region and Coronal Explorer (TRACE) satellite. (Courtesy NASA/TRACE)

Our Sun is an active star. Solar prominences are huge magnetic structures that separate from the solar surface and are filled with plasma. Prominences can last for several days and demonstrate the co-existence of a plasma with a magnetic field. Explosive emission of particles and radiation occurs in solar flares, which is the process of destroying active coronal loops. Figure 1.3 shows the evolution of a flare according to the Sweet-Parker model [16, 17]. The dipolar field of a coronal loop is partially connected to the interplanetary field. The elongated field lines contain magnetic energy, which can be released by reconnection of field lines. The plasma trapped inside the magnetic field is accelerated by the contracting field lines.

The largest explosive events on the Sun are coronal mass ejections (CMEs), which release on average 1.6×10^{12} kg of plasma moving at a speed between (200–2700) km s^{-1}. The frequency of CME events varies according to the 11-year sunspot cycle with typically one event per day at solar minimum and 5–6 events during solar maximum. As an example, the CME event of February 27, 2000 (during solar maximum conditions) is shown in Fig. 1.4. The observation was made with the Large Angle Spectrometric Coronograph (LASCO) aboard the SOHO satellite.[2] The central disk blocks direct light from the sun. The diameter of the sun is indicated by the white circle. The plasma bubble released in this CME event did not propagate towards the Earth. A new pair of satellites, NASA's Solar TErrestrial RElations Observatory (STEREO)[3] was launched in 2006 to observe, in three dimensions, plasma structures that may be heading towards the Earth. When such plasma bubbles hit the Earth's magnetosphere, magnetic storms can be triggered, which may lead to disruptions in power line grids by large induced currents and can damage communication satellites. CME's and the associated high energy particles are a major hazard for astronauts. The CME that hit the Earth on October 30, 2003, expanded the auroral zone, which (in Europe) usually has a southern boundary in

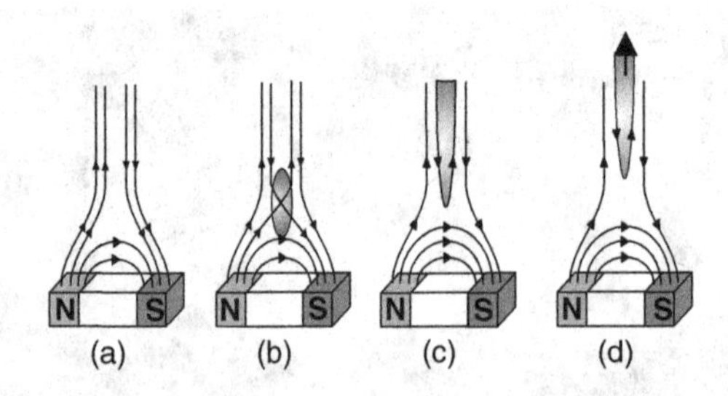

Fig. 1.3 Development of a solar flare in the Sweet-Parker model. (**a**) The dipolar field of a coronal loop connects to the interplanetary magnetic field. (**b**) By reconnection of antiparallel field lines the stress of the field lines is released. (**c, d**) The relaxing magnetic field accelerates the trapped plasma

---

[2] see: `http://lasco-www.nrl.navy.mil/`

[3] see: `http://www.nasa.gov/mission_pages/stereo/main/index.html`

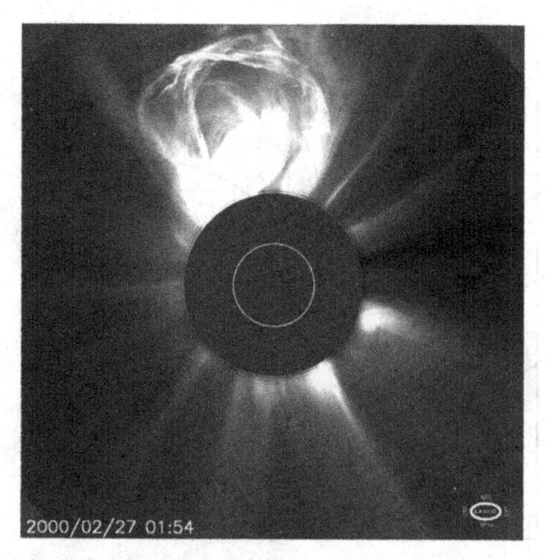

Fig. 1.4 Coronal mass ejection of February 27, 2000 as observed by the LASCO instrument aboard the SOHO satellite. (Courtesy NASA/SOHO)

mid-Scandinavia at $> 60°$ latitude, down to Lake Constance near the German–Swiss border at 47° latitude.

1.2.3 The Solar Wind

The space between Sun and Earth is filled with the plasma of the solar wind. This is a flow of charged particles from the Sun, whose existence was first conjectured, in 1908, by the Norwegian physicist Kristian Birkeland (1867–1917) [18]. Birkeland also recognized that this solar wind must comprise both positive ions and negative electrons. Ludwig Biermann, in 1951, inferred [19], that the pressure of the solar radiation on the molecules in the comet tail is by far insufficient to explain why comet tails always point away from the Sun. Rather, a solar *corpuscle flow* with velocities of the order of $10^6\,\mathrm{m\,s^{-1}}$ was necessary to deflect the comet tail. Eugene Parker [20] recognized that the solar magnetic field is "frozen" in the mass flow of the solar wind—an effect from "magneto-hydrodynamics", a novel concept introduced by Hannes Alfvén (1908–1995). Although the mass flow of the solar wind is radially outward, solar rotation shifts the footpoints of the particle flow azimuthally, which transforms a radial beam into an Archimedian spiral (Fig. 1.5). Experimental evidence of the existence of the solar wind was given by Konstantin Gringauz (1918–1993), who had designed the hemispherical retarding-potential ion detectors aboard the Soviet moon probes Luna 1 and Luna 2, which were both launched in 1959 [21].

The *quasistationary solar wind* describes plasma streams whose sources on the Sun exist for more than one day, often weeks and even months. There are two types

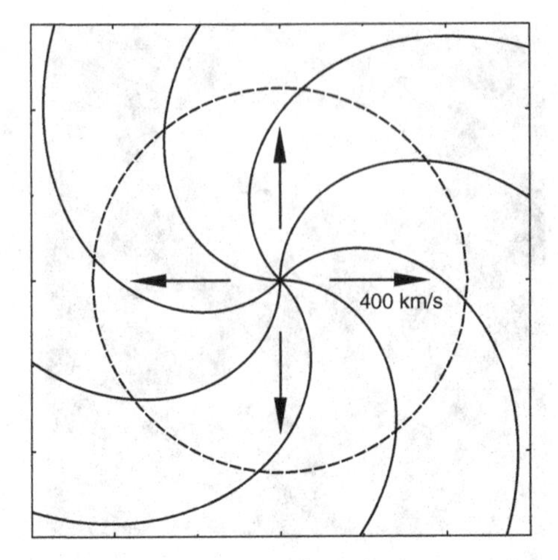

Fig. 1.5 The sun rotation shapes beams of solar wind, which emerge from distinct spots, into an Archimedian spiral. The motion of the solar wind is radially outward, but the magnetic field is trapped in the spiral arms. The Earth orbit is indicated by the dashed circle

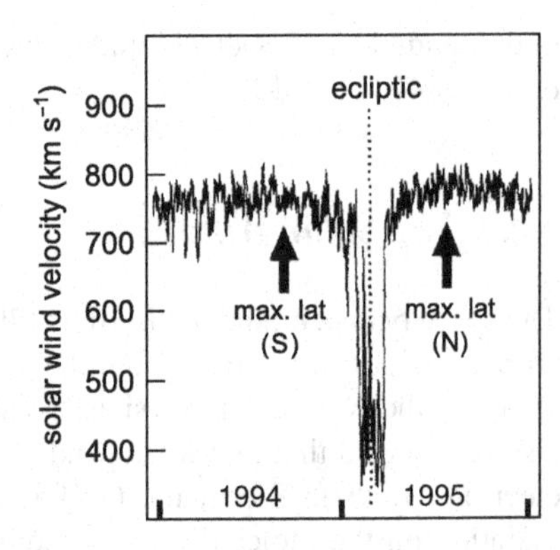

Fig. 1.6 The speed of the solar wind observed by the *Solar Wind Observations Over the Poles of the Sun* (SWOOPS) instrument aboard the ULYSSES spacecraft during its first passage. (Reprinted from [22] with kind permission from Springer Science+Business Media)

of plasma streams with distinct plasma properties, the *slow* solar wind with velocities below $450\,\mathrm{km\,s^{-1}}$ originating in the coronal streamer belt at low heliospheric latitudes, and the *fast* solar wind with velocities between 700 and $800\,\mathrm{km\,s^{-1}}$ flowing out of coronal holes at high heliospheric latitudes. These two types of solar

Table 1.2 Properties of the high-latitude solar wind [24] converted to conditions at 1 AU

Quantity	Fast SW	Slow SW	
v_{SW}	773	$\approx$ (300–500)	$\mathrm{km\,s^{-1}}$
Proton density	2.47×10^6	$(5\text{–}15) \times 10^6$	$\mathrm{m^{-3}}$
Proton temperature	1.86×10^5		K
Electron temperature	0.84×10^5		K
He^{++} ions/protons	0.044		

wind were detected by the ULYSSES[4] spacecraft during its first passage of the Sun (Fig. 1.6) [22, 23], when the 11-year solar activity cycle was at its minimum (Table 1.2).

1.2.4 Earth's Magnetosphere and Ionosphere

The interaction between the solar wind with the Earth gives rise to spectacular plasma phenomena in Nature. The Earth is protected by its magnetic field against the flow of energetic particles in the solar wind. Some of these particles can flow along magnetic field lines and hit the upper atmosphere at polar latitudes, where they cause the curtain-like aurora borealis or *Northern Lights*. These phenomena had already fascinated the Norwegian polar researcher Fridtjof Nansen (1861–1930). Nansen often illustrated his books with colored woodcuts displaying the aurora. Seen from space, the Northern Lights are located in bands forming an auroral oval about the magnetic North and South pole (see Fig. 1.7).

The magnetosphere is separated from the incoming solar wind by the bow shock. The dipolar field of the Earth is dramatically distorted by the impinging momentum flux of the solar wind and forms a long magnetotail on the night side (see Fig. 1.8). The similarity of the aurora borealis with a gas discharge was already recognized by another Norwegian physicist, Kristian Birkeland, who studied the relationship between auroral activity with fluctuations of the Earth's magnetic field.

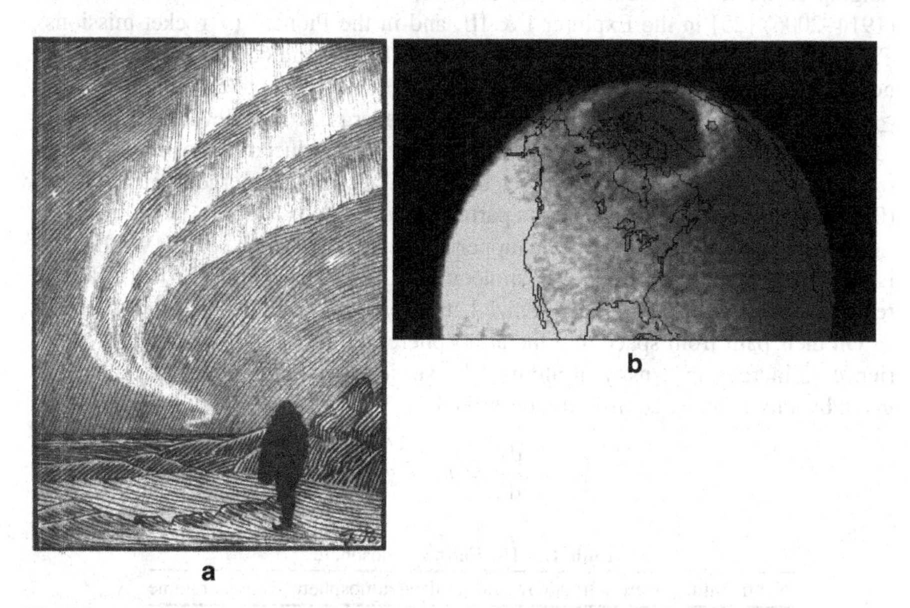

Fig. 1.7 (**a**) Aurora borealis, woodcut by Fridtjof Nansen (1911). (**b**) Auroral oval centered about the North magnetic pole as seen by the Dynamics Explorer 1 satellite (Courtesy NASA)

[4] Named after the mythical Greek seafarer Ulysses.

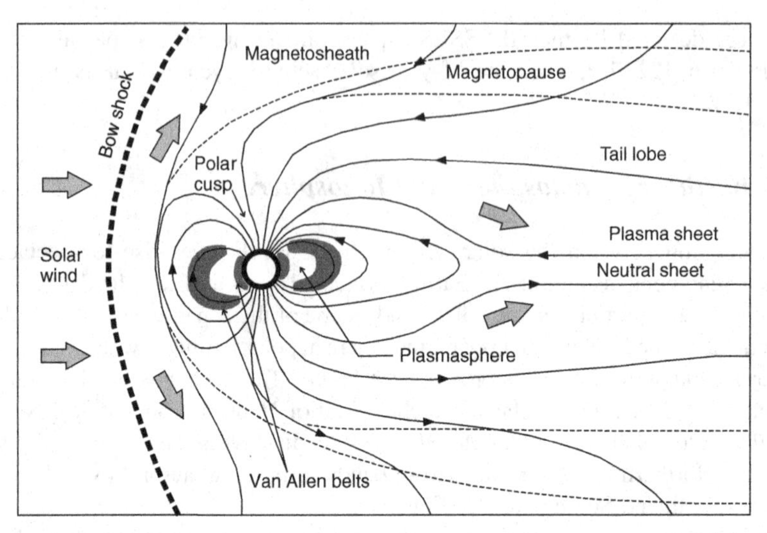

Fig. 1.8 The Earth's dipolar magnetic field is deformed into an elongated magnetsphere by the interaction with the solar wind

Active space research, beginning in the International geophysical year 1957, opened the way to new discoveries. In 1958–1959, two toroidal belts of energetic particles, ranging between 700 and 10,000 km altitude, were detected by James van Allen (1914–2006) [25] in the Explorer I & III, and in the Pioneer IV rocket missions. This inner belt and a second outer belt between 13,000 and 65,000 km altitude are now known as the *van Allen radiation belts*. The inner belt is filled with protons of $\geq$ 100 MeV and electrons of hundreds of keV energy. It is believed that the protons result from the beta decay of neutrons that are produced by cosmic rays hitting the upper atmosphere. The outer belt mainly contains energetic electrons of (0.1–10) MeV energy, protons, alpha particles and O^+ ions.

The ionosphere is that part of the upper atmosphere in which solar UV radiation is absorbed by ionizing atoms and molecules. The nomenclature for the different regions of the Earth's neutral and ionized atmosphere is compiled in Table 1.3.

On their path from space into the atmosphere, the incoming UV photons experience an increasing density of atoms. The vertical structure of the atmosphere is given by a hydrostatic equilibrium described by

$$-\frac{\mathrm{d}p}{\mathrm{d}h} = n_\mathrm{n} m_\mathrm{n} g, \tag{1.1}$$

Table 1.3 The Earth's atmosphere

Neutral atmosphere	Altitude regime	Ionized atmosphere	Altitude regime
Exosphere	> 500 km	F-layer	(120–500) km
Thermosphere	(85–500) km	E-layer	(90–120) km
Mesosphere	(45–85) km	D-layer	(50–90) km
Stratosphere	(12–45) km		
Troposphere	(0–12) km		

where $p = n_n k_B T$ is the gas pressure. For a region of constant temperature and uniform gravitational acceleration, this gives an exponential decay with altitude

$$n(h) = n_0 \exp\left(-\frac{h - h_0}{H_0}\right), \tag{1.2}$$

where $H_0 = k_B T (m_n g)^{-1}$ is the *scale height*. In Sidney Chapman's model of the ionospheric layers, the photoionization rate along the path of the UV flux at a given wavelength first rises because of the increasing atom density, goes through a maximum, but eventually dies out because nearly all photons of that wavelength have been absorbed.

This behavior can be seen in the electron density profile in Fig. 1.9a. There, the ionospheric F-layer is shown over the author's location. This density profile was calculated from the International Reference Ionosphere [26] model (IRI-2007).[5] The daytime profile for October 24, 2009 (2:00 pm local time) has a maximum density of $\approx 6 \times 10^{11}\,\text{m}^{-3}$ at 250 km altitude. In the F-layer, plasma is produced by photoionization of atomic oxygen by extreme UV photons in the 10 – 100 nm range. Note, that the maximum appears slightly above the rapid increase in atomic oxygen density (dotted curve). At night (10:00 pm local time), the density maximum is $\approx 1.4 \times 10^{11}\,\text{m}^{-3}$ at a higher altitude of 330 km. This vertical shift of the maximum is caused by the higher recombination rates at low altitudes, i.e., electrons at higher altitude have a longer time of survival after the production ended at sunset.

The E-layer between 90 and 120 km altitude is formed by photoionization of molecular oxygen by radiation in the (100–150) nm range, and by soft X-rays of (1–10) nm. The ion composition in the E-layer is mostly O_2^+ and NO^+, as shown

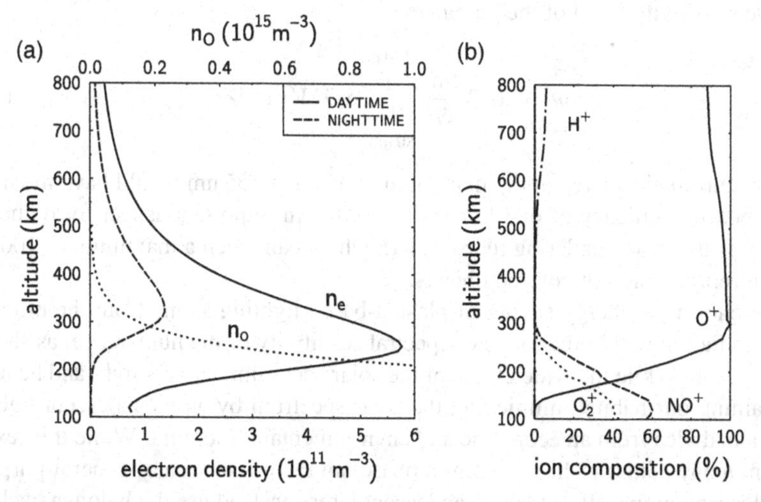

Fig. 1.9 (**a**) The electron density profile in the daytime and nighttime ionosphere given by IRI-2007 for the author's location (54.3N, 10.1E). For comparison the profile of neutral oxygen atoms from IRI-2007 is shown. (**b**) The ion composition in the ionosphere

[5] `http://iri.gsfc.nasa.gov/`

in Fig. 1.9b. At the mid-latitude location shown here, there is no clear separation between E-layer and F-layer. The D-layer is produced by the hydrogen Lyman-α line at 121.5 nm and by hard X-rays ($\lambda < 1$ nm).

1.3 Gas Discharges

Let us now switch to man-made cold plasmas that are produced by electric discharges. This is the realm of applied plasma science, which comprises fluorescent tubes, photographic flash tubes, plasma TVs, high-power arc lamps for data projectors or street illumination, and many industrial applications like etching of silicon wafers or silicon deposition on substrates for manufacturing solar cells and computer displays. They are all driven by an applied direct current (dc), alternating current (ac) or radio frequency (rf) voltage, which generates an electric gas breakdown and sustains the discharge.

1.3.1 Lighting

Lighting is one of the traditional domains for plasma applications. Electric arcs in high-pressure lamps are used for street lights and low-pressure discharges in fluorescent tubes for office and domestic lighting. In Table 1.4 various light sources are compared in terms of their efficacy given in lumens per watt electric input power.

Lumen is a unit to characterize the visible light flux Φ_v into the full solid angle 4π that originates from a radiated spectral power density $S(\lambda)$, weighted by the relative sensitivity $V(\lambda)$ of the human eye,

$$\Phi_v = 683 \frac{\mathrm{lm}}{\mathrm{W}} \int_{380\,\mathrm{nm}}^{780\,\mathrm{nm}} S(\lambda) V(\lambda)\, \mathrm{d}\lambda\,. \tag{1.3}$$

A monochromatic source at the maximum of $V(\lambda)$ at 555 nm would have the maximum possible efficacy of 683 lumens per watt. An important aspect for domestic lighting is the color rendering index (CRI), which can reach a maximum of 100 for faithful reproduction of colored objects.

The enormous energy saving of plasma-based lighting stems from the efficient use of radiation within the range of spectral sensitivity of the human eye, as shown in Fig. 1.10b. While the wide extent of the solar spectrum delivers light and heat for maintaining our habitat, mimicking the solar spectrum by an incandescent light is today a bad idea from an economic and environmental standpoint. While this text is written, many countries have begun to phase out the production of general-purpose incandescent lamps. High-tech incandescent lamps instead use the halogen cycle to diminish blackening of the glass by evaporated tungsten and save heating power by an internal coating that reflects the infrared part of the spectrum back to the filament. The efficacy reaches nearly twice that of general-purpose lamps.

Table 1.4 Comparison of the efficacy and colour rendering index (CRI) of various light sources

Lamp type	lm/W	CRI	Source
General-purpose incandescent lamp	9–15	100	a
Low-voltage halogen	12–19	100	a
Halogen with internal reflective coating	17–24	100	a
Compact fluorescent lamp (stick)	46–61	82	a
Compact fluorescent lamp (spiral)	60–67	82	a
Light emitting diodes (warm tone)	66	90	b
Light emitting diodes (cool)	105	70	b
T8 tube with electronic ballast	80–100	80–89	c
High-pressure mercury lamp	65–85	25–50	d
High-pressure sodium lamp	90–135	15–25	d
Ceramic metal halide lamp	65–115	70	b

[a]Jacob [27], [b]Philips data sheet, [c]Osram data sheet, [d]Report [28]

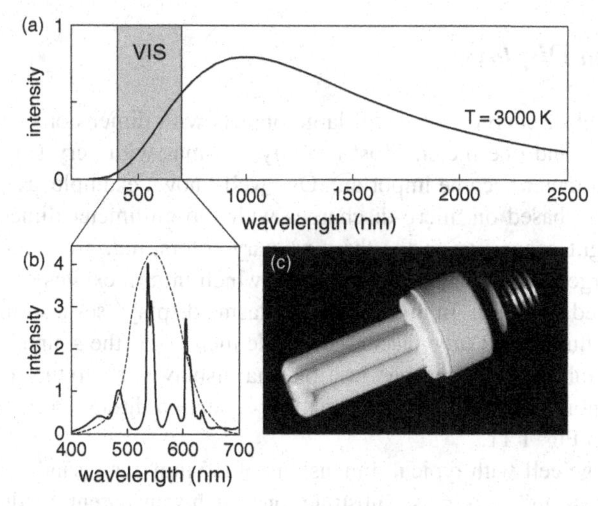

Fig. 1.10 (**a**) The spectrum of an incandescent lamp is represented by black-body radiation at $T =$ 3000 K. The shaded rectangle marks the visible spectral range. (**b**) The spectrum of a fluorescent tube with a modern tri-phosphor coating (solid line) in comparison with the eye-sensitivity curve $V(\lambda)$ (dashed line). (**c**) Compact fluorescent lamp

Most energy-efficient plasma-based light sources use the fact, that about 80% of the electric power of a low-pressure discharge in mercury vapour can be transformed to ultraviolet light, which can then be converted to visible light by fluorescent materials. Early fluorescent tubes used the mercury spectral lines at 435 nm and 546 nm in combination with the fluorescence of a halophosphor coating of the inner tube wall that contributed to the yellow and red part of the spectrum. Modern tri-phosphor coatings, see Fig. 1.10b, are better targeted to the eye-sensitivity curve $V(\lambda)$.

For office lighting, the CRI should be greater than 80. For domestic applications, customers prefer values greater than 90. Light-emitting-diodes are an emerging technology that has just reached break-even with fluorescent lamps regarding efficacy and color rendering and may overtake fluorescent lamps in some applications. This is true for back-lighting of computer screens and foreseeable for domestic applications. For street lights, efficacy was formerly of higher priority than color rendering. Nowadays, urban lighting is beginning to benefit both from higher efficacy and better color rendering by replacing high-pressure mercury lights by improved metal halide lamps. For high-power stadium lighting with more than 1 kW per luminaire there is presently no alternative to plasma lamps.

From an environmental point of view, the pros for a lighting technology based on mercury lamps lie in its efficacy and the reduced carbon footprint, the cons in the toxicity of mercury. The ban of mercury in plasma lamps for car headlights has already stimulated alternative mercury-free plasma lightsources [29], which, in the near future, may also replace high-pressure lamps for street lighting.

1.3.2 Plasma Displays

So far, technical discharges were still large objects with dimensions between several centimeters and one meter. Most recently, plasmas with very small scales of less than a millimeter became important. One well-known example are plasma displays. These are based on micro-discharges with sub-millimeter dimensions. The principle of light generation in the three primary colors, red, green and blue, is a plasma discharge that produces UV radiation, which in turn excites a phospor that emits the desired spectrum. In this sense, the plasma display uses a similar chain of processes as a fluorescent tube discharge. While mercury is the source of UV light in fluorescent tubes, the filling gas in a plasma display is a mixture of neon and xenon with xenon delivering the UV radiation. A section through a plasma display cell is shown in Fig. 1.11.

The discharge cell with typical dimensions of 0.5 mm has a sandwich structure with a front glass and back glass substrate, on which transparent conducting elec-

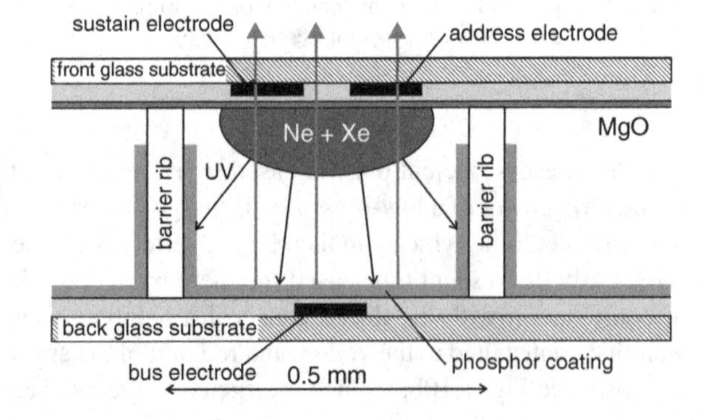

Fig. 1.11 A discharge cell in a plasma display. The conducting electrode layers and the MgO coating are transparent. The electrodes on the front glass substrate are actually oriented at right angle to form an address matrix in combination with the bus electrode

trodes are printed that form rows and columns of a display matrix. The cells can be addressed by applying a voltage pulse that is applied between the address electrode and bus electrode. After the discharge has fired a smaller voltage on a sustain electrode maintains the discharge. The discharge electrodes are not in contact with the plasma but embedded in a dielectric layer. Therefore, the discharge current is a displacement current that flows only for a short period and generates a short discharge flash. The number of subsequent flashes determines the brightness of that pixel.

The cathode of this discharge is made of a thin layer of magnesium-oxide. This material has the unique property that one impinging ion creates more than 30 secondary electrons that maintain the discharge. In this way, the discharge cell can be operated very efficiently at a very low discharge voltage ($\approx$ 95 V). Neighboring cells are separated by glass barrier-ribs. Three neighboring cells of different color form a pixel.

1.4 Dusty Plasmas

So far we have discussed the elementary mechanisms in a gas discharge and the way of their technical application. Let us now shortly digress to a field, where the complexity of plasma systems is the driver of research.

Dust in space is the material from which stars and planets are formed. The huge amount of dust in a galaxy can be seen in an edge-on view of the sombrero galaxy M104 and the distribution in the spiral arms becomes evident from a face view of the Whirlpool galaxy M51, as shown in Fig. 1.12. The collapse of a dust cloud, often triggered by strong stellar winds from nearby star clusters or a supernova, leads to the formation of newborn stars in protostellar disks.

In our planetary system, electrically charged dust is found in Saturn's rings. While the details of dust charging in the ring system are still under investigation, the collective interaction of the dust can be directly observed. During the passage of the Voyager 2 spacecraft in 1981, unexpected radial structures (spokes) were found in the B-ring (Fig. 1.13), which appear dark in backscattered light [30].

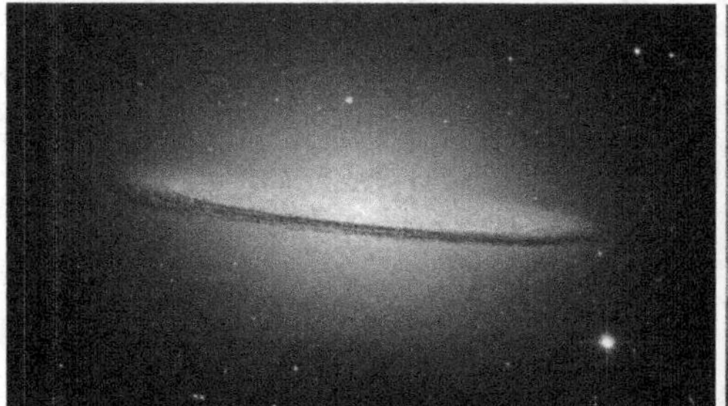

Fig. 1.12 (*left*) The Sombrero galaxy M104, seen edge-on by the Hubble Space Telescope, reveals huge amounts of dust in the galactic plane. (*right*) The Whirlpool galaxy M51 gives a face view that displays the dust distribution in the spiral arms. (Courtesy NASA and Hubble Heritage Team)

Fig. 1.13 Dark radial "spokes" were observed in Saturn's B-ring during the fly-by of the Voyager 2 spacecraft [30]. These structures are attributed to a collective motion of electrically-charged fine dust particles. (Courtesy NASA/JPL-Caltech)

After passing Saturn, Voyager 2 observed the same structures as bright features in forward scattered light. This is a clear hint that the spokes consist of micrometer-sized dust. These spokes are typically 10,000 km long and 2000 km wide and do not follow the Kepler motion of the ring particles.

One of the currently assumed models [31] assumes that the dust is electrostatically levitated above the ring plane. An initial transient event, such as a meteoritic impact or a high-energy auroral electron beam, could create a short-lived dense plasma that charges the boulders in the main ring to a negative potential. Dust grains on the surface of the boulder collect an extra electron and are repelled from the surface. Subsequently they leave the dense plasma cloud and are found in the ever-present background plasma environment. The spokes are now under detailed investigation by the Cassini spacecraft, which is in an orbit about Saturn.

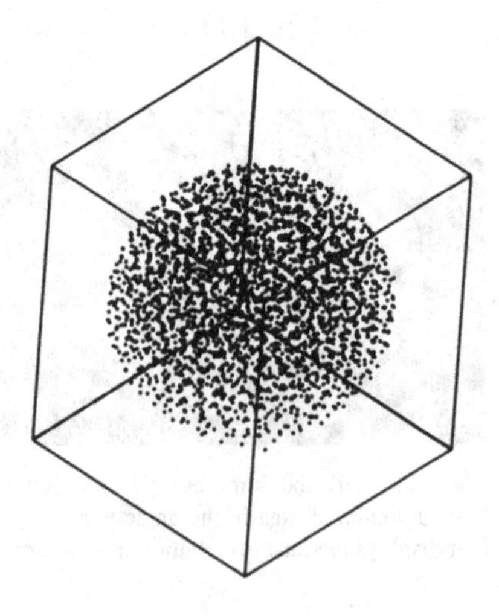

Fig. 1.14 Yukawa ball: a three-dimensional plasma crystal of charged dust particles with unusual shell structure. The image shows the positions of the dust particles obtained by scanning videomicroscopy

The field of dusty plasmas has grown rapidly since the 1990s. Dust charging, interaction forces, wave phenomena and phase transitions were studied. Dusty plasmas in the laboratory showed new physics, like the formation of two-dimensional [32–34] and three-dimensional plasma crystals [35] or spherical Yukawa balls [36], see Fig. 1.14. The high attractivity of this field of investigations lies in the high transparency of the dust clouds and the slow motion of the dust particles, which can be traced with fast video cameras. This is one of the rare occasions, where plasma phenomena can be studied by simultaneously observing the many-particle system at the "atomic level".

1.5 Controlled Nuclear Fusion

Our tour through plasma science finally returns to the hot plasmas of the stars. But now we are interested how to mimic the conditions in the interior of stars by hot plasmas confined in fusion reactors. Research on controlled nuclear fusion promises an energy source that could provide the worlds growing energy demand in the 21st century and beyond.

In the cold-war era after World War II, research on nuclear energy was done within secret programs. In the United States, the astrophysicist Lyman Spitzer (1914–1997) began building a *stellarator* device at Princeton University. Richard F. Post (1918–) was setting up a *mirror machine* at the University of California's Livermore laboratory. In the Soviet Union, the tokamak concept was introduced by Igor Tamm (1895–1971) and Andrei Sakharov (1921–1989). In 1956, the Soviet research on controlled nuclear fusion was unilaterally disclosed to Western scientists by Igor V. Kurchatov (1903–1960). In short time, the road to a peaceful use of nuclear fusion energy opened in 1958 at the *2nd Atoms for Peace Conference* in 1958, when scientists from around the world were allowed to share their results and laid the foundation for "one of the most closely collaborative scientific endeavours ever undertaken" [37]. The common goal of all these attempts is to use the energy resulting from the fusion of deuterium and tritium nuclei to operate a power plant. The reaction channels and associated energies are compiled in Table 1.5.

A significant yield of fusion reactions can only be expected at such kinetic energies of the fusion partners that overcome the Coulomb repulsion between the like-charged nuclei. Figure 1.15 shows the fusion cross sections as a function of the particle energy in the center-of-mass system. The figure uses the tabulated values from [38, 39]. The fusion reactions set in between 10 and 100 keV energy. Moreover, the cross section for the D–T reaction is found, at the same energy, much larger than that of the D–D or ^{3}He–D reaction. This is the reason why all present experiments for igniting a fusion reaction use D–T mixtures. Actual concepts for obtaining nuclear

Table 1.5 Fusion reactions of the hydrogen isotopes

$^2D + {}^2D \rightarrow {}^3T + p +$	4.0 MeV
$^2D + {}^2D \rightarrow {}^3He + n +$	3.3 MeV
$^2D + {}^3T \rightarrow {}^4He + n +$	17.6 MeV
$^2D + {}^3He \rightarrow {}^4He + p +$	18.3 MeV

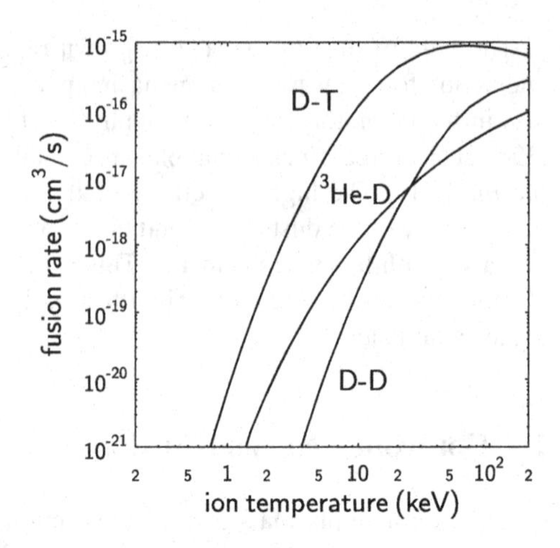

Fig. 1.15 The cross section for D–T, D–^{3}He and D–D fusion reactions as a function of the center-of-mass energy. The D–D cross section is the sum of both reaction channels

fusion are either based on magnetically-confined hot plasmas in so-called tokamak or stellarator devices, or on heating small pellets containing deuterium and tritium with ultra-intense laser beams.

1.5.1 A Particle Accelerator Makes No Fusion Reactor

Why can't we simply operate a particle accelerator as a fusion reactor? Obviously, today it is no big technical problem to accelerate ions to (0.1–1) MeV. Let us assume that we shoot a beam of tritium ions with the optimum energy into a solid target of deuterium ice, which may be a cube of 1 cm edge length that contains roughly 5.4×10^{19} deuterium atoms (Fig. 1.16). The probability p of hitting one of these target atoms is the ratio of the blocked area to the cross section of the cube, i.e., $p = 2.7 \times 10^{-4}$. This means, however, that 99.97% of the projectiles have *not* performed a fusion reaction. Let us further assume that the tritium beam represents an electric current of $I = 1\,\text{A}$, which is quite substantial at 100 keV energy. Then the cube is hit by $\mathrm{d}N_T/\mathrm{d}t = I/e = 6.3 \times 10^{18}$ tritium ions per second (e is the elementary charge). The product of this hit rate with the reaction probability and the fusion energy of 17 MeV gives a respectable fusion power of 4.6 kW per cubic centimeter. However, will this ion beam be able to penetrate a solid deuterium ice cube? Unfortunately, *no*. The interaction of the tritium ion beam with the electrons

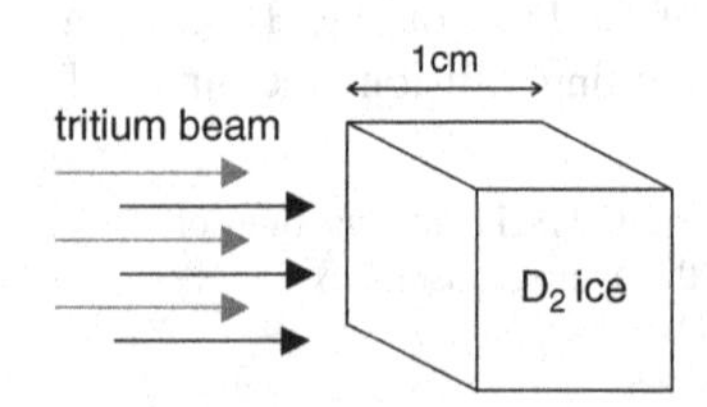

Fig. 1.16 Cartoon of the deuterium ice-cube with an impinging tritium ion beam

of the densely packed deuterium atoms leads to a rapid energy loss, which is of the order of 4×10^5 eV cm^{-1}. Hence, the initial energy of 100 keV will be completely lost as heat within the ice cube. Since no ion energy is left on the exit side, we would have to replenish the ion energy at a rate of 100 kV $\times$ 1 A = 100 kW, which is much more than we would gain from fusion.

This is why nuclear fusion uses a different concept. The trick is that the heat becomes not lost energy for the fusion processes. The *magnetic confinement fusion* approach starts with a hot gaseous plasma containing deuterium and tritium ions. Collisions between D^+ and T^+ ions, which do not lead to fusion, only scatter the collision partners but do not alter the heat content of the hot plasma. Admittedly, there is an energy leak by means of radiation losses (*Bremsstrahlung*), which are generated during the scattering process. However, different from the accelerator concept, where energy is dissipated in microseconds, the particle energy of the fusion partners in the hot plasma can be contained for fractions of a second. This is necessary to compensate for the lower density of the gaseous medium.

The other approach, *inertial confinement fusion* (ICF), which will be touched in Sect. 1.5.6, achieves nuclear fusion in a highly compressed D–T target that has a density of 300 g cm^{-3}, about 1500 times the density of D–T ice. The plasma is confined, for a short time of the order of a nanosecond, by its own inertia. This concept was originally developed by John Nuckolls, in 1957, before the invention of the laser. A full concept using lasers to compress the plasma was published in 1972 [40]. Alternatively, heavy-ion beams were suggested for ICF [41].

1.5.2 Magnetic Confinement in Tokamaks

The hot D–T plasma in a fusion device may be dilute but the energy yield from fusion will be substantial when we can confine the particles and their kinetic energy for a sufficiently long time. Such confinement can be achieved by means of strong magnetic fields in a so-called *tokamak* device. Then, each projectile has many repeated chances to collide with a fusion partner and the fusion yield is increased accordingly. A cut-away view of a tokamak is shown in Fig. 1.17a. First of all, a tokamak is a huge transformer, in which the plasma torus forms a single secondary winding. Therefore, the first impression of a tokamak comes from the iron-yokes of the transformer. The plasma itself is contained in a toroidal vacuum chamber, which is surrounded by magnetic field coils for the confinement of the charged particles. The principles of particle confinement and the reason for choosing a tokamak geometry will be outlined in Chap. 3.

1.5.3 Experiments with D–T Mixtures

While operating a tokamak as a power plant is still an ambitious goal for the near future, some important milestones on this road can already be considered as history.

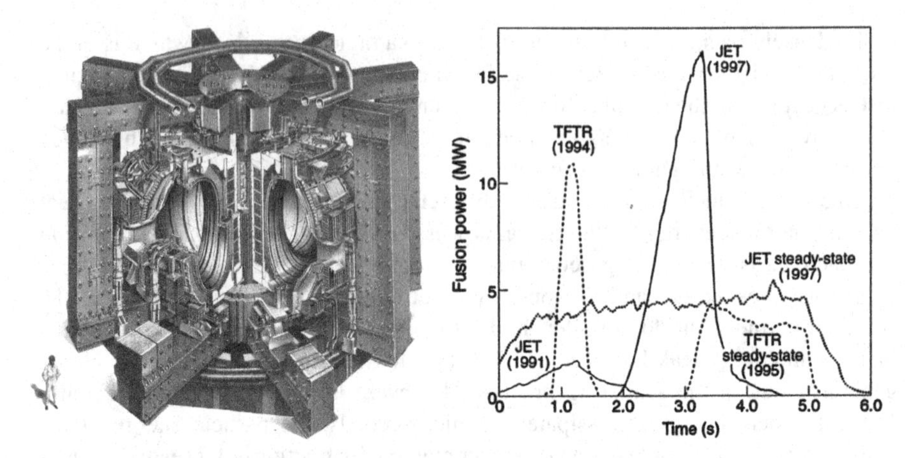

Fig. 1.17 (**a**) The JET tokamak is 12 m high and has a D-shaped plasma cross-section and a total plasma volume of 80–100 m^3 (Image: EFDA-JET). (**b**) Fusion power development in the D-T campaigns of JET and TFTR. (Graphic: EFDA/JET. Reprinted with permission from [42]. © 2006, American Nuclear Society)

In the 1990s, experiments with D–T mixtures were performed on the Tokamak Fusion Test Reactor (TFTR) at Princeton Plasma Physics Lab, USA, and on the Joint European Torus (JET) at Culham, UK, shown in Fig. 1.17a. These experiments aimed at demonstrating a thermonuclear fusion plasma close to the *break-even* point where the power production from nuclear fusion becomes comparable to the heating power of the plasma. Such experiments became feasible after the *high-confinement regime* (H-regime) of tokamak operation was discovered [43–45].

A preliminary fusion experiment with 10% tritium and 90% deuterium had been performed on JET in 1991 resulting in a fusion power output of about 1.7 MW [46]. Between the end of 1993 and the beginning of 1997, TFTR has been routinely operated in high-confinement D–T discharges resulting in a maximum fusion power output of 10.7 MW [47, 48]. A second D–T experimental campaign was performed on JET, in 1997, which resulted in the demonstration of a near-breakdown operation at $Q = P_{\text{fusion}}/P_{\text{heating}} = 0.62$ transiently and a maximum output power of 16 MW. Figure 1.17b shows a comparison of the fusion power development in the JET and TFTR D–T experiments [42].

1.5.4 The International Thermonuclear Experimental Reactor

The large-scale fusion experiments of the 1990s could only be performed on the scale of a big economy, like the TFTR in the USA, JET in Europe, or JT-60 in Japan. The next larger fusion reactor, however, requires joint efforts on a world scale. Plans to establish such a device date back to 1985. Since 1988 a planning group of 50 physicists and engineers from Europe, Japan, the former Soviet Union, and the United States worked on the design of a test reactor, which was presented in December 1990. The detailed planning for an *International Thermonuclear Experimental Reactor* (ITER) began in 1992 and the design report was presented to the ITER council in 1998. Because of budget constraints in the member states, the ITER design had to be cut back. In 2005 the ITER partners, which were joined by South Korea and China in 2003, decided the location for ITER to be Cadarache, France.

ITER is designed to deliver a fusion power of 500 MW. An artistic cut-away rendering of the device is shown in Fig. 1.18. Its magnet system comprises 18 superconducting toroidal and 6 poloidal field coils. The magnets are cooled with liquid helium at 4 K. The toroidal magnetic field can reach a maximum of 11.8 T and represents a total magnetic energy of 41 GJ. Besides the vacuum vessel, the magnetic field coils will be the biggest components with a total weight of 6540 tons. The mechanical and operational parameters are compiled in Table 1.6 [49].

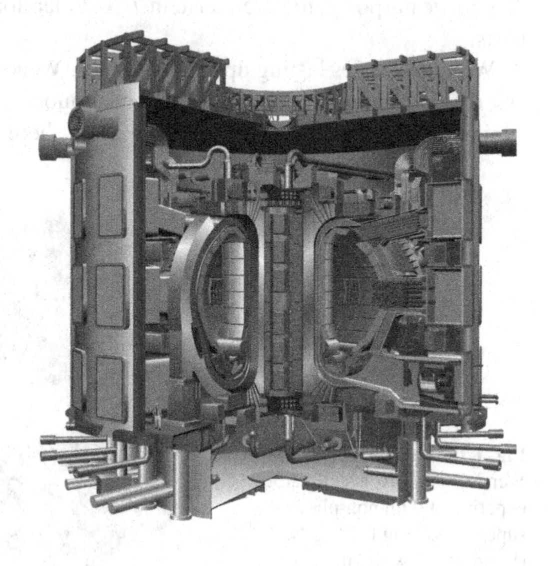

Fig. 1.18 Conceptual design of the International Thermonuclear Experimental Reactor (ITER). At a total height of 30 m, ITER is nearly 3 times larger than JET. The D-shaped vacuum vessel is surrounded by superconducting magnetic field coils. (Reproduced with permission. © ITER Organization)

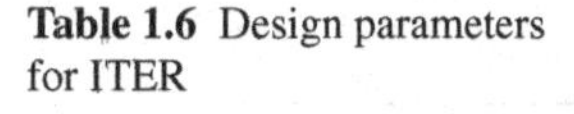

Table 1.6 Design parameters for ITER

Total radius	10.7 m
Total height	30 m
Plasma radius	6.2 m
Plasma volume	840 m^3
Plasma mass	0.5 g
Magnetic field	5.3 T
Maximum plasma current	15 MA
Heating power and current drive	73 MW
Fusion power	500 MW
Energy gain	10
Mean plasma temperature	2×10^8 K
Steady operation	> 400 s

1.5.5 Stellarators

There are two roads towards a fusion reactor with magnetic confinement, the tokamak and the stellarator. Most of the world's devices today are tokamaks. While part of the magnetic confinement in tokamaks is achieved by a strong electric current in the plasma, stellarators form the magnetic cage only by means of external field coils. Stellarators are therefore suitable for continuous operation. The plasma current in a tokamak is produced by a transformer, which limits operation to a pulsed mode unless other means of driving the plasma current are installed.

The Wendelstein 7-X stellarator, which is under construction in Greifswald, Germany, has the objective to show that stellarators are fundamentally suitable for operation of a power plant. After completion, it will be the world's largest stellarator-type fusion experiment (see Fig. 1.19 and Table 1.7). With 30 m^3 plasma volume it is still a small experiment compared to ITER's 840 m^3. The main plasma heating methods will be microwave, neutral particle and radio-frequency heating. For heating and for diagnostic purposes, the Wendelstein 7-X stellarator is equipped with more than 250 ports.

With discharges lasting up to 30 minutes, Wendelstein 7-X is to demonstrate the essential advantage of the stellarator: continuous operation with plasma conditions similar to those in the ITER tokamak. But there is no intention of going for an

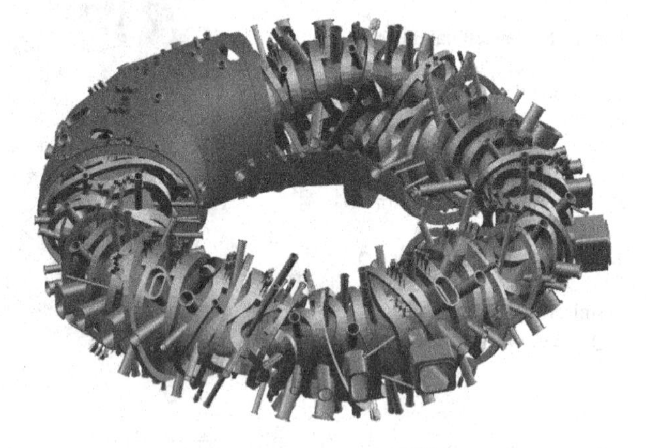

Fig. 1.19 Conceptual view of Wendelstein 7-X, a stellarator experiment with non-planar superconducting field coils. (Reproduced with kind permission. © IPP/MPG)

Table 1.7 Essential data of the Wendelstein 7-X stellarator

Major plasma radius	5.5 m
Minor plasma radius	0.53 m
Magnetic field	3 T
Discharge time	30 min
Plasma heating power	15 MW
Plasma composition	H, D
Plasma volume	30 m^3
Plasma temperature	$< 10^8$ K
Plasma density	$< 3 \times 10^{20}$ m^{-3}
Energy confinement time	0.15 s

energy-yielding plasma. This objective is reserved for ITER. If Wendelstein 7-X can experimentally confirm the good properties predicted from model calculations, a future demonstration power plant could also be based on the stellarator concept.

1.5.6 Inertial Confinement Fusion

Todays largest experiment for inertial confinement fusion is the *National Ignition Facility* (NIF), located at Lawrence Livermore National Laboratory in California [50]. While the primary goal of the NIF belongs to the National Security Program, various basic science objectives have been defined: (a) to demonstrate fusion ignition and energy gain for exploring laser fusion as a future energy source and (b) to generate matter under extreme pressures and temperatures for studying astrophysical processes in the laboratory, which are occurring in stars and supernovae.

With its 192 laser beams (Table 1.8), a nearly spherically-symmetric illumination of the target can be achieved. The laser beam lines are housed in two separate *laser bay* buildings, each 200 m long. The UV laser light is produced by frequency-tripling of the infrared neodymium-glass laser wavelength of 1053 nm. The laser beams end up in a 10 m-diameter target chamber and hit the inner wall of a small gold cylinder of about 5 mm diameter and 10 mm length with the 2 mm diameter D–T *pellet* at the center (Fig. 1.20). Construction of NIF began in 1997. Operation

Table 1.8 Design parameters for the NIF

Number of beam lines	192
Energy per beam line	≈ 10 kJ
Laser wavelength	351 nm
Pulse width	3.5 ns
Total energy per pulse (design)	1.8 MJ
Total power (design)	500 TW

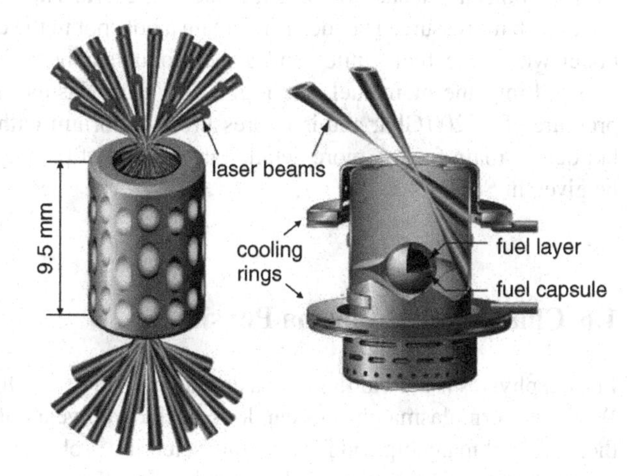

Fig. 1.20 Gold cylinder (*hohlraum*) for conversion of the incoming laser energy to X-rays. The hollow capsule of D–T ice is suspended in the center. The hohlraum is cooled to 18 K temperature. (Credit: Lawrence Livermore National Laboratory and Department of Energy)

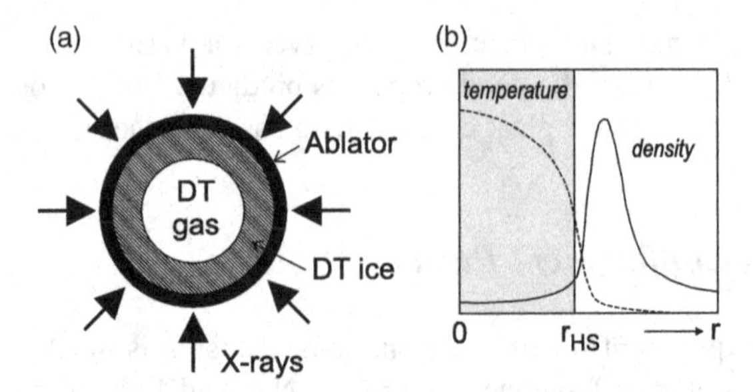

Fig. 1.21 (**a**) One of the present pellet designs for the NIF consists of a plastic hollow sphere with D–T ice condensed to the wall and a D–T gas filling of 30 bar pressure. The arrows indicate the X-rays heating up the ablator. (**b**) Development of a hot spot in the center of the compressed pellet

of a stack of four beam lines, between 2002 and 2004, demonstrated that the lasers are able to deliver an energy of 10 kW each. Combined operation of all 192 lasers began in February 2009, starting at a reduced test level of 80 kJ. Science experiments are scheduled for 2010 at a tenfold higher pulse energy.

The principle of laser fusion is inertial confinement. The fusion reaction has to burn a considerable part of the D–T fuel before the hot plasma has expanded. The ignition condition for laser fusion requires that the D–T gas be compressed to a density of about 1000–2000 times the density of D–T ice [51, 52]. Such a compression is achieved by shining intense laser radiation (direct drive) or X-rays (indirect drive) on a hollow sphere containing the D–T fuel. At typical intensities of $10^{15}\,\mathrm{W\ cm^{-2}}$, the outer ablation layer of the hollow sphere rapidly evaporates and, by momentum conservation, the fuel is driven inward. The capsule behaves as a spherical ablation-driven rocket. The pressure generated by the ablation is 100 Mbar or greater. Presently, indirect drive is preferred because of the achieved high homogeneity of compression.

One of the presently discussed pellet designs for the NIF uses an ablator made of hydrocarbons and a shell of D–T ice, see Fig. 1.21a. The volume is filled with D–T gas at 30 bar pressure. The idea is to create a hot spot in the center of the compressed pellet where the fuel ignites and a thermonuclear burn front propagates radially outward into the main fuel, see Fig. 1.21b. The plasma in the hot spot reaches a pressure of $\simeq$ 200 Gbar and is in pressure equilibrium with the surrounding colder but denser main fuel. A more detailed discussion of the physics of laser fusion will be given in Sect. 4.4.2.

1.6 Challenges of Plasma Physics

Plasma physics is a vivid field of basic and applied research in many subdisciplines. While modern plasma physics can look back on 80 years of success in many fields, there are still many big and far-reaching scientific problems that pose a challenge for the next young generation of plasma scientists. Some of these big and fascinating questions have been compiled in a recent analysis of the status and perspectives of plasma science [53].

- **Space and astrophysical plasmas**
  - What are the origins and the evolution of plasma structures throughout the magnetized universe?
  - How are particles accelerated throughout the universe?
  - How do plasmas interact with non-plasmas?
- **Low temperature plasmas**
  - How can plasmas be used in the next generation of energy-efficient light sources?
  - How can plasma methods be optimized for purifying drinking water and for other environmental problems?
  - To which extent can new materials or advanced nanoparticles and nanowires be tailored by plasma processes?
- **Plasma physics at high energy densities**
  - Can we achieve fusion ignition and, eventually, useful fusion energy from compressed and heated fusion plasma?
  - Can we generate, using intense short-pulse lasers, electric fields in the multi-GeV/cm range for accelerating charged particles to energies far beyond the present limits of standard accelerators?
  - Can we better understand some aspects of observed high-energy astrophysical phenomena, such as supernova explosions or galactic jets, by carrying out appropriately scaled experiments?
- **Basic plasma science**
  The fields of basic research at the present forefront of plasma science are:
  - Non-neutral plasmas and single-component plasmas
  - Ultracold neutral plasmas
  - Dusty plasmas
  - Laser produced and high energy density plasmas
  - Microplasmas at atmospheric pressure
  - Plasma turbulence and turbulent transport
  - Magnetic fields in plasmas
  - Plasma waves, structures and flows

1.7 Outline of the Book

Before starting with the physics of plasmas, some words about using this book are necessary.

Chapters 2–7 cover the typical subjects of introductory courses to plasma physics. These chapters can be used in parallel with an introductory course. Chapters 8–11 address more specialized topics covered in advanced courses. Each chapter concludes with a brief summary of the basics, which may help the reader to recapitulate the essentials of that chapter and may be helpful in preparing an exam. Problems at the end of each chapter (with worked-out solutions in the appendix) help delving deeper into the subject.

Starting from the definition of the plasma state (Chap. 2) in terms of quasineutrality and shielding, single particle motion (Chap. 3) in different field geometries or in time-varying fields is used as a first step into the realm of collisionless plasmas. A bottom-up approach is chosen to make the reader familiar with the sometimes strange behavior of plasmas without using too advanced mathematical methods. The basics of magnetic confinement illustrate the single particle model.

Chapter 4 introduces the concepts for collisional plasmas. The intention is to make the beginner familiar with the full span of plasma physics, which covers the cold collisional plasmas of gas discharges or the ionosphere as well as the hot collisionless plasmas in astrophysics or in fusion devices. Consequently, this chapter illustrates electron heating and ambipolar diffusion in the positive column of a gas discharge. At the same time, the heat balance of hot plasmas is used to introduce the concepts of magnetic and inertial confinement fusion. The question of cross-field currents is discussed in technical terms for an ion Hall thruster.

Fluid models and magnetohydrodynamics are introduced intuitively in Chap. 5. The differences between single particle drifts and fluid models is exemplified by the diamagnetic drift. The central concepts of isobaric surfaces, magnetic pressure and field line tension are introduced. The consequences of frozen-in magnetic fields are discussed for Alfvén waves and for the solar wind.

Plasma waves are discussed in Chap. 6 with a strong emphasis on a wide variety of diagnostic applications in laboratory or natural plasmas. The concept of a plasma as a dielectric medium is the guiding motive of this chapter.

Chapter 7 introduces the role of plasma boundaries and discusses space-charge sheaths. These concepts find immediate application for diagnostics in terms of Langmuir probes. Space-charge limited flow is a central concept with applications to ion extraction in thrusters.

The basic concepts of plasma instabilities are described in Chap. 8. The beam-plasma instability is discussed at some length in terms of plasma normal mode analysis to prepare the later discussion of Landau damping in the subsequent Chapter. The method of finding unstable modes is transferred to the Buneman instability. Macroscopic instabilities in current carrying plasmas and Rayleigh-Taylor modes are briefly discussed.

Chapter 9 uses a mathematically simplified introduction to the Vlasov model and Landau damping. The relationship between kinetic theory and fluid models is discussed. The physical processes behind Landau damping are presented in detail to resolve the paradox of collisionless damping. A complementary kinetic description by particle simulations is used to trace the instabilities into the non-linear regime where trapping occurs. Virtual cathode oscillations are discussed as an example for current interruption by space charge accumulation.

The new field of complex (dusty) plasmas is surveyed in Chap. 10 with emphasis on such phenomena that have no counterpart in classical plasmas, such as charge variability, ion drag force or plasma crystallization.

The final Chap. 11 makes the reader familiar with plasma generation. For this purpose a brief introduction is given to plasma discharge mechanisms in low-pressure dc discharges, parallel-plate rf discharges and inductively coupled plasmas.

Chapter 2
Definition of the Plasma State

"I can't tell you just now what the moral of that is, but I shall remember it in a bit."
"Perhaps it hasn't one", Alice ventured to remark.
"Tut, tut, child!" said the Duchess, "Everything 's got a moral, if only you can find it."

Lewis Carroll, Alice in Wonderland

The plasma state is a gaseous mixture of positive ions and electrons. Plasmas can be fully ionized, as the plasma in the Sun, or partially ionized, as in fluorescent lamps, which contain a large number of neutral atoms. In this section we will discuss the defining qualities of the plasma state, which result from the fact that we have a huge number of charged particles that interact by electric forces. In particular we will see that the plasma state is able to react in a *collective* manner. Therefore, the plasma medium is more then the sum of its constituents.

2.1 States of Matter

Before going deeper into definitions of the plasma state, let us recall the characteristic properties of a neutral gas. A gas is characterized by the number of particles per unit volume, which we call the *number density* n. The unit of n is m^{-3}. The motion of the particles (in thermodynamic equilibrium) is determined by the *temperature* T of the gas. In an ideal gas, the product of number density and temperature gives the pressure, $p = nk_{\mathrm{B}}T$, in which k_{B} is Boltzmann's constant.

We will use the same terminology for plasmas, but in the plasma state we have a mixture of two different gases, light electrons and heavy ions. Therefore, we have to distinguish the electron and ion gas by individual densities, n_{e} and n_{i}. Moreover, plasmas are often in a non-equilibrium state with different temperatures, T_{e} and T_{i} of electrons and ions. Such two-temperature plasmas are typically found in gas discharges. The solar plasma (in the interior and photosphere), on the other hand, is a good example for an isothermal plasma with $T_{\mathrm{e}} = T_{\mathrm{i}}$.

Plasmas exist in an environment that provides for a large number of ionization processes of atoms. These can be photoionization by an intense source of ultraviolet radiation or collisional ionization by energetic electrons. Impact ionization is the dominant process in gas discharges because of the ample supply of energetic

A. Piel, *Plasma Physics*, DOI 10.1007/978-3-642-10491-6_2,

Table 2.1 Ionization and recombination processes

$e + A \rightarrow A^+ + 2e$	Collisional ionization
$h\nu + A \rightarrow A^+ + e$	Photoionization
$A^+ + 2e \rightarrow A + e$	Three-body recombination
$A^+ + e \rightarrow A$	Two-body recombination

electrons. Photoionization is found in space plasmas where the electron and atom densities are low but a large number of ultraviolet (UV) photons may be present. These processes and their reciprocal processes can be written in terms of simple reaction equations, as summarized in Table 2.1.

Besides recombination by these volume processes, electrons and ions can effectively recombine at surfaces, which may be the walls of discharges or embedded microparticles. In thermodynamic equilibrium, each of these volume processes is balanced by the corresponding reciprocal process (i.e., photoionization and two-body recombination, or impact ionization and three-body recombination.) Because the ionization energy of neutral atoms lies between 3 and 25 eV, plasmas produced by impact ionization typically exist at high temperatures. Photoionized plasmas require short wavelength radiation, typically in the UV region. There are also situations, in hot and dilute plasmas, where collisional ionization is efficient but electrons are too few for three-body recombination. Then, a steady state can be reached, in which two-body recombination balances the impact ionization. The solar corona is an example for such a plasma that is in a non-thermodynamic equilibrium.

As a final remark, it is worth mentioning that some plasmas are not governed by local equilibria but by non-local processes. The properties of the solar wind at the Earth orbit, for example, are mostly determined by the emission process at the Sun's surface and by heating processes (e.g., shocks) during the propagation from Sun to Earth. We will see in Chap. 11 that a negative glow is also produced by electrons that have gained their energy at a different place.

2.1.1 The Boltzmann Distribution

Before we discuss the thermodynamic equilibrium of a plasma in more detail, it is meaningful to recall some elementary concepts of classical statistical mechanics. There, the relative population of different energy states is regulated by the Boltzmann factor. The relative population of the energy states W_i and W_k[1] is given by

$$\frac{n_i}{n_k} = \frac{g_i}{g_k} \exp\left(-\frac{W_i - W_k}{k_B T}\right) . \qquad (2.1)$$

g_i and g_k are the degeneracies of the states i and k, i.e., the number of substates with the same energy. The exponential of the form $\exp(-W/k_B T)$ determines how

[1] To avoid confusion with the electric field E we denote energies by the symbol W.

many atoms have overcome the energy barrier $W_i - W_k$ between the states i and k. Another example for a Boltzmann distribution is the Maxwell-Boltzmann velocity distribution of free particles

$$f_{\mathrm{M}}(v_x, v_y, v_z) = \frac{1}{Z} \exp\left(-\frac{m(v_x^2 + v_y^2 + v_z^2)}{2\,k_{\mathrm{B}}T}\right). \tag{2.2}$$

Z is a normalization factor. Here, the distribution of velocities is determined by the kinetic energy $W = m(v_x^2 + v_y^2 + v_z^2)/2$. The Maxwell distribution will be discussed in more detail in Sect. 4.1.

2.1.1.1 Derivation of the Boltzmann Distribution

The derivation of the Boltzmann distribution from statistical mechanics is given here for completeness. This paragraph may be skipped at first reading of this section.

We start with the concept of entropy, which attains a maximum value in a thermodynamic equilibrium. Already in 1866, the Austrian physicist Ludwig Boltzmann (1844–1906) introduced the logarithmic dependence of entropy on probability. The entropy of a classical system of N particles, having a total energy U, which can populate its different energy states W_i with N_i particles, is defined by

$$S = -k_{\mathrm{B}} \sum_i n_i \ln n_i\,. \tag{2.3}$$

Here, $n_i = N_i/N$ is the relative population of the energy state W_i. Letting S take a maximum value, we must take care of the constraining conditions

$$g(n_i) = \sum_i n_i W_i = U \quad \text{and} \quad h(n_i) = \sum_i n_i = 1\,. \tag{2.4}$$

A maximum with constraints is found by the method of Lagrange multipliers, which requires

$$\frac{\partial S}{\partial n_i} = \lambda \frac{\partial g}{\partial n_i} + \mu \frac{\partial h}{\partial n_i}\,. \tag{2.5}$$

Herefrom we immediately obtain

$$\begin{aligned} -\ln n_i - 1 &= \lambda W_i + \mu \\ n_i &= \exp\{-\mu - 1 - \lambda W_i\}\,. \end{aligned} \tag{2.6}$$

The two Lagrange-multipliers λ and μ are determined using the constraint $h = 1$

$$1 = \mathrm{e}^{-\mu-1} \sum_k \mathrm{e}^{-\lambda W_k}\,, \tag{2.7}$$

which gives

$$\mathrm{e}^{-\mu-1} = \left(\sum_k \mathrm{e}^{-\lambda W_k}\right)^{-1} \tag{2.8}$$

and finally

$$n_i = \frac{1}{Z}\,\mathrm{e}^{-\lambda W_i} \quad \text{and} \quad Z = \sum_k \mathrm{e}^{-\lambda W_k}\,. \tag{2.9}$$

The normalizing factor Z is called the partition function. The other Lagrange multiplier is found from the thermodynamic relationship $1/T = \partial S/\partial U$ and yields $\lambda = (k_\mathrm{B}T)^{-1}$ (cf. Problem 2.6). Then the relative population of the energy states is given by the Boltzmann distribution

$$n_i = \frac{1}{Z}\exp\left(-\frac{W_i}{k_\mathrm{B}T}\right). \tag{2.10}$$

The exponential $\exp(-W_i/k_\mathrm{B}T)$ is the *Boltzmann factor* and the Boltzmann distribution over energy states (2.1) follows immediately.

2.1.2 The Saha Equation

The Boltzmann factor Eq. (2.1) describes the distribution of the internal states of an atom or the free states of the Maxwell-Boltzmann gas. Now we seek for a thermodynamic description of the equilibrium between atoms and ions.

Thermal equilibrium conditions of a plasma are typically found in the interior of stars or in the electric arc discharges used for street and stadium illumination. The thermodynamic equilibrium state is characterized by the detailed balancing of each process with its reciprocal process. Here we consider the balance of electron impact ionization and three-body recombination

$$\mathrm{e} + \mathrm{A} \rightleftharpoons \mathrm{A}^+ + 2\mathrm{e}\,, \tag{2.11}$$

which can be quantified by the balance of the reaction rates

$$n_\mathrm{e}\, n_\mathrm{A}\, S(T) = n_\mathrm{e}^2\, n_{\mathrm{A}^+}\, R(T)\,. \tag{2.12}$$

Here, n_e is the electron density, n_A the neutral atom density and n_{A^+} the ion density. The rate coefficients, $S(T)$ for ionization and $R(T)$ for three-body recombination are only dependent on temperature. (Rate coefficients will be discussed in more detail in Sect. 4.2.3.) Therefore, Eq. (2.12) can be rearranged into a *mass action law*

$$\frac{n_e n_{A^+}}{n_A} = \frac{S(T)}{R(T)} =: f_{Saha}(T)$$
$$f_{Saha}(T) = \frac{2Z_{A^+}}{Z_A} \exp\left(-\frac{W_{ion}}{k_B T}\right) . \tag{2.13}$$

The function f_{Saha} was derived in 1920 by the Indian astrophysicist Megh Nad Saha (1893–1956) [54]. It contains the exponential $\exp(-W_{ion}/k_B T)$ that determines the probability to overcome the energy barrier W_{ion} and the *partition functions* $Z_A = \sum_k g_k \exp(-W_k/k_B T)$ of the atom [cf. Eq. (2.9)] and Z_{A^+} of the ion. The factor 2 is the degeneracy of the free electrons, which have two distinguishable spin states. The exponential in the Saha function has obvious similarities with the Boltzmann factor.

The Saha-equilibria between different ionization stages of an atom can be described in a similar manner. An example for the resulting ionization states of a free-burning argon arc discharge is shown in Fig. 2.1. Because this electric arc is operated in pressure equilibrium with the ambient air, the calculation was performed at constant pressure rather than at constant atom number. Ionization reaches a few percent at $T > 10,000$ K and full single ionization is established at $\approx 20,000$ K, where also the onset of double-ionization A^{++} is observed. Converting temperature to energy units $k_B T$, the onset of ionization in argon occurs at about 1 eV (corresponding to 11,600 K, see Sect. 4.1.3).

The high operating temperature of an arc discharge lamp leads to a high energy efficiency, because the maximum of the Planck curve for black-body radiation, as given by Wien's displacement law

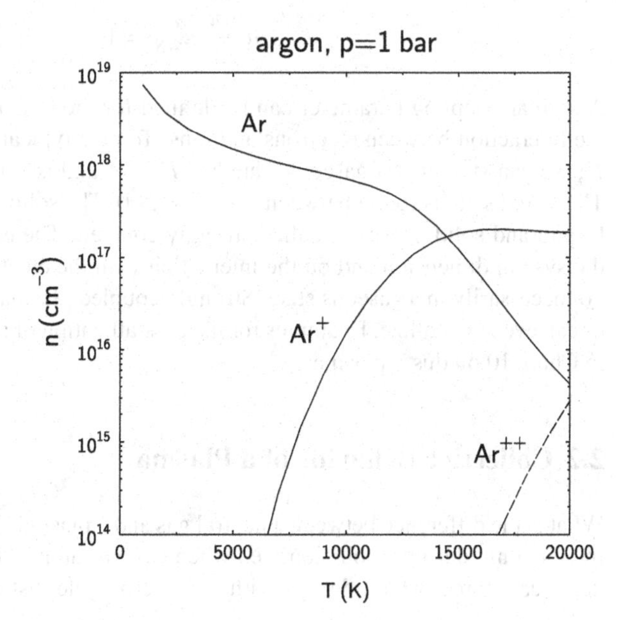

Fig. 2.1 Ionization states of an argon plasma in thermodynamic equilibrium at constant pressure calculated from Saha's equation

$$\lambda_{\max}(\mathrm{nm}) = \frac{2.898 \times 10^6}{T(K)}\,, \tag{2.14}$$

shifts to the blue end of the visible spectrum (414 nm at $T = 7000\,\mathrm{K}$). Arc discharge lamps used in data projectors have a sealed quartz discharge tube and develop operating pressures of (50–200) bar.

2.1.3 *The Coupling Parameter*

In the preceding paragraphs we have investigated how a rising temperature leads to population of excited atomic states, to ionization of atoms, and to equilibria with multiply ionized atoms. Now we focus our interest on the influence of particle density on the state of the plasma system. Then, the potential energy of the interacting particles becomes important.

The states of neutral matter, solid–liquid–gaseous, are determined by the degree of coupling between the atoms, which is described by the *coupling parameter* $\Gamma = W_{\mathrm{pot}}/k_{\mathrm{B}}T$, i.e., the ratio of the potential energy of nearest neighbors and the thermal energy. For the Coulomb interaction of singly charged ions, the coupling parameter becomes

$$\Gamma_{\mathrm{i}} = \frac{e^2}{4\pi\varepsilon_0 a_{\mathrm{WS}} k_{\mathrm{B}} T_{\mathrm{i}}}\,. \tag{2.15}$$

Here, a_{WS} is the Wigner-Seitz radius, a measure for the interparticle distance, defined by

$$n_{\mathrm{i}} \frac{4\pi}{3} a_{\mathrm{WS}}^3 = 1\,. \tag{2.16}$$

A similar coupling parameter can be defined for the interaction of the electrons or the interaction between electrons and ions. To give typical orders of magnitude for Γ, we can state that a gaseous state has $\Gamma < 1$ and is said to be weakly coupled. The liquid state is found between $1 < \Gamma < 200$. The solid phase exists at $\Gamma > 200$. Liquid and solid phase are called strongly coupled. The exact numbers depend on the system dimension and on the interaction with the electrons. Hence, a plasma is not necessarily in a gaseous state. Strongly coupled plasmas can behave like liquids or can even crystallize. Examples for the crystallization of a subsystem will be given in Chap. 10 on dusty plasmas.

2.2 Collective Behavior of a Plasma

What is the difference between a neutral gas and a plasma? In a neutral gas, particles interact only during a collision, i.e., when two gas atoms “feel” the short-range van der Waals force, which decays with the interparticle distance as r^{-6}. For most of the time, the gas atoms fly on a straight path independent of the other atoms. This is quite different in a plasma. The Coulomb force that describes the electrostatic interaction decays only slowly as r^{-2}, which makes it a long-range force. This means that each plasma particle interacts with a large number of other particles. Therefore, plasmas show a simultaneous response of many particles to an external stimulus. In this sense, plasmas show *collective behavior*, which means that the macroscopic result to an external stimulus is the cooperative response of many plasma particles. Mutual shielding of plasma particles or wave processes are examples of collective behavior.

2.2.1 Debye Shielding

The most important feature of a plasma is its ability to reduce electric fields very effectively. We can discuss this effect of *shielding* by placing a point-like extra charge $+Q$ into an infinitely large homogeneous plasma, which originally has equal densities of electrons and singly charged positive ions $n_{e0} = n_{i0}$. Let us assume that this extra charge $+Q$ is located at the origin of the coordinate system. We expect that electrons will be attracted and ions repelled by this extra charge as sketched in Fig. 2.2a. This gives rise to a net space charge in the vicinity of $+Q$, which tends to weaken the electric field generated by $+Q$.

The bending of the trajectory depends on the particle energy. The higher the energy of the electrons (ions) is, i.e., the higher the temperature of the electron (ion) gas is, the stiffer the trajectory becomes, as indicated in Fig. 2.2b. Therefore, a cold species of particles will be very effective in shielding the extra charge and we can conjecture that the size of the perturbed region is small, whereas the pertubation has a greater range for hotter electrons (ions).

Obviously, this shielding process is not static, but is governed by the thermal motion of the plasma electrons and ions. Therefore, we need a simple statistical description, for which we use the Boltzmann factor. For a quantitative description, we calculate the number of electrons and ions that are found at an enhanced electric potential in the vicinity of $+Q$. For a repulsive potential, the Boltzmann factor Eq. (2.1) gives the number of particles in a thermal distribution that have overcome

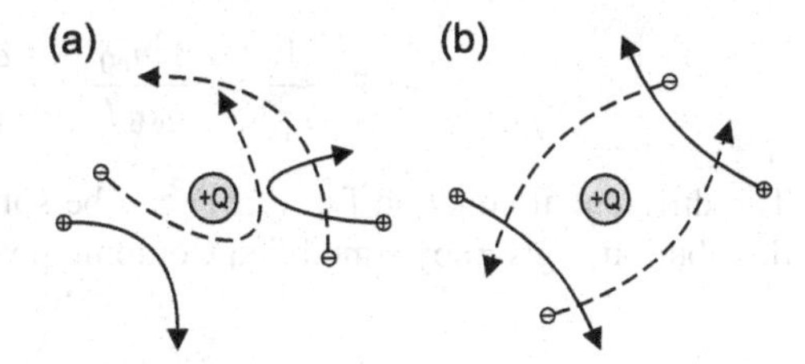

Fig. 2.2 (**a**) Shielding arises from a net attraction of electrons and repulsion of positive ions, leading to trajectory bending. (**b**) For higher energy the trajectories become stiffer and the shielding less efficient

a potential barrier Φ. For an attractive potential, the density can even become higher than the equilibrium value:

$$n_{\mathrm{e}}(\mathbf{r}) = n_{e0} \exp\left(+\frac{e\Phi(\mathbf{r})}{k_{\mathrm{B}}T_{\mathrm{e}}}\right),$$
$$n_{\mathrm{i}}(\mathbf{r}) = n_{i0} \exp\left(-\frac{e\Phi(\mathbf{r})}{k_{\mathrm{B}}T_{\mathrm{i}}}\right). \tag{2.17}$$

For simplicity, we assume that the perturbed potential Φ is small compared to the thermal energy, which allows us to expand the exponential and use only the first term in the Taylor-expansion

$$n_{\mathrm{e}}(\mathbf{r}) \approx n_{\mathrm{e0}}\left(1+\frac{e\Phi(\mathbf{r})}{k_{\mathrm{B}}T_{\mathrm{e}}}\right)$$
$$n_{\mathrm{i}}(\mathbf{r}) \approx n_{\mathrm{i0}}\left(1-\frac{e\Phi(\mathbf{r})}{k_{\mathrm{B}}T_{\mathrm{i}}}\right). \tag{2.18}$$

A self-consistent solution for the electric potential can then be obtained by using Poisson's equation

$$\Delta\Phi = -\frac{1}{\varepsilon_0}\left[Q\delta(\mathbf{r}) - en_{\mathrm{e}}(\mathbf{r}) + en_{\mathrm{i}}(\mathbf{r})\right] \tag{2.19}$$

and inserting the linearized densities from Eq. (2.18)

$$\frac{\partial^2\Phi}{\partial r^2} + \frac{2}{r}\frac{\partial\Phi}{\partial r} = -\frac{1}{\varepsilon_0}\left[Q\delta(r) - en_{\mathrm{e0}}\frac{e\Phi}{k_{\mathrm{B}}T_{\mathrm{e}}} - en_{\mathrm{i0}}\frac{e\Phi}{k_{\mathrm{B}}T_{\mathrm{i}}}\right]. \tag{2.20}$$

On the l.h.s. of this equation we have used the spherical symmetry of the problem, which makes Φ independent of angular variables. On the r.h.s. we have used the assumed neutrality of the unperturbed system, $n_{\mathrm{e0}} = n_{\mathrm{i0}}$. Rearranging all contributions that contain Φ to the left side, we obtain a differential equation of the Helmholtz type

$$\frac{\partial^2\Phi}{\partial r^2} + \frac{2}{r}\frac{\partial\Phi}{\partial r} - \frac{1}{\lambda_{\mathrm{D}}^2}\Phi = -\frac{Q}{\varepsilon_0}\delta(r). \tag{2.21}$$

The parameter λ_{D} has the dimension of a length and is defined by

$$\frac{1}{\lambda_{\mathrm{D}}^2} = \frac{e^2 n_{\mathrm{e0}}}{\varepsilon_0 k_{\mathrm{B}}T_{\mathrm{e}}} + \frac{e^2 n_{\mathrm{i0}}}{\varepsilon_0 k_{\mathrm{B}}T_{\mathrm{i}}}. \tag{2.22}$$

The differential equation Eq. (2.21) can be solved by assuming that the potential distribution is given by a modified Coulomb potential

$$\Phi(r) = \frac{a}{r}f(r). \tag{2.23}$$

Then, for all $r > 0$, the function $f(r)$ is a solution of the differential equation

$$f'' - \lambda_D^{-2} f = 0\,, \tag{2.24}$$

which gives $f_1(r) = \exp(-r/\lambda_D)$. A second solution, $f_2(r) = \exp(+r/\lambda_D)$, is unphysical because the perturbed field would increase indefinitely with distance r. The normalization constant a is obtained by applying Gauss' theorem to a small sphere around the origin

$$\oint \mathbf{D} \cdot d\mathbf{A} = 4\pi r^2 \varepsilon_0 E_r = Q\,. \tag{2.25}$$

Here we have assumed that for $r/\lambda_D \to 0$ the sphere only contains the extra charge Q but no space charge from the perturbed distributions of electrons and ions. From Eq. (2.23) we obtain

$$E_r = \frac{a}{r^2}\left(1 + \frac{r}{\lambda_D}\right) e^{-r/\lambda_D} \to \frac{a}{r^2}\,. \tag{2.26}$$

Hence, the normalization a is the same as for the Coulomb potential $a = Q\,(4\pi\varepsilon_0)^{-1}$. The complete solution

$$\Phi(r) = \frac{Q}{4\pi\varepsilon_0\, r^2}\, e^{-r/\lambda_D} \tag{2.27}$$

is called the *Debye-Hückel potential* after pioneering work of Pieter Debye (1884–1966) and Erich Hückel (1896–1980) on polarization effects in electrolytes [55]. A similar shielded Coulomb potential was later found in nuclear physics for interactions mediated by the exchange of a finite-mass particle like the pion by Nobel prize winner Hideki Yukawa (1907–1981). In classical weakly-coupled plasmas, the standard terminology is *Debye shielding* whereas the younger literature on strongly-coupled systems has a preference for *Yukawa-interaction*, a terminology also used in colloid science.

The parameter λ_D is the *Debye shielding length*, which describes the combined shielding action of electrons and ions. When we are interested in the individual contributions of electrons and ions we can define the *electron Debye length* λ_{De} and the *ion Debye length* λ_{Di} separately,

$$\lambda_{De} = \left(\frac{\varepsilon_0 k_B T_e}{n_{e0} e^2}\right)^{1/2} \qquad \lambda_{Di} = \left(\frac{\varepsilon_0 k_B T_i}{n_{i0} e^2}\right)^{1/2}\,. \tag{2.28}$$

Inspecting the dependence of the electron (ion) Debye length on temperature, we see that our initial conjecture is confirmed. The shielding length increases when the temperature rises, i.e., the size of the perturbed region becomes larger. The dependence on density $\propto n_{e0}^{-1/2}$ $(n_{i0}^{-1/2})$ means that an increasing number of shielding particles makes the shielding more efficient and diminishes the size of the perturbed volume.

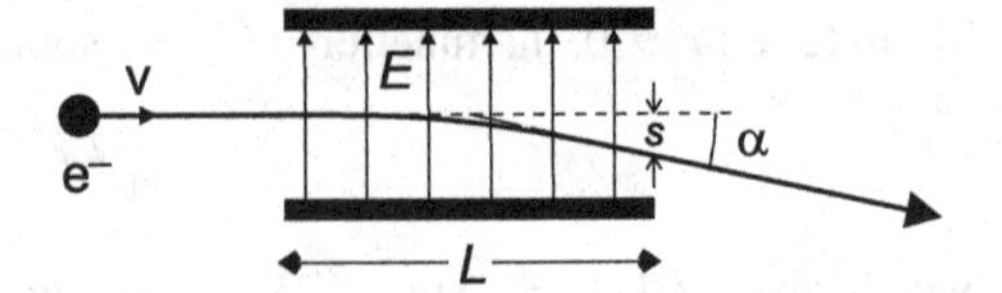

Fig. 2.3 Deflection of an electron beam by a transverse electric field E in a traditional cathode ray oscilloscope tube

Yet, why is the Debye length independent of the particle mass? We can gain insight into this property from a mathematically simpler situation, in which the electric field is homogeneous. Consider the cathode ray tube of a traditional oscilloscope. There, an electron of mass m, charge q and energy W enters the space between two deflection plates of length L (Fig. 2.3) and performs a free-fall motion in the electric field. Hence, the trajectory in the space between the deflection plates is a parabola.

The initial velocity is $v = (2W/m)^{1/2}$ and the electron needs a transit time $\tau = L/v$ to traverse the deflection plates. In this time, it has fallen a distance

$$s = \frac{qE}{2m}\tau^2 = \frac{qEL^2}{4W}, \tag{2.29}$$

which is independent of the mass m but depends on the energy W. Noting that the tangent to the parabola at the exit point intersects the undeflected orbit at $x = L/2$, the deflection angle becomes $\alpha = \arctan(2s/L)$. This independence of mass is the reason, why a transverse electric field can be used as an *energy filter* to sort out particles of same energy independent of their mass.

λ_D is often called the *linearized Debye length*. The reader may also recognize an analogy between the structure of Eq. (2.22) and the parallel circuit of two resistors in electricity:

$$\frac{1}{\lambda_D^2} = \frac{1}{\lambda_{De}^2} + \frac{1}{\lambda_{Di}^2} \quad \leftrightarrow \quad \frac{1}{R_{total}} = \frac{1}{R_1} + \frac{1}{R_2}. \tag{2.30}$$

In the shielding process, electrons and ions work in parallel. Attracting electrons and repelling ions both results in a net negative charge in the vicinity of the extra charge. Similar to the total resistance R_{total} of the parallel circuit, which is smaller than any of the two resistors R_1 and R_2, the linearized Debye length is smaller than λ_{De} and λ_{Di}. A comparison between a Coulomb potential and a shielded potential is shown in Fig. 2.4. For $r > \lambda_D$, the potential decays much faster than a Coulomb potential.

Summarizing, the perturbed electric potential around an extra charge Q decays exponentially for $r > \lambda_D$. This observation has two consequences. When we require that a cloud of electrons and ions behaves as a plasma, the cloud must have a size of several Debye lengths. Moreover, any deviation from equal densities of electrons and ions tends to be smoothed by Debye shielding. Therefore, a plasma has the natural tendency to become quasineutral.

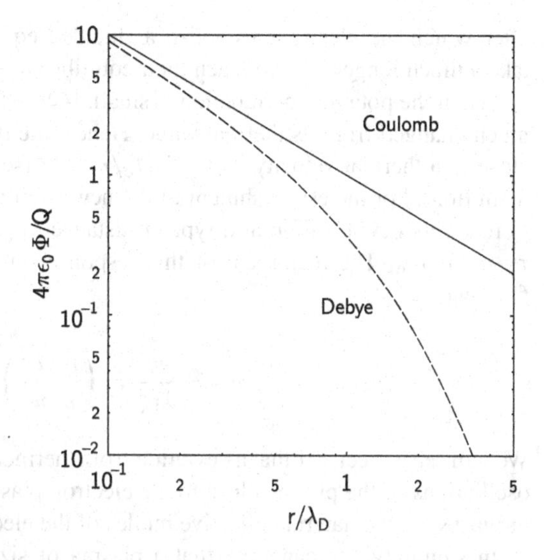

Fig. 2.4 Comparison of a Coulomb and shielded (Debye - Hückel or Yukawa) potential. Note the stronger decay for $r/\lambda_D > 1$

2.2.2 *Quasineutrality*

As we have learned in the preceding paragraph, a plasma is not a strictly neutral mixture of electrons and positive ions but deviations from neutrality can develop on short scales, the Debye length. Therefore, we define that on length scales larger than the Debye length the plasma must be quasineutral

$$\left|\sum_j Z_j\, e\, n_{i0,j} - n_{e0}\, e\right| \ll n_{e0}\, e\,. \tag{2.31}$$

Here, the sum is extended over all (positive) ion species j of charge number Z_j. The charge number is a positive quantity, hence negative ions will have a minus-sign before the charge number. For a single ion species of charge $q = +e$ we often use the short-hand notation of the quasineutrality condition $n_{i0} = n_{e0}$. In this way, quasineutrality can be used as a defining quality of a *classical* plasma, in Langmuir's parlance, to distinguish the plasma region from space-charge regions.

There are, however, systems consisting of one polarity of charges only, like electrons or ions in potential traps, which show similar collective behavior as plasmas. In these *nonneutral* plasmas (cf. Sect. 3.1.6), as they are called, the potential trap takes the role of the neutralizing species of the other polarity.

2.2.3 *Response Time and Plasma Frequency*

The response to an external electric perturbation was established by the combined action of many particles. Therefore, Debye shielding is one example for collective behavior of a plasma. The second aspect of collective behavior is the time scale, after which the electrons establish a shielded equilibrium. The heavier ions will take a much longer time to reach their equilibrium positions.

When the potential pertubation is small, $|e\Phi| \ll k_B T$, the electron energy is not much changed from its thermal value. Hence, the typical electron velocity remains close to a thermal velocity $v_e \approx (k_B T_e/m_e)^{1/2}$ (see Sect. 4.1 for a more thorough definition). For the establishment of the new equilibrium, the electron must be able to reach its new position at a typical distance λ_{De}. This time can be estimated as $\tau \approx \lambda_{De}/v_e$. The reciprocal of this response time is called the *electron plasma frequency*

$$\omega_{pe} = \frac{v_e}{\lambda_{De}} = \left(\frac{n_{e0}e^2}{\varepsilon_0 m_e}\right)^{1/2} . \tag{2.32}$$

We will see in Sect. 8.1 that a deviation from thermodynamic equilibrium may excite oscillations of the plasma close to the electron plasma frequency. Such oscillations or waves are the natural collective modes of the electron gas.

In summary, we can state that a plasma of size L must be sufficiently large, i.e., $L \gg \lambda_D$, to behave in a collective manner. This would disqualify a typical candle flame as a plasma because it is too small though it may have some ionization. Although not so obvious, a plasma must also exist for a period of time larger then the response time, $T \gg \omega_{pe}^{-1}$, to behave in a collective manner.

2.3 Existence Regimes

Plasmas are found in a huge parameter space, which covers seven orders of magnitude in temperature and twenty-five orders of magnitude in electron density (see Fig. 2.5). The temperature scale refers to the electron temperature. Typical examples are marked for astrophysical situations, some technical plasmas and the regime of controlled nuclear fusion.

2.3.1 Strong-Coupling Limit

The usual definition of a weakly coupled or *ideal* plasma is the requirement that there are many electrons inside the electron Debye sphere. This ensures that Debye shielding is a collective process and that the statistical derivation of a Debye length was correct. For this purpose we define the number of electrons inside the electron Debye sphere as the *plasma parameter* N_{De}

$$N_{De} = \frac{4\pi}{3}\lambda_{De}^3 n_e . \tag{2.33}$$

Because of the different temperatures, we can have different coupling states of electrons and ions $N_{De} \neq N_{Di}$. We will see in Chap. 10 that the dust system can be

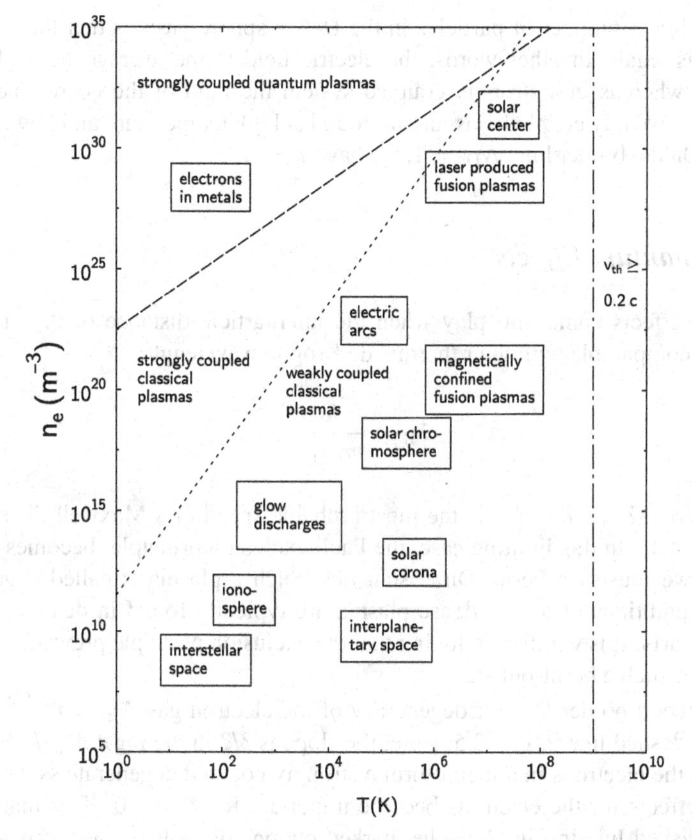

Fig. 2.5 Existence diagram of various plasmas. The dotted line marks the border for strong coupling, the dashed line the onset of quantum effects. Relativistic effects play a role for $T > 10^9$ K

strongly coupled because of the high charge number on a dust grain whereas the electron and ion gas remain weakly coupled.

The border line between weakly and strongly coupled plasmas is defined by $N_{\mathrm{De}} = 1$ (see dotted line in Fig. 2.5), which gives the equation of the border line

$$n_{\mathrm{e}} = \left(\frac{4\pi\varepsilon_0}{3e^2}\right)^2 T_{\mathrm{e}}^3\,. \tag{2.34}$$

This line has the slope 3 in the $\log n$–$\log T$ representation. From Eqs. (2.15) and (2.16) we obtain a relation between the plasma parameter N_{Di} and the coupling parameter Γ_{i} (see Problem 2.5)

$$\Gamma_{\mathrm{i}} = \frac{1}{3} N_{\mathrm{Di}}^{-2/3}\,. \tag{2.35}$$

Hence, a larger number of particles in the Debye sphere ensures that the coupling strength is small. In other words, the electric field is the average field of many particles, whereas in a strongly coupled system the field of the nearest neighbor dominates. Weakly coupled plasmas are found at high temperature and low electron density. On the border line, $N_{\mathrm{Di}} = 1$, we have $\Gamma = 1/3$.

2.3.2 *Quantum Effects*

Quantum effects come into play when the interparticle distance of the electrons becomes comparable with their thermal de Broglie wavelength

$$\lambda_{\mathrm{B}} = \frac{h}{m_{\mathrm{e}} v_{\mathrm{Te}}} . \tag{2.36}$$

Here, $v_{\mathrm{Te}} = (2k_{\mathrm{B}}T_{\mathrm{e}}/m_{\mathrm{e}})^{1/2}$ is the most probable speed of a Maxwell distribution (see Sect. 4.1). In this limiting case, the Pauli exclusion principle becomes important and we must use Fermi-Dirac statistics. Such a plasma is called degenerate and the conditions of a cold dense plasma are typically found in dead stars, like White Dwarfs. It is worth mentioning that the exclusion principle prevents the final collapse of such a burnt-out star.

The second border line for degeneracy of the electron gas, $\lambda_{\mathrm{B}} = n_{\mathrm{e}}^{-1/3}$ is also shown as dashed line in Fig. 2.5. Here, the slope is 3/2 in the $\log n$–$\log T$ diagram. Note that the electrons in a metal form a strongly coupled degenerate system. Relativistic effects for the electrons become important for $T > 10^9$ K as marked by the dot-dashed line in Fig. 2.5. The marked regions of typical plasmas can all be treated by non-relativistic models. This simplifies the plasma models in the subsequent chapters.

The Basics in a Nutshell

- Plasmas are quasineutral: $n_{\mathrm{e}} = \sum_k Z_k n_k$.
- Quasineutrality can be violated within a Debye length λ_{D}

$$\lambda_{\mathrm{D}} = \frac{\lambda_{\mathrm{De}}\lambda_{\mathrm{Di}}}{(\lambda_{\mathrm{De}}^2 + \lambda_{\mathrm{Di}}^2)^{1/2}} , \quad \lambda_{\mathrm{De,Di}} = \left(\frac{\varepsilon_0 k_{\mathrm{B}} T_{\mathrm{e,i}}}{n_{\mathrm{e,i}}^2}\right)^{1/2} .$$

- Quasineutrality can be established by the electrons within $\tau = \omega_{\mathrm{pe}}^{-1}$, with the plasma frequency

$$\omega_{\mathrm{pe}} = \left(\frac{n_{\mathrm{e}} e^2}{\varepsilon_0 m_{\mathrm{e}}}\right)^{1/2} .$$

- The coupling parameter Γ determines the state of each plasma component (electrons, ions, dust)

$$\Gamma = \frac{q^2}{4\pi \varepsilon_0 a_{WS}^2 k_B T} .$$

Γ may be different for the components, depending on the individual temperatures and densities. A gaseous phase is found for $\Gamma \ll 1$, the liquid state for $1 < \Gamma < 180$ and the solid phase for $\Gamma > 180$.

Problems

2.1 Prove that the electron Debye length can be written as

$$\lambda_{De} = 69\,\mathrm{m} \left[\frac{T(\mathrm{K})}{n_e(\mathrm{m}^{-3})} \right]^{1/2}$$

2.2 Calculate the electron and ion Debye length
(a) for the ionospheric plasma ($T_e = T_i = 3000\,\mathrm{K}$, $n = 10^{12}\,\mathrm{m}^{-3}$).
(b) for a neon gas discharge ($T_e = 3\,\mathrm{eV}$, $T_i = 300\,\mathrm{K}$, $n = 10^{16}\,\mathrm{m}^{-3}$).

2.3 Consider an infinitely large homogeneous plasma with $n_e = n_i = 10^{16}\,\mathrm{m}^{-3}$. From this plasma, all electrons are removed from a slab of thickness $d = 0.01\,\mathrm{m}$ extending from $x = -d$ to $x = 0$ and redeposited in the neighboring slab from $x = 0$ to $x = d$. (a) Calculate the electric potential in this double slab using Poisson's equation. What are the boundary conditions at $x = \pm d$? (b) Draw a sketch of space charge, electric field and potential for this situation. What is the potential difference between $x = -d$ and $x = d$?

2.4 Show that the equation for the shielding contribution (2.24) results from (2.21) and (2.23).

2.5 Derive the relationship between the coupling parameter for ion-ion interaction Γ Eqs. (2.15) and N_D (2.33) under the assumption that $T_e = T_i$.

2.6 Show that the second Lagrange multiplier in Eq. (2.6) is $\lambda = (k_B T)^{-1}$.
Hint: Start from

$$\frac{1}{T} = \frac{\partial S}{\partial \lambda} \frac{\partial \lambda}{\partial U}$$

and use $\sum n_i = 1$.

Chapter 3
Single Particle Motion in Electric and Magnetic Fields

'Twas brillig, and the slithy toves
Did gyre and gimble in the wabe.

Lewis Carroll, Jabberwocky

Plasmas belong to two different categories, unmagnetized and magnetized. The plasma in a fluorescent tube is unmagnetized, because the motion of electrons and ions is determined by electric fields and collisions, and the Earth magnetic field is too weak to bend the trajectories. The ionosphere, the magnetosphere, the solar wind, the interstellar medium and the solar surface are examples for natural magnetized plasmas. There, the motion of the particles is strongly affected by the magnetic field.

This chapter is focused on the motion of individual charged particles in given electric and magnetic fields. Of particular importance is the quest for magnetic confinement of plasmas. The inhomogeneity and curvature of magnetic field lines, or the variation of the fields in time cause complex particle motion. The model of *single particle motion* neglects the influence of particle currents on the electric and magnetic fields. In this respect, the model is still incomplete. Nevertheless, from an understanding of particle motion the reader will gain insight into the basic properties of a plasma that is subjected to electromagnetic fields.

3.1 Motion in Static Electric and Magnetic Fields

3.1.1 Basic Equations

The starting point for establishing the single-particle model is Newton's equation[1] for the motion of a particle of mass m and charge q in a given electric field $\mathbf{E}$ and magnetic field $\mathbf{B}$

$$m\dot{\mathbf{v}} = q(\mathbf{E} + \mathbf{v} \times \mathbf{B}) \,, \tag{3.1}$$

[1] Sometimes called Newton-Lorentz equation

A. Piel, *Plasma Physics*, DOI 10.1007/978-3-642-10491-6_3,

in which the dot represents the time derivative at the position of the particle. This equation can be solved in rigid mathematical terms only for simple cases, e.g., homogeneous and stationary fields.

3.1.2 Cyclotron Frequencies

Let us first consider the case of a homogeneous and stationary magnetic field $\mathbf{B} = (0, 0, B_z)$ and a vanishing electric field $\mathbf{E} = 0$. The magnetic field is chosen as z-axis because of the cylindrical symmetry about the B-field direction. Then, we obtain Newton's equation of motion in cartesian coordinates as

$$\begin{aligned}\dot{v}_x &= +v_y \frac{q}{m} B_z \\ \dot{v}_y &= -v_x \frac{q}{m} B_z \\ \dot{v}_z &= 0 .\end{aligned} \tag{3.2}$$

By combining the equations for the x and y-motion we obtain the differential equation for a harmonic oscillator

$$\ddot{v}_{x,y} = -\left(\frac{q B_z}{m}\right)^2 v_{x,y} . \tag{3.3}$$

This harmonic oscillator describes a periodic motion at a frequency

$$\omega_\mathrm{c} = \frac{|q|}{m} B_z , \tag{3.4}$$

which we call the *cyclotron frequency*. Inserting numbers for q, B, and m we find the cyclotron frequency of an electron in a magnetic field of 1 T at $\omega_\mathrm{ce} = 1.759 \times 10^{11}\,\mathrm{s}^{-1} = 2\pi \times 27.99$ GHz. At the same magnetic field, the proton cyclotron frequency is $\omega_\mathrm{cp} = 9.579 \times 10^7\,\mathrm{s}^{-1} = 2\pi \times 15.25$ MHz.

In the x-y plane, a particle with perpendicular velocity $v_\perp$ performs a circular orbit with the gyroradius or Larmor radius, named after the Irish physicist Joseph Larmor (1857–1942),

$$r_\mathrm{L} = \frac{v_\perp}{\omega_\mathrm{c}} . \tag{3.5}$$

When the initial velocity v_z along the magnetic field is nonzero, the orbit becomes a helix of constant pitch about the magnetic field direction (z). The motion about a magnetic field line is referred to as gyromotion or gyroorbit.

The sense of rotation about the magnetic field depends on the sign of the particle's charge. Electrons move in a *right-handed*, positive ions in a *left-handed* orbit (see Fig. 3.1).

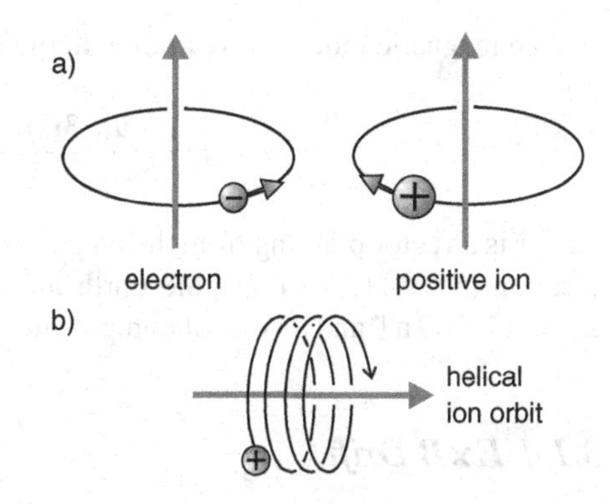

Fig. 3.1 (**a**) Gyro motion of electrons and ions. Note that electrons perform a right-handed motion about the magnetic field (Consider the thumb of your right hand representing the magnetic field direction, then the fingers give the sense of electron motion). (**b**) Helix orbit of an ion

When the ions (electrons) can perform complete gyroorbits, the ions (electrons) are called *magnetized*. This is the case, when the gyroorbits are not interrupted by collisions. A condition for this to happen is that the ion (electron) collision frequency (cf. Sect. 4.2.2) is smaller than the ion (electron) cyclotron frequency. Usually, this condition can be better fulfilled by electrons than by ions. A more detailed discussion of the interplay of gyromotion and collisions can be found in Sect. 4.3.4.

The ions in a gas discharge plasma in the presence of the Earth's magnetic field can be considered as unmagnetized. This may be unrelated to the frequency of collisions but is rather a consequence of the size of the gyroradius, which is larger than the diameter of the discharge tube.

3.1.3 The Earth Magnetic Field

In the immediate neighborhood of the Earth, the magnetic field has the shape of a dipole field (Fig. 3.2). The source of this field can be represented by a magnetic dipole at the Earth's center with a magnetic moment of $|\mathbf{M}| = 7.3 \times 10^{22}\,\mathrm{A\,m^2}$. This dipole is tilted from the axis of rotation leading to a deviation of the magnetic poles from the geographic poles. The Earth magnetic field is generated by electric currents in the Earth's core. The general shape of the distorted Earth magnetic field under the influence of the solar wind was shown in Fig. 1.6.

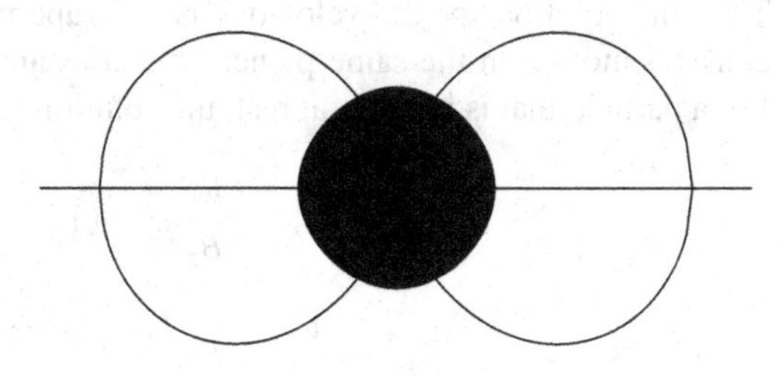

Fig. 3.2 The unperturbed dipole field near the Earth. The horizontal line marks the equatorial plane

The magnetic induction $\mathbf{B}(\mathbf{r})$ of a dipole field is given by the expression

$$\mathbf{B}(\mathbf{r}) = \frac{\mu_0}{4\pi} \frac{3\mathbf{r}(\mathbf{r}\cdot\mathbf{M}) - r^2\mathbf{M}}{r^5} \,. \tag{3.6}$$

Here $\mathbf{r}$ is a vector pointing from the magnetic dipole to the field point. At the author's location (54.3° N.; 10.1° E), the Earth magnetic field has a horizontal component $B_\mathrm{h} = 17,700\,\mathrm{nT}$ and a vertical component $B_\mathrm{v} = -46,150\,\mathrm{nT}$.

3.1.4 E×B Drift

When we now allow for a stationary and homogeneous electric field, we can choose the orientation of our coordinate system, without loss of generality, to have electric and magnetic field in the x-y plane, $\mathbf{E} = (E_x, 0, E_z)$ and $\mathbf{B} = (0, 0, B_z)$. In this case, Newton's equation of motion reads

$$\begin{aligned} \dot{v}_x &= \frac{q}{m}\left(E_x + v_y B_z\right) \\ \dot{v}_y &= \frac{q}{m}(-v_x B_z) \\ \dot{v}_z &= \frac{q}{m} E_z \,. \end{aligned} \tag{3.7}$$

The motion along the magnetic field is now accelerated but independent of the motion in the x-y plane. According to the principle of superposition of motions, we can consider both effects separately. For the x-y plane, the motion can again be decomposed

$$\begin{aligned} \ddot{v}_x &= -\omega_\mathrm{c}^2 v_x \\ \ddot{v}_y &= -\omega_\mathrm{c}^2 (v_y + E_x/B_z) \,. \end{aligned} \tag{3.8}$$

Again, we find a harmonic oscillation in x-direction, but the motion in y-direction is more complex. In a moving frame of reference, $\tilde{v}_y = v_y + E_x/B_z$, which moves at a constant velocity $-E_x/B_z$ in negative y-direction, we obtain a simple harmonic motion

$$\ddot{\tilde{v}}_y = -\omega_c^2 \tilde{v}_y \,. \tag{3.9}$$

Thus the solution for the velocities is the superposition of a circular orbit and a constant motion in the same plane. This constant motion is called the E×B-drift. For a particle that is initially at rest, the solution reads

$$\begin{aligned} v_x &= \frac{E_x}{B_z} \sin \omega_\mathrm{c} t \\ v_y &= \frac{E_x}{B_z} \left[\cos \omega_\mathrm{c} t - 1\right] \,. \end{aligned} \tag{3.10}$$

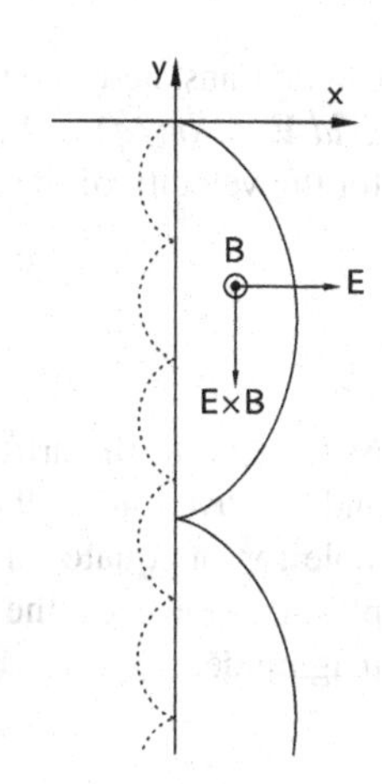

Fig. 3.3 Cycloidal trajectory resulting from the superposition of gyro-motion and E×B-drift. The electric field is oriented along the x-axis, the magnetic field is perpendicular to the x–y plane. Note that ions (full line) and electrons (dashed line) have the same drift direction. An artificial electron mass was assumed here for clarity, $m_e = 0.3\, m_{ion}$

A typical trajectory is shown in Fig. 3.3. Mathematically, the trajectory is a cycloid. For a positive particle starting at $t = 0$ in the origin, the trajectory is described by

$$x = \frac{E_x}{B_z \omega_c} [1 - \cos(\omega_c t)] , \quad y = \frac{E_x}{B_z \omega_c} [\sin(\omega_c t) - \omega_c t] \tag{3.11}$$

The E×B-drift can also be understood from energy considerations. On the high-potential side, the kinetic energy is small, which makes the instantaneous gyroradius small. On the low-potential side, the ion has gained kinetic energy from the electric field, which makes the gyro-radius larger. The combination of these two effects results in a cycloidal motion.

It is a peculiarity of the E×B-drift that negative electrons and positive ions experience the same sign of the drift velocity. This is a consequence of the fact that the applied electric field force $q\mathbf{E}$ and the resulting Lorentz force $q\mathbf{v} \times \mathbf{B}$ both depend on the sign of q, which cancels in the result. This effect can also be seen in Fig. 3.3. In vector notation, the E×B drift velocity is given by

$$\mathbf{v}_E = \frac{\mathbf{E} \times \mathbf{B}}{B^2} . \tag{3.12}$$

3.1.5 Gravitational Drift

When we consider the ionospheric plasma at the magnetic equator, we find a similar situation to crossed electric and magnetic fields. Here, the force of gravity, $m\mathbf{g}$, is perpendicular to the (horizontal) magnetic field lines. Neglecting collisions, which indeed are important in the lower ionosphere, Newton's equation of motion

$$m\dot{\mathbf{v}} = m\mathbf{g} + q\mathbf{v} \times \mathbf{B} \tag{3.13}$$

can be translated into the case of E×B motion by introducing an *equivalent electric field* $\mathbf{E} = (m/q)\mathbf{g}$. Without solving Eq. (3.13) we can immediately give the result for the velocity of the gravitational drift

$$\mathbf{v}_g = \frac{m}{q} \frac{\mathbf{g} \times \mathbf{B}}{B^2} . \tag{3.14}$$

Note that now the drift velocity depends on mass and charge. In particular, electrons and positive ions will drift in opposite directions. The gravitational drift is responsible for an equatorial net electric current that is driven by the weight force on the plasma. However, the collisionless approximation is too crude to give its correct magnitude.

3.1.6 Application: Confinement of Nonneutral Plasmas

A nonneutral plasma consists of only one sort of charged particles, often positive ions. Fig. 3.4 shows a typical magnetic trap of the Penning-Malmberg type, which is suitable for trapping electrons or ions. These traps use strong magnetic fields of $|B| > 1\,\mathrm{T}$. A review of experiments with this device can be found in [56].

The axial magnetic field B provides magnetic confinement by having the ions gyrate about the magnetic field line. The axial confinement is achieved by electric fields from the positively-biased outer cylinders that repels the ions towards the center. The ion cloud represents a region of positive space charge. From Poisson's equation in cylindrical geometry

$$\frac{1}{r}\frac{\partial}{\partial r}(rE_r) = \frac{n_i e}{\varepsilon_0} \tag{3.15}$$

we obtain $E_r = \frac{1}{2} n_i e r \varepsilon_0^{-1}$, i.e., the electric field increases linearly from the center to the edge of the ion cloud. Hence, the E×B velocity increases in the same manner, which means that the cloud rotates as a rigid body with an angular frequency $\omega = E(rB)^{-1} = \frac{1}{2} n_i e (\varepsilon_0 B)^{-1}$. Ions can be cooled to milli-Kelvin temperatures by a technique called laser-doppler cooling [57]. Such devices can be used to trap antiprotons for a sufficiently long time to recombine with positrons from a radioactive source to form antihydrogen [58–60].

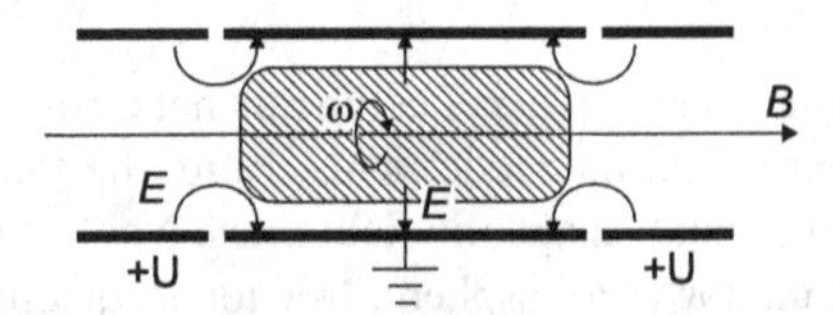

Fig. 3.4 The Penning–Malmberg trap for confining a nonneutral plasma of positive ions (*hatched area*) uses three cylindrical tube electrodes aligned with a strong magnetic field B

3.2 The Drift Approximation

In this Section, approximate solutions are sought for the case of inhomogeneous and curved magnetic fields. We will discuss the influence of inhomogeneity and curvature in separate steps although these two aspects are intertwined by Maxwell's equations.

3.2.1 The Concept of a Guiding Center

We have seen that the effect of an external force on a gyrating particle can be described as a net *drift motion* that is superimposed on the gyromotion. We will apply this idea to the motion in inhomogeneous magnetic fields. For this purpose, we assume that the true particle orbit can be decomposed into a circular orbit about a local *guiding center* and a drift motion of the guiding center (see Fig. 3.5). For the drift motion we calculate a net force, which is the average over one gyro-period. This net force is then converted into an equivalent electric field—as in the case of the gravitational drift—and the drift velocity is obtained from Eq. (3.12).

Such an approximation requires that the gradient of the magnetic field is small. This can be expressed by the requirement that the change of the magnetic field across one gyroradius is small compared to the magnitude of the magnetic field at the guiding center

$$r_{\mathrm{L}} \frac{\partial B_z}{\partial r} \ll B_z \,. \tag{3.16}$$

The guiding center approximation is in fact more than a simple Taylor expansion of the fields. The resulting expressions have a wider range of applicability than expected from the requirement of Eq. (3.16).

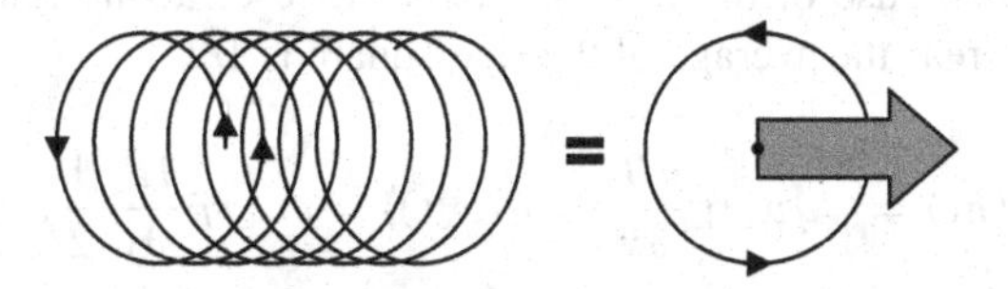

Fig. 3.5 The concept of a guiding center decomposes the actual cycloidal orbit into a circular motion about the guiding center and a drift motion of the guiding center

3.2.2 Gradient Drift

In a first step, we assume that the magnetic field is inhomogeneous, but that the field lines are straight and parallel. The influence of field line curvature will be discussed separately below. The particle experiences a Lorentz force $\mathbf{F} = q\mathbf{v} \times \mathbf{B}$. In the

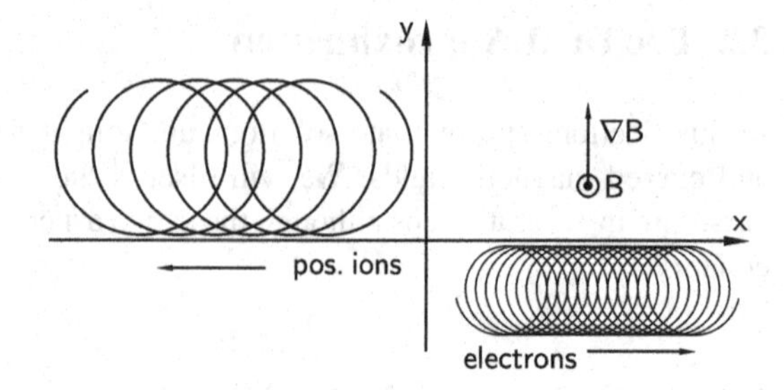

Fig. 3.6 Gradient drift of electrons and ions. This drift is charge sensitive. Note that the instantaneous curvature of the trajectory is smaller in regions of stronger magnetic field

geometry shown in Fig. 3.6, the y-component of this force is given by

$$F_y = -qv_x B_z(y)\,, \tag{3.17}$$

where $B_z(y)$ is the true magnetic field at the position of the particle. This can be estimated from the field at the guiding center by Taylor expansion, $B_z(y) = B_0 + y(t)(\partial B_z/\partial y)$, yielding

$$F_y = -qv_\perp \sin(\omega_c t)\left[B_0 \pm r_L \sin(\omega_c t)\frac{\partial B_z}{\partial y}\right]. \tag{3.18}$$

$v_\perp$ is the orbit velocity of the gyrating particle in a plane perpendicular to the magnetic field. The upper sign in this expression corresponds to positive, the lower sign to negative particles. When we put B_0 outside the brackets, the small expansion parameter becomes visible

$$F_y = -qv_\perp \sin(\omega_c t) B_0 \left[1 \pm \frac{r_L(\partial B_z/\partial y)}{B_0}\sin(\omega_c t)\right]. \tag{3.19}$$

According to the recipe given above, we now need to average the force over one gyroperiod and make use of the fact that the average of a sine function over one period is zero whereas the average of the sine-square is 1/2

$$\langle F_y\rangle = \mp q v_\perp r_L \frac{\partial B_z}{\partial y}\langle \sin^2(\omega_c t)\rangle = e v_\perp r_L \frac{\partial B_z}{\partial y}\frac{1}{2}. \tag{3.20}$$

The resulting average force is independent of the sign of the charge. However, the corresponding equivalent electric field $E = (1/q)\langle F_y\rangle$ is charge sensitive. Hence, we obtain a drift velocity

$$\mathbf{v}_{\nabla B} = \frac{1}{q}\frac{\langle \mathbf{F}\rangle \times \mathbf{B}}{B^2} = \pm\frac{1}{2} v_\perp r_L \frac{\mathbf{B}\times\nabla|\mathbf{B}|}{B^2}. \tag{3.21}$$

This is the velocity of the *gradient drift*. The charge dependence leads to charge separation and to the formation of a net current across the magnetic field.

3.2.3 Curvature Drift

In the second step, we now consider curved field lines with a constant radius of curvature R_c. At the same time, we neglect the gradient of the magnetic field, which we have already discussed in the preceding paragraph. The curvature drift is an effect of motion along the field line, where the particle experiences a centrifugal force F_c from the curvature

$$\langle \mathbf{F}_c \rangle = \frac{m v_z^2}{R_c} \mathbf{e}_R \,. \tag{3.22}$$

Here, v_z is the parallel velocity and $\mathbf{e}_R$ the unit vector in radial direction. We have retained the average over a gyroperiod for compatibility with calculations above. This reflects the idea that the particle experiences a constant net curvature during one gyroorbit. This force leads to a drift velocity

$$\mathbf{v}_R = \frac{1}{q} \frac{\mathbf{F}_c \times \mathbf{B}}{B^2} = \frac{m v_z^2}{q B^2} \frac{\mathbf{R}_c \times \mathbf{B}}{R_c^2} \,. \tag{3.23}$$

3.2.4 The Toroidal Drift

The expression for the curvature drift Eq. (3.23) looks quite different from the gradient drift as long as we do not know how to relate the radius of curvature of a field line to the gradient of the magnitude of $\mathbf{B}$. We assume that the field lines of interest are generated by currents that flow outside the considered volume. These may be the currents in magnetic field coils that confine a plasma, or the dynamo currents in the Earth's core that generate Earth's magnetic field.

Because of the separation of the current region from the field region, we can assume the magnetic field being irrotational

$$(\nabla \times \mathbf{B})_z = \frac{\partial B_r}{\partial s} - \frac{\partial B_\theta}{\partial r} = 0 \,. \tag{3.24}$$

From Fig. 3.7 we obtain the relation

$$\frac{\mathrm{d}s}{R_c} = -\frac{\mathrm{d}B_r}{B_\theta} \quad \text{order :} \quad \frac{1}{R_c} = -\frac{1}{B_\theta} \frac{\partial B_r}{\partial s} \,. \tag{3.25}$$

Using Eq. (3.24) we find

$$\frac{1}{R_c} = -\frac{1}{B_\theta} \frac{\partial B_\theta}{\partial r} \,, \tag{3.26}$$

which means that the radius of curvature is the inverse of the logarithmic derivative of the magnetic field strength. Introducing the vector $\mathbf{R}_c$, we can rewrite the

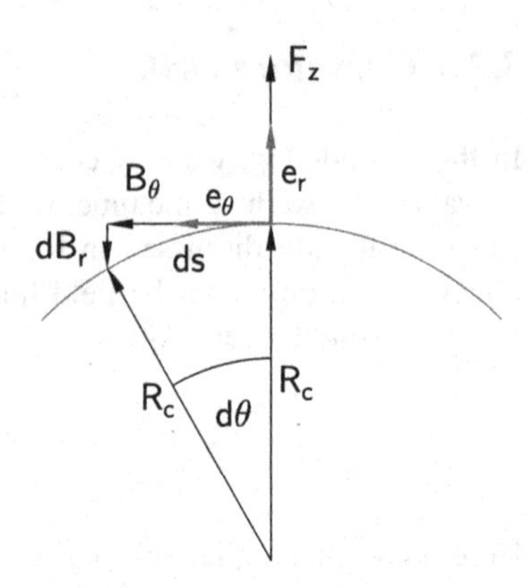

Fig. 3.7 The relationship between magnetic field gradient and radius of field line curvature

magnetic field gradient as

$$\frac{\nabla |B|}{|B|} = -\frac{\mathbf{R}_c}{R_c^2}\,. \tag{3.27}$$

This expression allows us to calculate the additional contribution of the curvature drift

$$\mathbf{v}_{\nabla B} = \frac{1}{2}\frac{m}{q}\frac{v_\perp^2}{R_c^2 B^2}\mathbf{R}_c \times \mathbf{B}\,. \tag{3.28}$$

The total drift in inhomogeneous curved magnetic field finally becomes the sum of Eqs. (3.23) and (3.28),

$$\mathbf{v}_R + \mathbf{v}_{\nabla B} = \frac{m}{q}\left(v_z^2 + \frac{1}{2}v_\perp^2\right)\frac{\mathbf{R}_c \times \mathbf{B}}{R_c^2 B^2}\,. \tag{3.29}$$

This sum is often called the *toroidal drift* because of its role in magnetic confinement in torus-like configurations. Since both contributions depend on the square of the particle velocities the two effects lead always to an increased total drift.

3.3 The Magnetic Mirror

In this Section we consider a situation where the gradient of the magnetic field is parallel to the field direction (see Fig. 3.8). For mathematical simplicity we assume a bundle of straight field lines with rotational symmetry about a central field line. The guiding center is assumed to move along this central field line. We are allowed to neglect the curvature of the field lines, which would lead to a curvature drift as discussed above.

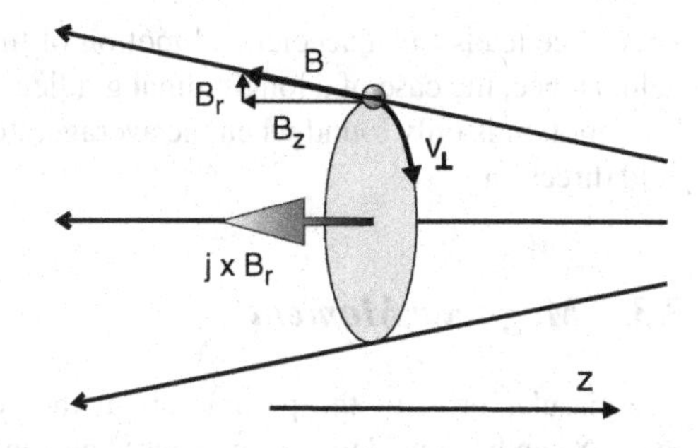

Fig. 3.8 The magnetic field lines with a longitudinal gradient form a magnetic mirror

3.3.1 Longitudinal Gradient

In such a type of inhomogeneous magnetic field, a charged particle experiences a constant net Lorentz force $q(\mathbf{v}_\perp \times \mathbf{B}_r)$ that has its origin in the radial component of the magnetic field. The force vector is oriented along the central field line and points in the direction of a weaker field. Note that the sign of the charge cancels because of the reversed sense of gyration for negative particles. This force acts to decelerate and eventually reflect a particle that has originally moved into the region of stronger field. Therefore, this field geometry is called a *magnetic mirror*.

We obtain the radial part of the magnetic field from the vanishing of the divergence of the magnetic induction

$$0 = \nabla \cdot \mathbf{B} = \frac{1}{r}\frac{\partial}{\partial r}(r B_r) + \frac{\partial}{\partial z} B_z \, . \tag{3.30}$$

When we prescribe the longitudinal gradient $\partial B_z/\partial z$ at $r = 0$, and assume this as approximately constant, we can integrate Eq. (3.30)

$$r B_r = -\int_0^r r \frac{\partial B_z}{\partial z} \mathrm{d}r \approx -\frac{1}{2} r^2 \left[\frac{\partial B_z}{\partial z}\right]_{r=0} \tag{3.31}$$

and obtain the radial magnetic field as

$$B_r \approx -\frac{1}{2} r \left[\frac{\partial B_z}{\partial z}\right]_{r=0} \, . \tag{3.32}$$

Then, the net Lorentz force in z-direction, acting on the ring current with a gyroradius r_L, is

$$\langle F_z \rangle = -\frac{1}{2} q v_\perp r_\mathrm{L} \left[\frac{\partial B_z}{\partial z}\right]_{r=0} \, . \tag{3.33}$$

This force leads to an accelerated motion of the guiding center along the magnetic field. Hence, the case of a longitudinal gradient does *not* lead to a new drift velocity. Drift motion is only found when the averaged force is perpendicular to the magnetic field direction

3.3.2 Magnetic Moment

The circular orbit of the particle about the central field line in the geometry of Fig. 3.8 can be considered as an electric current. This ring current has an associated magnetic moment μ, which is the product of the current I flowing at the edge of a circular disk with Larmor radius r_L and the area A of this disk. The current is given by the charge q performing a revolution in the gyroperiod T:

$$|\mu| = IA = \frac{|q|}{T}\pi r_{\mathrm{L}}^2 = |q|\frac{\omega_{\mathrm{c}}}{2\pi}\pi\left(\frac{v_\perp}{\omega_{\mathrm{c}}}\right)^2 = \frac{mv_\perp^2}{2B} = \frac{W_\perp}{B}\,. \tag{3.34}$$

The magnetic moment is a vector that is antiparallel to the ambient magnetic field B (Fig. 3.9) and—according to Lenz's rule—weakens the external field. With this definition of the magnetic moment, we can rewrite Eq. (3.33)

$$\langle F_z \rangle = -\mu \frac{\partial B_z}{\partial z}\,. \tag{3.35}$$

This shows that the gyrating particle experiences a force like a piece of diamagnetic matter in an inhomogeneous magnetic field. The diamagnetism results from the left-handed motion of a positive ion, which creates a magnetic dipole that is antiparallel to the acting magnetic field. The same is true for electrons, which have the opposite charge and the opposite sense of gyration.

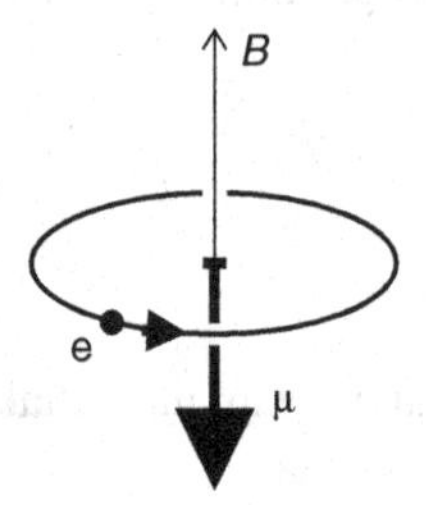

Fig. 3.9 The magnetic moment μ of a gyroorbit is antiparallel to the magnetic field. This makes a magnetized plasma diamagnetic

3.4 Adiabatic Invariants

It is shown in classical mechanics that the action integral over a periodic orbit, $\oint p\,\mathrm{d}q$, is a conserved quantity of the system. This concept can be extended to weak gradients, in which the orbit is nearly periodic. The associated action integrals then are no longer strict invariants but become *adiabatic invariants*.

3.4.1 The Magnetic Moment as First Invariant

When we assume that the diamagnetic force in Eq. (3.35) is a valid description, we obtain an energy relation for the motion of the guiding center by multiplying with v_z

$$\begin{aligned} m\dot{v}_z &= -\mu \frac{\partial B}{\partial z} \\ m v_z \dot{v}_z &= -\mu \frac{\partial B}{\partial z} \frac{\mathrm{d}z}{\mathrm{d}t} \\ \frac{\mathrm{d}}{\mathrm{d}t}\left(\frac{1}{2} m v_z^2\right) &= -\mu \frac{\mathrm{d}B}{\mathrm{d}t} . \end{aligned} \tag{3.36}$$

Here, $\mathrm{d}B/\mathrm{d}t$ is the change in the magnetic field, which the guiding center experiences by moving along the central field line. For the guiding center, we have no radial magnetic field and therefore can drop the index z. A time-invariant magnetic field does not alter the kinetic energy, as can be seen from $\mathbf{F}\cdot\mathrm{d}\mathbf{s} = q(\mathbf{v}\times\mathbf{B})\cdot\mathbf{v}\,\mathrm{d}t = 0$, because the Lorentz force is always perpendicular to the trajectory. The change in kinetic energy can then be written as

$$0 = \frac{\mathrm{d}}{\mathrm{d}t}\left(\frac{1}{2} m v_z^2 + \frac{1}{2} m v_\perp^2\right) = \frac{\mathrm{d}}{\mathrm{d}t}\left(\frac{1}{2} m v_z^2 + \mu B\right) . \tag{3.37}$$

Combining Eqs. (3.36) and (3.37) one obtains

$$-\mu \frac{\mathrm{d}B}{\mathrm{d}t} + \frac{\mathrm{d}}{\mathrm{d}t}(\mu B) = 0 \tag{3.38}$$

and finally

$$\frac{\mathrm{d}\mu}{\mathrm{d}t} = 0 . \tag{3.39}$$

Hence, the magnetic moment is conserved to the same degree of accuracy as the diamagnetic force gave a sufficiently accurate description of the motion of the guiding center.

3.4.2 The Mirror Effect

The adiabatic invariance of the magnetic moment can be used to calculate the confinement properties of a magnetic mirror field. A natural magnetic mirror is given by the Earth's magnetic dipole field. Following a field line from the equator towards the pole, one observes the increase in field line density and hence in magnetic flux density (see Fig. 3.10a). Therefore, charged particles can be trapped between the mirrors at North and South pole.

Magnetic mirrors can also be formed in the laboratory, e.g., between a set of circular magnetic field coils, as shown in Fig. 3.10b. When the distance of the coils

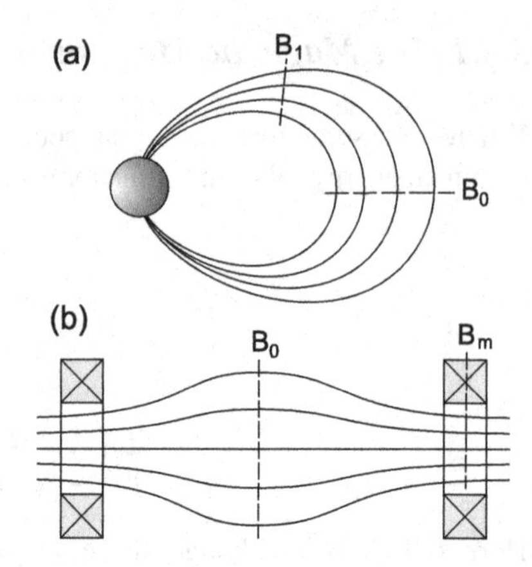

Fig. 3.10 (**a**) Mirror action of the Earth's dipole field. Following a field line from the equator to the pole the magnetic field increases $B_1 > B_0$. (**b**) Magnetic mirror created in the laboratory by a set of field coils

is larger than their radius, the magnetic field becomes inhomogeneous along the axis of the system, as can be seen by the high density of field lines at the coil position and the lower density in the midplane. We will now discuss the confinement of a charged particle that moves along the central field line of this system.

At any position, the velocity components of the particle are v_z along the axis and $v_\perp = (v_x^2 + v_y^2)^{1/2}$, which is the speed of gyromotion. In the symmetry plane of the mirror the magnetic field is $B = B_0$ and there the particle has initial velocities $v_{\perp 0}$ and v_{z0}. The maximum magnetic field $B = B_{\mathrm{m}}$ is found in the vicinity of the mirror coils. The motion of the particle is governed by the conservation of energy and the adiabatic invariance of the magnetic moment, which can be written as

$$\begin{aligned} v_\perp^2 + v_z^2 &= v_{\perp 0}^2 + v_{z0}^2 = v_0^2 \\ v_\perp^2/B &= v_{\perp 0}^2/B_0 \,. \end{aligned} \tag{3.40}$$

When the particle moves into regions of higher magnetic field, its parallel energy is consumed by the diamagnetic force. At the same time, the gyrofrequency increases, which leads to a larger kinetic energy of the gyromotion. A reflection by the magnetic mirror occurs when the energy of parallel motion becomes zero at any position before B attains its maximum value B_{m}. Solving Eq. (3.40) for $v_z = 0$ and setting $v_{\perp 0} = v_0 \sin\theta$ gives

$$0 = v_z^2 = v_0^2 \left(1 - \frac{B}{B_0}\sin^2(\theta)\right) . \tag{3.41}$$

This means that the stopping point of a particle is only dependent on the starting angle θ with respect to the magnetic field. It is independent of the magnitude of the initial velocity v_0. All particles with a starting angle $\theta > \theta_{\mathrm{m}}$ are confined, while the particles with $\theta < \theta_{\mathrm{m}}$ can overcome the mirror point B_{m} and form the *loss cone* in velocity space. The angle of the loss cone is

$$\theta_{\mathrm{m}} = \arcsin\left(\sqrt{B_0/B_{\mathrm{m}}}\right) . \tag{3.42}$$

The quantity $R_\mathrm{m} = B_\mathrm{m}/B_0$ is called the *mirror ratio*, which defines the confinement quality of a mirror machine. A large mirror ratio is equivalent to a small loss-cone angle.

In Fig. 3.11 a mirror field is shown that is produced by a pair of circular currents, each of magnitude I, at positions $z = \pm L/2$ and radius R. From Biot and Savart's law, the magnetic field on the axis of a current ring is

$$B(z) = \frac{\mu_0}{2} \frac{IR^2}{(R^2 + z^2)^{3/2}} \,. \tag{3.43}$$

Hence the total mirror field becomes $B_\mathrm{tot} = B(z - L/2) + B(z + L/2)$. For $R = 0.15\,\mathrm{m}$, $L = 0.3\,\mathrm{m}$, $I = 500\,\mathrm{A}$ the mirror ratio becomes $B_\mathrm{m}/B_0 = 1.48$ and the half-angle of the loss-cone $\theta = 55.3°$.

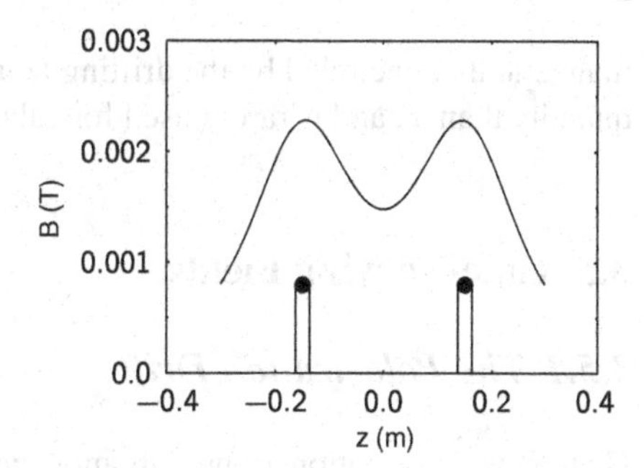

Fig. 3.11 Mirror field of an arrangement of two ring currents. The positions and radius of the ring currents are indicated

3.4.3 The Longitudinal and the Flux Invariant

According to the three degrees of freedom of motion, we can define exactly three adiabatic invariants. The first is the magnetic moment, which corresponds to the periodic gyromotion. In a magnetic mirror, trapped particles are bouncing back and forth, on a slower time scale, between the reflection points, which can be seen in Fig. 3.12. There, an energetic proton is trapped in the dipole field of the Earth. With this secondary periodic moment, we associate the *second adiabatic invariant* J, also called the *longitudinal invariant*,

$$J = \int v_\parallel \mathrm{d}l \,. \tag{3.44}$$

The integral is taken between the reflection points. The invariant J is more fragile than the fairly robust magnetic moment μ.

The third periodic motion, on an even longer time scale, is associated with the toroidal drift of this bouncing trajectory in the curved dipole field (see Fig. 3.12), which leads to a circular motion in the equatorial plane. Associated with this slow periodic drift motion is the third or *flux invariant* Φ, which represents the total

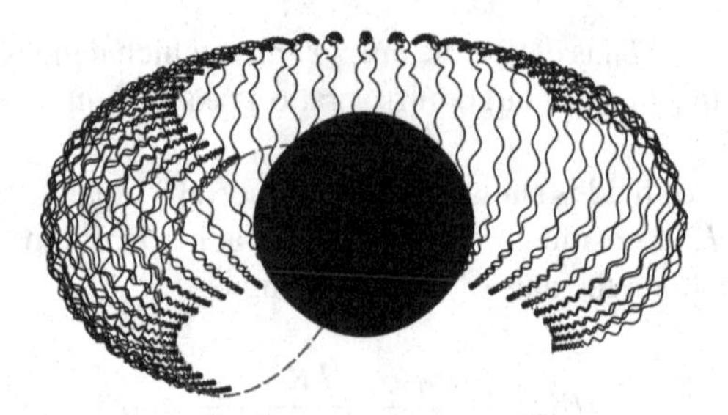

Fig. 3.12 Mirror effect and particle drifts in the Earth's dipole field. A 10 MeV proton with a pitch angle of $\theta = 30°$ starting at $3R_E$ is trapped in the Earth magnetic field. The initial field line, on which the particle motion started, is shown by the dashed curve. The particle performs a hierarchy of three periodic motions: gyration about the field line, bouncing between the mirror points, and a slow (toroidal) drift in the equatorial plane

magnetic flux encircled by the drifting bounce-trajectory. Φ is an even more fragile quantity than J, and is rarely used for calculations.

3.5 Time-Varying Fields

3.5.1 The Polarization Drift

Up to now, only stationary but inhomogeneous fields have been considered. We will now allow a time-varying electric field $\mathbf{E} = (E_x(t), 0, 0)$, which is again assumed perpendicular to the stationary magnetic field $\mathbf{B} = (0, 0, B_z)$. For simplicity, we consider the case when the electric field is increasing at a constant rate $\mathrm{d}E_x/\mathrm{d}t =$ const. The equation of motion in the perpendicular plane

$$\dot{\mathbf{v}} = \frac{q}{m}(\mathbf{E}(t) + \mathbf{v} \times \mathbf{B}) \tag{3.45}$$

can be decoupled and yields for the x direction:

$$\ddot{v}_x = -\omega_c^2 \left[v_x \mp \frac{1}{\omega_c} \frac{\dot{E}_x}{B_z} \right] . \tag{3.46}$$

The upper sign corresponds to positive ions, the lower to electrons. Transforming to a moving frame of coordinates $\tilde{v}_x = v_x \mp \dot{E}_x/\omega_c B_z$, one obtains the familiar circular orbits in this moving frame. Hence, the trajectory is the superposition of gyromotion and a *polarization drift* v_p in the direction of the electric field,

$$v_p = \pm \frac{1}{\omega_c} \frac{\dot{E}_x}{B_z} , \tag{3.47}$$

which has a constant speed because of the assumed constant rate $\dot{E}_x$. At the same time, there is a time-dependent E×B-drift

$$v_E = -E_x(t)/B_z\,. \qquad (3.48)$$

The polarization drift can also be considered as a switch-on effect of the plasma, which reflects the inertia of the particles. This can be seen most clearly when the electric field is suddenly switched on at $t = 0$, as shown in Fig. 3.13a. At $t = 0$ a particle, which is initially at rest, first moves in the direction of the electric field until, with increasing speed, the Lorentz force bends its trajectory into a perpendicular direction. This effect has opposite signs for positive ions and electrons, which leads on average to a net displacement of the charge by one gyroradius in plus or minus x-direction and represents a net polarization of the plasma.

For a linearly increasing electric field, cf. Fig. 3.13b, we obtain an ion polarization drift in the electric field direction. The electron polarization drift has the opposite sign. Hence, an increasing electric field gives rise to a polarization current. Note that the polarization drift is based on the motion of the guiding center. Since the E×B drift has now a linearly increasing speed, the trajectory of the guiding center becomes a parabola, as shown by the dashed line in Fig. 3.13b.

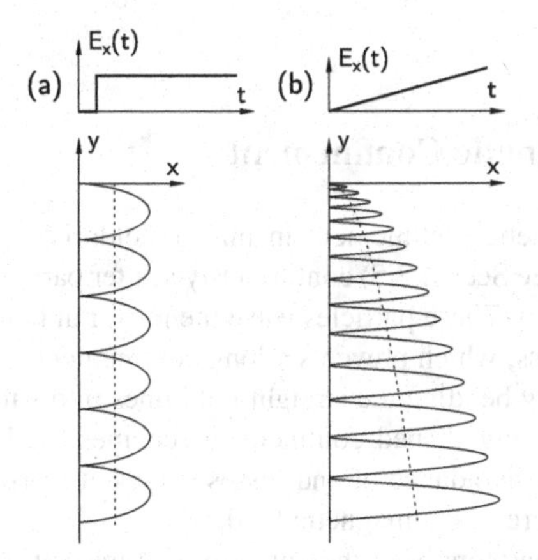

Fig. 3.13 (**a**) Polarization by the sudden switch-on of an electric field perpendicular to a static magnetic field. The ion trajectory is on average displaced by a gyroradius. (**b**) Polarization drift of positive ions in a linearly increasing electric field and a perpendicular magnetic field. Note that in both diagrams the scaling in y-direction is compressed with respect to the x-scaling. The motion of the guiding center is indicated by dashed lines

3.5.2 Time-Varying Magnetic field

Here, we consider a gyrating particle with Larmor radius $r_L = v_\perp/\omega_c$ in a homogeneous magnetic field $\mathbf{B}(t)$ that is slowly increasing. The time-varying magnetic field induces a loop voltage $\Delta U = \pi r_L^2 \, \mathrm{d}B/\mathrm{d}t$ along the gyroorbit of the particle (Fig. 3.14). Hence, in one gyroperiod $\Delta t = 2\pi/\omega_c$, the particle gains additional kinetic energy $\Delta W_\perp = q\Delta U$. Then the energy gain rate is

$$\frac{\mathrm{d}W_\perp}{\mathrm{d}t} = q\Delta U \frac{\omega_c}{2\pi} = \frac{1}{B}\frac{\mathrm{d}B}{\mathrm{d}t} W_\perp \,. \tag{3.49}$$

This gives the relation

$$\frac{\mathrm{d}W_\perp}{W_\perp} = \frac{\mathrm{d}B}{B} \,, \tag{3.50}$$

which can be integrated to give $W_\perp/B = \mu = \text{const}$. Therefore, in a slowly time-varying magnetic field the magnetic moment μ is conserved.

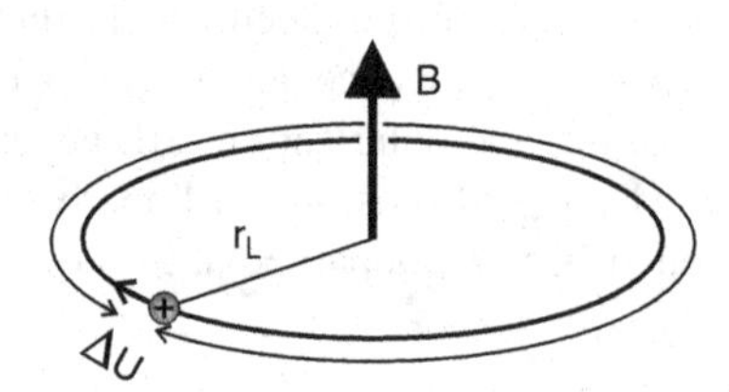

Fig. 3.14 Orbit of a positive ion in a magnetic field that increases slowly in time. During each orbit the ion gains energy from the induced loop voltage ΔU

3.6 Toroidal Magnetic Confinement

The problem of magnetic confinement in mirror fields is unsatisfactory because Coulomb collisions (see Sect. 4.2.5) continuously scatter particles into the loss-cone region of velocity space. These particles leave the mirror at both ends and represent an intolerably large loss, which prevents a long confinement time. It suggests itself to avoid these losses by bending the straight field lines into a torus, which removes the end losses. These ring-shaped confinement schemes are known as *tokamaks*[2] and *stellarators*.[3] The avoidance of end losses comes at a prize, namely the inhomogeneity and curvature of the magnetic field.

Tokamaks and stellarators have the common feature that the toroidal magnetic field is created by external field coils, as shown for a simple torus in Fig. 3.15. In this Section we will discuss the basic ideas of plasma confinement in terms of the single particle model.

The toroidal field is generated by field coils, which have a winding density n/l per unit length that is greater at the inner edge of the torus than on the outer edge.

---

[2] Tokamak is a Russian acronym meaning *toroidal chamber with magnetic field coils*

[3] Lyman Spitzer (1914–1997) originally devised a *figure-eight stellarator*. The name alludes to the Latin word *stella* for star

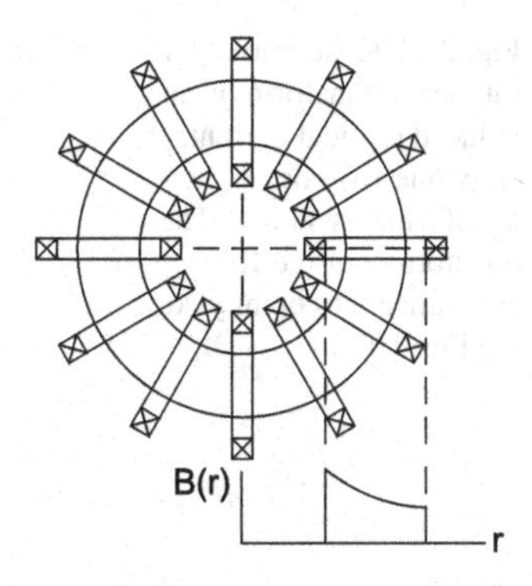

Fig. 3.15 Simple torus with field coils that generate the toroidal magnetic field. The inset shows that the toroidal magnetic field strength decays as $B_t \propto 1/r$

Hence, the magnetic field will be radially inhomogeneous, as can be seen by applying Ampere's law

$$\oint \mathbf{H} \cdot \mathrm{d}\mathbf{s} = 2\pi r H_t(r) = nI . \tag{3.51}$$

Here, the integration follows a field line of radius r which encircles the total current nI, when n is the number of windings. This means that the toroidal magnetic flux density $B_t = \mu_0 H_t$ decreases radially as

$$B_t = \mu_0 \frac{nI}{2\pi r} . \tag{3.52}$$

In this inhomogeneous and curved magnetic field, charged particles experience the combined toroidal drift Eq. (3.29), which is dependent on the sign of the charge and effects charge separation in vertical direction. This charge separation leads to the establishment of a vertical electric field that is responsible for a secondary E×B drift driving both ions and electrons radially outwards (see Fig. 3.16).

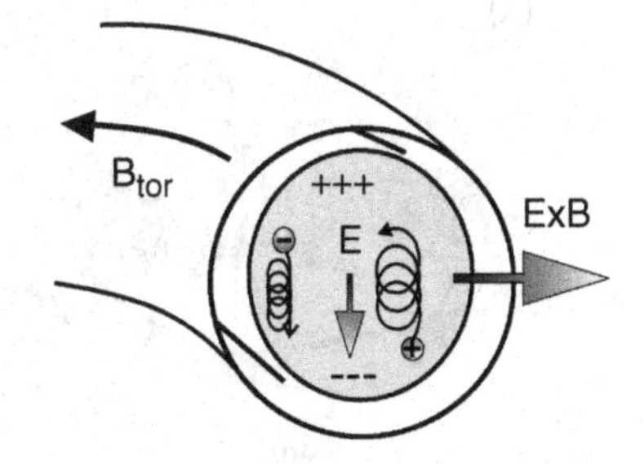

Fig. 3.16 The toroidal drift leads to charge separation and a subsequent particle loss by E×B drift

3.6.1 The Tokamak Principle

The net outward drift of the particles in a simple magnetized torus can be compensated by twisting of the toroidal field lines in such a way that a field line on the

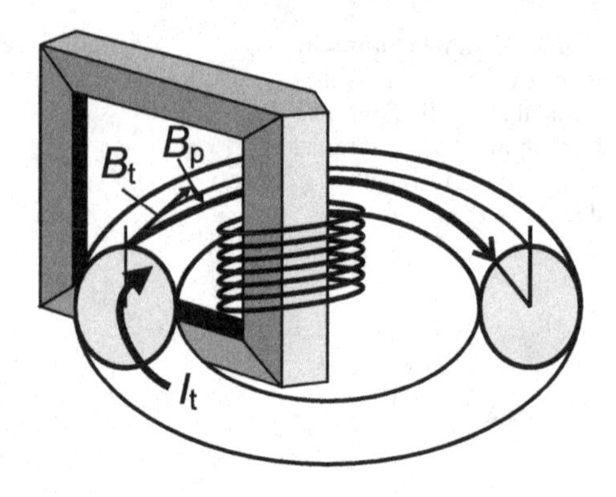

Fig. 3.17 Generation of a rotational transform by an induced toroidal current. Only one yoke of the transformer is shown. The transformer of the JET experiment has eight yokes, see Fig. 1.17**a**

outside of the torus moves to the inside of the torus after a few revolutions about the major axis of the torus. Such a *rotational transform* of the magnetic field can be achieved by superimposing a *poloidal magnetic field* B_{p}. In a tokamak, such a poloidal field is generated by inducing a toroidal current I_{t} into the plasma ring, which forms a one-turn secondary of a huge transformer. Fig. 3.17 shows the principle of a tokamak and the twisting of the field lines.

While tokamaks have been successfully used to demonstrate energy gain by fusion reactions (cf. Sect. 1.5), their operation is limited to some ten seconds because of the necessity to ramp-up the magnetic field in the transformer to drive the toroidal current by induction, which is ultimately limited by the saturation of the transformer's iron core. A mode of quasi-steady state operation can be achieved, in principle, by driving the toroidal current by other means, for example by the radiation pressure of intense radio waves or microwaves [61, 62].

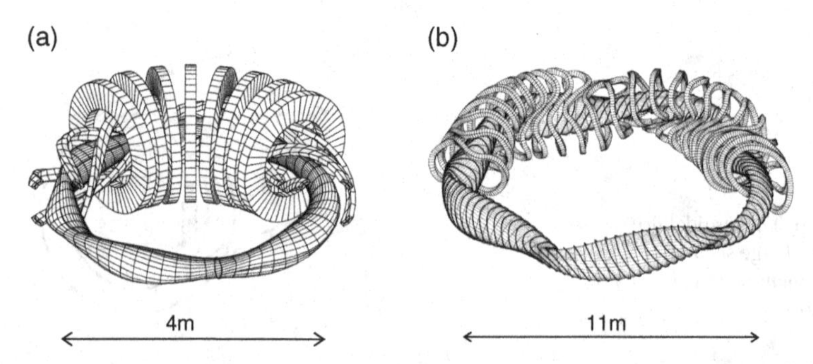

Fig. 3.18 (**a**) Field coil arrangements for the Wendelstein 7-A stellarator with planar toroidal field coils and two pairs of helical windings. (**b**) Wendelstein 7-X stellarator with superconducting modular field coils that simultaneously produce toroidal and poloidal fields. (Reproduced with permission. © IPP/MPG.)

3.6.2 The Stellarator Principle

In a stellarator, the rotational transform is produced by external currents. This makes the stellarator attractive because it allows steady-state operation. In a classical stellarator, the poloidal field is generated by pairs of conductors that are wound in a helix around the torus. Figure 3.18a shows such an arrangement for the Wendelstein 7-A stellarator [63]. The plasma has an elliptical cross-section with sudden bends and changes in the orientation of the major axis.

Modern stellarator concepts use non-planar field coils, which produce both a toroidal and a poloidal magnetic field. The most recent development is the stellarator Wendelstein 7-X, which is under construction in Greifswald, Germany. A sketch of the arrangement of the superconducting modular coils is shown in Fig. 3.18b The vacuum vessel and the cryostat are omitted.

3.6.3 Rotational Transform

The angle ι (iota), by which a magnetic field line is twisted after one revolution about the torus is easily understood for a tokamak and can be estimated as follows. Let us denote the *major radius* of the torus by R and the *minor radius* of the plasma by a. Further, r is a radial coordinate measuring from the center of the plasma. Assuming that the current density is approximately homogeneous in the plasma, we have a current density flowing in toroidal direction

$$j_\mathrm{t} = \frac{I_\mathrm{t}}{\pi a^2}, \tag{3.53}$$

where I_t is the total toroidal current. The poloidal magnetic field H_p follows from Ampere's law

$$2\pi r H_\mathrm{p} = I(r) = \pi r^2 j_\mathrm{t}. \tag{3.54}$$

The poloidal magnetic flux density B_p then increases with r

$$B_\mathrm{p} = \mu_0 \frac{I_\mathrm{t}}{2\pi a^2} r. \tag{3.55}$$

The poloidal angle, by which a field line is twisted around the torus can be estimated in the following way.

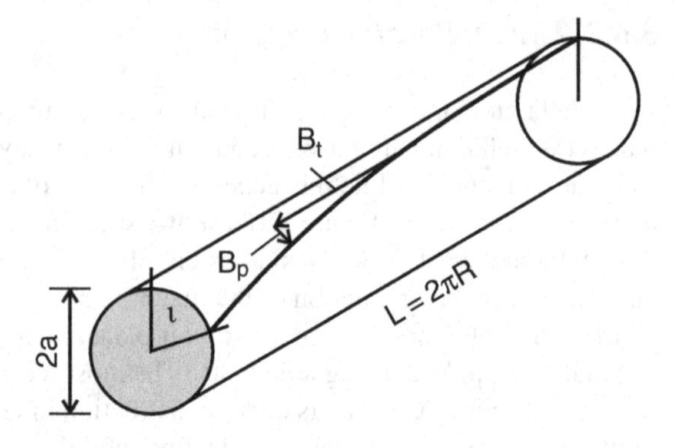

Fig. 3.19 The rotational transform $\iota(r)$ is estimated by straightening a torus into a cylinder

Consider the torus being cut and bent into a cylinder of length $L = 2\pi R$ and radius r as shown in Fig. 3.19. Then, the arc segment $r\,\iota(r)$ is given by

$$r\,\iota(r) = 2\pi R\frac{B_{\mathrm{p}}(r)}{B_{\mathrm{t}}} = 2\pi R\frac{\mu_0 I_{\mathrm{t}}}{2\pi a^2 B_{\mathrm{t}}}r$$

$$\text{and} \quad \iota(r) = \frac{\mu_0 I_{\mathrm{t}} R}{a^2 B_{\mathrm{t}}}, \tag{3.56}$$

which is independent of r. Hence, the transformation angle is the same for all radial positions and the magnetic field is unsheared. In a real tokamak, however, the current density peaks in the center of the plasma and the magnetic field is sheared. The *rotational transform* is defined as $\iota/2\pi$, and its reciprocal value $q = 2\pi/\iota$ is called the *safety factor*, which is the number of toroidal revolutions a field line has to make to complete one poloidal revolution.

Many detailed investigations of plasma confinement were made in stellarators, which can be operated in a steady state mode. The field lines can be visualized by producing a localized electron beam that travels along the field line and hits a small phosphor screen. Moving the phosphor screen in a section of the torus, the trajectory of the electron beam is intersected after different numbers of revolutions. The totality of these luminous dots can be recorded photographically with a long time exposure, as shown in Fig. 3.20a. This kind of representation is known as a Poincaré section.

When $\iota/2\pi \neq n/m$, i.e., an irrational number, the field lines form a set of nested *magnetic surfaces*, see Fig. 3.20a. This topology changes when a rational value of the rotational transform is approached, here $\iota/2\pi = 1/2$. Figure 3.20b shows that the magnetic surfaces breaks up and forms two additional magnetic islands. At rational values of the rotational transform, the plasma confinement is deteriorated, as was shown in early experiments on the Wendelstein IIa stellarator (Fig. 3.21). There, the plasma density generally increased with the applied magnetic field, but pronounced minima appeared at $\iota/2\pi = 1/3, 1/4, 1/5\ldots$.

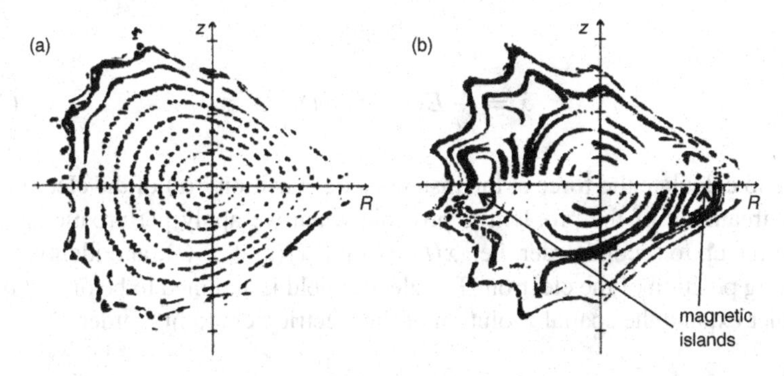

Fig. 3.20 Visualization of the rotational transform by an electron beam and a small movable phosphor screen in the Wendelstein 7-AS stellarator. (**a**) Nested magnetic surfaces for $\iota/2\pi = 0.47$. (**b**) Magnetic islands form at $\iota/2\pi = 1/2$. (Reprinted with permission from [64]. © 1993, IAEA.)

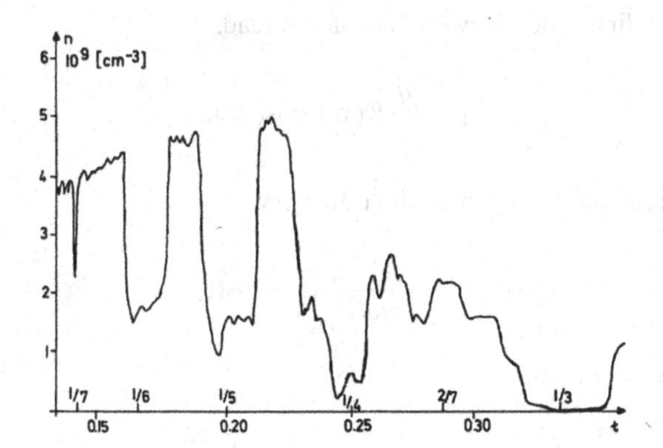

Fig. 3.21 Plasma density in the Wendelstein IIa stellerator for a toroidal magnetic field of 0.6 T and varying rotational transform. The electron density breaks down at rational values $\iota/2\pi = 1/3, 1/4, 1/5 \ldots$. (Reprinted with permission from [63]). © IPP/MPG

3.7 Electron Motion in an Inhomogeneous Oscillating Electric Field

Up to this point, we have only discussed the motion of charged particles in slowly varying magnetic fields. In this last Section, we will study the motion of electrons in an inhomogeneous oscillating electric field. This is a simplified approach to understand the behavior of electrons in radio-frequency or laser fields.

3.7.1 The Ponderomotive Force

For simplicity, we use a non-relativistic description of electron motion and consider only an inhomogeneity along the field direction x, which gives the equation of motion as

$$\ddot{x} = \frac{q}{m} E(x)\cos(\omega t)\,. \tag{3.57}$$

We have to calculate the force at the instantaneous position $x(t)$ of the electron. An analytic treatment of the problem is possible when we apply perturbation theory, retain terms up to second order, i.e., $x(t) = x_0 + x_1(t) + x_2(t)$, and identify x_0 as the starting position of the electron. The electric field is assumed to be of first order. We further expand the spatial evolution of the electric field to first order

$$E(x) \approx E(x_0) + x_1(t)\frac{\mathrm{d}}{\mathrm{d}x}E(x_0)\,. \tag{3.58}$$

Note that the product $x_1 E'$ is already of second order, so no contribution from x_2 is needed. Then, in first order, Newton's equation reads

$$\ddot{x}_1 = \frac{q}{m}E(x_0)\,\cos(\omega t), \tag{3.59}$$

from which we immediately obtain the trajectory

$$x_1(t) = -\frac{q}{m\omega^2}E(x_0)\cos(\omega t)\,. \tag{3.60}$$

In second order, we obtain

$$\ddot{x}_2 = \frac{q}{m}x_1(t)\frac{\mathrm{d}}{\mathrm{d}x}E(x_0) = -\frac{q^2}{m^2\omega^2}E(x_0)\frac{\mathrm{d}}{\mathrm{d}x}E(x_0)\cos^2(\omega t)\,. \tag{3.61}$$

Noting that $\cos^2(\alpha) = \frac{1}{2}[1 - \sin(2\alpha)]$, the electron experiences a fast acceleration at 2ω, which we are not interested in, and a mean acceleration

$$\langle \ddot{x}_2 \rangle = -\frac{q^2}{2m^2\omega^2}E(x_0)\frac{\mathrm{d}}{\mathrm{d}x}\left[E^2(x_0)\right] = -\frac{q^2}{4m^2\omega^2}\frac{\mathrm{d}}{\mathrm{d}x}\left[E^2(x_0)\right], \tag{3.62}$$

in which the average is taken over the fast time scale and the mean particle position is replaced by $x_2(t)$. The associated force

$$F_{\mathrm{p}} = -\frac{q^2}{4m\omega^2}\frac{\mathrm{d}}{\mathrm{d}x}\left[E^2(x_0)\right] \tag{3.63}$$

is called the *ponderomotive force*, which drives the electron into regions of diminishing field amplitude. Note that we do not end up with a new type of constant drift velocity.

The Basics in a Nutshell

- The complex trajectory of a charged particle in a magnetic field has been decomposed into a hierarchy of (periodic) motions

  1. gyration about the field line at the cyclotron frequency,
  2. periodic bouncing between mirror points,
  3. curvature and gradient drift, which can lead to a very slow periodic motion about the axis of the magnetic mirror.

- Each of these periodic motions is associated with an adiabatic invariant, which has a decreasing degree of conservation: the magnetic moment, the longitudinal invariant, and the flux invariant. Therefore, in the guiding center model, the real particle is replaced by a small ring current with an associated magnetic moment.
- The guiding center of this ring current performs various types of drift motion

E×B drift	$\mathbf{v}_E = (\mathbf{E} \times \mathbf{B})/B^2$
Gravitational drift	$\mathbf{v}_g = (m/q)(\mathbf{g} \times \mathbf{B})/B^2$
Gradient drift	$\mathbf{v}_{\nabla B} = (m/q)(\frac{1}{2}v_\perp^2/R_c^2)(\mathbf{R}_c \times \mathbf{B})/B^2$
Curvature drift	$\mathbf{v}_R = (m/q)(v_z^2/R_c^2)(\mathbf{R}_c \times \mathbf{B})/B^2$
Polarization drift	$\mathbf{v}_p = (m/q)(\partial \mathbf{E}/\partial t)/B^2$

- In a tokamak, the twist of the confining magnetic field is effected by the toroidal current, which is induced by a big transformer with the plasma torus as secondary winding. The rotational transform of the field lines counteracts the losses arising from plasma drifts.
- In a stellarator, the rotational transform is effected by external helical currents. Modern stellarators use modular coils which produce both the confining magnetic field and the rotational transform.

Problems

3.1 Consider a cylindrical straight wire of radius a with a homogeneous distribution of current density inside. Use Ampere's law to derive the azimuthal magnetic field $H_\varphi(r)$ for $r < a$ and $r \geq a$.

3.2 Consider now a cylindrical discharge tube, in which the plasma density profile and the associated current distribution is parabolic:

$$j(r) = j_0 \left(1 - \frac{r^2}{a^2}\right) .$$

What is the magnetic field distribution $H_\varphi(r)$ for $r < a$ in this case?

3.3 (a) What is the electron cyclotron frequency resulting from the Earth magnetic field at the author's location? (c.f. Sect. 3.1.3)
(b) What is the gyroradius of an electron with 10 eV kinetic energy in this field?

3.4 (a) The magnetic field created by a dipole of magnetic moment $\mathbf{M} = M\mathbf{e}_z$ reads in cartesian coordinates:

$$\mathbf{B}(\mathbf{r}) = \frac{\mu_0}{4\pi} \frac{3\mathbf{r}(\mathbf{r} \cdot \mathbf{M}) - r^2\mathbf{M}}{r^5} .$$

Find the corresponding components B_r and B_θ in spherical coordinates (r, θ).
(b) In the equatorial ionosphere the horizontal component of the Earth magnetic field is approximately 30 μT. Calculate the dipole moment at the Earth center that would generate such a magnetic field.

3.5 (a) Calculate the gradient of the Earth magnetic field at the magnetic equator at an altitude of 500 km and the radius of curvature of a magnetic field line, $R_c = |B_\theta/(\mathrm{d}B_\theta/\mathrm{d}r)|$.
(b) What is the speed of the gradient drift and curvature drift for electrons, which have 3 eV kinetic energy in parallel and perpendicular motion?

3.6 Determine the trajectory $[x(t), y(t)]$ of an electron in crossed fields $\mathbf{B} = (0, 0, B_z)$ and $\mathbf{E} = (E_x, 0, 0)$, when the electron is initially at rest, $\mathbf{v}(t = 0) = 0$.

3.7 The vector of the magnetic field is tangent to the field line. Therefore, the differential equation for a magnetic field line is

$$\frac{\mathrm{d}\mathbf{s}}{\mathrm{d}t} = \mathbf{e}_B .$$

Here, $\mathbf{s} = (x, y, z)$ is a point on the field line and t a parameter, which makes tick-marks along the trajectory. Write the defining equation for the field line in components, eliminate t, and show that the equation for a magnetic field line in the x–z plane reads

$$\frac{\mathrm{d}z}{\mathrm{d}x} = \frac{B_z}{B_x} .$$

Solve this differential equation for the dipole field given in Problem 3.4 by separating the variables and show that the field line is given as

$$z(x) = \sqrt{x_0^{2/3} x^{4/3} - x^2} \, ,$$

where x_0 marks the intersection of the field line with the x-axis.

Chapter 4
Stochastic Processes in a Plasma

"When I use a word", Humpty Dumpty said, in rather a scornful tone, "it means just what I choose it to mean—neither more nor less."
"The question is", said Alice, "whether you can make words mean so many different things."

Lewis Carroll, Through the Looking-Glass

The description of the hot gas of electrons and ions forming a plasma involves a number of non-deterministic or stochastic processes that require a statistical description. The plasma constituents have a wide spread of velocities and perform collisions between the charged particles, or with the gas atoms of the parent gas. The behavior of the plasma as a whole can no longer be reduced to the deterministic motion of individual particles in prescribed fields. Rather, the large number of particles introduces uncertainties that force us to describe the plasma by average quantities. For example, the average motion depends on macroscopic quantities like temperature and density gradients, which generate particle fluxes or electric currents. This Section discusses the stochastic motion of particles and introduces simple statistical concepts to describe typical transport processes in gas discharges. To illustrate the concepts, typical applications are given in gas discharges, in ion thrusters designed for spacecrafts, or in the heat balance for nuclear fusion.

4.1 The Velocity Distribution

4.1.1 The Maxwell Velocity Distribution in One Dimension

In thermodynamic equilibrium the particles forming a gas attain a Maxwell velocity distribution, which reads in one spatial dimension

$$f_{\mathrm{M}}^{(1)}(v_x) = a \exp\left(-\frac{m v_x^2}{2k_{\mathrm{B}}T}\right) . \tag{4.1}$$

Here, $f_{\mathrm{M}}^{(1)}(v_x)\mathrm{d}v_x$ is the number of particles per volume with velocities between v_x and $v_x + \mathrm{d}v_x$. $k_{\mathrm{B}} = 1.38 \times 10^{-23}\,\mathrm{J\,K^{-1}}$ is Boltzmann's constant and T the

A. Piel, *Plasma Physics*, DOI 10.1007/978-3-642-10491-6_4,

thermodynamic temperature. a is a normalization factor. The particle density n is given by the integral of f_{M} over all velocities

$$n = \int_{-\infty}^{\infty} f_{\mathrm{M}}^{(1)}(v_x)\mathrm{d}v_x \,. \tag{4.2}$$

Hence, the normalization constant a is

$$a = n\left(\frac{m}{2\pi k_{\mathrm{B}}T}\right)^{1/2} . \tag{4.3}$$

The width of the distribution function is determined by the temperature as shown in Fig. 4.1a. Defining a characteristic velocity v_T by

$$v_T = (2k_{\mathrm{B}}T/m)^{1/2} \,, \tag{4.4}$$

we see that the Maxwellian $f_{\mathrm{M}}^{(1)}(v_x) \propto \exp[-(v_x/v_T)^2]$ takes half its maximum value at $v_{1/2} = (\ln 2)^{1/2}v_T = 0.833v_T$. This characteristic velocity v_T should not be confused with the mean thermal speed defined below.

The one-dimensional Maxwell distribution can easily be generalized to three spatial dimensions by taking the product of velocity distributions for each spatial direction and renormalizing to the particle density

$$\begin{aligned} f_{\mathrm{M}}^{(3)}(\mathbf{v}) &= a' f_{\mathrm{M}}^{(1)}(v_x) f_{\mathrm{M}}^{(1)}(v_y) f_{\mathrm{M}}^{(1)}(v_z) \\ &= n\left(\frac{m}{2\pi k_{\mathrm{B}}T}\right)^{3/2} \exp\left(-\frac{m(v_x^2+v_y^2+v_z^2)}{2k_{\mathrm{B}}T}\right) . \end{aligned} \tag{4.5}$$

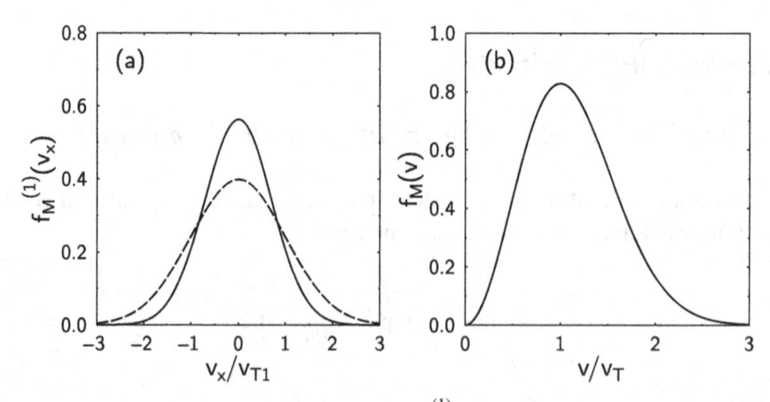

Fig. 4.1 (**a**) One-dimensional Maxwell distribution $f_{\mathrm{M}}^{(1)}(v_x)$ for two different temperatures. *Solid line*: T_1, *dashed line*: $T_2 = 2T_1$. (**b**) Maxwell distribution $f_{\mathrm{M}}(v)$ of particle speeds

Again, $f_{\rm M}^{(3)}(\mathbf{v})\mathrm{d}v_x\mathrm{d}v_y\mathrm{d}v_z$ is the density of particles which have a velocity vector with components lying between $v_x \ldots v_x + \mathrm{d}v_x$, $v_y \ldots v_y + \mathrm{d}v_y$, and $v_z \ldots v_z + \mathrm{d}v_z$.

4.1.2 The Maxwell Distribution of Speeds

When we are not interested in the distribution of a specific velocity component of a gas but want to describe the distribution of particle speeds without caring for the orientation of motion, we can define the Maxwell distribution of particle speeds $v = |\mathbf{v}|$

$$f_{\rm M}(v) = 4\pi v^2 n \left(\frac{m}{2\pi k_{\rm B}T}\right)^{3/2} \exp\left(-\frac{mv^2}{2k_{\rm B}T}\right) . \tag{4.6}$$

The additional factor $4\pi v^2$ arises from the fact that the volume element in three-dimensional velocity space is a thin sheet on the surface of a sphere. Hence, for $v \ll v_T$ the distribution of speeds rises like v^2 because of the increasing number of combinations of v_x, v_y, and v_z that lead to the same speed $v = (v_x^2 + v_y^2 + v_z^2)^{1/2}$, see Fig. 4.1b. The Maxwellian of speeds takes a maximum at v_T (see Problem 4.1). In other words, the characteristic velocity v_T is the most probable speed of the gas. For larger speeds, the Maxwellian distribution of speeds decays because the exponential decreases much more rapidly than the growth of the spherical surface in velocity space.

4.1.3 Moments of the Distribution Function

The mean thermal speed of a gas is defined as the first moment of the distribution of speeds

$$v_{\rm th} = \frac{1}{n}\int_0^\infty f_{\rm M}(v)v\,\mathrm{d}v = \left(\frac{8k_{\rm B}T}{\pi m}\right)^{1/2} , \tag{4.7}$$

which is 13% larger than v_T. Likewise, the mean kinetic energy of a gas is defined by the second moment of the distribution of speeds

$$\langle W_{\rm kin}\rangle = \frac{1}{n}\frac{m}{2}\int_0^\infty f_{\rm M}(v)v^2\mathrm{d}v = \frac{3}{2}k_{\rm B}T . \tag{4.8}$$

The evaluation of these two integrals is left to the reader (see Problems 4.2 and 4.3).

The relation between mean kinetic energy and temperature is often a source of confusion in plasma physics. There, it is common practice to convert temperature units into energy units by the relation $W = k_B T$. In this language, 1 eV = 11,600 K, which brings typical temperatures in gas discharges to small numbers of a few eV. Nevertheless, from Eq. (4.8), an electron gas of 10 eV temperature has a mean kinetic energy of 15 eV.

4.1.4 Distribution of Particle Energies

For many calculations we need the particle distribution function on an energy scale $W = \frac{1}{2}mv^2$. For this purpose, we must perform a transformation from speed to energy, which conserves the number of particles in a certain velocity and energy interval, respectively, $f_M(v)\mathrm{d}v = F_M(W)\mathrm{d}W$. With $\mathrm{d}W = mv\,\mathrm{d}v$, we obtain

$$F_M(W) = n\frac{2}{\sqrt{\pi}}\frac{1}{(k_B T)^{3/2}} W^{1/2} \exp\left(-\frac{W}{k_B T}\right). \tag{4.9}$$

For gas discharges, the terminology *electron energy distribution function* (EEDF) and *electron energy probability function* (EEPF) is used. For a Maxwellian, the EEDF $F(W)$ can be identified with $F_M(W)$, but in most gas discharges $F(W)$ is a non-Maxwellian distribution. The EEPF is defined as

$$g(W) = W^{-1/2} F(W). \tag{4.10}$$

In a semilog plot, $\log[g(W)]$ vs. W, we obtain a straight line when the distribution is Maxwellian (see Fig. 4.2). Note that the slope of the EEPF and EEDF in the semilog plots are slightly different. Therefore, the EEPF must be used for directly reading the temperature.

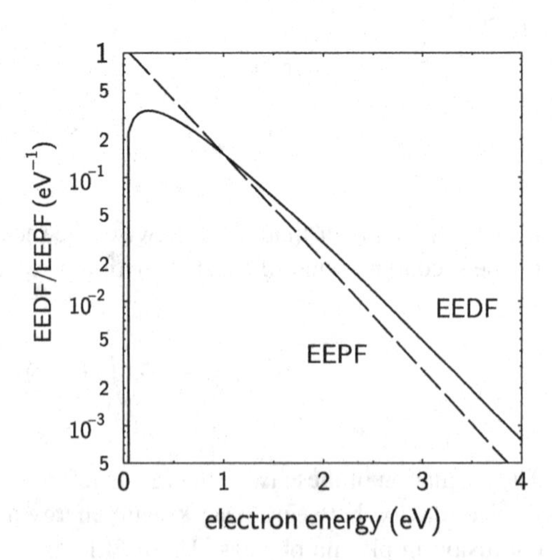

Fig. 4.2 Comparison of EEDF (*solid line*) and EEPF (*dashed line*) for a Maxwellian distribution

We will see in Sect. 7.5.1 that the EEPF is the immediate result from evaluating the second derivative of a Langmuir probe characteristic. The particle density can be obtained from the integral of the EEDF

$$n_e = \int_0^\infty F_M(W)dW \tag{4.11}$$

and the *effective temperature*

$$\frac{3}{2}k_B T_e = \frac{1}{n_e}\int_0^\infty W F_M(W)dW\,. \tag{4.12}$$

4.2 Collisions

In this section, we first introduce the concept of atomic cross sections, which define the collisions between charged plasma particles and atoms in gas discharges. Later, we discuss Coulomb collisions between charged particles, which are primarily important for fully ionized hot plasmas.

4.2.1 Cross Section

The probability that an electron hits an atom can be described by a geometrical quantity, the cross section of an assumed atomic sphere. Such a classical picture of fine particles hitting a target is also valid in quantum mechanics when the de Broglie wavelength of the electron is small compared to the size of the atom. Collisions between atoms, or between an ion and atom, can often be approximated by collisions of two billiard balls with radii r_1 and r_2. The concept of cross section assumes that the projectile is point-like and the target is assigned the effective cross section. Hence, a collision between billiard balls is defined by the sum of the collision radii $\sigma = \pi(r_1 + r_2)^2$, as shown in Fig. 4.3.

The real situation for electron-atom collisions is more complex. The fact that the cross section for elastic electron collisions in argon shows a minimum at small energies (*Ramsauer effect*) can be understood as an interference phenomenon when

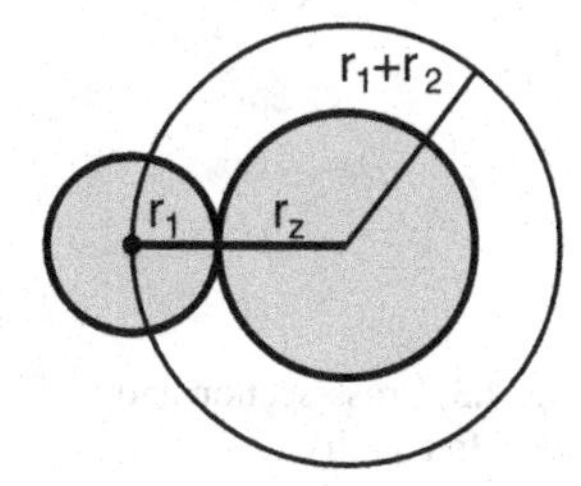

Fig. 4.3 Collision between two spheres of radius r_1 and r_2. The collision is described by a point-like particle hitting an effective cross section $\sigma = \pi(r_1 + r_2)^2$

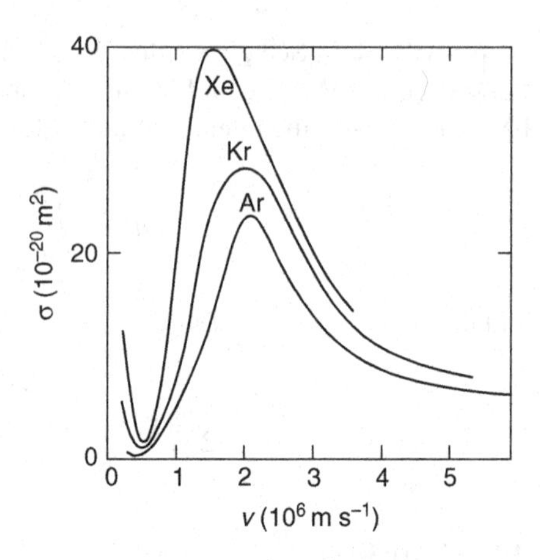

Fig. 4.4 Total cross-section for collisions of electrons with noble gas atoms. At low energy, the atoms show the so-called Ramsauer minimum at small collision energies (Data taken from [65])

the electron's de Broglie wavelength matches the atom size. The total collision cross section for electrons in the noble gases Ar, Kr and Xe [65] is shown in Fig. 4.4.

4.2.2 Mean Free Path and Collision Frequency

For calculating the collision probability we assume a directed stream of point-like particles with a velocity v that penetrate a gas volume. The targets have a cross section σ (Fig. 4.5). We consider a cylinder of entrance area A and length $\Delta z = v\Delta t$. This volume contains N_a atoms acting as target spheres and the density of these targets is $n_a = N_a(A\Delta z)^{-1}$. The differential probability for a collision in this infinitesimal volume is then the ratio of blocked area $N_a\sigma$, i.e., the shadow of the particles at the exit plane, to the total area A.

$$\Delta w = \frac{N_a\sigma}{A} = n_a\sigma\,\Delta z\,. \tag{4.13}$$

The macroscopic probability to traverse a distance z without any collision is the product of the differential probabilities to suffer no collision in any of the intermediate steps

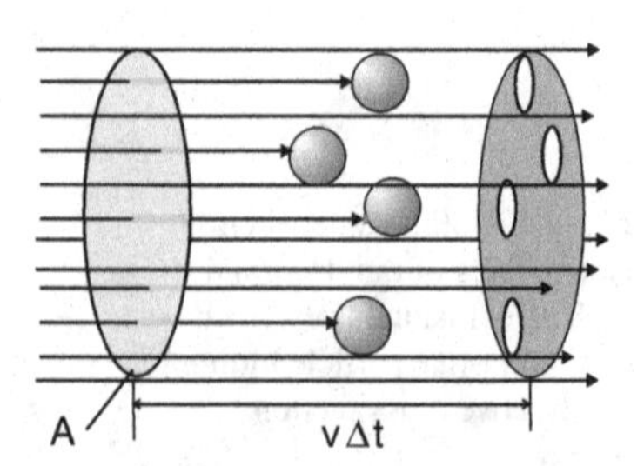

Fig. 4.5 Cross-section and mean free path

$$w(z) = \lim_{\Delta z \to 0} \prod_{i=1}^{z/\Delta z} (1 - n_a \sigma \Delta z) = \lim_{\Delta z \to 0} (1 - n_a \sigma \Delta z)^{z/\Delta z}$$
$$= \exp(-n_a \sigma z) = \exp(-z/\lambda_{\mathrm{mfp}}) \,. \tag{4.14}$$

We call the quantity $\lambda_{\mathrm{mfp}} = (n_a \sigma)^{-1}$ the *mean free path*. Because λ_{mfp} is inversely proportional to the density of targets the mean free path for plasma particles in the neutral gas background scales as $\lambda_{\mathrm{mfp}} \propto p_{\mathrm{gas}}^{-1}$. The collision frequency is defined as

$$\nu_{\mathrm{coll}} = \sigma \, v \, n_{\mathrm{a}} \tag{4.15}$$

and is the reciprocal of the mean free time τ_{coll} between two interactions.

Pressure Units

The international unit of gas pressure is the Pascal (1 Pa = 1 Nm^{-2}). Despite many attempts of introducing metric SI units, there are still different units in use. The conversion between these units is compiled here:

1 bar	= 10^5 Pa
1 torr	= 133 Pa
1 Pa	= 7.52 mtorr
1 bar	= 0.9869 atm (phys. atmosphere)
1 atm	= 760 torr

In the older literature, gas discharge conditions are often referred to the atom density at 1 torr pressure and $T = 273$ K, which is $n_{\mathrm{a}} = 3.54 \times 10^{22}$ m^{-3}.

4.2.3 *Rate Coefficients*

Up to now, we have only considered the collisions of a monoenergetic particle hitting a target. In a gas discharge, we may be interested in the number of ionizing collisions per second that are caused by a Maxwellian electron distribution with a temperature T_{e}. This leads to the definition of rate coefficients for ionization or excitation.

The *rate coefficient* for ionization by electrons is defined as the average of $\sigma_{\mathrm{ion}} v$ over the actual *velocity probability function*, $(1/n_{\mathrm{e}}) f_{\mathrm{M,e}}(v)$, which is normalized to unity,

$$\langle \sigma v \rangle = \frac{1}{n_{\mathrm{e}}} \int_0^\infty \sigma_{\mathrm{ion}}(v) v \, f_{\mathrm{e}}(v) \, \mathrm{d}v \,, \tag{4.16}$$

and has the dimension volume per second. Rate coefficients for other processes are defined in a similar way. The average number of ionizing events by a single electron is obtained as the *ionization frequency*

$$\nu_{\text{ion}} = n_{\text{a}} \langle \sigma_{\text{ion}} \nu \rangle = \frac{n_{\text{a}}}{n_{\text{e}}} \int_0^\infty f_{\text{M,e}}(\nu) \sigma_{\text{ion}}(\nu) \nu \text{d}\nu \,. \tag{4.17}$$

The lower bound of the latter integral can be chosen at the ionization threshold W_{ion}.
The total number of ionization processes per volume and second is

$$S_{\text{ion}} = n_{\text{e}} n_{\text{a}} \langle \sigma_{\text{ion}} v \rangle \,. \tag{4.18}$$

For ionization of atoms by electron collisions, we need not worry about the proper relative velocity of the collision partners, which is determined by the electron velocity in the laboratory frame because the electron mass is much smaller than the atom mass. The situation is different for collision partners of comparable masses m_1 and m_2. Then, the reduced mass $m_{\text{r}} = m_1 m_2 (m_1 + m_2)^{-1}$ must be inserted in the distribution function $f_{\text{M}}(v)$ that is used in (4.17).

Typical electron temperatures in low-pressure gas discharges are a few electron volts only, say 3 eV in an argon discharge, which has an ionization energy of 15.8 eV. This means that the ionization is mostly due to electrons in the tail of the distribution function. The ionization rate is therefore determined by the overlap of the decreasing Maxwellian and the increasing ionization cross section, as indicated by the shaded region in Fig. 4.6. Because electrons in this overlap regime loose energy by the ionization process, the true distribution function will be depleted above the ionization threshold. Therefore, in accurate calculations, one has to use such a self-consistent distribution function.

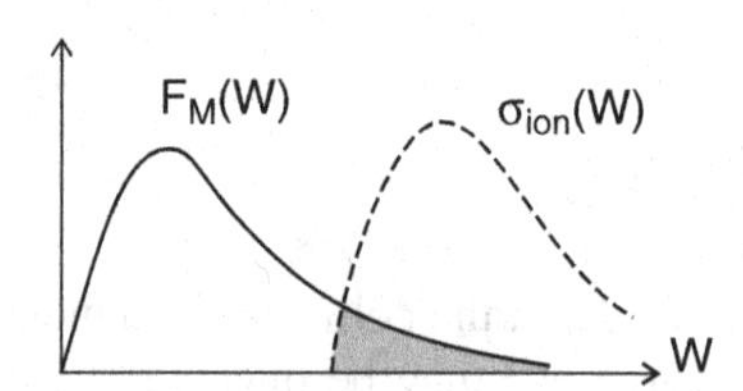

Fig. 4.6 The ionization frequency is determined by the overlap of the distribution function $F_{\text{M}}(W)$ and the ionization cross section $\sigma_{\text{ion}}(W)$. Only the shaded tail of the distribution function contributes to ionization

4.2.4 Inelastic Collisions

Inelastic collisions of electrons and ions can lead to excitation or ionization of atoms. Here, we are only interested in a comparison of the cross sections for ionization and elastic scattering. Experimental data for the ionization cross sections for helium and argon atoms are shown in Fig. 4.7 [66].

Below the ionization energy W_{ion} (He: 24.6 eV, Ar: 15.8 eV) the ionization cross section is zero; it rapidly rises to take a flat maximum at about 4 W_{ion} and decays like $W^{-1/2}$ at high projectile energies. Generally, the ionization cross section is two orders of magnitude smaller than the cross section for elastic collisions. The ionization frequency in a plasma can be calculated from Eq. (4.17).

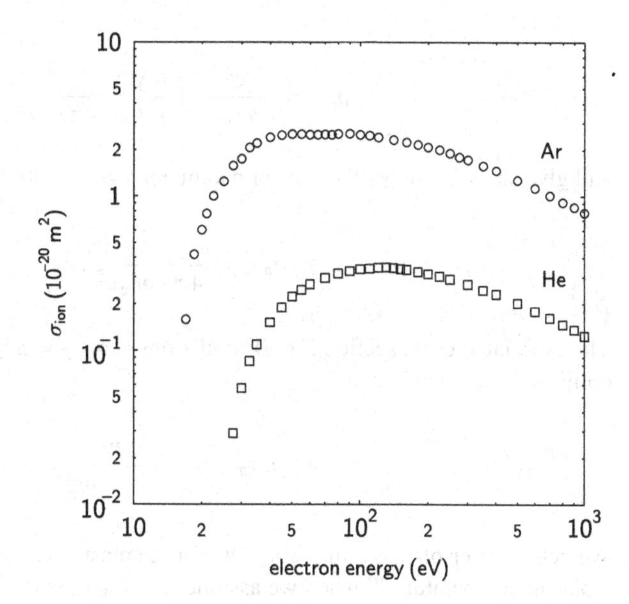

Fig. 4.7 Ionization cross-sections for helium and argon atoms from tabulated values published in [66]

4.2.5 Coulomb Collisions

Collisions between charged particles play an important role in fully ionized plasmas. There, the scattering of electrons by ions is responsible for the resistivity of a hot plasma. Collisions between electrons do not change the total momentum of the electron gas and therefore do not affect the conductivity. However they determine classical electron diffusion.

In this paragraph, a simplified description of Coulomb collisions is presented. The trajectory of an electron in the field of an isolated ion is shown in Fig. 4.8. The impact parameter b is assumed to be smaller than the Debye length, which justifies the approximation by a Coulomb force. For simplicity, we split the hyperbolic orbit into three sections, the two asymptotes with impact parameter b and a nearly circular arc. The time of interaction can be estimated as $\tau \approx b/v$. For large deflection angles, the change in momentum is roughly equal to the initial momentum $\Delta p \approx p$, and can be estimated from the Coulomb force and the interaction time, $\Delta p \approx F_{\mathrm{C}}\tau$. These considerations yield

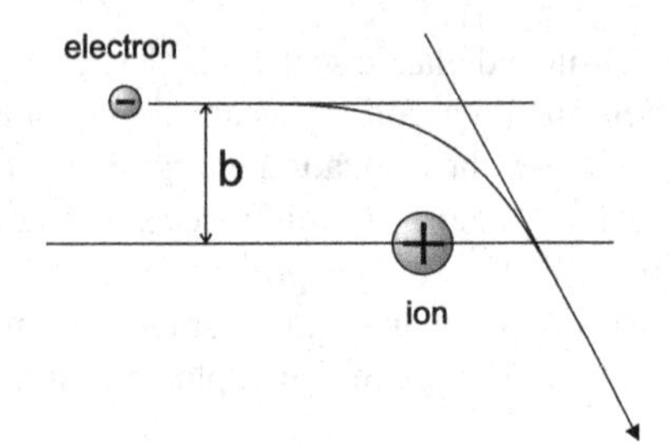

Fig. 4.8 Geometry of an electron-ion collision

$$m_{\mathrm{e}}v = \frac{e^2}{4\pi\varepsilon_0 b^2}\left(\frac{b}{v}\right) = \frac{e^2}{4\pi\varepsilon_0 bv} \tag{4.19}$$

and give an estimate for the impact parameter $b_{\pi/2}$ for 90° scattering

$$b_{\pi/2} = \frac{e^2}{4\pi\varepsilon_0 m_{\mathrm{e}} v^2}\,. \tag{4.20}$$

The associated cross section for 90° collisions is $\sigma_{\pi/2} = \pi b_{\pi/2}^2$ and the electron-ion collision frequency becomes

$$\nu_{\mathrm{ei}} = n\sigma_{\pi/2}v = \frac{ne^4}{16\pi\varepsilon_0^2 m_{\mathrm{e}}^2 v^3}\,. \tag{4.21}$$

We get an order-of-magnitude estimate for the plasma resistivity, $\eta = m_{\mathrm{e}}\nu_{\mathrm{ei}}/ne^2$, at a plasma temperature T when we assume $v = (k_{\mathrm{B}}T/m_{\mathrm{e}})^{1/2}$

$$\eta = \frac{\pi e^2 m_{\mathrm{e}}^{1/2}}{(4\pi\varepsilon_0)^2(k_{\mathrm{B}}T)^{3/2}}\,. \tag{4.22}$$

This rough estimate already shows that the resistivity is independent of the electron density and scales with $T^{-3/2}$. These simple arguments were based on large deflection angles. A more thorough treatment of the problem leads to the *Spitzer resistivity* [67]

$$\eta_{\mathrm{s}} = \frac{\pi e^2 m_{\mathrm{e}}^{1/2}}{(4\pi\varepsilon_0)^2(k_{\mathrm{B}}T)^{3/2}}\ln\Lambda\,. \tag{4.23}$$

that contains a correction factor $\ln(\Lambda) \approx \ln(\lambda_{\mathrm{D}}/b_{\pi/2}) = \ln(4\pi N_{\mathrm{D}})$, called the *Coulomb logarithm* that is related to the number of particles N_{D} in a Debye sphere.

A hot plasma of $T = 10\,\mathrm{keV}$ has a resistivity $\eta = 5\cdot 10^{-10}\,\Omega\,\mathrm{m}$, which is lower than the resistivity of copper $\eta_{\mathrm{Cu}} = 2\cdot 10^{-8}\,\Omega\,\mathrm{m}$. This explains why a hot plasma behaves like a nearly perfect conductor. We will discuss Coulomb collisions in more detail in Chap. 10 on dusty plasmas.

4.3 Transport

Transport processes in plasmas comprise mobility-limited motion, electric currents described by conductivity, and free or ambipolar diffusion. Here, only steady-state processes will be discussed. The section ends with a discussion of the influence of a magnetic field on mobility.

4.3.1 Mobility and Drift Velocity

In a gas discharge with a low degree of ionization, the motion of electrons and ions is governed by the applied electric field and collisions with the atoms of the background gas. Most of the electron collisions are elastic. Therefore, we will neglect ionizing collisions in the calculation of friction forces. Because of the equal mass of positive ions and atoms of the parent gas, the momentum exchange between the heavy particles is very efficient. Besides elastic scattering, the process of charge exchange plays an important role, in which a moving ion captures an electron and leaves a slow ion behind. In the momentum balance this process is equivalent to a head-on collision in a billiards game.

A cartoon of electron motion in a gas background is shown in Fig. 4.9. In the collision between a light electron with a heavy atom the momentum transfer is small. Rather, the incoming electrons experience a random redirection of their momentum. The trajectory is a sequence of parabolic segments. Since we have no diagnostic to follow the trajectories of individual electrons, we must be content with evaluating the average motion of an ensemble of electrons.

The equation of motion for an individual electron can be written as

$$m_e \dot{\mathbf{v}}_e = -eE + \sum_k m_e \Delta \mathbf{v}_k \delta(t - t_k)\,. \tag{4.24}$$

Here $m_e\, \Delta \mathbf{v}_k$ is the momentum loss in the kth collision. By averaging this equation over many collisions we obtain the mean drift velocity $\langle \mathbf{v}_e \rangle$. Then, the sum on the r.h.s. of Eq. (4.24) becomes $m_e \langle \Delta \mathbf{v}_e \rangle \tau_{\mathrm{coll}}^{-1}$, which represents the average momentum

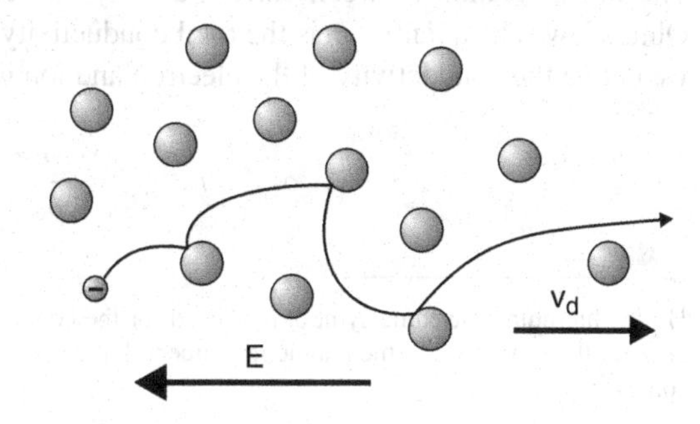

Fig. 4.9 Cartoon of an electron trajectory in a homogeneous electric field. The trajectory is interrupted by elastic collisions with neutral atoms

loss per unit time. $\tau_{\rm coll}$ is the mean free time between two collisions defined in Eq. (4.15).

The elastic scattering of electrons on atoms is almost isotropic [68]. Therefore, on average, the electron loses its mean momentum $m_{\rm e}\bar{\mathbf{v}}_{\rm e}$ and we can write the equation of motion for an *average electron*

$$m\dot{\bar{v}} = -eE - m\bar{v}\nu_{\rm m} \,. \tag{4.25}$$

This average electron now moves in $-E$-direction. The quantity $\nu_{\rm m} = 1/\tau_{\rm coll}$ is the effective collision frequency for momentum transfer. Because of the one-dimensional motion, the vector symbol was dropped. The solution of this equation of motion

$$\bar{v}(t) = -\frac{eE}{m\nu_{\rm m}}\left[1 - {\rm e}^{-\nu_{\rm m}t}\right] + v(0){\rm e}^{-\nu_{\rm m}t} \tag{4.26}$$

has two parts: the first describes the approach to a terminal velocity

$$v_{\rm d} = -\frac{e}{m\nu_{\rm m}}E = -\mu_{\rm e}E \,, \tag{4.27}$$

the second the loss of memory on the initial velocity v_0. The terminal velocity $v_{\rm d}$ is called the *drift velocity*, which is established when the electric field force is balanced by the friction force. The mobilities of electrons and ions are defined as

$$\mu_{\rm e} = \frac{e}{m_{\rm e}\nu_{\rm m,e}} \quad ; \quad \mu_{\rm i} = \frac{e}{m_{\rm i}\nu_{\rm m,i}} \,. \tag{4.28}$$

4.3.2 Electrical Conductivity

The drift velocity of electrons and ions can be used to define the electric current density

$$j = j_{\rm e} + j_{\rm i} = n[(-e)v_{\rm de} + ev_{\rm di}] = ne(\mu_{\rm e} + \mu_{\rm i})E = \sigma E \,. \tag{4.29}$$

The linear relation between current density and electric field is the equivalent to Ohm's law. The quantity σ is the total conductivity[1] of the gas discharge. Likewise we define the conductivity of the electron and ion gas

$$\sigma_{\rm e,i} = ne\mu_{\rm e,i} = \frac{ne^2}{m_{\rm e,i}\nu_{\rm m}} \,. \tag{4.30}$$

---

[1] In the literature, the same symbol σ is used for the conductivity and the collision cross section, or μ for the mobility and the magnetic moment, but confusion is unlikely because of the different context

This concept for the conductivity of a gas discharge is applicable at a low degree of ionization n_e/n_a. For a typical gas pressure of 1 mbar and room temperature the atom density is $n_a = 2.8 \times 10^{22}\,\mathrm{m}^{-3}$ whereas a typical electron density can be $n_e \leq 10^{19}\,\mathrm{m}^{-3}$.

4.3.3 Diffusion

The average electron and ion motion in a gas discharge is determined by electric field forces and by density gradients. The latter type of net motion is called diffusion. What is diffusion on a microscopic scale? Let us consider a situation with a hot electron gas that has initially a density gradient in negative x-direction (see Fig. 4.10).

Since the electron thermal speed is much higher than that of the gas atoms we assume that the gas atoms are at rest. Because of the density gradient it is evident that per unit time more electrons will move to the right than to the left. This gives rise to a net down-hill motion. However, this electron motion is inhibited by collisions with the gas atoms. This combination of electron thermal motion and friction with the neutral gas is described by a diffusion coefficient D. Again, we can only describe the average motion in terms of a relation between the density gradient and a resulting particle flux $\boldsymbol{\Gamma}_{\mathrm{e,i}}$

$$\boldsymbol{\Gamma}_{\mathrm{e,i}} = n_{\mathrm{e,i}}\bar{\mathbf{v}}_{\mathrm{e,i}} = -D\nabla n_{\mathrm{e,i}}\,, \tag{4.31}$$

which is known as *Fick's law*. Such relations were originally developed for neutral gases, in which the motion of particles is determined by collisions with other particles of the same species. In that situation, diffusion is the result from a greater number of collision with neighboring particles of the same kind from the left than from the right, which on average gives a net force directed down-hill.

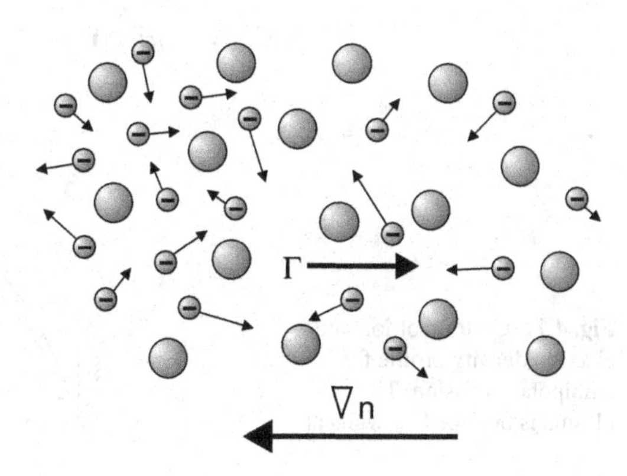

Fig. 4.10 Cartoon of electron diffusion in an electron density gradient

4.3.3.1 Ambipolar Diffusion

The situation for plasma electrons is quite different, because diffusion does not mean that the electrons collide with other electrons. This effect can be neglected for weakly coupled plasmas. Rather, as described above, the net motion is only determined by their thermal motion and the inhomogeneous density distribution. In this way, electron diffusion is similar to drift motion with the electron temperature—together with the density gradient—providing the driving force. The same considerations can be applied to ions.

Einstein had shown that diffusion coefficient and mobility are related by the temperature of the gas

$$\frac{D}{\mu} = \frac{k_{\mathrm{B}}T}{e}\,. \tag{4.32}$$

This relation quantifies the arguments given above that electron diffusion in a neutral gas background with a density gradient is driven by the temperature and inhibited by electron-neutral collisions. Because the diffusion of electrons and ions leads to different values of the particle fluxes, which would lead to unequal densities of electrons and ions, the plasma reacts by forming a space charge electric field E. This field reduces the electron diffusion and accelerates the ion diffusion until the two fluxes reach a common value and the plasma remains macroscopically neutral. This final state is called *ambipolar diffusion* when electrons and ions are lost at the same rate and E is called the *ambipolar electric field*.

Figure 4.11 shows schematically how electron and ion density profiles in a plasma look like under the influence of ambipolar diffusion. The difference between the two profiles is exaggerated, for clarity. In the plasma center, a surplus of ions is expected that generates a positive plasma potential in the plasma center because electrons have the tendency to leave the system faster than ions. Therefore, a slight surplus of electrons is found in the outer plasma region. The corresponding space charge field E that accelerates the ions but slows down the electrons, is indicated.

The particle fluxes for this diffusion process are given by

$$\boldsymbol{\Gamma}_{\mathrm{e,i}} = \pm n\mu_{\mathrm{e,i}}\mathbf{E} - D_{\mathrm{e,i}}\nabla n\,. \tag{4.33}$$

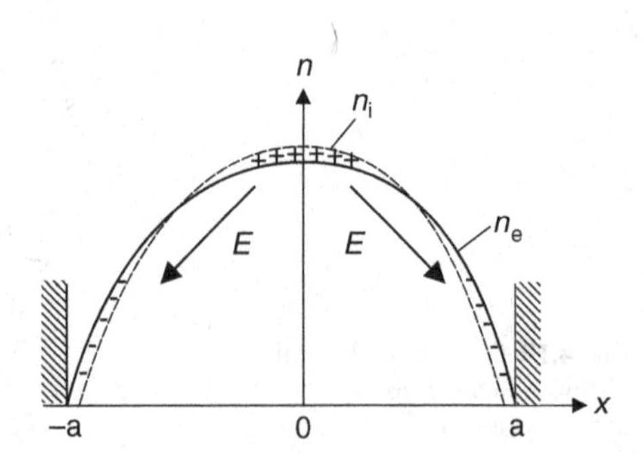

Fig. 4.11 Cartoon of ion and electron density profile for ambipolar diffusion. The plasma is bounded by walls at $x = \pm a$

$D_{\mathrm{e,i}}$ are the diffusion coefficients for electrons and ions. The upper sign refers to electrons, the lower to ions. For equal electron and ion fluxes (in the one-dimensional geometry of Fig. 4.11) we obtain the set of equations

$$\Gamma_{\mathrm{a}} = -n\mu_{\mathrm{e}}E - D_{\mathrm{e}}\frac{\mathrm{d}n}{\mathrm{d}x}$$
$$\Gamma_{\mathrm{a}} = +n\mu_{\mathrm{i}}E - D_{\mathrm{i}}\frac{\mathrm{d}n}{\mathrm{d}x}, \tag{4.34}$$

from which we can determine the ambipolar flux

$$\Gamma_{\mathrm{a}} = -D_{\mathrm{a}}\frac{\mathrm{d}n}{\mathrm{d}x} \tag{4.35}$$

and define the ambipolar diffusion coefficient

$$D_{\mathrm{a}} = \frac{D_{\mathrm{i}}\mu_{\mathrm{e}} + D_{\mathrm{e}}\mu_{\mathrm{i}}}{\mu_{\mathrm{e}} + \mu_{\mathrm{i}}}. \tag{4.36}$$

The inequality $D_{\mathrm{i}} < D_{\mathrm{a}} < D_{\mathrm{e}}$ shows that the ion diffusion is accelerated while the electron diffusion is reduced. Finally, the ambipolar electric field is given by

$$E = -\frac{D_{\mathrm{e}} - D_{\mathrm{i}}}{\mu_{\mathrm{e}} + \mu_{\mathrm{i}}}\frac{1}{n}\frac{\mathrm{d}n}{\mathrm{d}x}. \tag{4.37}$$

Using $D_{\mathrm{e}} \gg D_{\mathrm{i}}$, $\mu_{\mathrm{e}} \gg \mu_{\mathrm{i}}$ and Eq. (4.32) the ambipolar field can be estimated as

$$E_{\mathrm{a}} \approx -\frac{k_{\mathrm{B}}T_{\mathrm{e}}}{e\ell}, \tag{4.38}$$

where $\ell = n/(\mathrm{d}n/\mathrm{d}x)$ is a characteristic length scale.

The ambipolar field is the typical response of a collisional plasma to the initial imbalance of losses by an electron flux that exceeds the ion flux. The ambipolar field holds the more diffusive electrons back and pulls the less diffusive ions towards a higher speed (see Fig. 4.12).

Fig. 4.12 The ambipolar field ties a mobile species to a less mobile one and secures that both agree on the same speed

4.3.3.2 Diffusion in Cylindrical Geometry

A typical example for diffusion problems is the formation of a plasma density profile in a homogeneous cylindrical glass tube of length $L \gg a$, where a is the tube radius. In gas discharge physics, this situation is found in the positive column—a region of quasineutral plasma that is characterized by the balance of plasma production and radial ambipolar diffusion with subsequent recombination at the wall. The governing equation for the steady-state problem is the extended equation of continuity with a source term by ionization

$$\frac{\partial n}{\partial t} + \nabla \cdot (n\mathbf{v}) = \nu_{\mathrm{ion}}\, n\,. \tag{4.39}$$

Here, we have neglected the small difference between n_{e} and n_{i}, which gives rise to the ambipolar field and set $n_{\mathrm{e}} \approx n_{\mathrm{i}} = n$. It is further assumed that the electron temperature is the same everywhere. This justifies to ascribe a fixed ionization frequency to each electron. Therefore, the number of ionization processes per volume and second can be written as the product of electron density and ionization frequency.

For steady state solutions, we can drop the term $\partial n/\partial t$. Combining with Fick's law for ambipolar diffusion,

$$n\mathbf{v} = -D_{\mathrm{a}} \nabla n\,, \tag{4.40}$$

and assuming that the diffusion coefficient D_{a} is constant, we obtain in cylindrical coordinates

$$\frac{\partial^2 n}{\partial r^2} + \frac{1}{r}\frac{\partial n}{\partial r} + \frac{\nu_{\mathrm{ion}}}{D_{\mathrm{a}}} n = 0\,. \tag{4.41}$$

Here, we have assumed that, in steady state and in the long-tube limit, there is no dependence on the coordinates φ and z. Equation (4.41) is the special case ($m = 0$) of Bessel's differential equation $x^2 y'' + x y' + (x^2 - m^2) y = 0$. Requiring that the plasma density has to be positive and must vanishes at the inner tube wall, $r = a$, we obtain the radial density profile as

$$n(r) = n(0) J_0\left(\frac{2.405 r}{a}\right). \tag{4.42}$$

$p_1 = 2.405$ is the first zero of the Bessel function $J_0(x)$. The differential equation has a second solution, $Y_0(x)$, which has a singularity at $r = 0$ and can be discarded here. The Bessel functions $J_0(x)$ and $Y_0(x)$ are shown in Fig. 4.13. Inserting the solution Eqs. (4.42) in (4.41) yields a relation between the physical parameters ν_{ion} and D_{a}

$$\nu_{\text{ion}} = D_{\text{a}} \left(\frac{2.405}{a} \right)^2 , \tag{4.43}$$

which defines the steady-state condition $\nu_{\text{ion}} = \nu_{\text{loss}}$. This means that the frequency at which a plasma particle is lost by diffusion to the wall, is given by the r.h.s. of Eq. (4.43). We can interpret $\ell = a/2.405$ as the characteristic length for the Bessel profile Eq. (4.42).

In summary, the plasma in a cylindrical discharge tube has a radial density profile with a peak in the center, as shown in Fig. 4.14. This can be understood because the number of ionization processes $S_{\text{ion}} = n_{\text{e}}\nu_{\text{ion}}$ is maximum there whereas losses occur only at the wall. The plasma potential in the center is the integral of the ambipolar electric field, which points radially outward, and is more positive than the wall potential. This type of discharge plasma is found, for example, in fluorescent tubes or in neon display tubes.

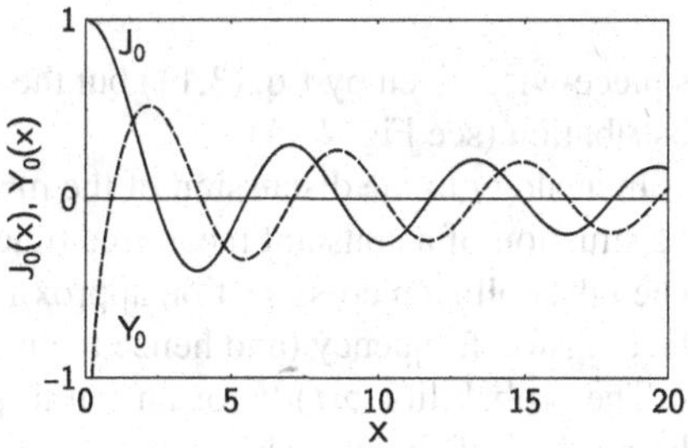

Fig. 4.13 The Bessel functions of first and second kind, $J_0(x)$ and $Y_0(x)$. The Bessel functions are the "cousins" of cosine and sine for cylindrical geometries

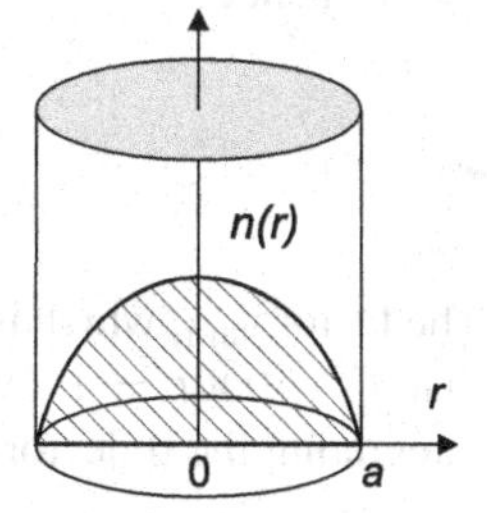

Fig. 4.14 The density profile in a cylindrical tube attains its maximum on the axis and becomes zero at the wall $r = a$

4.3.4 Motion in Magnetic Fields in the Presence of Collisions

In Sect. 3.1.4 we have seen that, without collisions, an ion in crossed electric and magnetic fields performs an E×B drift at right angle to the electric field. In the following we will see that current flow across the magnetic field requires collisions. The trajectory of an ion in crossed electric and magnetic fields under the action of collisions can be discussed thoroughly when we assume that a collision leads to the total transfer of momentum like in charge-exchange collisions. Then, the trajectory

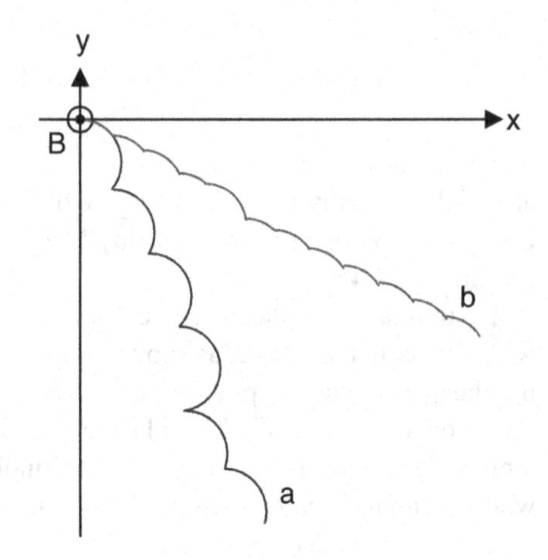

Fig. 4.15 Typical ion trajectories in crossed fields. The electric field is oriented in x-direction. (a) $\nu_{m,i}/\omega_{ci} < 1$ (b) $\nu_{m,i}/\omega_{ci} \approx 1$. For increasing collisionality $\nu_{m,i}$, the trajectory from the $E \times B$-direction to the E-direction

is piece-wise given by Eq. (3.11) but the length of the segments follows a statistical distribution (see Fig. 4.15).

In analogy to the discussion of the mean free path in Eq. (4.14), we discuss here the situation of a constant mean free time between collisions. This concept is valid, when the collision cross section approximately decreases as $\sigma \propto v^{-1}$, which makes the collision frequency (and hence the mean time between collisions) constant.

The probability $p(t)\mathrm{d}t$ for an ion to perform a collision after a given time t is the product of the probability to have no collision for $0 < t' < t$, which decays exponentially, and the probability to have a collision between t and $t + \mathrm{d}t$, which is $\nu_{m,i}\mathrm{d}t$, hence

$$p(t) = \nu_{m,i} \exp(-\nu_{m,i} t) . \tag{4.44}$$

The factor $\nu_{m,i}$, which is the ion collision frequency for momentum transfer, ensures that $\int_0^\infty p(t)\mathrm{d}t = 1$. We obtain the mean displacement between two collisions by integrating the trajectory Eq. (3.11) over the probability distribution (4.44)

$$\Delta x = \frac{E_x \nu_{m,i}}{\omega_{ci} B_z} \int_0^\infty [1 - \cos(\omega_{ci} t)] \exp(-\nu_{m,i} t)\mathrm{d}t ,$$

$$\Delta y = \frac{E_x \nu_{m,i}}{\omega_{ci} B_z} \int_0^\infty [\sin(\omega_{ci} t) - \omega_{ci} t] \exp(-\nu_{m,i} t)\mathrm{d}t . \tag{4.45}$$

The evaluation of the integrals is straightforward and yields the average particle velocities

$$\bar{v}_x = \nu_{\mathrm{m,i}}\Delta x = \frac{E_x}{B_z}\frac{\omega_{\mathrm{ci}}/\nu_{\mathrm{m,i}}}{1+(\omega_{\mathrm{ci}}/\nu_{\mathrm{m,i}})^2} = \frac{\mu_{\mathrm{i}}E_x}{1+(\omega_{\mathrm{ci}}/\nu_{\mathrm{m,i}})^2}$$
$$\bar{v}_y = \nu_{\mathrm{m,i}}\Delta y = -\frac{E_x}{B_z}\frac{(\omega_{\mathrm{ci}}/\nu_{\mathrm{m,i}})^2}{1+(\omega_{\mathrm{ci}}/\nu_{\mathrm{m,i}})^2} \tag{4.46}$$

Here, $\mu_{\mathrm{i}} = e(m\nu_{\mathrm{m,i}})^{-1}$ is the ion mobility. The resulting velocities $\bar{v}_x$ and $\bar{v}_y$ are plotted in Fig. 4.16 as a function of the *Hall parameter*, $\omega_{\mathrm{ci}}/\nu_{\mathrm{m,i}}$, which describes the number of gyro periods between two collisions. The Hall effect in solid matter was discovered, in 1879, by Edwin Hall (1855–1938). When $\omega_{\mathrm{ci}}/\nu_{\mathrm{m,i}} \gg 1$, the ions experience only few collisions and the velocity $\bar{v}_y$ approaches the E×B velocity while $\bar{v}_x \to 0$. This is the limit of a magnetized plasma. In the opposite limit, $\omega_{\mathrm{ci}}/\nu_{\mathrm{m,i}} \gg 1$, the ion motion is preferentially in x-direction and approaches the collisional result $\bar{v}_x = \mu_{\mathrm{i}}E_x$ of the unmagnetized plasma.

Instead of particle velocities we can also consider current densities, which lead to the ion conductivity tensor in a collisional magnetized plasma

$$\begin{pmatrix} j_x \\ j_y \\ j_z \end{pmatrix} = \sigma_{\mathrm{i}} \begin{pmatrix} \frac{1}{1+(\omega_{\mathrm{ci}}/\nu_{\mathrm{m,i}})^2} & \frac{\omega_{\mathrm{ci}}/\nu_{\mathrm{m,i}}}{1+(\omega_{\mathrm{ci}}/\nu_{\mathrm{m,i}})^2} & 0 \\ \frac{-\omega_{\mathrm{ci}}/\nu_{\mathrm{m,i}}}{1+(\omega_{\mathrm{ci}}/\nu_{\mathrm{m,i}})^2} & \frac{1}{1+(\omega_{\mathrm{ci}}/\nu_{\mathrm{m,i}})^2} & 0 \\ 0 & 0 & 1 \end{pmatrix} \cdot \begin{pmatrix} E_x \\ E_y \\ E_z \end{pmatrix} . \tag{4.47}$$

Here, $\sigma_{\mathrm{i}} = ne^2/(m_{\mathrm{i}}\nu_{\mathrm{m,i}})$ is the ion conductivity in the unmagnetized plasma. The current in electric field direction is called the *Pedersen current* and the cross-field current the *Hall current*. The current along the magnetic field is the same as in the unmagnetized case. Similar expressions can be derived for electrons, in which the ratio $\omega_{\mathrm{ce}}/\nu_{\mathrm{m,e}}$ determines the direction of the current.

In the ionosphere, the conductivity parallel to the field lines is several orders of magnitude higher than the Hall and Pedersen conductivities. Therefore, magnetic field lines connecting the ionosphere with the magnetosphere can be considered as wires that transport current between these regions. Moreover, magnetic field

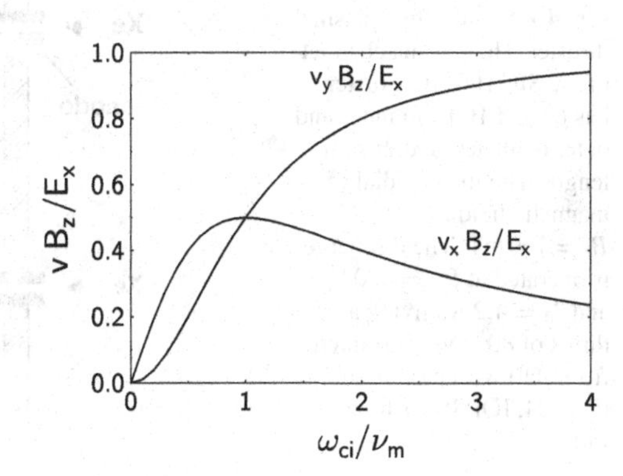

Fig. 4.16 Normalized ion velocities in a crossed field situation with collisions

lines become essentially equipotentials because the perpendicular components of the electric field become larger than the parallel electric fields.

In summary, the ion motion in crossed electric and magnetic fields under the influence of collisions is the proper model to discuss the transition from unmagnetized ions to magnetized ions. The essential parameter is the ion Hall parameter $\omega_{ci}/\nu_{m,i} = \mu_i B$, which determines the amount of Hall motion. When the Hall parameter is small, collisions dominate over gyromotion and the ions are unmagnetized. When it becomes large, the ions are magnetized. The same terminology applies to the electrons. When both electrons and ions are magnetized, the entire plasma is called magnetized. The following example shows that a plasma with magnetized electrons but unmagnetized ions has interesting new properties.

4.3.5 Application: Cross-Field Motion in a Hall Ion Thruster

Modern ion thrusters for small spacecrafts make use of the plasma Hall effect described in the previous Section. Such a device (as described in the review [69] and in references therein) was used for ESA's Earth-to-Moon mission SMART-1 [70, 71]. The principle of this device is shown in Fig. 4.17. The plasma is created inside an annular gap that contains a metallic anode through which the propellant gas (xenon) is introduced. The walls of the plasma gap are insulated with ceramics. The most important part of this device is the magnetic circuit, consisting of magnets and ferromagnetic discs (hatched areas), which produces a localized transverse magnetic field near the exit of the plasma channel. A dc discharge is operated by applying a high voltage between an electron gun and the anode. The electron gun

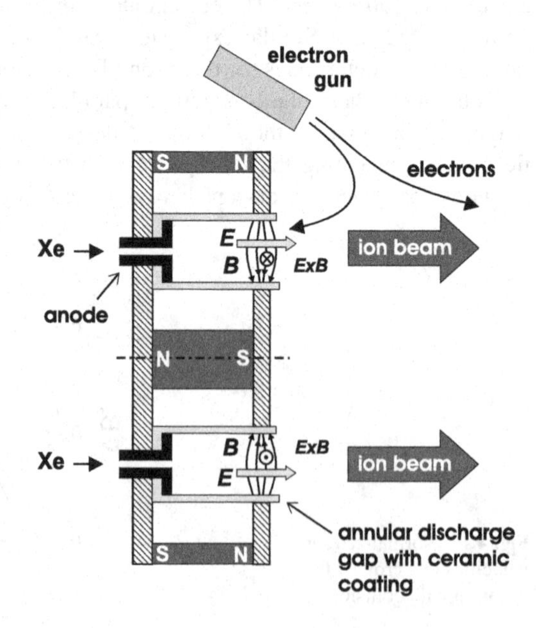

Fig. 4.17 Hall-effect plasma thruster. The plasma channel of the SPT100ML thruster has 69 and 100 mm inner and outer diameter, and 25 mm length. The mean radial magnetic field is $B_r = 160$ mT. The discharge is operated at $U_d = 300$ V and $I_d = 4.2$ A, giving a thrust of 80 mN. (Reprinted from [69] with permission. © 2004, IOP Publishing Ltd.)

has a dual purpose: it delivers primary electrons for the discharge process and acts as neutralizer for the exiting ion beam.

The transverse magnetic field in the discharge gap of $B_r = 160\,\mathrm{mT}$ guarantees that the electron Hall parameter $\omega_{ce}/\nu_{m,e}$ becomes very large. Therefore, the electron motion is preferentially in E×B direction (indicated by the symbols ⊙ and ⊗) and the electrons are practically confined in a ring. These electrons have a speed $v_{e,\varphi} = 2.5 \times 10^6\,\mathrm{m\,s^{-1}}$, which can be explained by E×B motion when we assume that the discharge voltage of 300 V drops over a distance of 7.5 mm, which compares well with the thickness of the magnetized region. The mean electron speed corresponds to an energy of 17.7 eV, compared to an ionization energy for xenon of 12.1 eV. The number of ionization processes per volume and second, $S_i = n_{Xe}\langle\sigma_{ion}v\rangle$, attains large values under these conditions, leading to 90–95% ionization of the propellant gas.

The axial motion of electrons, which constitutes part of the discharge current, is not fully understood yet because the cross-field mobility of the electrons resulting from electron-neutral collisions is too low to describe the actual electron current. It was conjectured [69] that an anomalous collision frequency results from microinstabilities, which were the subject of recent investigations [72]. Although the anomalous collision frequency will reduce the effective electron Hall parameter, the electron cross-field resistivity will be still high enough that practically the entire voltage drop of the discharge occurs across the magnetized plasma layer.

The ion motion is nearly unaffected by the magnetic field in the discharge gap. Here, we do not use the argument that the ion-neutral collision frequency exceeds the ion cyclotron frequency, which applies to extended plasma regions that are much larger than the size of the gyroorbit. Rather, we see that a xenon ion in the anodic part of the plasma gap has essentially wall temperature. As soon as it enters the magnetized region at the exit of the gap, the ion is accelerated by the high electric field $E_z \approx 4 \times 10^4\,\mathrm{V\,m^{-1}}$ and performs a short section of a gyroorbit of radius $r_L = E(B\omega_{ci})^{-1} = 204\,\mathrm{m}$. Since this radius is very much larger than the thickness of the magnetized region, the ion orbit is practically straight, and the ion gains the entire potential energy from the potential drop and leaves the thruster with 300 eV energy. The acceleration to such an exhaust velocity of about $2 \times 10^4\,\mathrm{m\,s^{-1}}$ across the magnetized plasma region has been verified by spatially-resolved laser induced fluorescence measurements on Xe^+ ions [73].

A beam of charged particles cannot simply be blown into ambient space, because it represents a space charge $\rho = j_i/v$, which repels other ions that leave the beam source at a later time. Such fundamental questions of space-charge limited flow will be discussed in Sects. 9.2 and 9.4.4. Therefore, the ion space charge is neutralized by a source of electrons near the exhaust of a thruster. In the Hall thruster, the electron gun delivers these electrons.

In summary, the Hall ion thruster is an example for a plasma where different Hall parameters lead, on the one hand, to E×B motion of the electrons for efficient ionization and, on the other hand, to localization of the potential drop in the magnetized region and subsequent acceleration of ions in the axial electric field.

4.4 Heat Balance of Plasmas

In this Section, we will discuss the heat balance in different plasma systems. The simplest example is the heating of electrons in a low-pressure gas discharge, which gives a first insight, why the electron temperature can be much higher than the temperature of the heavy particles. The other two examples address the conditions to realize controlled nuclear fusion in magnetic confinement fusion experiments and in inertial confinement fusion.

4.4.1 Electron Heating in a Gas Discharge

So far we were only interested in the mean motion of the electron (or ion) gas which resulted in the electric current and the diffusive flux. Now we are looking at the random motion of electrons, especially at the energy gain and loss processes, and the resulting electron temperature.

Between two elastic collisions, electrons gain kinetic energy in the electric field at a rate

$$\left.\frac{\mathrm{d}W}{\mathrm{d}t}\right|_{\mathrm{gain}} = -eEv_{\mathrm{d}} = \frac{e^2}{m_{\mathrm{e}}\nu_{\mathrm{m}}}E^2 \,. \tag{4.48}$$

From conservation of momentum and energy in an elastic collision between a light and heavy particle, in which the light particle is deflected by an angle θ, we find (Problem 4.4) that the energy loss ΔW of the light particle is given by

$$\left.\frac{\Delta W}{W}\right|_{\mathrm{loss}} = 2\frac{m_{\mathrm{e}}}{m_{\mathrm{a}}}[1 - \cos(\theta)] \,. \tag{4.49}$$

This energy loss has to be averaged for an isotropic distribution of scattering angles θ. Because of this isotropy, the proper average is calculated in spherical coordinates

$$\left\langle\frac{\Delta W}{W}\right\rangle = \frac{1}{4\pi}\int_0^{2\pi}\mathrm{d}\varphi\int_0^{\pi}\mathrm{d}\theta\,\sin\theta\,\frac{2m_{\mathrm{e}}}{m_{\mathrm{a}}}(1-\cos\theta) = \frac{2m_{\mathrm{e}}}{m_{\mathrm{a}}} \,. \tag{4.50}$$

The result shows, how small the fractional energy loss is. Therefore, the mean kinetic energy (i.e., the electron temperature) will rise until the energy loss rate equals the energy gain Eq. (4.48). While the real energy loss rate is determined by elastic and inelastic collisions, we contend ourselves here with discussing the more frequent elastic losses

$$\left.\frac{\mathrm{d}W}{\mathrm{d}t}\right|_{\mathrm{loss}} = n\langle\Delta W\rangle\nu_{\mathrm{m}} \,. \tag{4.51}$$

In balancing gain and loss rates, we must bear in mind that the electron temperature is not only contained in $\langle W \rangle = (3/2)k_B T_e$ but also in the collision frequency $\nu_m \approx v_T/\lambda_{mfp}$. Using Eqs. (4.48) and (4.51) this leads to the gain-loss balance

$$\langle W \rangle \nu_m^2 = \frac{e^2}{m_e} \frac{m_a}{2m_e} E^2 \tag{4.52}$$

in which the l.h.s. is proportional to $(k_B T_e)^2$. Finally, we obtain an estimate for the equilibrium electron temperature

$$k_B T_e = 3^{-1/2} \left(\frac{m_a}{2m_e} \right)^{1/2} eE\lambda_{mfp} \, . \tag{4.53}$$

As we had expected, the electron temperature is proportional to the energy gain between two collisions, which is proportional to $E\,\lambda_{mfp}$, the product of the electric field and the mean free path. However, the proportionality factor is not the atom-electron mass ratio, as one could have conjectured from Eq. (4.49), but only the square root of that expression. Besides increasing the temperature, we had also to account for the decrease of gain rate and increase of loss rate with the square root of temperature. The proportionality $T_e \propto E\lambda_{mfp}$ is known in gas discharge physics as the fundamental law, $T_e = f(E/p)$, where the gas pressure p represents the atom density.

In summary, the electron temperature is not simply given by the energy gain $E\,\lambda_{mfp}$ between two collisions. Rather, the collision is mainly responsible for changing the direction of the momentum vector and only a small fraction $\propto m_e/m_a$ of the energy is lost in this collision. So, in a sequence of free flights between two collisions, the kinetic energy of the electron gets higher and higher. A steady state is reached when the rate of energy loss equals the energy gain as described by Eq. (4.53). This shows that the electron temperature is high because the cooling mechanism is inefficient.

An everyday experience for a flow equilibrium that is governed by a loss rate, which depends on the reached level of a quantity, is shown in Fig. 4.18. When a bath tub is filled with a constant water current I_{in}, the water level h is determined by Bernoulli's law that gives the speed of outflow as $v = (2gh)^{1/2}$ and the water

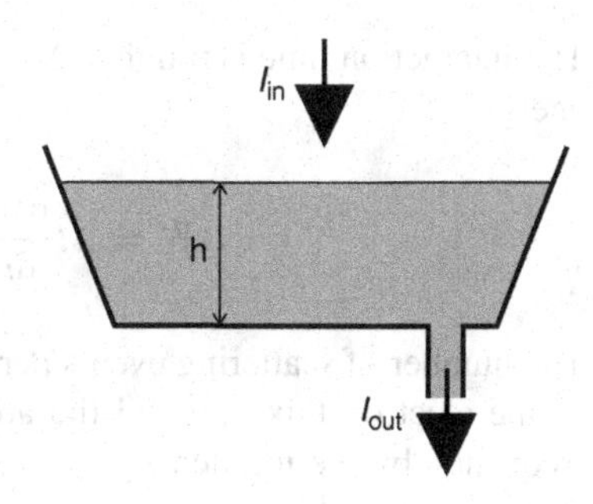

Fig. 4.18 The bath tub analogy for a flow equilibrium that is determined by the loss rate (see text)

outflow current through an opening of cross-section A becomes $I_{\mathrm{out}} = Av$. Hence, $h = I_{\mathrm{in}}^2/(2gA^2)$. The water level rises to the filling height h until the outflow rate reaches the inflow rate. By the same principle, the electron temperature rises until the gain-loss equilibrium is reached.

4.4.2 *Ignition of a Fusion Reaction: The Lawson Criterion*

The establishment of a steady-state nuclear fusion reaction requires that there is more energy produced per unit time by fusion reactions than is lost by radiation processes, and by the loss of energetic particles to the walls. Otherwise, the plasma temperature would decrease and the fusion reaction would extinguish. In a hot plasma of $T > 10\,\mathrm{keV}$, all deuterium and tritium atoms will be fully ionized. The main source of radiation in a dilute plasma, as used in magnetic fusion experiments, is *Bremsstrahlung* that arises from Coulomb collisions of electrons with ions.

4.4.2.1 Bremsstrahlung

Bremsstrahlung is well known from X-ray tubes, where electrons of (20–100) keV energy hit a solid target (e.g., a copper or tungsten anode). These energetic electrons can penetrate deep into the electron shells of the atom and are deflected by the strong electric field of the atomic nucleus. The deflection is an accelerated motion, which leads to radiation of the electrons. The non-relativistic expression for the radiated energy per unit time at an acceleration a is [74]

$$\frac{\mathrm{d}W}{\mathrm{d}t} = \frac{e^2}{6\pi\varepsilon_0 c^3}\,a^2\,. \tag{4.54}$$

In the case of a hot plasma, the acceleration is given by the Coulomb force of the ion of charge number Z at an impact parameter b (see Fig. 4.19). The characteristic value for the acceleration is then given by

$$a \approx \frac{Ze^2}{4\pi\varepsilon_0 b^2 m_{\mathrm{e}}}\,. \tag{4.55}$$

The interaction time is roughly $\Delta t = 2b/v$, which gives an estimate of the radiated energy

$$\Delta W = \Delta t\frac{\mathrm{d}W}{\mathrm{d}t} = \frac{4}{3}\frac{Z^2e^6}{(4\pi\varepsilon_0)^3 m_{\mathrm{e}}^2 c^3 b^3 v}\,. \tag{4.56}$$

The number of scattering events per second for a single ion is given by the product of the electron flux $n_{\mathrm{e}}v$ and the area of the ring $2\pi b\mathrm{d}b$. When we multiply this frequency by the ion density n_{i}, we obtain the event rate per volume and second.

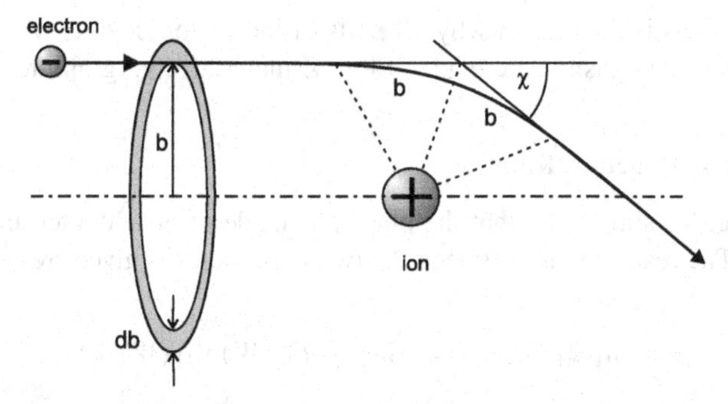

Fig. 4.19 Scattering geometry for an electron-ion collision. The trajectory is approximated by straight lines and two curved segments of length b. The impact parameter b defines a ring of area $2\pi b db$ for deflection by an angle χ

Multiplying again by the radiated energy ΔW of a single event, we obtain the energy loss rate per volume from collisions with an impact parameter between b and $b + db$,

$$\mathrm{d}P_{\mathrm{br}} = \frac{8\pi}{3} \frac{Z^2 e^6 n_\mathrm{e} n_\mathrm{i}}{(4\pi\varepsilon_0)^3 m_\mathrm{e}^2 c^3} \frac{db}{b^2} . \tag{4.57}$$

To obtain the total radiation power by Bremsstrahlung per unit volume, we have to integrate this expression over all meaningful impact parameters. Obviously, there is a singularity at $b = 0$. However, at the smallest scales, the electron can no longer be treated as a point charge. At atomic scales, the electron shows its wave character, which is described by the *de Broglie wavelength*

$$\lambda_\mathrm{B} = \frac{h}{m_\mathrm{e} v} \approx \frac{h}{(3 k_\mathrm{B} T_\mathrm{e} m_\mathrm{e})^{1/2}} . \tag{4.58}$$

In the last step we have used $\frac{1}{2} m_\mathrm{e} v^2 = \frac{3}{2} k_\mathrm{B} T_\mathrm{e}$ to estimate the electron speed. Integrating Eq. (4.57) from $b_{\mathrm{min}} = \lambda_\mathrm{B}$ to infinity, we obtain

$$P_{\mathrm{br}} = \frac{8\pi^2}{\sqrt{3}} \frac{(k_\mathrm{B} T_\mathrm{e})^{1/2}}{(4\pi\varepsilon_0)^3 m_\mathrm{e}^{3/2} c^3 h} n_\mathrm{e} n_\mathrm{i} Z^2 . \tag{4.59}$$

This approximate result gives the correct dependence on the physical quantities. A precise treatment, assuming a Maxwellian distribution of the electrons and quantum mechanical corrections, yields an increase by a factor of the order 2 [75].

In an ideal fusion plasma, there will be only deuterium and tritium ions from the filling gas and α-particles or ^{3}He nuclei from the fusion reaction, which lead to electron scattering. Impurity ions with a high Z-number, however, can dramatically increase the energy losses by Bremsstrahlung because of the Z^2 dependence in Eq. (4.59). This is the reason why all parts of the fusion device, which may get into contact with the plasma, are made of low-Z material, e.g., graphite.

4.4.2.2 Nuclear Reaction Rate

Let us assume, for simplicity, that the plasma is made of 50% deuterium and 50% tritium ions. The reaction rate between the two ion species is given by

$$S_{\mathrm{DT}} = n_{\mathrm{D}} n_{\mathrm{T}} \langle \sigma_{\mathrm{DT}} v \rangle = n_{\mathrm{D}} n_{\mathrm{T}} \int F'_{\mathrm{M}}(W) \sigma_{\mathrm{DT}}(W)\, \mathrm{d}W\,. \tag{4.60}$$

The expression is similar to the ionization rate Eq. (4.18), but we have to bear in mind that this calculation requires the proper relative velocity of reaction partners, which have similar mass. $F'_{\mathrm{M}}(W)$ is the energy distribution function Eq. (4.9), but normalized to unity. This can be done by using the reduced mass of the deuterium–tritium system in the Maxwell distribution. The reaction rates in Fig. 4.20 are calculated in this way [38]. The power from this reaction is $P_{\mathrm{DT}} = S_{\mathrm{DT}} Q_{\mathrm{DT}}$. Here, $Q_{\mathrm{DT}} = 17.6\,\mathrm{MeV}$ is the sum of the energies in the α-particle and in the neutron.

The D–T reaction attains appreciable values for ion temperatures of the order of 10 keV. Although the maximum of the D–T fusion cross-section (Fig. 1.10) is only reached at $\approx 70\,\mathrm{keV}$, the plasma temperature can be considerably below this value. We had observed a similar tendency for the electron temperature and ionization threshold in Sect. 4.2.3.

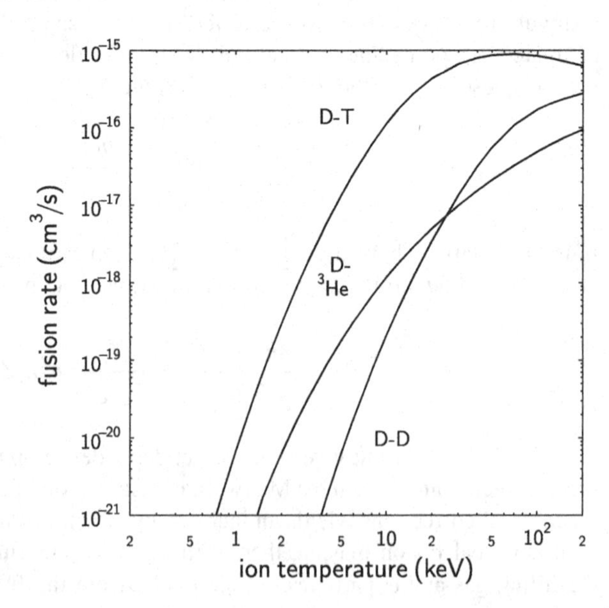

Fig. 4.20 Rate of fusion processes as a function of ion temperature (from [38])

4.4.2.3 Lawson's Energy Balance

The thermal energy content (per unit volume) of a hot isothermal ($T_e = T_D = T_T =: T$) and quasineutral ($n_e = n_D + n_T$) D–T fusion plasma is

$$W_{th} = \frac{3}{2} k_B \left(n_e T_e + n_D T_D + n_T T_T \right) = 3 n_e k_B T \,. \tag{4.61}$$

The heat loss per unit volume and per second can be expressed in terms of an energy confinement time τ_E

$$P_H = \frac{W_{th}}{\tau_E} \,. \tag{4.62}$$

The conditions for maintaining the plasma temperature were first analyzed by John D. Lawson (1923–2008) [76], who assumed that the total power leaving the plasma, $P_{br} + P_H + P_{DT}$, can be first converted to electricity and then to heating power (e.g., using radio waves) with an overall efficiency η, which then replenishes the heat loss and Bremsstrahlung loss. This gives the balance

$$P_{br} + P_H = \eta (P_{br} + P_H + P_{DT}) \,. \tag{4.63}$$

Inserting the functional dependence of the various terms on the temperature and using $n_D = n_T = n_e/2$, we obtain

$$A_{br} n_e^2 (k_B T)^{1/2} + \frac{3 n_e k_B T}{\tau_E} = \frac{\eta}{1-\eta} \frac{1}{4} n_e^2 \langle \sigma v \rangle Q_{DT} \tag{4.64}$$

$$n_e \tau_E = \frac{3 k_B T}{\frac{\eta/4}{1-\eta} \langle \sigma v \rangle Q_{DT} - A_{br} (k_B T)^{1/2}} \,. \tag{4.65}$$

Because $\langle \sigma v \rangle$ is only a function of temperature, we can plot a diagram $n_e \tau_E = f(T)$ (cf. Fig. 4.20). For this purpose, we use the coefficient for Bremsstrahlung $A_{br} = 5.35 \times 10^{-37}\,\mathrm{Wm^{-3}}$ which is valid for $Z = 1$ and when $k_B T$ is given in keV [75]. The fusion reactivity $\langle \sigma v \rangle$ is taken from [38].

The curves in Fig. 4.21 can be interpreted as follows: Assuming an efficiency $\eta = 0.3$, a minimum value of the product $n_e \tau_E = 7 \times 10^{19}\,\mathrm{m^{-3}\,s^{-1}}$ must be reached. At the same time, the plasma temperature must be ≈ 25 keV. These two conditions are known as the Lawson criterion for steady state operation. A higher efficiency leads to a lower value of $n_e \tau_E$.

4.4.2.4 Ignition

So far we have assumed that the α-particles are escaping from the plasma and are part of the heat loss P_H. When the α-particles, which have an energy of

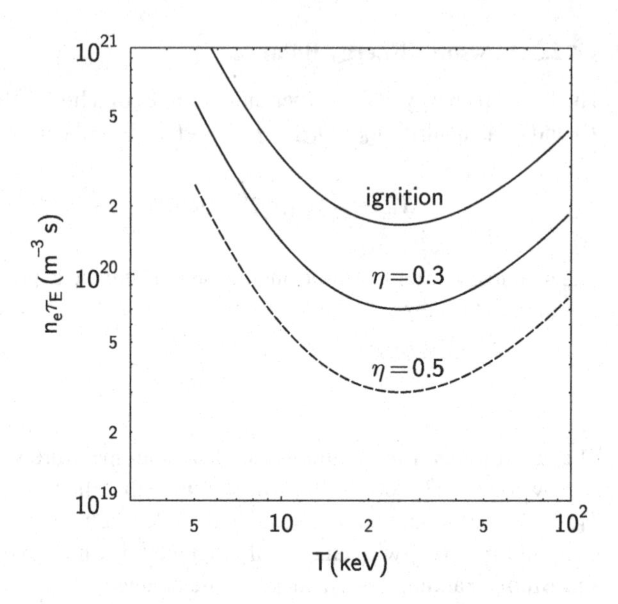

Fig. 4.21 Lawson's criterion for steady state operation at $\eta = 0.5$ and $\eta = 0.3$. The ignition curve corresponds to $\eta = 0.154$

$Q_\alpha = 3.52\,\text{MeV}$, can be confined, this energy can be used to compensate for the heat losses, which defines the ignition condition

$$P_{\text{br}} + P_{\text{H}} = P_\alpha = \frac{1}{4}n_{\text{e}}^2\langle\sigma v\rangle Q_\alpha\,. \tag{4.66}$$

One can easily show that this condition is equivalent to Lawson's condition (4.63) for $\eta = 0.154$. The corresponding ignition curve is also shown in Fig. 4.20. Ignition requires $n_{\text{e}}\tau_{\text{E}} > 1.5 \times 10^{20}\,\text{m}^{-3}\,\text{s}$.

4.4.2.5 The Fusion Triple Product

The definition of the Lawson parameter $n_{\text{e}}\tau_{\text{E}}$ may lead to the conjecture that the fusion yield could be increased by feeding more neutral gas into the reactor. While the electron density would generally rise in such an attempt, the plasma temperature would decay. The reason for this behavior lies in the fact that the plasma pressure $p = 2n_{\text{e}}k_{\text{B}}T$ determines the plasma losses. Therefore, a fusion plasma is operated at a constant pressure. A useful figure of merit for a magnetic fusion experiment is the so-called *triple product* $n_{\text{e}}\tau_{\text{E}}T$, which can also be identified as the product of plasma pressure and energy confinement time. The evolution of this figure of merit is shown in Fig. 4.22 [77]. Today's most advanced magnetic fusion experiments, e.g., the Japanese tokamak JT-60U, have reached their optimum performance. The next step can only be achieved in a bigger machine like ITER.

A different way to look at the triple product comes from the observation that in the interesting regime the fusion reactivity scales approximately as $\langle\sigma v\rangle \propto T^2$. Then the ratio of fusion power and heat loss becomes

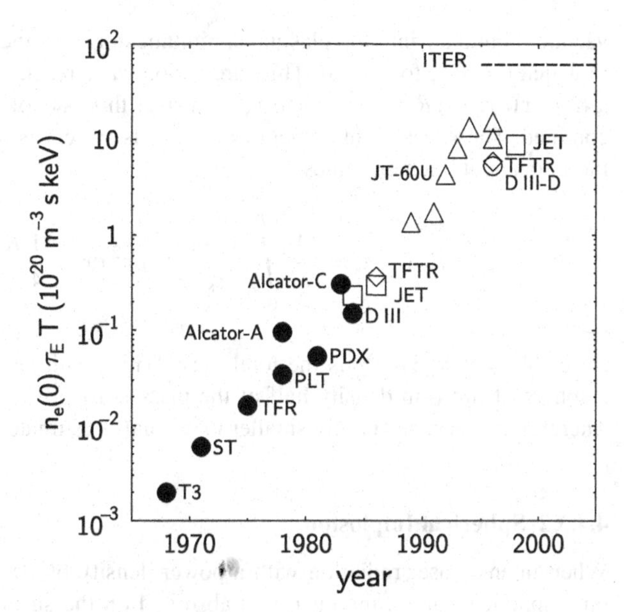

Fig. 4.22 The evolution of the fusion triple product towards the reactor regime. The design goal for the ITER-experiment is given as dashed line. A fusion reactor will have $n_e \tau_E T \approx 10^{22}\,\mathrm{m^{-3}\,s\,keV}$ (Reprinted with permission from [77]. © 2005, European Physical Society)

$$\frac{P_{\mathrm{fus}}}{P_{\mathrm{H}}} \propto \frac{n_e^2 T^2}{n_e T/\tau_E} = n_e T \tau_E \,. \tag{4.67}$$

Long energy confinement times became possible by operating the plasma in the so-called high-confinement mode (H-mode) [43]. A description of H-mode physics is beyond the scope of this introduction. The interested reader can find more on this subject in [45].

4.4.3 Inertial Confinement Fusion

The ideas behind the Lawson criterion can be applied to two different scenarios. In magnetic confinement fusion, the plasma density is low ($\approx 10^{20}\,\mathrm{m^{-3}}$) and the energy confinement time is long ($\tau_E \approx 1$ s). Inertial confinement fusion (ICF) uses the opposite concept with a high plasma density and a short confinement time. The idea is to burn a great part of the D-T content of a small pellet before the plasma has significantly expanded. Modern ICF concepts have been described in reviews [51, 78] and in a tutorial [79].

4.4.3.1 Inertial Confinement Time

The confinement time for a homogeneous sphere of hot plasma can be estimated as follows: Let us assume that the high-pressure plasma of radius R is surrounded by a vacuum. Then a rarefaction wave will propagate at the sound speed c_s from the vacuum boundary into the plasma, communicating that there is a vacuum out there to which it is free to expand. This rarefaction wave reaches a radial position r after a time $\tau(r) = (R - r)/c_s$, and upon arrival this part of the plasma is no longer confined. The global confinement time for this system is then the mass-average of these local confinement times

$$\tau_c = \frac{1}{M} \int_0^R \frac{R-r}{c_s} 4\pi \rho r^2 \, dr = \frac{1}{4} \frac{R}{c_s} . \tag{4.68}$$

Here, $M = (4\pi/3)\rho R^3$ is the total mass. This result seems plausible because in a sphere of uniform density half of the mass is found in the outer 20% of radius. Therefore, τ_c is substantially smaller than a naïve estimate R/c_s.

4.4.3.2 Spherical Implosion

When intense laser radiation with a power density of 10^{14}–10^{15} W m^{-2} impinges on a spherical target, the energy is absorbed on the surface to generate a plasma of (2–3) keV temperature and a few hundred megabars pressure. Currently discussed fusion targets have a design as shown in Fig. 4.23. The outermost layer (ablator) is made of plastic foam or a low-Z material like beryllium. On the inside, a layer of D–T ice is deposited with about 80 μm thickness. The volume is filled with D–T gas with a mass density of 0.3 mg cm^{-3}, corresponding to 30 bar pressure at room temperature. The intention is to heat and compress the central gas filling to fusion temperature of about 5 keV while the surrounding main fuel stays dense and relatively cold. This concept assumes self-ignition in the central hot spot.

The pressure from the ablated surface material accelerates the outer shell of the target towards the center. The acceleration mechanism is the same as for rocket

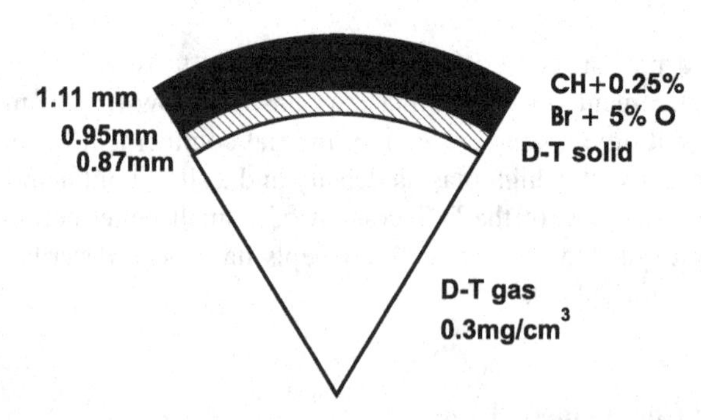

Fig. 4.23 NIF pellet design. A plastic microsphere with the main fuel as a hollow shell of D–T ice. The volume is filled with D–T gas at 30 bar normal pressure to form a hot spot. (Reprinted with permission from [52]. © 2004, American Institute of Physics)

propulsion. When the accelerated fuel collides in the center, compression and heating occurs. In inertial-confinement fusion experiments, the radius of the compressed fuel is about 1/30 of the original pellet radius. The achievable final velocity of the imploding shell of solid D–T can be estimated from the rocket equation

$$v_{\mathrm{shell}} = v_{\mathrm{exhaust}} \ln\left(\frac{m_{\mathrm{ablator}}}{m_{\mathrm{DT}}}\right) . \tag{4.69}$$

Typical implosion velocities vary from $(2\text{–}4)\times 10^5\,\mathrm{m\,s^{-1}}$ [51].

4.4.3.3 Ignition

In ICF, ignition occurs when the α-particles from the D–T fusion can deposit their energy inside the hot spot more rapidly than the heat content is lost by plasma expansion. Assuming that the density of the hot spot is high enough to stop all α-particles by collisions, we can use the Lawson condition for ignition, $n_\mathrm{e}\tau_\mathrm{E} > 1.5\times 10^{20}\,\mathrm{m^{-3}\,s}$ and insert the confinement time τ_c from (4.68) to obtain

$$n_\mathrm{e} R_{\mathrm{hs}} = (1.5 \times 10^{20}\,\mathrm{m^{-3}\,s}) \times 4 c_\mathrm{s} . \tag{4.70}$$

This is a condition for the product of plasma density and radius to achieve ignition. Introducing the mass density $\rho_\mathrm{m} = (5/2) n_\mathrm{e} m_\mathrm{p}$, m_p being the proton mass, this defines a critical value for the product $(\rho_\mathrm{m} R)_{\mathrm{crit}} \approx 0.4\,\mathrm{g\,cm^{-2}}$, which is sufficient to stop the α-particles [78]. Unfortunately, the ignition condition is not yet sufficient for an optimum use of the D–T fuel.

4.4.3.4 Burn Fraction

For an efficient fusion process, a considerable fraction (usually 1/3) of the main fuel must be burned during the confinement time. The rate, at which tritium (and deuterium) react in a 50% D/50% T fuel is

$$\frac{\mathrm{d}n_\mathrm{T}}{\mathrm{d}t} = \frac{\mathrm{d}n_\mathrm{D}}{\mathrm{d}t} = -n_\mathrm{D} n_\mathrm{T} \langle \sigma v \rangle_{\mathrm{DT}} . \tag{4.71}$$

Inserting the total fuel density $n = 2n_\mathrm{T} = 2n_\mathrm{D}$, the burn process obeys

$$\frac{\mathrm{d}n}{\mathrm{d}t} = -\frac{1}{2} n^2 \langle \sigma v \rangle_{\mathrm{DT}} , \tag{4.72}$$

which can be easily integrated from $t = 0$ to τ_c with the result

$$\frac{1}{n} - \frac{1}{n_0} = \frac{\tau_c}{2} \langle \sigma v \rangle_{\mathrm{DT}} , \tag{4.73}$$

where n_0 is the initial density of the compressed fuel. We define the burn fraction $f_b = 1 - (n/n_0)$ and introduce the initial mass density $\rho_{\mathrm{m}0} = (5/2) n_0 m_\mathrm{p}$, m_p being the proton mass. After simple algebraic manipulations we obtain

$$f_b = \frac{\rho_{\mathrm{m}0} R}{\rho_{\mathrm{m}0} R + g(T)} , \tag{4.74}$$

with $g(T) = 20 m_\mathrm{p} c_s / \langle \sigma v \rangle_{DT}$. $g(T)$ takes a minimum of $6\,\mathrm{g\,cm^{-2}}$ at 30 keV. To achieve a burn fraction of 1/3, we need $\rho_{\mathrm{m}0} R \approx 3\,\mathrm{g\,cm^{-2}}$, which is an order of magnitude higher than the mass per area for ignition.

In retrospect, we can ask why we need a highly compressed target for ICF. The functional dependence on the mass per area, $\rho_\mathrm{m} R$, might suggest to work with solid D–T at a density of $\rho_\mathrm{m} = 0.21\,\mathrm{g\,cm^{-3}}$ and a corresponding larger radius of $R \approx 14.3\,\mathrm{cm}$. The need for compression becomes clear, when we consider the total mass in a sphere of radius R that fulfills the constraint $\rho R \approx 3\,\mathrm{g\,cm^{-2}}$

$$M_\mathrm{s} = \frac{4}{3} \pi \rho_\mathrm{m} R^3 = \frac{4}{3} \pi \frac{(\rho_\mathrm{m} R)^3}{\rho_\mathrm{m}^2} . \tag{4.75}$$

This shows that the mass of D–T fuel shrinks as ρ_m^{-2}. The energy yield of the DT–fuel is $\varepsilon_{\mathrm{DT}} = 17.6\,\mathrm{MeV}/(5 m_\mathrm{p}) = 3.4 \times 10^{11}\,\mathrm{J/g}$. The mentioned $R = 14\,\mathrm{cm}$ ball of D–T ice represents about 2.5 kg fusionable material with an explosive yield of about 70 kilotons TNT [79]. Compressing the radius by a factor of 10 would increase the density by a factor of 1000 and reduce the fuel content by 10^{-6} to 2.5 mg—equivalent to 70 kg TNT, or $\approx 300\,\mathrm{MJ}$—which can be handled in a fusion reactor. For a power plant design, about 5 shots per second can be envisaged with a total thermal power of $\approx 1.5\,\mathrm{GW}$. The NIF capsule shown in Fig. 4.23 contains only about 1 mg D–T in view of the power handling capacity of the NIF target chamber.

4.4.3.5 Lawson Criterion for ICF

We have seen above that the balance equations of ICF are based on characteristic $\rho_\mathrm{m} R$ values rather than on the $n_\mathrm{e} \tau_\mathrm{E}$ criterion of magnetic fusion devices. In the spirit of (4.70), we can rewrite the characteristic $\rho_\mathrm{m} R = 3\,\mathrm{g\,cm^{-2}}$ for a burn fraction of 1/3 in terms of number density and confinement time $n\tau_c = 2 \times 10^{21}\,\mathrm{m^{-3}\,s}$ [79]. Thus the Lawson criterion for ICF is typically a factor of 20 higher than that of magnetic fusion due to the inefficiencies in assembling the fuel that ICF has to overcome.

The Basics in a Nutshell

- The various definitions of a Maxwell distribution are

1-dimensional	$f_{\mathrm{M}}^{(1)}(v_x) = n\left(\frac{m}{2\pi k_{\mathrm{B}}T}\right)^{1/2} \exp\left(-\frac{mv_x^2}{2k_{\mathrm{B}}T}\right)$
3-dimensional	$f_{\mathrm{M}}^{(3)}(\mathbf{v}) = n\left(\frac{m}{2\pi k_{\mathrm{B}}T}\right)^{3/2} \exp\left(-\frac{m(v_x^2+v_y^2+v_z^2)}{2k_{\mathrm{B}}T}\right)$
distribution of speed	$f_{\mathrm{M}}(v) = 4\pi v^2 n\left(\frac{m}{2\pi k_{\mathrm{B}}T}\right)^{3/2} \exp\left(-\frac{mv^2}{2k_{\mathrm{B}}T}\right)$
distribution of energy	$F_{\mathrm{M}}(W) = n\frac{2}{\sqrt{\pi}}\frac{1}{(k_{\mathrm{B}}T)^{3/2}} W^{1/2} \exp\left(-\frac{W}{k_{\mathrm{B}}T}\right)$

- The various definitions of "thermal velocity" are:

mean thermal velocity	$v_{\mathrm{th}} = \left(\frac{8k_{\mathrm{B}}T}{\pi m}\right)^{1/2}$
most probable velocity	$v_T = \left(\frac{2k_{\mathrm{B}}T}{m}\right)^{1/2}$

- The mean free path is $\lambda_{\mathrm{mfp}} = (n\sigma)^{-1}$ and the collision frequency $\nu_{\mathrm{c}} = v/\lambda_{\mathrm{mfp}}$.
- The number of collision events in a hot gas per volume and second is given by the rate coefficient $\langle\sigma v\rangle$, in which the angle brackets denote averaging over the distribution function.
- The Coulomb collision frequency decreases for rising temperature as $\nu_{ei} \propto T^{-3/2}$ and is independent of plasma density.
- Transport of plasma particles is accomplished by electric fields or density gradients. The individual transport coefficients are the mobilities $\mu_{\mathrm{e,i}} = e/(m_{\mathrm{e,i}}\nu_{\mathrm{e,i}})$ and the diffusion coefficients $D_{\mathrm{e,i}} = \mu_{\mathrm{e,i}}k_{\mathrm{B}}T_{\mathrm{e,i}}/e$.
- The global transport coefficients are the electrical conductivity $\sigma = ne(\mu_{\mathrm{e}} + \mu_{\mathrm{i}})$ and the ambipolar diffusion coefficient $D_{\mathrm{a}} = (D_{\mathrm{i}}\mu_{\mathrm{e}} + D_{\mathrm{e}}\mu_{\mathrm{i}})/(\mu_{\mathrm{e}} + \mu_{\mathrm{i}})$.
- In the presence of a magnetic field, the transport coefficients become tensors that link the velocity to the force. This applies to mobility, conductivity and diffusivity. The Pedersen conductivity is the diagonal element and the Hall conductivity the off-diagonal element of the conductivity tensor.

Problems

4.1 Show that the maximum of the Maxwell distribution function $f_M(|v|)$ is found at v_T.

4.2 Prove that the mean thermal velocity in Eq. (4.7) is $v_{\mathrm{th}} = [(8k_{\mathrm{B}}T)/(\pi m)]^{1/2}$.

4.3 Prove that the mean kinetic energy in Eq. (4.8) is $(\mathrm{m}/2)\langle v^2\rangle = (3/2)k_\mathrm{B}T$.

4.4 Derive (4.49) from conservation of energy and momentum by considering the scattering of a light particle on a heavy particle. Hint: In this limit the modulus of momentum of the scattered electron is the same as the momentum before the collision.

4.5 Solve the integral in Eq. (4.50).

4.6 Solve the integrals in Eq. (4.45) and derive (4.46).

4.7 Show that the velocities in Eq. (4.46) can also be derived from the force balance between friction and total Lorentz force

$$m_\mathrm{i}\nu_\mathrm{m}\mathbf{v} = e(\mathbf{E} + \mathbf{v} \times \mathbf{B})\,.$$

4.8 Assume that the radial electron density profile in a long cylindrical discharge tube of radius a is parabolic

$$n_\mathrm{e}(r) = n(0)\left[1 - \frac{r^2}{a^2}\right].$$

Determine the equivalent electron density of a homogeneous density distribution that would give the same current.

4.9 Perform the intermediate steps for proving the statement that the ignition line for fusion is equivalent to $\eta = 0.154$ in the Lawson curves.

Chapter 5
Fluid Models

"The time has come," the Walrus said,
"To talk of many things:
Of shoes—and ships—and sealing wax—
Of cabbages—and kings—
And why the sea is boiling hot—
And whether pigs have wings."
Lewis Carroll, Through the Looking-Glass

In the single-particle model (Chap. 3) the motion of the particles was derived from fixed external electric and magnetic fields. This approach is very useful to obtain a first insight into the richness of plasma motion, which results in a host of particle drifts. The major drawback of this model is the neglect of the modification of the fields by the electric currents represented by these drifts. The present chapter on fluid models attempts to overcome this weakness.

The self-consistency of a plasma model is an important aspect. Only in such models (Fig. 5.1) phenomena can be described where a magnetic field is apparently frozen in the highly conductive plasma, such as in solar prominences. The Swedish physicist and Nobel prize winner Hannes Alfvén (1908–1995) had recognized this cooperative action of plasma and magnetic field and had predicted that a new type of magnetohydrodynamic waves should exist, which are now named Alfvén waves.

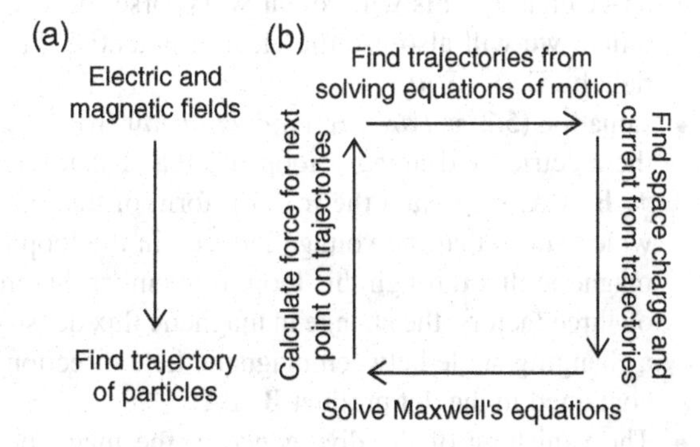

Fig. 5.1 (**a**) A plasma model with prescribed forces. (**b**) A self-consistent plasma model

A. Piel, *Plasma Physics*, DOI 10.1007/978-3-642-10491-6_5,

5.1 The Two-Fluid Model

The huge number of particles makes it impossible to solve Newton's equation for each of these particles. Therefore, we will seek a description similar to hydrodynamics, in which the motion of fluid elements is studied instead of tracing the individual molecules. In particular, to make the plasma model realistic, a self-consistent description for the plasma motion and the electromagnetic fields will be used.

5.1.1 Maxwell's Equations

The starting point for the fluid model is a proper combination of Maxwell's equations with the particle currents and space charges. Maxwell's equations are given in a form that uses the electric field **E** and the magnetic field **B**. For **B**, we will use the names *magnetic induction, magnetic field* and *magnetic flux density* synonymously. Because they appear in the equation of motion (3.1), **E** and **B** are the natural field quantities.

$$\nabla \cdot \mathbf{E} = \frac{\rho}{\varepsilon_0} \tag{5.1}$$

$$\nabla \times \mathbf{E} = -\frac{\partial \mathbf{B}}{\partial t} \tag{5.2}$$

$$\nabla \cdot \mathbf{B} = 0 \tag{5.3}$$

$$\nabla \times \mathbf{B} = \mu_0 \left(\mathbf{j} + \varepsilon_0 \frac{\partial \mathbf{E}}{\partial t} \right) . \tag{5.4}$$

Here, ρ is the total charge density and $\mathbf{j}$ the current density carried by particles. These two quantities contain the action of the plasma motion on the fields. Let us shortly recall the physical contents of the set of Maxwell's equations:

- The relationship (5.1) is *Poisson's equation*, which links the electric field to the space charge. This will be our workhorse for electrostatic problems in plasmas. Often, we will also use the electric potential Φ, which is linked to the electric field by $\mathbf{E} = -\nabla\Phi$.
- Equation (5.2) is *Faraday's induction law* in differential form. When we integrate the electric field along a loop of area $\mathbf{A}$ that encircles a magnetic flux $\Phi_{\mathrm{m}} = \oint_{\mathbf{A}} \mathbf{B} \cdot \mathrm{d}\mathbf{A}$, we obtain the integral form of the induction law, $U_{\mathrm{ind}} = -\mathrm{d}\Phi_{\mathrm{m}}/\mathrm{d}t$, which states that the voltage induced in the loop is the (negative) change of the magnetic flux through this loop. Remember that the induced voltage can depend on three factors: the change in magnetic flux density **B**, the change of area $|\mathbf{A}|$ and a changing angle between magnetic field direction and the area normal, which is contained in the dot product $\mathbf{B} \cdot \mathrm{d}\mathbf{A}$.
- The vanishing of the divergence of the magnetic flux density **B** in (5.3) is an experimental fact that there are no magnetic monopoles in ordinary matter.
- *Ampere's law* (5.4) states that the curl of the magnetic flux density **B** is determined by the conduction current $\mathbf{j}$ *and* the displacement current $\varepsilon_0 \partial\mathbf{E}/\partial t$. For

calculating the stationary magnetic field of a straight wire, we will use the integral form of Ampere's law, $\oint \mathbf{B} \cdot d\mathbf{s} = \mu_0 I$.

At this point, we need to discuss why we do not use the magnetic field strength **H**, or the dielectric displacement **D**. The single particle model had the important result that a gyrating particle has a magnetic moment, which is antiparallel to the direction of **B**. Therefore, a plasma is a diamagnetic medium. Moreover, the magnetic moment is either a conserved quantity, when we try to change the magnetic flux density, or it scales as $\propto 1/B$, when we prescribe the particle energy. Therefore, we cannot expect any proportionality $\mathbf{B} = \mu_r \mu_0 \mathbf{H}$ that is typical of ferromagnetic materials. Therefore, introducing **H** would not help simplifying our models.

We had also seen in Sect. 3.5.2 that electric polarization of a plasma does only appear in time-varying fields. Any static polarization charges can only exist at the plasma surface, but the resulting electric field will be shielded in the plasma interior. Hence, **D** is also no suitable quantity to describe static situations.

On the other hand, we will explicitly use the concept of a plasma as a dielectric medium in connection with plasma waves. When we take the simplified picture for electron waves in a low-temperature plasma, where an ion is essentially at rest and the electron reacts to the oscillating electric field, we can group the plasma particles into pairs of electrons and ions that form local oscillating dipoles.

5.1.2 The Concept of a Fluid Description

There is one essential difference between hydrodynamics and plasma fluid models. In hydrodynamics, the molecules of the liquid are strongly coupled. This means that the molecules are continuously colliding with their neighbors. A pair of particles will only slowly drift apart by diffusion. Hence, it is meaningful to partition the liquid into macroscopic fluid elements, which contain many molecules that stay close together for a long time. These fluid elements move along streamlines of the flow pattern.

In an ideal plasma, however, the electrons and ions do not experience their nearest neighbors. This means that Coulomb collisions are rare. Rather, the electrons and ions follow the forces from the average electric and magnetic fields that are produced by many other particles. Therefore, we can partition the plasma into small cells but this does not imply that the particles will stay inside their cells for an extended time. The electrons and ions will typically leave a cell of size ℓ after a transit time $T_t \approx \ell / v_{th}$ while particles from neighboring cells enter this volume. Therefore, we can use these cells as a kind of bank account to keep a gain and loss record of the total number of particles in such a cell, or the total momentum, or the heat content. We will see that this approach gives us a kind of hydrodynamic description, but the analogy to real liquids has its limitations.

Depending on the situation, we can arbitrarily choose a description with cells that are fixed in a resting frame of reference, or we can transform to a moving frame of reference that follows the mean flow velocity of the plasma.

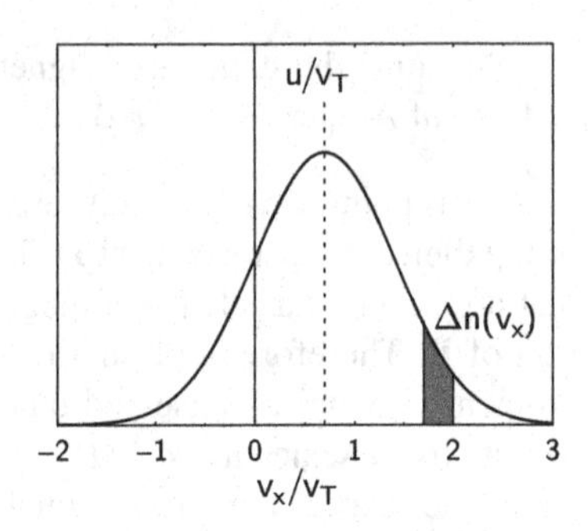

Fig. 5.2 Shifted Maxwellian with a mean drift velocity u. Thee group of particles in the interval $[v_x, v_x + \Delta v_x]$ contains $\Delta n(v_x)$ particles

In the following, electrons and ions are assumed to form two independent fluids that penetrate each other (two-fluid model). The defining properties of a plasma for the fluid description are the densities n_e and n_i of electrons and ions, the temperatures T_e and T_i, as well as the streaming velocities $\mathbf{u}_e$ and $\mathbf{u}_i$.

A streaming electron population can be described by a shifted distribution function (Fig. 5.2), which for simplicity is assumed to be Maxwellian. However, the arguments given below apply to arbitrary distributions. In one dimension, the shifted Maxwellian has the form

$$f_M(v_x) = n\left(\frac{m}{2\pi k_B T}\right)^{1/2} \exp\left(-\frac{m(v_x - u_x)^2}{2k_B T}\right), \tag{5.5}$$

with a mean drift velocity $\mathbf{u} = (u_x, 0, 0)$ that defines the x-direction.

5.1.3 The Continuity Equation

The balance for the number of particles in a fixed cell of size $\Delta V = \Delta x \Delta y \Delta z$ is discussed for the one-dimensional flow described by (5.5). The number of particles inside the interval $[x, x+\Delta x]$ is $N = nA\Delta x$ with $A = \Delta y \Delta z$. The incident particle flux is $I_N = nAu_x$. When this flux is decelerated or accelerated inside the cell by external forces, the flux on the exit side is larger or smaller.

Accordingly, the number of particles in the cell is diminished or increased (Fig. 5.3)

$$-\frac{\partial N}{\partial t} = I_N(x + \Delta x) - I_N(x) \approx \frac{\partial I_N}{\partial x}\Delta x\,. \tag{5.6}$$

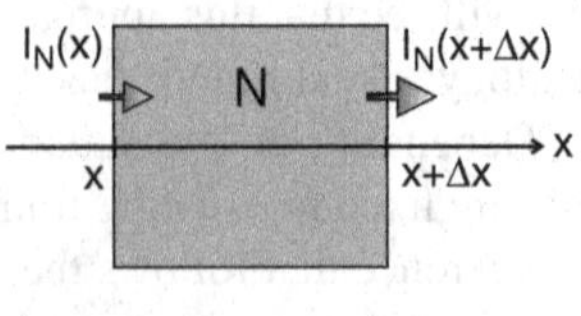

Fig. 5.3 Definitions used to derive the continuity equation

In the last step, we have Taylor-expanded the particle flux and retained only the differential change of the flux. Dividing by $\Delta V = A\Delta x$ and taking the limit $\Delta V \to 0$ gives

$$\frac{\partial n}{\partial t} + \frac{\partial (n\, u_x)}{\partial x} = 0\,. \tag{5.7}$$

This result can easily be generalized to a three-dimensional flow pattern, which results in the *continuity equation*

$$\frac{\partial n}{\partial t} + \nabla \cdot (n\mathbf{u}) = 0\,. \tag{5.8}$$

This balance equation describes the conservation of the number of particles in the flow. When particles are generated or annihilated inside the cell, say by ionization or recombination, the zero on the right hand side is replaced by a net production rate S (see Sect. 4.2.3).

The continuity equation can be easily generalized to an equation for the conservation of charge by introducing the charge density $\rho = \sum_\alpha n_\alpha q_\alpha$ and the current density $\mathbf{j} = \sum_\alpha n_\alpha q_\alpha \mathbf{u}_\alpha$

$$\frac{\partial \rho}{\partial t} + \nabla \cdot \mathbf{j} = 0\,. \tag{5.9}$$

5.1.4 Momentum Transport

The net force in the balance of the considered cell is a result of the sum of all forces acting on the particles within the cell plus the export and import of momentum by particles that leave and enter the cell. The starting point of our calculation is Newton's equation for the force acting on a single particle

$$m\frac{\mathrm{d}\mathbf{v}}{\mathrm{d}t} = q(\mathbf{E} + \mathbf{v} \times \mathbf{B})\,. \tag{5.10}$$

Here, $\mathrm{d}/\mathrm{d}t$ is the derivative calculated at the position of the point-like particle. The correct momentum balance for a many-particle system can be obtained by multiplying (5.10) with the density n. However, in an inhomogeneous flow, the time derivative has to be calculated according to the rules of hydrodynamic flow

$$\frac{\mathrm{d}\mathbf{u}}{\mathrm{d}t} = \frac{\partial \mathbf{u}}{\partial t} + \frac{\partial \mathbf{u}}{\partial x}\frac{\mathrm{d}x}{\mathrm{d}t} + \frac{\partial \mathbf{u}}{\partial y}\frac{\mathrm{d}y}{\mathrm{d}t} + \frac{\partial \mathbf{u}}{\partial z}\frac{\mathrm{d}z}{\mathrm{d}t}\,. \tag{5.11}$$

The vector $(\mathrm{d}x/\mathrm{d}t, \mathrm{d}y/\mathrm{d}t, \mathrm{d}z/\mathrm{d}t)$ is just the velocity $\mathbf{u}$ of the cell. This leads to the compact notation

$$\frac{\mathrm{d}\mathbf{u}}{\mathrm{d}t} = \frac{\partial \mathbf{u}}{\partial t} + (\mathbf{u} \cdot \nabla)\mathbf{u}\,, \tag{5.12}$$

in which $\mathbf{u} \cdot \nabla$ represents the *convective derivative*, which describes the change of a quantity originating from the motion of the flow. To gain an insight into this quantity, consider a man in a boat that is driven by the flow of a river from a narrow region with rapid flow speed to a wide reach of slow speed. Although the flow pattern is continuous, and does not change in time, the experience of a subject following the flow is a change in velocity. Hence, the correct balance of the internal forces for a fluid element is

$$nm\left[\frac{\partial \mathbf{u}}{\partial t} + (\mathbf{u} \cdot \nabla)\mathbf{u}\right] = nq(\mathbf{E} + \mathbf{u} \times \mathbf{B})\,. \tag{5.13}$$

We now need to sum up the surface forces that arise from particles entering and leaving the fluid cell. For this purpose we consider the particle exchange through the cell surface sketched in Fig. 5.4.

The calculation is presented for the x-direction only. The cell boundaries are at x_0 and $x_0 + \Delta x$. Further, we select a group of velocities between v_x and $v_x + \Delta v_x$. The particle flux represented by this group of velocities is

$$\Delta I_N(v_x) = \Delta n(v_x) v_x \Delta y \Delta z\,. \tag{5.14}$$

The number density $\Delta n(v_x)$ of this particle group is related to the distribution function $f(\mathbf{v})$ by

$$\Delta n(v_x) = \Delta v_x \iint f(v_x, v_y, v_z)\mathrm{d}v_y \mathrm{d}v_z\,, \tag{5.15}$$

In analogy to the definition of particle flux, we introduce the momentum flux that is carried by the group of particles around v_x

$$\Delta I_\mathrm{P} = (m v_x)\Delta n(v_x)|v_x|\Delta y \Delta z\,. \tag{5.16}$$

The momentum flux is the momentum transported through a boundary per unit time. The factor $|v_x|$ is a measure for the rate at which the particles pass through the

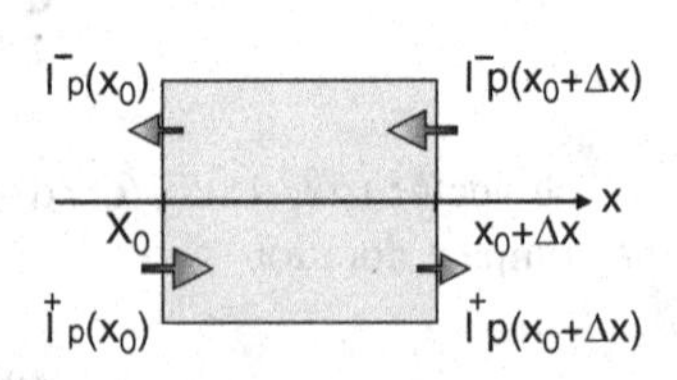

Fig. 5.4 Calculation of pressure forces

boundary, and therefore a positive quantity. The gain and loss balance for the interval $[x_0, x_0 + \Delta x]$ can hence be written as

$$\text{Gain at } x_0: \quad I_{\mathrm{P}}^{+}(x_0) = \sum_{v_x>0} \Big[\Delta n(v_x)(mv_x)|v_x|\Big]_{x_0} \Delta y \Delta z \tag{5.17}$$

$$\text{Loss at } x_0: \quad I_{\mathrm{P}}^{-}(x_0) = \sum_{v_x<0} \Big[\Delta n(v_x)(mv_x)|v_x|\Big]_{x_0} \Delta y \Delta z \tag{5.18}$$

$$\text{Gain at } x_0 + \Delta x_0: I_{\mathrm{P}}^{-}(x_0 + \Delta x) = \sum_{v_x<0} \Big[\Delta n(v_x)(mv_x)|v_x|\Big]_{x_0+\Delta x} \Delta y \Delta z \tag{5.19}$$

$$\text{Loss at } x_0 + \Delta x_0: I_{\mathrm{P}}^{+}(x_0 + \Delta x) = \sum_{v_x>0} \Big[\Delta n(v_x)(mv_x)|v_x|\Big]_{x_0+\Delta x} \Delta y \Delta z\,. \tag{5.20}$$

The upper index $\pm$ describes the sign of the velocity. Gain and loss by particles moving to the left represent negative values. The net gain of momentum per unit time then becomes

$$\frac{\partial P_x}{\partial t} = I_{\mathrm{P}}^{+}(x_0) - I_{\mathrm{P}}^{+}(x_0 + \Delta x) + I_{\mathrm{P}}^{-}(x_0 + \Delta x) - I_{\mathrm{P}}^{-}(x_0)\,. \tag{5.21}$$

By Taylor expanding the momentum flux and replacing negative velocities by $|v_x| = -v_x$, we can combine the result as

$$\frac{\partial P_x}{\partial t} = -m \sum_{v_x=-\infty}^{\infty} \left([\Delta n(v_x) v_x^2]_{x_0+\Delta x} - [\Delta n(v_x) v_x^2]_{x_0} \right) \tag{5.22}$$

$$= -m \frac{\partial}{\partial x} \left(n \langle v_x^2 \rangle \right) \Delta x \Delta y \Delta z \tag{5.23}$$

and $n\langle v_x^2 \rangle = \int f(v_x) v_x^2 \mathrm{d}v_x$. The next step is to split the particle velocities into a mean flow u_x and a random thermal motion $\tilde{v}_x$

$$v_x = u_x + \tilde{v}_x\,. \tag{5.24}$$

Then, we obtain the momentum balance as

$$\frac{\partial}{\partial t}(nmu_x) = -m \frac{\partial}{\partial x} \left[n \left(\langle u_x^2 \rangle + 2u_x \langle \tilde{v}_x \rangle + \langle \tilde{v}_x^2 \rangle \right) \right]. \tag{5.25}$$

For a one-dimensional Maxwellian we know that $(1/2)\, m \langle \tilde{v}_x^2 \rangle = (1/2)\, k_{\mathrm{B}} T$. By definition, the average of the random motion is $\langle \tilde{v}_x \rangle = 0$. Hence, the momentum balance becomes

$$\frac{\partial}{\partial t}(nmu_x) = -\frac{\partial}{\partial x} \left[nmu_x^2 + nk_{\mathrm{B}} T \right]. \tag{5.26}$$

This is the correct balance for a fixed volume in space. On the r.h.s. of (5.26) we find the stagnation pressure nmu_x^2 and the kinetic pressure $p = nk_BT$. Evaluating the derivatives on both sides and using the continuity Eq. (5.8) we obtain

$$nm\left(\frac{\partial u_x}{\partial t} + u_x\frac{\partial u_x}{\partial x}\right) = -\frac{\partial p}{\partial x}\,. \tag{5.27}$$

In this representation, the fluid element is considered to follow the flow, which can be identified by the convective derivative on the l.h.s. of the equation. Generalizing this one-dimensional result to three dimensions, and adding the volume forces, the final result gives the *momentum transport equation*

$$nm\left(\frac{\partial \mathbf{u}}{\partial t} + (\mathbf{u}\cdot\nabla)\mathbf{u}\right) = nq(\mathbf{E}+\mathbf{u}\times\mathbf{B}) - \nabla p\,. \tag{5.28}$$

5.1.5 Shear Flows

In the previous paragraph, we have calculated the momentum exchange between neighboring cells along the mean flow. This could be summed up into a new net volume force, the pressure gradient. Now, we focus our attention on the momentum exchange across the flow (Fig. 5.5). Because of their random thermal motion, particles passing the boundaries at y and $y+\Delta y$ belong to populations that have different mean flow velocities.

The calculation is quite similar to that of the previous paragraph, but now we define a *shear stress* tensor P_{ij}

$$P_{ij} = nm\langle\tilde{v}_i\tilde{v}_j\rangle\,, \tag{5.29}$$

which involves the random thermal velocities that are responsible for the momentum exchange between neighboring cells. P_{ij} replaces the scalar pressure. Instead of the pressure gradient we now have the divergence of the shear stress tensor. Shear flows are associated with viscosity, which, however, is negligible in many plasmas.

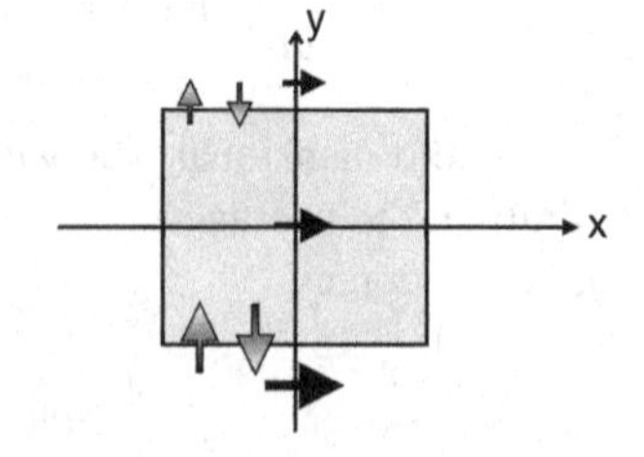

Fig. 5.5 Momentum transport in a shear flow. The *black horizontal arrows* mark the mean local velocity in the flow. The *shaded arrows* indicate the momentum exchange by particles traversing the boundary at y and $y+\Delta y$

Summary of the Two-Fluid Model

The two-fluid model of a plasma includes two individual momentum transport equations for electrons and ions

$$n_\mathrm{e}m_\mathrm{e}\left[\frac{\partial \mathbf{u}_\mathrm{e}}{\partial t}+(\mathbf{u}_\mathrm{e}\cdot\nabla)\mathbf{u}_\mathrm{e}\right]=-n_\mathrm{e}e\,(\mathbf{E}+\mathbf{u}_\mathrm{e}\times\mathbf{B})-\nabla p_\mathrm{e}$$
$$n_\mathrm{i}m_\mathrm{i}\left[\frac{\partial \mathbf{u}_\mathrm{i}}{\partial t}+(\mathbf{u}_\mathrm{i}\cdot\nabla)\mathbf{u}_\mathrm{i}\right]=+n_\mathrm{i}e\,(\mathbf{E}+\mathbf{u}_\mathrm{i}\times\mathbf{B})-\nabla p_\mathrm{i}\,. \tag{5.30}$$

The connection with space charge ρ and current density $\mathbf{j}$ is established by

$$\rho = n_\mathrm{i}e - n_\mathrm{e}e$$
$$\mathbf{j} = n_\mathrm{i}e\,\mathbf{u}_\mathrm{i} - n_\mathrm{e}e\,\mathbf{u}_\mathrm{e}\,. \tag{5.31}$$

Both fluids obey individual equations of continuity

$$\frac{\partial n_\mathrm{e}}{\partial t}+\nabla\cdot(n_\mathrm{e}\mathbf{u}_\mathrm{e})=0$$
$$\frac{\partial n_\mathrm{i}}{\partial t}+\nabla\cdot(n_\mathrm{i}\mathbf{u}_\mathrm{i})=0\,. \tag{5.32}$$

Together with Maxwell's equations,

$$\nabla\cdot\mathbf{E}=\frac{\rho}{\varepsilon_0} \tag{5.33}$$
$$\nabla\times\mathbf{E}=-\frac{\partial\mathbf{B}}{\partial t} \tag{5.34}$$
$$\nabla\cdot\mathbf{B}=0 \tag{5.35}$$
$$\nabla\times\mathbf{B}=\mu_0\left(\mathbf{j}+\varepsilon_0\frac{\partial\mathbf{E}}{\partial t}\right)\,. \tag{5.36}$$

we now have a complete self-consistent fluid model of the plasma.

5.2 Magnetohydrostatics

As a first application of the two-fluid model we will inspect slowly evolving plasma situations, in which the non-linear term $\mathbf{u}\cdot\nabla\mathbf{u}$ can be neglected. Then, the momentum transport for electrons and ions including gravitational forces and friction between the electron and ion fluid is given by

$$
\begin{aligned}
nm_{\mathrm{i}}\frac{\partial \mathbf{u}_{\mathrm{i}}}{\partial t} &= ne(\mathbf{E}+\mathbf{u}_{\mathrm{i}}\times\mathbf{B})-\nabla p_{\mathrm{i}}+nm_{\mathrm{i}}\mathbf{g}+n\nu_{\mathrm{ei}}m_{\mathrm{e}}(\mathbf{u}_{\mathrm{e}}-\mathbf{u}_{\mathrm{i}})\\
nm_{\mathrm{e}}\frac{\partial \mathbf{u}_{\mathrm{e}}}{\partial t} &= -ne(\mathbf{E}+\mathbf{u}_{\mathrm{e}}\times\mathbf{B})-\nabla p_{\mathrm{e}}+nm_{\mathrm{e}}\mathbf{g}+n\nu_{\mathrm{ei}}m_{\mathrm{e}}(\mathbf{u}_{\mathrm{i}}-\mathbf{u}_{\mathrm{e}})\,. \quad (5.37)
\end{aligned}
$$

The momentum exchange between the electron and ion fluid is described by a collision frequency ν_{ei} and the mean exchanged momentum per volume, $nm_{\mathrm{e}}(\mathbf{u}_{\mathrm{e}}-\mathbf{u}_{\mathrm{i}})$.

Instead of solving the pair of fluid equations, it is useful to transform these equations into a set of new variables that describe the mean mass motion $\mathbf{v}_{\mathrm{m}}$ and the relative motion $\propto \mathbf{j}$ of the two fluids. This approach is similar to splitting a two-particle problem into center-of-mass motion and relative motion. The mean mass motion is described by

$$
\rho_{\mathrm{m}}\frac{\partial \mathbf{v}_{\mathrm{m}}}{\partial t} = \mathbf{j}\times\mathbf{B}-\nabla p+\rho_{\mathrm{m}}\mathbf{g} \quad (5.38)
$$

with the mass density $\rho_{\mathrm{m}} = n(m_{\mathrm{i}}+m_{\mathrm{e}})$, total pressure $p = p_{\mathrm{e}}+p_{\mathrm{i}}$ and the mean mass velocity

$$
\mathbf{v}_{\mathrm{m}} = \frac{(m_{\mathrm{i}}\mathbf{u}_{\mathrm{i}}+m_{\mathrm{e}}\mathbf{u}_{\mathrm{e}})}{m_{\mathrm{e}}+m_{\mathrm{i}}}\,. \quad (5.39)
$$

Note that now the Lorentz force $\mathbf{j}\times\mathbf{B}$ acts on the total current density. Moreover, the mass motion is not affected by the friction between electron and ion fluid because it does not change the total momentum, but leads only to redistribution between electron and ion fluid.

When $\partial\mathbf{v}_{\mathrm{m}}/\partial t = 0$, (5.38) defines the static equilibria of a magnetized plasma, which are defined by the force balance

$$
0 = \mathbf{j}\times\mathbf{B}-\nabla p+\rho_{\mathrm{m}}\mathbf{g}\,. \quad (5.40)
$$

This framework is called *magnetohydrostatics*. In the next two paragraphs we will discuss two simple applications of this concept.

5.2.1 Isobaric Surfaces

Let us shortly return to the problem of toroidal confinement. Neglecting gravitational forces as small compared to the magnetic forces, we define the *magnetohydrostatic equilibrium* by

$$
\mathbf{j}\times\mathbf{B} = \nabla p\,. \quad (5.41)
$$

By taking the dot product with $\mathbf{B}$ on the both sides of the equation, the dot product vanishes yielding $0 = \mathbf{B}\cdot\nabla p$, i.e., $\mathbf{B}$ and ∇p are perpendicular to each other.

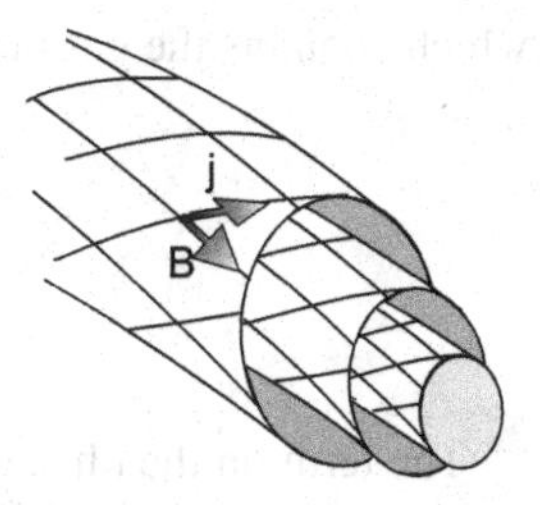

Fig. 5.6 Nested magnetic surfaces in a tokamak. Each surface is spanned by a set of magnetic field lines and current stream lines. The force $\mathbf{j} \times \mathbf{B}$ points inward, balancing the pressure gradient

The same is true for $\mathbf{j}$ and ∇p. Therefore, the vectors $\mathbf{B}$ and $\mathbf{j}$ must lie in a plane of constant pressure. The magnetic field lines and the current streamlines span a *magnetic surface*, which is also an *isobaric surface*. Figure 5.6 shows that the force $\mathbf{j} \times \mathbf{B}$ is directed inward and balances the pressure force.

5.2.2 Magnetic Pressure

The relationship between current density and magnetic induction follows from Ampere's law

$$\nabla \times \mathbf{B} = \mu_0 \mathbf{j}\,, \tag{5.42}$$

which yields

$$\mathbf{j} \times \mathbf{B} = \frac{1}{\mu_0}(\nabla \times \mathbf{B}) \times \mathbf{B} = -\frac{1}{\mu_0}\mathbf{B} \times (\nabla \times \mathbf{B})\,. \tag{5.43}$$

This expression can be evaluated by using the vector identity for arbitrary vectors $\mathbf{a}$ and $\mathbf{b}$,

$$\mathbf{a} \times (\nabla \times \mathbf{b}) = (\nabla \mathbf{b}) \cdot \mathbf{a}_{\mathrm{c}} - (\mathbf{a} \cdot \nabla)\mathbf{b}\,. \tag{5.44}$$

Then, we obtain a tensor $\nabla \mathbf{B}$ with components $(\nabla \mathbf{B})_{ij} = \partial B_j / \partial x_i$. The symbol $\mathbf{a}_{\mathrm{c}}$ means that $\mathbf{a}$ is held constant in the differentiation by the ∇-operator on its left side. Finally, we can use $(\nabla \mathbf{B}) \cdot \mathbf{B} = (1/2)\nabla(\mathbf{B} \cdot \mathbf{B})$ and obtain

$$\mathbf{j} \times \mathbf{B} = -\frac{1}{2\mu_0}\nabla(B^2) + \frac{1}{\mu_0}(\mathbf{B} \cdot \nabla)\mathbf{B}\,. \tag{5.45}$$

In the term $(\mathbf{B} \cdot \nabla)\mathbf{B}$ we recognize the analogy to the convective derivative in fluid motion discussed in Sect. (5.1.4). Here, the derivative describes the change of $\mathbf{B}$ (regarding magnitude and orientation) along a field line. Combining (5.41) and (5.45), we obtain a pressure balance

$$\nabla(p + p_{\mathrm{mag}}) = \frac{(\mathbf{B} \cdot \nabla)\mathbf{B}}{\mu_0}\,, \tag{5.46}$$

which contains the gas pressure p and a new *magnetic pressure*

$$p_{\mathrm{mag}} = \frac{B^2}{2\mu_0}\,. \tag{5.47}$$

The term on the r.h.s. of (5.46) describes a force that arises from the mechanical tension of a magnetic field line, which leads to a net force per unit volume when the field line is curved. Let us shortly discuss this curvature force. For this purpose we assume a curved flux tube of curvature radius R_c as shown in Fig. 5.7. The mechanical tension $\mathcal{T}$ produces a pair of forces acting on the left and right end of a small flux tube element of length $\mathrm{d}s = R_c\mathrm{d}\theta$ and cross-section A. Because of the curvature, the force vectors are tilted by $\pm\mathrm{d}\theta/2$. Whereas the horizontal part of the two forces cancel each other, the vertical components add up and give a radial net force, $\mathrm{d}F_r = 2\mathrm{d}F = \mathcal{T}A\mathrm{d}\theta$. This is the same principle as used in calculating the restoring force for a vibrating string.

This mechanical consideration can be related to the term on the r.h.s. of (5.46) in the following way:

$$\frac{(\mathbf{B}\cdot\nabla)\mathbf{B}}{\mu_0} = \frac{B}{\mu_0}\frac{\mathrm{d}\mathbf{B}}{\mathrm{d}s} = \frac{B^2}{\mu_0}\frac{\mathrm{d}\mathbf{e}_\theta}{\mathrm{d}s} = \frac{B^2}{\mu_0}\frac{\mathrm{d}\mathbf{e}_\theta}{R_c\mathrm{d}\theta} = -\frac{B^2}{\mu_0 R_c}\mathbf{e}_r\,. \tag{5.48}$$

In the first step we have specified that the change of $\mathbf{B}$ occurs along the field line segment $\mathrm{d}s$. In the second step we make the assumption that the magnetic field only changes orientation but maintains its magnitude. The third step uses the definition of $\mathrm{d}s$ and in the last step, we have used the rotation of the system of coordinates when we follow the field line. For this specific geometry, the result describes a net volume force in negative radial direction. Multiplying with the volume $A\,\mathrm{d}s$ we obtain the same result as from our geometrical consideration in Fig. 5.7. The value of the elastic stress results from Maxwell's stress tensor, which associates a magnetic field with an isotropic magnetic pressure $B^2/(2\mu_0)$ and an additional tension of the field line $\mathcal{T} = -B^2/\mu_0$ acting only along the field line. This completes the discussion of the curvature force.

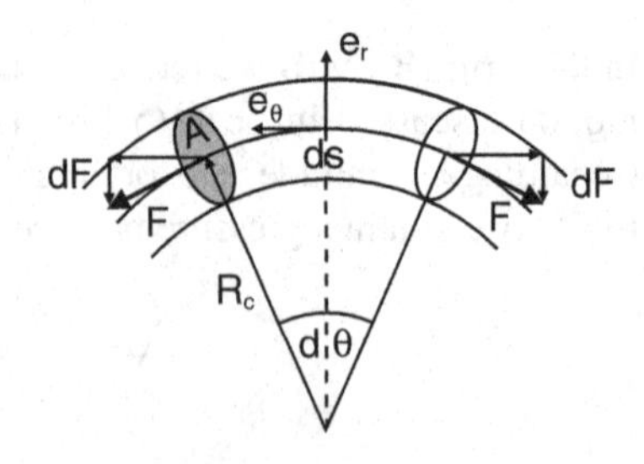

Fig. 5.7 Action of a mechanical tension of the field lines in a curved flux tube. For a short segment ds of the flux tube the tension produces a radial net force $dF_r = -\mathcal{T} A\,d\theta$

5.2.3 Diamagnetic Drift

In this paragraph, we will address the question what the microscopic interpretation of the current density $\mathbf{j}$ in the fluid model is. A steady drift motion can be derived in the fluid description of the electrons given by (5.28) when we drop the inertial forces and form the vector product with the magnetic flux density

$$0 = \mathbf{E} \times \mathbf{B} + (\mathbf{u} \times \mathbf{B}) \times \mathbf{B} - \frac{1}{nq} \nabla p \times \mathbf{B} . \tag{5.49}$$

Decomposing the mean velocity into components perpendicular and parallel to the magnetic field direction, $\mathbf{u} = \mathbf{u}_\perp + \mathbf{u}_\parallel$, and evaluating the double vector product $(\mathbf{u} \times \mathbf{B}) \times \mathbf{B} = (\mathbf{u}_\perp \cdot \mathbf{B})\mathbf{B} - B^2 \mathbf{u}_\perp$, we obtain the plasma motion across the magnetic field

$$\mathbf{u}_\perp = \frac{\mathbf{E} \times \mathbf{B}}{B^2} - \frac{\nabla p \times \mathbf{B}}{qnB^2} = \mathbf{v}_E + \mathbf{v}_D . \tag{5.50}$$

The first term on the r.h.s. is the well-known E×B drift motion. The second term is called the *diamagnetic drift* $\mathbf{v}_\mathrm{D}$. While the former is based on a drift of the guiding centers—as we had derived in the single particle model—the latter drift does not require any motion of the guiding centers, as can be seen from the cartoon in Fig. 5.8a. The superposition of the ring currents represented by gyrating particles that have an inhomogeneous distribution of guiding centers gives a net electric current. The same net current arises at the surface of a finite size magnetoplasma, as shown in Fig. 5.8b.

Instead of being generated by an inhomogeneous density of guiding centers, such net currents are also produced by temperature gradients, which affect the gyroradius and the orbit velocity of the particles. Both effects are covered by the pressure gradient that determines the diamagnetic drift. As a rule, the pressure gradient does not produce a motion of the guiding centers.

Comparing with the description in the single particle model, we can state that the magnetic moment associated with each gyrating particle contributes to a diamagnetic magnetization of the plasma, i.e., a reduction of the magnetic flux density. The magnetization is the product of the individual magnetic moment, $\mu \propto W_\perp$, see

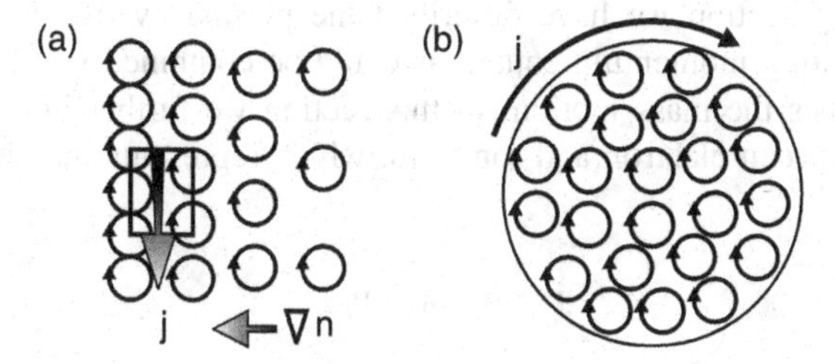

Fig. 5.8 (**a**) The inhomogeneous distribution of guiding centers gives rise to the diamagnetic drift which represents a net current j. (**b**) Surface current of a finite size homogeneneous magnetoplasma

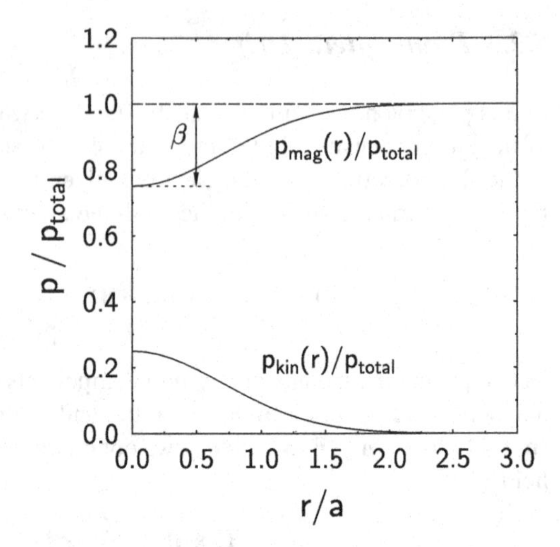

Fig. 5.9 Magnetic pressure and kinetic pressure in the cross-section of a hot plasma column with effective radius a. The diamagnetic currents weaken the magnetic induction. The decrease of magnetic pressure in the center is described by the β factor

(3.34), and the density of guiding centers. Hence, the diamagnetic magnetization is proportional to the particle pressure. This is just what the fluid model says when it balances pressure gradient and the Lorentz force from the diamagnetic current in (5.28).

Applying this understanding of the diamagnetic current to a magnetically confined fusion device, we can start from the pressure balance $p_{\mathrm{kin}} + p_{\mathrm{mag}} = p_{\mathrm{total}} =$ const, which results from (5.46), when we neglect the curvature force. Since the kinetic pressure vanishes in the cool outer layers at the plasma surface, the magnetic field inside the plasma is weakened by the diamagnetism which increases with plasma kinetic pressure $p_{\mathrm{mag}} = p_{\mathrm{total}} - p_{\mathrm{kin}}$, and takes a minimum in the center of the plasma (see Fig. 5.9). This decrease of magnetic confinement in the plasma center is described by the ratio β of the kinetic pressure at the center to the total pressure at the surface, which is given by the magnetic pressure,

$$\beta = \frac{p_{\mathrm{kin}}(0)}{p_{\mathrm{total}}} . \tag{5.51}$$

5.3 Magnetohydrodynamics

In the preceding section we have described the plasma by two interpenetrating fluids. The resulting momentum equations could be combined into a single equation that describes the mass motion. In this section we further introduce the relative motion between electron and ion fluid, which represents the electric current density

$$\mathbf{j} = ne(\mathbf{u}_{\mathrm{i}} - \mathbf{u}_{\mathrm{e}}) . \tag{5.52}$$

Such a single-fluid model of mass motion and electric current flow is called magnetohydrodynamics (MHD).

5.3.1 The Generalized Ohm's Law

A dynamic equation for the spatio-temporal evolution of the current results from the momentum (5.37) after multiplying the ion equation by m_e and the electron equation by m_i, and subtracting the equations:

$$nm_i m_e \frac{\partial}{\partial t}(\mathbf{u}_i - \mathbf{u}_e) = ne(m_e + m_i)\mathbf{E} + ne(m_e\mathbf{u}_i + m_i\mathbf{u}_e) \times \mathbf{B}$$
$$-m_e \nabla p_i + m_i \nabla p_e + n(m_e + m_i)\nu_{ei} m_e(\mathbf{u}_e - \mathbf{u}_i) \,. \quad (5.53)$$

This equation can be simplified by neglecting m_e in the sum of the masses. The mixed term

$$m_e\mathbf{u}_i + m_i\mathbf{u}_e = m_i\mathbf{u}_i + m_e\mathbf{u}_e + m_i(\mathbf{u}_e - \mathbf{u}_i) + m_e(\mathbf{u}_i - \mathbf{u}_e) \quad (5.54)$$
$$= \frac{1}{n}\rho_m\mathbf{v}_m - (m_i - m_e)\frac{1}{ne}\mathbf{j} \quad (5.55)$$

can be decomposed into contributions from mass motion and current density, which results in

$$\frac{m_i m_e}{e}\frac{\partial \mathbf{j}}{\partial t} = e\rho_m\left(\mathbf{E} + \mathbf{v}_m \times \mathbf{B} - \frac{\nu_{ei} m_e}{ne^2}\mathbf{j}\right)$$
$$-m_i\mathbf{j} \times \mathbf{B} - m_e\nabla p_i + m_i\nabla p_e \,. \quad (5.56)$$

As long as we are interested in slowly varying phenomena, we can set $\partial\mathbf{j}/\partial t = 0$ and neglect terms of the order of m_e/m_i. In this way we obtain the *generalized Ohm's law*

$$\mathbf{E} + \mathbf{v}_m \times \mathbf{B} = \eta\mathbf{j} + \frac{1}{ne}(\mathbf{j} \times \mathbf{B} - \nabla p_e) \,. \quad (5.57)$$

Here, $\eta = \nu_{ei} m_e/ne^2$ is the plasma resistivity that arises from Coulomb collisions between electrons and ions. The l.h.s. of (5.57) is the correct electric field in the moving reference frame. This electric field balances the voltage drop ηj by the resistivity, the contribution from the Hall effect $\mathbf{j} \times \mathbf{B}/(ne)$, and the electron pressure term $-\nabla p_e/ne$.

5.3.2 Diffusion of a Magnetic Field

As an application of the generalized Ohm's law, we consider a plasma that is moving at a velocity $\mathbf{v}_m$, and at an arbitrary angle to the magnetic field direction. We start from

$$\mathbf{E} + \mathbf{v}_m \times \mathbf{B} = \eta\mathbf{j} = \frac{\eta}{\mu_0}\nabla \times \mathbf{B} \,, \quad (5.58)$$

where the Hall and pressure term were neglected for simplicity. By taking the curl we have

$$\nabla \times \mathbf{E} + \nabla \times (\mathbf{v}_m \times \mathbf{B}) = \frac{\eta}{\mu_0} \nabla \times (\nabla \times \mathbf{B}) \,. \tag{5.59}$$

Using Faraday's induction law, we obtain a differential equation that links the magnetic field to the mass motion

$$-\frac{\partial \mathbf{B}}{\partial t} - \frac{\eta}{\mu_0} \nabla \times (\nabla \times \mathbf{B}) = -\nabla \times (\mathbf{v}_m \times \mathbf{B}) \,. \tag{5.60}$$

Let us first consider the situation with a plasma at rest, $\mathbf{v}_m = 0$. In this case, (5.58) attains the mathematical shape of a time-dependent diffusion equation

$$-\frac{\partial \mathbf{B}}{\partial t} + D_B \Delta \mathbf{B} = 0 \,, \tag{5.61}$$

which describes the diffusion of magnetic field lines in a conducting medium. The magnetic diffusion coefficient is $D_B = \eta / \mu_0$. We can estimate the diffusion time τ_B by setting $\mathbf{B}(t) \propto \exp(-t/\tau_B)$ and replacing the Laplacian by the square of a characteristic scale length $\Delta \mathbf{B} \approx \mathbf{B}/\ell^2$

$$\tau_B = \frac{\mu_0 \ell^2}{\eta} \,. \tag{5.62}$$

With decreasing resistivity the diffusion time gets longer and longer. This consideration is not restricted to plasmas. For the conditions in the metallic core of the Earth we obtain $\tau_B \approx 10^4$ years. Not surprisingly, this is just the time for the observed reversal of the Earth magnetic field. Even a copper sphere of 1 m diameter has a long diffusion time of $\approx 10\,\text{s}$.

The relative importance of the diffusion term and the flow term in (5.60) can be estimated by a similar dimensional analysis

$$\frac{\eta}{\mu_0} \nabla \times (\nabla \times \mathbf{B}) \approx \frac{\eta}{\mu_0} \frac{B}{\ell^2} \,, \qquad \nabla \times (\mathbf{v}_m \times \mathbf{B}) \approx \frac{v_m B}{\ell} \,. \tag{5.63}$$

This leads to the definition of the *magnetic Reynolds number*

$$R_m = \frac{\mu_0 v_m \ell}{\eta} \,, \tag{5.64}$$

which characterizes the ratio of mass flow to magnetic diffusion.

5.3.3 The Frozen-in Magnetic Flux

The conductivity of a hot plasma is many times larger than that of metals. Therefore, we can describe hot laboratory plasmas or astrophysical plasmas by the concept of infinitely large conductivity (zero resistivity). This theory is named *ideal magnetohydrodynamics* and is reached for $R_{\mathrm{m}} \to \infty$. From (5.60), this limit gives us the relationship

$$\frac{\partial \mathbf{B}}{\partial t} = \nabla \times (\mathbf{v}_{\mathrm{m}} \times \mathbf{B})\,. \tag{5.65}$$

Using the identity

$$\nabla \times (\mathbf{v}_{\mathrm{m}} \times \mathbf{B}) = (\mathbf{B} \cdot \nabla)\mathbf{v}_{\mathrm{m}} - (\mathbf{v}_{\mathrm{m}} \cdot \nabla)\mathbf{B} + \mathbf{v}_{\mathrm{m}} \underbrace{(\nabla \cdot \mathbf{B})}_{=0} - \mathbf{B}(\nabla \cdot \mathbf{v}_{\mathrm{m}}) \tag{5.66}$$

and the continuity (5.8) in the form

$$\nabla \cdot \mathbf{v}_{\mathrm{m}} = -\frac{1}{\rho_{\mathrm{m}}}\left(\frac{\partial \rho_{\mathrm{m}}}{\partial t} + (\mathbf{v}_{\mathrm{m}} \cdot \nabla)\rho_{\mathrm{m}}\right) = -\frac{1}{\rho_{\mathrm{m}}}\frac{\mathrm{d}\rho_{\mathrm{m}}}{\mathrm{d}t} \tag{5.67}$$

we obtain the relation

$$\frac{\mathrm{d}\mathbf{B}}{\mathrm{d}t} = (\mathbf{B} \cdot \nabla)\mathbf{v}_{\mathrm{m}} + \frac{\mathbf{B}}{\rho_{\mathrm{m}}}\frac{\mathrm{d}\rho_{\mathrm{m}}}{\mathrm{d}t}\,. \tag{5.68}$$

We can further use the identity

$$\frac{\mathrm{d}}{\mathrm{d}t}\left(\frac{\mathbf{B}}{\rho_{\mathrm{m}}}\right) = \frac{1}{\rho_{\mathrm{m}}}\frac{\mathrm{d}\mathbf{B}}{\mathrm{d}t} - \frac{\mathbf{B}}{\rho_{\mathrm{m}}^2}\frac{\mathrm{d}\rho_{\mathrm{m}}}{\mathrm{d}t}\,, \tag{5.69}$$

which results in the theorem of Truesdell [80]

$$\frac{\mathrm{d}}{\mathrm{d}t}\left(\frac{\mathbf{B}}{\rho_{\mathrm{m}}}\right) = \left(\frac{\mathbf{B}}{\rho_{\mathrm{m}}} \cdot \nabla\right)\mathbf{v}_{\mathrm{m}}\,. \tag{5.70}$$

The quantity B/ρ_{m} can be considered as the number of field lines per unit mass of the plasma. When the mass flow is strictly perpendicular to the magnetic field, the r.h.s. vanishes. Hence, B/ρ_{m} becomes a conserved quantity. This means that the mass motion can only occur together with the magnetic field. In other words, the magnetic flux is frozen into the plasma. When the mass motion has a field-aligned component, the r.h.s. describes the (B/ρ_{m})-weighted rate of change of the mass flow velocity along the field line. This slipping along the field line for inhomogeneous flows allows a change of B/ρ_{m}.

5.3.4 The Pinch Effect

A very efficient technique to produce a hot plasma without caring for long confinement times is the concept of a pinch discharge. With high pulsed currents, ranging from 10 kA to several MA, a plasma can be magnetically confined and heated to millions of degrees. Two different geometries are using the pinch effect (see Fig. 5.10), the Z-Pinch and the Θ-pinch. The names give the direction of current flow in cylindrical coordinates. In the Z-pinch the plasma is in contact with metallic electrodes, which may be responsible for plasma contamination. The Θ-pinch has no electrodes. Rather, the current I_θ in an external coil of only one winding induces an opposing surface current $-I_\theta$ of the plasma cylinder. Both currents create an axial magnetic field B_z in the gap. In both cases the plasma is squeezed into a narrow cylinder under the action of the magnetic pressure at the plasma surface.

The equilibrium of a Z-pinch can be described by the balance of kinetic pressure in the center and magnetic pressure at its surface, $r = a$,

$$nk_{\mathrm{B}}(T_{\mathrm{e}} + T_{\mathrm{i}}) = \frac{B(a)^2}{2\mu_0}\,. \tag{5.71}$$

The magnetic field at the plasma surface is calculated from the total current by using Ampere's law

$$B(a) = \frac{\mu_0 I}{2\pi a}\,. \tag{5.72}$$

From these two equations we can find the relationship between the temperature on the discharge axis, the radius of the compressed plasma, and the total current

$$(T_{\mathrm{e}} + T_{\mathrm{i}}) \propto \frac{I^2}{a^2}\,, \tag{5.73}$$

which is known as the Bennett-relation, named after Willard Bennett, who already, in 1932, studied the plasma pinch effect [81]. Note that the temperature increases with the square of the discharge current. Modern Z-pinch experiments [82, 83] start

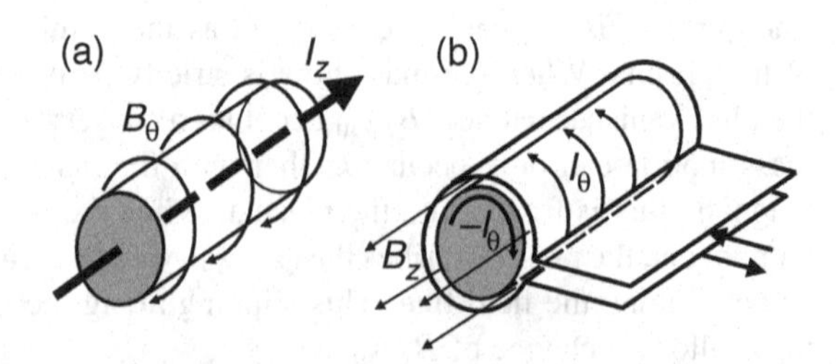

Fig. 5.10 (**a**) Z-pinch und (**b**) Θ-pinch. Magnetic self-confinement by the pinch effect is achieved by pulsed high-current discharges

from a cylindrical wire-cage that vaporizes into a plasma and implodes under the magnetic pressure from several ten MA current.[1]

5.3.5 Application: Alfvén Waves

The concept of frozen-in magnetic flux can be best demonstrated by studying low-frequency waves of a magnetized plasma. Such MHD waves were predicted by Alfvén [84, 85] and were first generated in liquid metals [86, 87] before they were demonstrated in magnetized plasmas, e.g., [88–90].

5.3.5.1 The Shear Alfvén Wave

We know from introductory mechanics courses that simple harmonic waves of a fluctuating quantity, e.g., the gas density n in a sound wave, are described by a second-order differential equation of the type

$$\left[\frac{\partial^2 n}{\partial t^2} - v^2 \frac{\partial^2 n}{\partial z^2}\right] = 0\,. \tag{5.74}$$

The quantity v is the phase velocity of the wave and wave solutions have the form

$$n(x,t) = \hat{n}\sin[k(x \pm vt)]\,. \tag{5.75}$$

Here, $k = 2\pi/\lambda$ is the wavenumber. The $\pm$ sign indicates that waves can propagate in $\pm x$ direction because the wave equation only depends on v^2.

The starting point of our calculation is the ideal MHD ($\eta = 0$), which ensures that an internal magnetic field cannot leave the plasma by magnetic diffusion. The momentum equation is written in the simplified form

$$\rho_{\mathrm{m}} \frac{\partial \mathbf{v}_{\mathrm{m}}}{\partial t} = \mathbf{j} \times \mathbf{B} \tag{5.76}$$

and the evolution of the magnetic field is given by (5.65) and (5.66) as

$$\frac{\partial \mathbf{B}}{\partial t} = (\mathbf{B}\cdot\nabla)\mathbf{v}_{\mathrm{m}} - (\mathbf{v}_{\mathrm{m}}\cdot\nabla)\mathbf{B} - \mathbf{B}(\nabla\cdot\mathbf{v}_{\mathrm{m}})\,. \tag{5.77}$$

Since we are not interested in sound waves of the plasma, the additional assumption of an incompressible flow $\nabla\cdot\mathbf{v}_{\mathrm{m}} = 0$ is made, which gives $\rho_{\mathrm{m}} = \mathrm{const}$. Sound waves will be discussed separately in Sect. 6.5.3. The linear wave analysis assumes that the

[1] For recent experiments, see Sandia National Lab's website `http://zpinch.sandia.gov/`

magnetic field and the mass velocity can be decomposed into a homogeneous and stationary equilibrium (subscript 0) and a wavelike perturbation (subscript 1).

$$\mathbf{B} = \mathbf{B}_0 + \mathbf{B}_1$$
$$\mathbf{v}_\mathrm{m} = \mathbf{v}_0 + \mathbf{v}_1 \,. \tag{5.78}$$

The magnetic field $\mathbf{B}_0 = (0, 0, B_0)$ defines the z-direction and the plasma is assumed to be at rest, $\mathbf{v}_0 = 0$. Then the perturbed quantities are described by

$$\rho_\mathrm{m} \frac{\partial \mathbf{v}_1}{\partial t} = \frac{1}{\mu_0} (\nabla \times \mathbf{B}_1) \times \mathbf{B}_0 \tag{5.79}$$
$$\frac{\partial \mathbf{B}_1}{\partial t} = (\mathbf{B}_0 \cdot \nabla) \mathbf{v}_1 \,. \tag{5.80}$$

Here, we have dropped the second-order term $(\mathbf{v}_\mathrm{m} \cdot \nabla)\mathbf{B}$. We now seek for perpendicular pertubations of the magnetic field $\mathbf{B}_1 = (B_{1x}, 0, 0)$, which describe a local transverse displacement of a field line in x-direction. For calculating the streaming velocity, we must decompose the double vector product $(\nabla \times \mathbf{B}_1) \times B_0 = (\mathbf{B}_0 \cdot \nabla)\mathbf{B}_1 - (\nabla \mathbf{B}_1) \cdot \mathbf{B}_0$. Since $\mathbf{B}_1$ has only an x-component, and $\mathbf{B}_0$ is oriented in z-direction, the expression $(\nabla \mathbf{B}_1) \cdot \mathbf{B}_0$ vanishes. Likewise, $(\mathbf{B}_0 \cdot \nabla)\mathbf{B}_1 = B_0(\partial B_{1x}/\partial z)\mathbf{e}_x$ and the acceleration of the mass is also in x-direction. In this way, we obtain the coupled set of equations

$$\rho_\mathrm{m} \frac{\partial v_{1x}}{\partial t} = \frac{B_0}{\mu_0} \frac{\partial B_{1x}}{\partial z}$$
$$\frac{\partial B_{1x}}{\partial t} = B_0 \frac{\partial v_{1x}}{\partial z} \,, \tag{5.81}$$

which can be combined into a wave equation for either of the perturbed quantities

$$\left[\frac{\partial^2}{\partial t^2} - v_\mathrm{A}^2 \frac{\partial^2}{\partial z^2} \right] v_{1x} = 0$$
$$\left[\frac{\partial^2}{\partial t^2} - v_\mathrm{A}^2 \frac{\partial^2}{\partial z^2} \right] B_{1x} = 0 \,. \tag{5.82}$$

This equation describes a transverse wave that propagates along the magnetic field line (see Fig. 5.11). This is the shear-Alfvén wave. A different shear-wave in a solid is described in Fig. 10.37b. The propagation velocity v_A is the Alfvén speed

$$v_\mathrm{A} = \left(\frac{B_0^2}{\mu_0 \rho_\mathrm{m}} \right)^{1/2} \,. \tag{5.83}$$

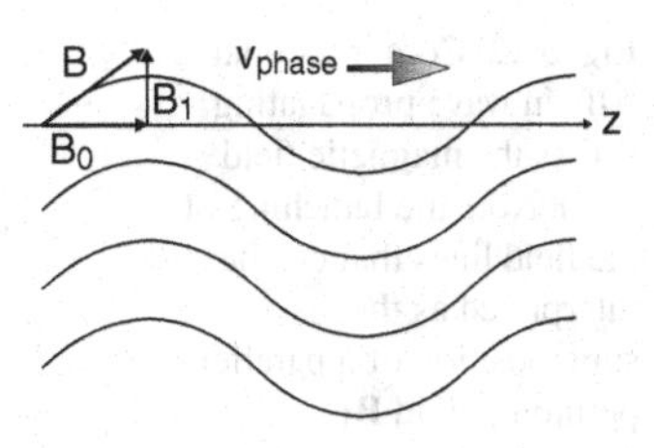

Fig. 5.11 The deformed field-line pattern of a transverse Alfvén wave propagating along the magnetic field $\mathbf{B}_0$

Although the Alfvén speed contains the term B_0^2/μ_0, it is misleading to associate this with the magnetic pressure. The transverse wave, like any shear wave, conserves the volume between the field lines and there is no compression of the plasma or of the bundle of magnetic field lines. Hence, the wave must be driven by a different mechanism.

The force, which a magnetic field exerts on a certain volume, is obtained by integrating the Maxwell stress tensor [74] over the surface of that volume

$$F_\alpha = \oint S_{\alpha\beta}\, \mathrm{d}A_\beta \tag{5.84}$$

and the magnetic part of the stress tensor is defined as

$$S_{\alpha\beta} = \frac{\mu_0}{2}\begin{pmatrix} -B_0^2 & 0 & 0 \\ 0 & -B_0^2 & 0 \\ 0 & 0 & +B_0^2 \end{pmatrix} = -p_{\mathrm{mag}}\begin{pmatrix} 1 & 0 & 0 \\ 0 & 1 & 0 \\ 0 & 0 & 1 \end{pmatrix} + \begin{pmatrix} 0 & 0 & 0 \\ 0 & 0 & 0 \\ 0 & 0 & 2p_{\mathrm{mag}} \end{pmatrix}. \tag{5.85}$$

The Maxwell stress can be decomposed into an isotropic magnetic pressure p_{mag} and an opposing tension of magnitude $\mathscr{T} = 2p_{\mathrm{mag}}$ along the field line. Then, the propagation velocity can be rewritten as

$$v_{\mathrm{A}} = \left(\frac{\mathscr{T}}{\rho_{\mathrm{m}}}\right)^{1/2}, \tag{5.86}$$

which shows the similarity of the Alfvén wave mechanism to a plucked string. Here, the tension $\mathscr{T}$ of the magnetic field line replaces the mechanical tension, which restores the string to its resting position. This was already discussed for the curvature force in Sect. 5.2.2. Moreover, the inertia of the string is replaced by the mass density ρ_{m} of the plasma. This means that the mass remains attached to the field line as predicted by the concept of frozen-in flux.

5.3.5.2 The Compressional Alfvén Wave

For completeness, it should be mentioned at the end that there is a different type of Alfvén wave, which propagates across the magnetic field direction and involves compression of the magnetic field (see Fig 5.12). When effects from gas pressure

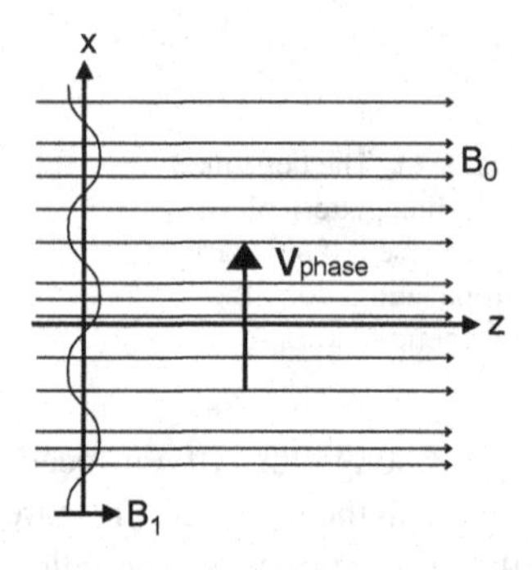

Fig. 5.12 Compressional Alfvén wave propagating across the magnetic field lines. Note the bunching of the field lines that can be interpreted as the superposition of a parallel pertubing field $\mathbf{B}_1$

can be neglected ($p_{\text{kin}} \ll p_{\text{mag}}$) the phase velocity becomes again $v_\phi = v_\text{A}$, but this time the analogy with a sound wave, $c_\text{s} = (\gamma p/\rho)^{1/2}$, is justified. In this situation, the magnetic pressure takes the role of gas pressure and $\gamma = 2$ reflects that there are two degrees of freedom corresponding to the two directions perpendicular to the magnetic field. When the kinetic pressure cannot be neglected, the compressional Alfvén wave becomes a magnetosonic wave, whose propagation speed is determined by $v_\varphi = (v_\text{A}^2 + c_\text{s}^2)^{1/2}$.

5.3.6 Application: The Parker Spiral

The solar wind is a highly conducting medium. Therefore, the magnetic field is frozen into the mass flow of the expanding plasma. In Sect. 1.2.3 we had seen that the rotation of the Sun shapes the mass flow into an Archimedian spiral, as shown in Fig. 1.5, which is named the *Parker spiral* in honor of Eugene Parker who first described this structure by MHD [20]. Here, we will now consider the consequences for the interplanetary magnetic field.

If the Sun did not rotate, the solar wind would simply expand in flux tubes formed by radial magnetic field lines, see Fig. 5.13. Because of flux conservation in this spherical geometry, the mass density would decrease as $(r/r_\odot)^{-2}$. In the same way,

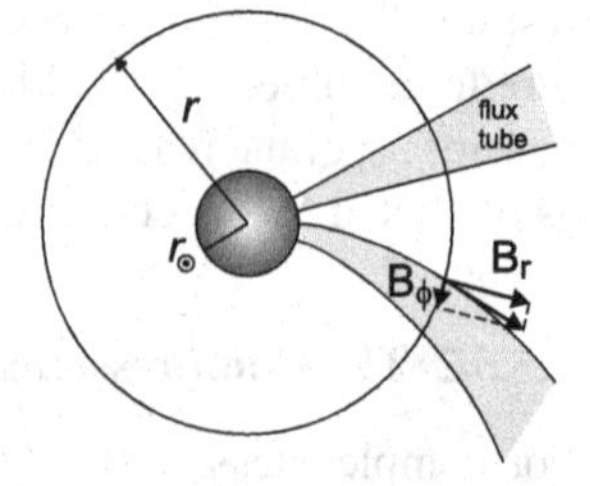

Fig. 5.13 Cartoon of a hypothetical purely radial magnetic flux tube for a non-rotating Sun and the flux tubes forming the Parker spiral

the magnetic flux conservation would give $B_r \propto (r/r_\odot)^{-2}$. This property does also apply in the presence of solar rotation, because the formation of an Archimedian spiral is simply a transformation to a rotating coordinate system, which does not affect the number of field lines traversing a spherical shell of radius r. In addition to the radial magnetic field component B_r, there will also be an azimuthal component B_φ in the solar equatorial plane.

The condition for a frozen-in magnetic field (5.65) for a stationary flow becomes $0 = \nabla \times (\mathbf{u} \times \mathbf{B})$, which reads in spherical coordinates

$$0 = \frac{1}{r}\frac{\partial}{\partial r}[r(u_\varphi B_r - u_r B_\varphi)] , \tag{5.87}$$

or $r(u_\varphi B_r - u_r B_\varphi) = \text{const}$. At the surface of the Sun we have $u_\varphi = \omega_\odot r_\odot$ and the magnetic field there has only a radial component $B_r = B_0$, which results in

$$r(u_\varphi B_r - u_r B_\varphi) = \omega_\odot r_\odot^2 B_0 . \tag{5.88}$$

This relation gives us the azimuthal magnetic field in the spiral-shaped flux tube

$$B_\varphi = \frac{r(u_\varphi B_r - \omega_\odot r_\odot^2 B_0)}{r u_r} = -\frac{\omega_\odot (r - r_\odot)}{u_r} B_r . \tag{5.89}$$

In the last step we have used $B_r = B_0 (r_\odot / r)^2$. For large distances, $r \gg r_\odot$, we have $\omega_\odot r \gg u_\varphi$. Then the azimuthal component of the magnetic field drops off as $B_\varphi \propto r^{-1}$ whereas the radial component decreases as $B_r \propto r^{-2}$. This means that the magnetic field direction changes from radial near the Sun to azimuthal at the orbits of the outer planets. At 1 AU, the inclination is $\arctan(B_\varphi / B_r) \approx 45°$. The intensity of the magnetic field decreases as

$$B(r) = B_0 \frac{r_\odot^2}{r^2} \left[1 + \left(\frac{\omega_\odot (r - r_\odot)}{u_r} \right)^2 \right]^{1/2} . \tag{5.90}$$

So far, we were only interested in the development of the magnetic field components along an individual flux tube that forms one branch of the Parker spiral. Since the Sun cannot be a magnetic monopole with outgoing field lines of same polarity, it is not surprising that under quiet conditions, the Parker spiral has two or four sectors of alternating polarity, which are stable for many solar rotation periods. At solar maximum conditions, the sector structure is complex and characterized by a large number of transient disturbances.

The Basics in a Nutshell

- The fluid models treat the electrons and ions as fluids and seek self-consistency of the problems by combining the fluid equations with the set of Maxwell's equations:

Faraday's induction law	$\nabla \times \mathbf{E} = -\partial \mathbf{B}/\partial t$
Ampere's law	$\nabla \times \mathbf{B} = \mu_0 \left(\mathbf{j} + \varepsilon_0 \partial \mathbf{E}/\partial t \right)$
Poisson's law	$\nabla \cdot \mathbf{E} = \rho/\varepsilon_0$
no magnetic monopoles	$\nabla \cdot \mathbf{B} = 0$

- The two-fluid model is based on separate equations for electrons and ions and describes the continuity and momentum flow of the fluids:

continuity	$\partial n/\partial t + \nabla \cdot (n\mathbf{u}) = 0$
momentum tansport	$nm \left(\frac{\partial \mathbf{u}}{\partial t} + (\mathbf{u} \cdot \nabla)\mathbf{u} \right) = nq(\mathbf{E} + \mathbf{u} \times \mathbf{B}) - \nabla p$

- The MHD-equations describe the mass transport and the electric current in a single fluid:

momentum transport	$\rho_{\mathrm{m}} \frac{\partial \mathbf{v}_{\mathrm{m}}}{\partial t} = \mathbf{j} \times \mathbf{B} - \nabla p + \rho_{\mathrm{m}} \mathbf{g}$
generalized Ohm's law	$\mathbf{E} + \mathbf{v}_{\mathrm{m}} \times \mathbf{B} = \eta \mathbf{j} + \frac{1}{ne}(\mathbf{j} \times \mathbf{B} - \nabla p_{\mathrm{e}})$

- The diamagnetic drift is a net effect in an inhomogeneous distribution of guiding centers. A net electric current is established without motion of the guiding centers.
  The diamagnetic drift velocity is $\mathbf{v}_{\mathrm{D}} = -[\nabla p \times \mathbf{B}](qnB^2)^{-1}$.
- A magnetic field exerts an isotropic magnetic pressure $p_{\mathrm{mag}} = B_0^2(2\mu_0)^{-1}$ and has a field line tension $\mathcal{T} = 2p_{\mathrm{mag}}$.
- When the plasma is an ideal conductor, the magnetic field is frozen in the plasma. The combined motion of plasma and magnetic field leads to Alfvén waves, which propagate at the Alfvén speed $v_{\mathrm{A}} = B_0(\mu_0 \rho_{\mathrm{m}})^{-1/2}$.

Problems

5.1 (a) Consider the pressure equilibrium in a Z-pinch that has been compressed by its self-generated magnetic field to a radius of $100\,\mu$m. What is the magnetic pressure at the surface of the pinch, when the total current amounts to 10 kA? How compares this to atmospheric pressure?
(b) Assume that the plasma inside the pinch is homogeneous and has $T_{\mathrm{e}} = T_{\mathrm{i}}$ and density $n_{\mathrm{e}} = 10^{24}\,\mathrm{m}^{-3}$. What is the temperature inside this plasma that is necessary to balance the magnetic pressure by gas kinetic pressure?

5.2 Calculate the magnetic field B that is necessary to produce a magnetic pressure at the surface of a magnetically confined fusion that is 4 times the kinetic pressure in the plasma center, when the central density is $n_e = 2\times10^{20}\,\mathrm{m}^{-3}$ and the temperature $T = 20\,\mathrm{keV}$. This corresponds to $\beta = 25\%$.

5.3 What is the Alfvén speed in a fusion plasma with deuterium ions of $n_i = 10^{20}\,\mathrm{m}^{-3}$ density at a typical magnetic field of $B = 3\,\mathrm{T}$?

5.4 The ionospheric F-layer has a plasma density of $n = 10^{12}\,\mathrm{m}^{-3}$ and consists mainly of O^+-ions.
(a) What is the Alfvén speed at a typical magnetic field of $B = 3\cdot10^{-5}\,\mathrm{T}$?
(b) Compare this result with the ion sound speed at a temperature $T_e = T_i = 3000\,\mathrm{K}$.

5.5 For the Parker spiral, draw a log-log plot of the normalized magnetic field $B(r)/B_0$ and its components, B_r/B_0 and B_φ/B_0, vs. the normalized radial position $r/r_\odot$. Assume $u_r = 4\times10^5\,\mathrm{m\,s}^{-1}$ and a solar rotation period of 27 d. Mark the position of the Earth's orbit in this plot.

5.6 A method to determine the temperature of a hot magnetized plasma column is based on measuring the change in magnetic flux when the plasma is switched off. This can be done by a *diamagnetic loop* of N windings, which is wound around the (non-conducting) cylindrical vessel of radius R that is assumed to contain the plasma column. Faraday's induction law gives $\Delta\Phi_{\mathrm{mag}} = -N\int U_{\mathrm{ind}}\,\mathrm{d}t$. Hence, the time integral of the voltage pulse from the diamagnetic loop gives the change in magnetic flux. To derive a relation between plasma temperature and integrated loop voltage, we assume that $T_e = T_i = \mathrm{const}$. The density profile is approximated by a Gaussian $n(r) = n_0\exp[-(r/a)^2]$ with $a^2 \ll R^2$. Use the pressure equilibrium $p_{\mathrm{kin}}(0)+p_{\mathrm{mag}}(0) = p_{\mathrm{mag}}(R)$ and calculate the total change in magnetic flux $\Delta\Phi_{\mathrm{mag}}$ from its vacuum value. Show that $\Delta\Phi_{\mathrm{mag}} \approx -\frac{1}{2}\pi a^2 n_0 B_0\beta$ in the limit $\beta \ll 1$ with β from (5.51).

Chapter 6
Plasma Waves

"What is the use of a book", thought Alice, "without pictures or conversations?"

Lewis Carroll, Alice in Wonderland

The interest in wave propagation in plasmas has different roots. One of these was the reflection of electromagnetic waves by the ionosphere [91]. Stimulated by Guglielmo Marconi's (1874–1937) experiments on long-distance radio in 1901, Oliver Heaviside (1850–1925) [92] and, independently, Arthur Edwin Kennelly (1861–1939) [93] postulated, in 1902, that the Earth's atmosphere at high altitude must contain an electrically conducting layer that reflects radio waves like a mirror. Many decades before, in 1839, Carl Friedrich Gauss (1777–1855) had conjectured that the fluctuations of the Earth magnetic field might be related to electric currents in the high atmosphere. The quantitative investigation of the ionoshere with radio waves began in the years 1924–1927 with the vertical sounding experiments in the U.S. of Gregory Breit (1899–1981) and Merle Antony Tuve (1901–1982) [94, 95], and in Great-Britain by Edward V. Appleton (1892–1965) [96], which proved the existence of a conducting atmospheric layer, now named the ionosphere, in the altitude regime of (100–500) km. At the same time, Irving Langmuir, at the General Electric Laboratories, discovered high-frequency fluctuations in gas discharges, now known as Langmuir oscillations [97].

At the introductory level of this book we are interested in the classification of the fundamental wave types in a plasma, which elucidate the diverse mechanisms that lead to wave phenomena. At the same time we will discuss the application of various wave types for plasma diagnostics. There is a number of modern text books [98–101] that give a comprehensive description of plasma waves.

6.1 Maxwell's Equations and the Wave Equation

In this Section, the interaction of the plasma particles with an electromagnetic wave is investigated in terms of a new concept in which the plasma is considered as a dielectric medium. For this purpose, the linear response of the plasma particles to the wave field is included in the dielectric constant of the plasma medium, which

A. Piel, *Plasma Physics*, DOI 10.1007/978-3-642-10491-6_6,

determines the propagation speed and polarization of the plasma waves. This model is developed step by step starting from Maxwell's equations.

6.1.1 Basic Concepts

Plasma waves are described by the set of Maxwell's equations

$$\nabla \times \mathbf{E} = -\frac{\partial \mathbf{B}}{\partial t} \tag{6.1}$$

$$\nabla \times \mathbf{B} = \mu_0 \left(\mathbf{j} + \varepsilon_0 \frac{\partial \mathbf{E}}{\partial t} \right) \tag{6.2}$$

and a proper equation of motion for the plasma species that establishes the relation between the alternating electric current $\mathbf{j}(\mathbf{E})$

$$\mathbf{j} = ne(\mathbf{v}_\mathrm{i} - \mathbf{v}_\mathrm{e}) \,, \tag{6.3}$$

and the electric field. The simplest case is the description of the plasma particles in the model of single-particle motion. The velocities $\mathbf{v}_\mathrm{e,i}$ are solutions of Newton's equation that is expanded by an additional friction force that is described by a collision frequency ν_m for momentum loss,

$$m\,(\dot{\mathbf{v}} + \nu_\mathrm{m}\mathbf{v}) = q(\mathbf{E} + \mathbf{v} \times \mathbf{B}) \,. \tag{6.4}$$

In warm plasmas, we could include pressure effects by solving the MHD equations for the variable $\mathbf{j}$. Other effects related to the distribution function of velocities, e.g., Landau damping, will be discussed in Chap. 9.

For discussing the propagation properties of the waves, we make the additional simplifying assumption that, at a chosen angular frequency ω, the relation between the alternating current $\mathbf{j}(\omega)$ and the electric field strength at that frequency $\mathbf{E}(\omega)$ is linear or can be linearized by suitable approximations

$$\mathbf{j}(\omega) = \sigma(\omega) \cdot \mathbf{E}(\omega) \,. \tag{6.5}$$

Here, $\sigma(\omega)$ is the frequency-dependent conductivity. Taking the curl in the induction law (6.1), we obtain the wave equation

$$\begin{aligned} \nabla \times (\nabla \times \mathbf{E}) &= -\nabla \times \frac{\partial \mathbf{B}}{\partial t} \\ &= -\frac{\partial}{\partial t}(\nabla \times \mathbf{B}) \\ &= -\mu_0 \varepsilon_0 \frac{\partial^2 \mathbf{E}}{\partial t^2} - \mu_0 \frac{\partial \mathbf{j}}{\partial t} \,. \end{aligned} \tag{6.6}$$

With $\mu_0\varepsilon_0 = 1/c^2$, the wave equation for the electric field takes the form

$$\nabla \times (\nabla \times \mathbf{E}) + \frac{1}{c^2}\frac{\partial^2 \mathbf{E}}{\partial t^2} = -\mu_0 \frac{\partial \mathbf{j}}{\partial t} . \tag{6.7}$$

6.1.2 Fourier Representation

The wave equation has solutions that are plane monochromatic waves of the form

$$\begin{aligned}
\mathbf{E} &= \hat{\mathbf{E}} \exp[\mathrm{i}(\mathbf{k}\cdot\mathbf{r} - \omega t)] \\
\mathbf{B} &= \hat{\mathbf{B}} \exp[\mathrm{i}(\mathbf{k}\cdot\mathbf{r} - \omega t)] \\
\mathbf{j} &= \hat{\mathbf{j}} \exp[\mathrm{i}(\mathbf{k}\cdot\mathbf{r} - \omega t)] .
\end{aligned} \tag{6.8}$$

Here, $\mathbf{k}$ is the wave vector, which describes the direction of wave propagation. The magnitude of the wave vector is related to the wavelength by $k = 2\pi/\lambda$. The wave amplitudes $\hat{\mathbf{E}}$ and $\hat{\mathbf{j}}$ are complex quantities, which give us a simple way to include a phase shift between current density and electric field. Both are functions of frequency and wavenumber, e.g., $\hat{E} = \hat{E}(\omega, \mathbf{k})$. Using this plane wave representation, we can establish simple substitution rules for the differential operations in the wave equation

$$\nabla \times \mathbf{E} \rightarrow \mathrm{i}\mathbf{k} \times \hat{\mathbf{E}} , \quad \nabla \cdot \mathbf{E} \rightarrow \mathrm{i}\mathbf{k} \cdot \hat{\mathbf{E}} , \quad \frac{\partial}{\partial t}\mathbf{E} \rightarrow -\mathrm{i}\omega\hat{\mathbf{E}} . \tag{6.9}$$

In this way Maxwell's equations (6.1) and (6.2) can be rewritten in terms of a set of algebraic relations between the complex wave amplitudes

$$\mathrm{i}\mathbf{k} \times \hat{\mathbf{E}} = \mathrm{i}\omega\hat{\mathbf{B}} \tag{6.10}$$

$$\mathrm{i}\mathbf{k} \times \hat{\mathbf{B}} = -\mathrm{i}\omega\varepsilon_0\mu_0\hat{\mathbf{E}} + \mu_0\hat{\mathbf{j}}_0 . \tag{6.11}$$

Here, the term $\exp[\mathrm{i}(\mathbf{k}\cdot\mathbf{r} - \omega t)]$ describing the phase evolution in space and time could be dropped.

6.1.3 Dielectric or Conducting Media

Since we have assumed a linear relation between the alternating current and the electric field, we can give different interpretations to the current density. When we consider the plasma as a dielectric medium, we can think of the wiggling motion of electrons and ions as a polarization current, which can be combined with the vacuum displacement current $\varepsilon_0(\partial\mathbf{E}/\partial t)$. In the limit of very high frequencies only the electrons will oscillate about their mean position while the much heavier ions will be immobile. Hence, we are allowed to consider the plasma as a set of dipoles formed by pairs consisting of an electron oscillating about a corresponding ion at rest. Such a medium is characterized by the dielectric displacement

$$\hat{\mathbf{D}}(\omega) = \varepsilon_0 \varepsilon(\omega) \hat{\mathbf{E}}(\omega) . \tag{6.12}$$

Noting that for a given frequency ω, the displacement current can be considered as the sum of the vacuum displacement current plus the conduction current,

$$\frac{\partial \mathbf{D}}{\partial t} = \varepsilon_0 \frac{\partial \mathbf{E}}{\partial t} + \mathbf{j} = \varepsilon_0 \varepsilon(\omega) \frac{\partial \mathbf{E}}{\partial t} . \tag{6.13}$$

This gives us a relation between the dielectric function $\varepsilon(\omega)$ and the electric conductivity $\sigma(\omega)$

$$\varepsilon(\omega) = 1 + \frac{\mathrm{i}}{\omega \varepsilon_0} \sigma(\omega) . \tag{6.14}$$

In the following, we will call $\varepsilon(\omega)$ the dielectric function when the frequency dependence is considered. For a specific value of ω, we will name $\varepsilon(\omega)$ the dielectric constant for that frequency.

In an unmagnetized plasma, $\sigma(\omega)$ and $\varepsilon(\omega)$ are simple scalar functions of the wave frequency ω. A magnetized plasma, however, is anisotropic because of the different motion along and across the magnetic field. Therefore, dielectric function and conductivity then become tensors, see Sect. 3.2 of Appendix.

$$\boldsymbol{\varepsilon}_\omega = \mathbf{I} + \frac{\mathrm{i}}{\omega \varepsilon_0} \boldsymbol{\sigma}_\omega . \tag{6.15}$$

Here, $\mathbf{I}$ is the unit tensor. This means that the electric field vector $\mathbf{E}$ and the electric current vector $\mathbf{j}$ may be no longer parallel to each other. Moreover, collisions of the plasma particles make the plasma a *lossy* dielectric medium and $\varepsilon(\omega)$ is in general a complex function.

In conclusion, there are two different views of the plasma medium. For weak losses, the plasma behaves mostly as a dielectric and is described by a dielectric function (or tensor) $\varepsilon(\omega)$, in which the real part is dominant over the imaginary part. When the collisions are frequent, the plasma behaves mainly as a conductor and is described by a complex conductivity $\sigma(\omega)$, in which the imaginary part represents phase shifts resulting from inertial effects. In the following, we will be mostly interested in cases, where the plasma waves are weakly damped. Then, the dielectric tensor elements are mostly real quantities. Therefore, we will prefer the dielectric description of a plasma.

6.1.4 Phase Velocity

We define the phase φ of a monochromatic wave by

$$\varphi = \mathbf{k} \cdot \mathbf{r} - \omega t\,. \tag{6.16}$$

A point of constant phase (think of wave crests or troughs) within a wave moves with a velocity that is defined by the constancy of the phase

$$0 = \frac{\mathrm{d}\varphi}{\mathrm{d}t} = \mathbf{k} \cdot \frac{\mathrm{d}\mathbf{r}}{\mathrm{d}t} - \omega\,. \tag{6.17}$$

Hence, the *phase velocity* is defined as

$$\mathbf{v}_\varphi = \frac{\omega}{k^2}\,\mathbf{k}\,. \tag{6.18}$$

The phase velocity is a vector with the magnitude $v_\varphi = \omega/k$ and the vector has the same orientation as the wave vector $\mathbf{k}$.

6.1.5 Wave Packet and Group Velocity

Let us now consider the propagation of two waves, which is the simplest case of a wave packet. The frequencies and wave numbers are assumed to be close to each other, i.e., $|\omega_1 - \omega_2| \ll (\omega_1 + \omega_2)/2$, $|k_1 - k_2| \ll (k_1 + k_2)/2$. For simplicity, we also assume that the waves have the same amplitude and propagate in x-direction. Then, the wave packet is the superposition of the two sine waves

$$E(x, t) = \sin(k_1 x - \omega_1 t) + \sin(k_2 x - \omega_2 t)\,. \tag{6.19}$$

Using the theorem for addition of sines we obtain

$$E(x, t) = 2 \sin\left(\frac{k_1 + k_2}{2} x - \frac{\omega_1 + \omega_2}{2} t\right) \cos\left(\frac{k_1 - k_2}{2} x - \frac{\omega_1 - \omega_2}{2} t\right)\,. \tag{6.20}$$

This gives us the well known interference pattern (Fig. 6.1), in which we find a sine term that describes a rapid oscillation at the arithmetic mean of the two frequencies and wavenumbers and a cosine term that describes the envelope of the signal. The phase velocity of this signal is given by the sine term as

$$v_\varphi = \frac{(\omega_1 + \omega_2)/2}{(k_1 + k_2)/2} = \frac{\bar{\omega}}{\bar{k}}\,. \tag{6.21}$$

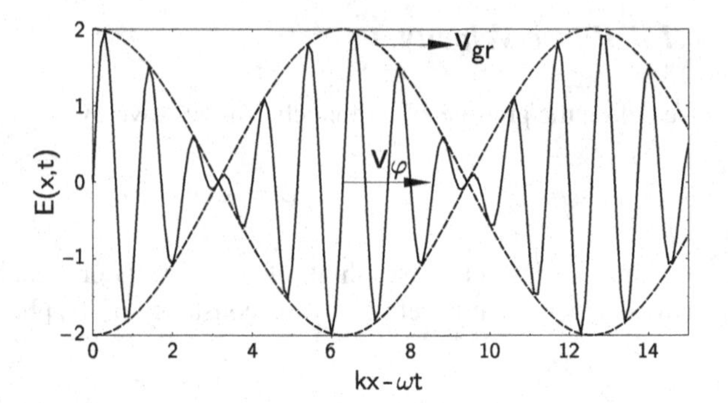

Fig. 6.1 Interference of two sine waves of same amplitude. The instantaneous phase propagates at the phase velocity v_φ, the envelope has a different propagation velocity, the group velocity v_{gr}

The envelope moves at a different velocity, the *group velocity*, which results from the phase evolution of the cosine term

$$v_{gr} = \frac{(\omega_1 - \omega_2)/2}{(k_1 - k_2)/2} = \frac{\Delta\omega}{\Delta k} . \tag{6.22}$$

In the more general case of an extended wave packet consisting of many *wavelets* one can show that the group velocity is given by the expression

$$\mathbf{v}_{gr} = \left(\frac{\partial\omega}{\partial k_x}, \frac{\partial\omega}{\partial k_y}, \frac{\partial\omega}{\partial k_z}\right) = \nabla_k \omega = \frac{d\omega}{d\mathbf{k}} . \tag{6.23}$$

The analogy to the simpler case given above is evident. The group velocity has the magnitude $v_{gr} = d\omega/dk$.

Why have group velocity and phase velocity different values? In Fig. 6.2a the relationship between frequency ω and wavenumber k is shown for a dispersive medium. The phase velocity is constructed by choosing a point (ω, k) on the dispersion curve and evaluating $\tan\alpha = \omega/k = v_\varphi$. The tangent to the dispersion branch at this point (ω, k) has a different slope, $\tan\beta = d\omega/dk = v_{gr}$. Obviously phase velocity and group velocity are different in this example. A non-dispersive medium is characterized by the equality of phase and group velocity, as shown in Fig. 6.2b. Obviously, $v_{gr} = v_\varphi$ can only be achieved by a dispersion relation that is represented by a straight line through the origin.

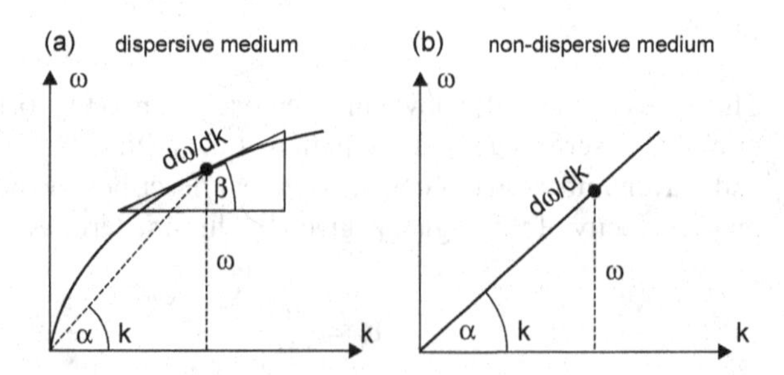

Fig. 6.2 (**a**) The phase velocity is the quotient $\omega/k = \tan(\alpha)$. The slope of the tangent to the function $\omega(k)$ is the group velocity, $d\omega/dk = \tan(\beta)$. (**b**) A non-dispersive medium has $v_{gr} = v_\varphi$

In an anisotropic medium, such as a magnetized plasma, the direction of the group velocity is not necessarily parallel to the phase velocity. There are exotic situations, e.g., for Whistler waves, where phase velocity and group velocity can become even perpendicular to each other [99, 100].

6.1.6 Refractive Index

In optics the refractive index of a transparent medium is defined as the ratio of the speed of light in vacuum to the speed in that medium. This concept can be applied in a similar manner to electromagnetic waves in a plasma. Hence, we define the refractive index as

$$\mathcal{N} = \frac{kc}{\omega}\,. \tag{6.24}$$

Because of the proportionality of $\mathcal{N}$ and k, we can also define a refractive index vector $\mathbf{N} = (c/\omega)\mathbf{k}$. Obviously, this is the complement to the phase velocity because it points in the direction of wave propagation but has a magnitude $\propto v_\varphi^{-1}$. As in optics, the concept of refractive index is useful for wave refraction, ray tracing, or interferometry.

6.2 The General Dispersion Relation

In this Section, we discuss the wave equation in Fourier representation. Using the vector identity $\mathbf{k}\times(\mathbf{k}\times\hat{\mathbf{E}}) = (\mathbf{kk} - k^2\mathbf{I})\hat{\mathbf{E}}$, the homogeneous wave equation for the Fourier amplitudes (6.7) can be transformed into one of the following forms

$$\left\{\mathbf{kk} - k^2\boldsymbol{I} + \frac{\omega^2}{c^2}\boldsymbol{I} + \mathrm{i}\omega\mu_0\boldsymbol{\sigma}(\omega)\right\}\cdot\hat{\mathbf{E}} = 0 \tag{6.25}$$

$$\left\{\mathbf{kk} - k^2\boldsymbol{I} + \frac{\omega^2}{c^2}\boldsymbol{\varepsilon}(\omega)\right\}\cdot\hat{\mathbf{E}} = 0\,. \tag{6.26}$$

Here, the dyadic product $\mathbf{kk}$ of the wave vectors is defined as the tensor

$$\mathbf{kk} = \begin{pmatrix} k_xk_x & k_xk_y & k_xk_z \\ k_yk_x & k_yk_y & k_yk_z \\ k_zk_x & k_zk_y & k_zk_z \end{pmatrix}. \tag{6.27}$$

Equations (6.25) or (6.26) represent a homogeneous linear system of equations for the electric field vector. This can be explicitely written for the dielectric model as

$$\begin{pmatrix} k_x k_x - k^2 + \frac{\omega^2}{c^2}\varepsilon_{xx} & k_x k_y + \frac{\omega^2}{c^2}\varepsilon_{xy} & k_x k_z + \frac{\omega^2}{c^2}\varepsilon_{xz} \\ k_y k_x + \frac{\omega^2}{c^2}\varepsilon_{yx} & k_y k_y - k^2 + \frac{\omega^2}{c^2}\varepsilon_{yy} & k_y k_z + \frac{\omega^2}{c^2}\varepsilon_{yz} \\ k_z k_x + \frac{\omega^2}{c^2}\varepsilon_{zx} & k_z k_y + \frac{\omega^2}{c^2}\varepsilon_{zy} & k_z k_z - k^2 + \frac{\omega^2}{c^2}\varepsilon_{zz} \end{pmatrix} \cdot \begin{pmatrix} \hat{E}_x \\ \hat{E}_y \\ \hat{E}_z \end{pmatrix} = 0\,. \tag{6.28}$$

Non-vanishing solutions for $\mathbf{E} \neq 0$ are only possible when the determinant of the matrix is zero. This determinant condition defines an implicit relation between frequency and wave number, which we will name the *dispersion relation*,

$$0 = D(\omega, \mathbf{k}) = \det\left[\mathbf{kk} - k^2 \boldsymbol{I} + \frac{\omega^2}{c^2}\boldsymbol{\varepsilon}(\omega)\right]\,. \tag{6.29}$$

In many cases, the relationship between ω and $\mathbf{k}$, which is defined by the zeroes $D(\omega, \mathbf{k}) = 0$, can be written in an explicit form, $\omega(\mathbf{k})$. As a rule, this relation has multiple branches. The explicit form is also called the dispersion relation of a wave.

In summary, (6.28) describes all possible wave modes of a plasma. The specific properties of the plasma are encoded in the elements of the dielectric tensor. Unmagnetized plasmas are isotropic media, for which the dielectric tensor reduces to a scalar dielectric function. It is the magnetic field that introduces anisotropy and requires a description by a tensor (cf. Sect. 6.6).

6.3 Waves in Unmagnetized Plasmas

Here, we investigate the wave modes in a plasma without the influence of a magnetic field. We will first consider very high frequency waves, for which the ion motion, because of the much larger ion inertia, can be neglected. This can be immediately seen from Newton's equation

$$m\frac{\mathrm{d}\mathbf{v}}{\mathrm{d}t} = q\hat{\mathbf{E}}\mathrm{e}^{\mathrm{i}(\mathbf{k}\cdot\mathbf{r}-\omega t)}\,, \tag{6.30}$$

from which the alternating current at the angular frequency ω becomes

$$\hat{\mathbf{j}} = nq\hat{\mathbf{v}} = \mathrm{i}\frac{ne^2}{\omega m}\hat{\mathbf{E}}\,. \tag{6.31}$$

Obviously the ion current is smaller by a factor $m_\mathrm{e}/m_\mathrm{i}$ than the electron current. At high frequencies the ions only act as an immobile neutralizing charge background.

In the last paragraph, we will introduce low-frequency electrostatic waves, where the ion motion becomes important.

6.3.1 Electromagnetic Waves

As a first example we study electromagnetic waves in the limit of a cold plasma, i.e., we neglect pressure effects. Further we neglect collisions of the electrons with neutrals. The coordinate system is chosen with the wave vector in x-direction, $\mathbf{k} = (k_x, 0, 0)$. From (6.31) we know that the directions of electric current and electric field are parallel. Hence, the conductivity tensor has only diagonal elements that have the same value.

$$\sigma_{xx} = \sigma_{yy} = \sigma_{zz} = \mathrm{i}\frac{ne^2}{\omega m} \tag{6.32}$$

and the dielectric tensor has only the components

$$\varepsilon_{xx} = \varepsilon_{yy} = \varepsilon_{zz} = 1 + \frac{\mathrm{i}}{\omega\varepsilon_0}\sigma_{yy} = 1 - \frac{\omega_{\mathrm{pe}}^2}{\omega^2}\,. \tag{6.33}$$

In the latter expression we have introduced the *electron plasma frequency*

$$\omega_{\mathrm{pe}} = \left(\frac{ne^2}{\varepsilon_0 m_{\mathrm{e}}}\right)^{1/2}\,. \tag{6.34}$$

The electron plasma frequency had already been introduced in Sect. 2.2 as the reciprocal of the electron reponse time. Here, it is the natural frequency of the electron gas, as we will see below. Taking into account that for the chosen geometry we have $k_x k_x - k^2 = 0$ and $k_y = k_z = 0$, the wave (6.26) becomes

$$\begin{pmatrix} \frac{\omega^2}{c^2}\left(1-\frac{\omega_{\mathrm{pe}}^2}{\omega^2}\right) & 0 & 0 \\ 0 & -k^2+\frac{\omega^2}{c^2}\left(1-\frac{\omega_{\mathrm{pe}}^2}{\omega^2}\right) & 0 \\ 0 & 0 & -k^2+\frac{\omega^2}{c^2}\left(1-\frac{\omega_{\mathrm{pe}}^2}{\omega^2}\right) \end{pmatrix} \cdot \begin{pmatrix} \hat{E}_x \\ \hat{E}_y \\ \hat{E}_z \end{pmatrix} = 0\,. \tag{6.35}$$

Obviously, the problem has a cylindrical symmetry about the x-direction, which manifests itself by the identical response in y and z-direction. We can now distinguish three cases:

- longitudinal waves: $\hat{E}_x \neq 0$ but $\hat{E}_y = \hat{E}_z = 0$.
- transverse waves: $\hat{E}_x = \hat{E}_z = 0$ but $\hat{E}_y \neq 0$,
- or $\hat{E}_x = \hat{E}_y = 0$ but $\hat{E}_z \neq 0$.

The transverse waves are twofold degenerate corresponding to the two possible directions of polarization in y or z-direction. The case of longitudinal waves will be postponed to Sect. 6.5.1. Here, we will focus on the transverse waves. For this purpose we set $\hat{E}_x = 0$ and retain only the middle line in the set of (6.35),

$$\left(-k^2 + \frac{\omega^2 - \omega_{\rm pe}^2}{c^2} \right) \hat{E}_y = 0 . \tag{6.36}$$

Since $\hat{E}_y \neq 0$ we conclude that the factor in parantheses must vanish, yielding

$$\omega^2 = \omega_{\rm pe}^2 + k^2 c^2 . \tag{6.37}$$

The same result is obtained from the last line of (6.35) because of the degeneracy. The explicit form of the dispersion relation for the transverse wave becomes

$$\omega = \left(\omega_{\rm pe}^2 + k^2 c^2 \right)^{1/2} . \tag{6.38}$$

Since we have $\mathbf{k}$ in x-direction and $\hat{\mathbf{E}}$ in y-direction the vector product $\mathbf{k} \times \hat{\mathbf{E}}$ is nonzero and the induction law (6.10) gives an associated wave magnetic field. Therefore, the transverse wave is an electromagnetic mode. When we consider the limit of vanishing electron density, the electron plasma frequency goes to zero and the wave dispersion takes the limiting form $\omega = kc$, which is the light wave in vacuum. The transverse mode in an unmagnetized plasma is therefore the light wave modified by the presence of the plasma as a dielectric medium.

The wave dispersion of the electromagnetic wave is shown in Fig. 6.3. The transverse wave is only propagating for $\omega > \omega_{\rm pe}$. Therefore, we call the electron plasma frequency the *cut-off frequency* for the electromagnetic mode. In the limit of very high frequencies the dispersion approaches the light wave in vacuum. This case is different from the limit of vanishing plasma density because, with increasing

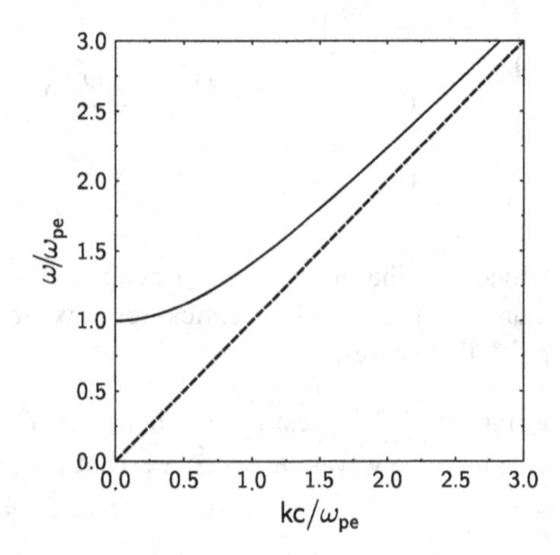

Fig. 6.3 Dispersion relation for electromagnetic waves in an unmagnetized plasma. Wave propagation is only possible for frequencies larger than the plasma frequency. For $\omega \gg \omega_{\rm pe}$ the wave dispersion approaches the light wave in vacuum $\omega = kc$

frequency, the electron inertia leads to a reduction of the electron current, $\hat{j} \propto \omega^{-1}$, so in the end the electron current has a vanishing influence on the wave.

A surprising result for the electromagnetic mode is the fact that the phase velocity

$$v_\varphi = \left(\frac{\omega_{\mathrm{pe}}^2}{k^2} + c^2 \right)^{1/2} \tag{6.39}$$

is higher than the speed of light This is not in conflict with Einstein's theory of relativity, because the phase is neither transporting energy nor information. Rather, energy is transported at the group velocity

$$v_{\mathrm{gr}} = \frac{kc^2}{\left(\omega_{\mathrm{pe}}^2 + k^2 c^2 \right)^{1/2}}, \tag{6.40}$$

which is always smaller than the speed of light. Other examples for electromagnetic waves with $v_\varphi > c$ are found, e.g., for the wave propagation in microwave wave-guides.

6.3.2 The Influence of Collisions

So far, we have assumed that the electron motion in the wave field is unaffected by friction with the neutral gas. This approximation is certainly valid when the wave frequency (for instance that of a laser beam) is much higher than the collision frequency. This conjecture is supported by Newton's equation in Fourier notation

$$m\,(-\mathrm{i}\omega + \nu_{\mathrm{m}})\,\hat{v} = q\,\hat{E}\,, \tag{6.41}$$

which shows that the resulting electron velocity can be decomposed into a real and imaginary response factor w.r.t. the electric field

$$\hat{v} = \left[\frac{\nu_{\mathrm{m}}}{\omega^2 + \nu_{\mathrm{m}}^2} + \frac{\mathrm{i}\omega}{\omega^2 + \nu_{\mathrm{m}}^2} \right] \frac{q}{m} \hat{E}\,. \tag{6.42}$$

The in-phase response, which is due to the electron collisions, corresponds to the action of a resistor. The imaginary part of the response represents a current that is 90° lagging behind the voltage. The lag is caused by the electron inertia, and this part of the system behaves like an inductance.

We can use a simple trick to write down the dispersion relation with collisions by noting that (6.41) becomes identical with the collisionless limit, when we rewrite it in terms of an *effective mass* m^*

$$-\mathrm{i}\omega m^* \hat{v} = q\hat{E}, \qquad m^* = m\left(1 + \mathrm{i}\frac{\nu_{\mathrm{m}}}{\omega}\right). \tag{6.43}$$

Hence, we only have to replace the real electron mass in the electron plasma frequency by m_e^*. By doing so, we find a complex wavenumber for any given real frequency. This is the proper approach to describe how a wave penetrates into a plasma. The complex wavenumber then reads

$$k = \frac{1}{c}\left(\omega^2 - \frac{\omega_{pe}^2}{1 + i(\nu_m/\omega)}\right)^{1/2} . \tag{6.44}$$

The resulting complex dispersion relation $k(\omega)$ is shown in Fig. 6.4. The real part of the wavenumber gives the spatial phase evolution while the imaginary part describes the damping of the wave amplitude. Note that the imaginary and real part of k intersect at the plasma frequency. For wave frequencies below the plasma frequency, the collisionality makes the plasma a resistive medium. This explains, why we can generate a plasma with radio frequency (e.g., at 13.56 MHz) even if the electron plasma frequency is much higher. The collisionality gives the proper dissipation of wave energy that leads to Joule heating of the electron gas. On the other hand, collisional damping becomes negligible when $\omega > 2\omega_{pe}$. Hence, a weakly collisional plasma can still be analyzed in terms of its refractive index as long as the wave frequency is sufficiently higher than the cut-off frequency.

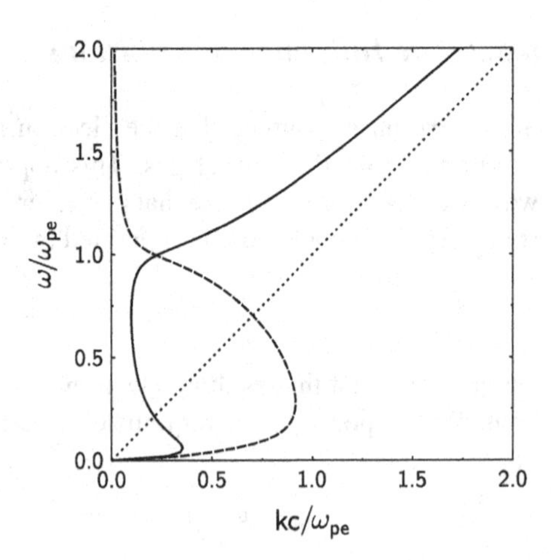

Fig. 6.4 Complex dispersion relation for electromagnetic waves in a weakly collisional $\nu_m/\omega_{pe} = 0.1$ unmagnetized plasma. The real part of the wavenumber (*solid line*) is dominant when $\omega > \omega_{pe}$. The imaginary part of the wavenumber (*dotted line*) shows that the wave can now penetrate into the plasma for frequencies $\omega < \omega_{pe}$. The dashed line again represents the speed of light, which is approached when $\omega \gg \omega_{pe}$

6.4 Interferometry with Microwaves and Lasers

Because the unmagnetized plasma is an isotropic medium, its dielectric properties are described by a dielectric constant rather than by a tensor

$$\varepsilon = 1 - \omega_{pe}^2/\omega^2 . \tag{6.45}$$

The dielectric constant depends on the electron density and is a function of the wave frequency ω. The refractive index $\mathcal{N}$ of this plasma is then determined by

$$\mathcal{N}^2 = \frac{c^2}{v_\varphi^2} = \frac{k^2 c^2}{\omega^2} = \varepsilon(\omega)\,. \tag{6.46}$$

Hence, by measuring the refractive index, e.g., by means of an interferometer, we can deduce the electron density. For a collisionless plasma, the refractive index $\mathcal{N}$ is smaller than unity and becomes zero at the electron plasma frequency and imaginary for lower frequencies. In this way, an electromagnetic wave is reflected at the surface of a plasma, when the wave frequency is lower than the electron plasma frequency. This explains why a thin silver layer on a glass mirror, by means of the free electrons of the silver atoms in the conduction band, can reflect visible light but becomes transparent in the UV range. As discussed above in Sect. 6.3.2, a weakly collisional plasma would allow wave penetration with an exponential decay $\exp(-k_\mathrm{I} x)$ described by the imaginary part of k. (See also the discussion of the skin effect in Sect. 11.3.1.

Instead of considering a given plasma with a fixed density for various wave frequencies, we can also study a plasma with a density gradient and a wave with a fixed frequency. Then the point where the local plasma frequency (which corresponds to the electron density at that point) agrees with the wave frequency defines the *cut-off density* n_co. Likewise, we can consider a non-stationary plasma, which is switched on at $t = 0$ and which, after a short while, reaches the cut-off density. Thus, the cut-off density can be connected with an inhomogeneous plasma as well as with the temporal evolution of plasma density, or both. Therefore, the crudest method of plasma density diagnostics is to look, whether the plasma allows for wave transmission or not.

6.4.1 Mach-Zehnder Interferometer

A more elegant way, however, is the measurement of the refractive index with the aid of an interferometer. This can be done, depending on plasma density, with coherent radiation sources in the microwave, infrared or visible regime. A typical arrangement for interferometry with microwaves or lasers is a Mach-Zehnder interferometer (Fig. 6.5). The typical cut-off densities that correspond to the wave frequencies are compiled in Table 6.1.

Interferometry is based on the phase difference between the wave that penetrates the plasma and the reference wave, which takes a way of the same length in air. This is practically accomplished by splitting the wave into two identical branches and recombining the two on a detector where the two waves interfere. The *optical path* through the plasma is the product of the geometrical length L and the refractive

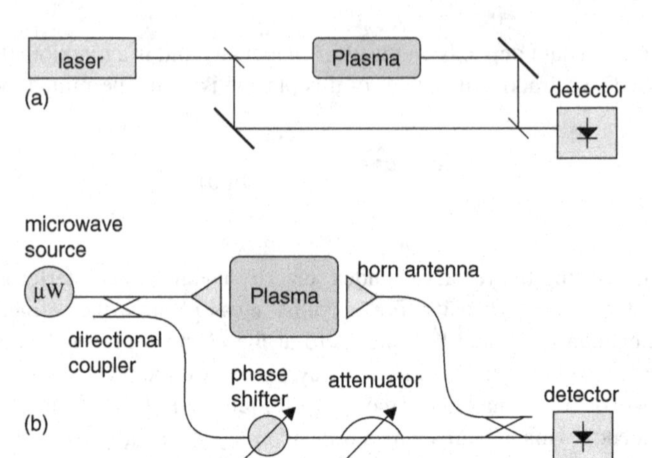

Fig. 6.5 (**a**) Laser interferometer in Mach-Zehnder arrangement, (**b**) microwave interferometer. The optical arrangement uses partially-reflecting and fully-reflecting mirrors. The analog to a partially reflecting mirror is the directional coupler for microwaves

Table 6.1 Cut-off densities for microwave and laser interferometers

Source	Wavelength λ	Frequency f	Cut-off-density $n_{\mathrm{co}}(\mathrm{m}^{-3})$
Microwave	3 cm	10 GHz	1.2×10^{18}
	8 mm	37 GHz	1.7×10^{19}
	4 mm	75 GHz	7.0×10^{19}
HCN-laser	337 μm	890 GHz	9.8×10^{21}
CO_2 laser	10.6 μm	28 THz	9.9×10^{24}
He-Ne laser	3.39 μm	88 THz	9.7×10^{25}
	0.633 μm	474 THz	2.8×10^{27}

index $\mathcal{N}$. At a wavelength λ the difference between the optical path in the plasma and the same path in vacuum is

$$\Delta\varphi = 2\pi \frac{(\mathcal{N} - 1)L}{\lambda} < 0\,, \tag{6.47}$$

where we have assumed that the plasma has a homogeneous density and a corresponding constant refractive index. In an inhomogeneous plasma, the phase difference is

$$\Delta\varphi = \frac{2\pi}{\lambda} \int [\mathcal{N}(x) - 1]\,\mathrm{d}x\,. \tag{6.48}$$

This means that interferometry can only determine the path-averaged refractive index $\bar{\mathcal{N}} = (1/L)\int \mathcal{N}(x)\mathrm{d}x$. The reduction to index profiles requires additional procedures like tomography. When the refractive index is not too close to zero, i.e., the plasma density not too close to the cut-off density, we can expand the refractive index into a Taylor series

$$\mathcal{N}=\sqrt{1-\omega_{\mathrm{pe}}^2/\omega^2}\approx 1-\frac{1}{2}\frac{\omega_{\mathrm{pe}}^2}{\omega^2}=1-\frac{1}{2}\frac{n}{n_{\mathrm{co}}}\,, \tag{6.49}$$

with the cut-off density

$$n_{\mathrm{co}}=\frac{\varepsilon_0 m_{\mathrm{e}}\,\omega^2}{e^2}\,. \tag{6.50}$$

Then the phase shift can be written as

$$\Delta\varphi\approx -\pi\,\frac{L}{\lambda}\,\frac{n}{n_{\mathrm{co}}}\,. \tag{6.51}$$

Because the cut-off density decreases with λ^{-2}, the phase shift becomes proportional to the wavelength and not proportional to its inverse, as (6.51) might suggest at first glance. For density diagnostics at low electron densities we therefore need long-wavelength lasers or microwaves. The maximum wavelength, on the other hand, is limited by the geometric optics approximation, which requires that the plasma dimensions are large compared to the wavelength. This conflict limits the achievable sensitivity of interferometers.

A practical example for interferometry in a time-varying plasma is shown in Fig. 6.6a. The dynamic response of the detector circuit is too slow to resolve the interferometer *fringes* during the build-up of the plasma by a strong current pulse.

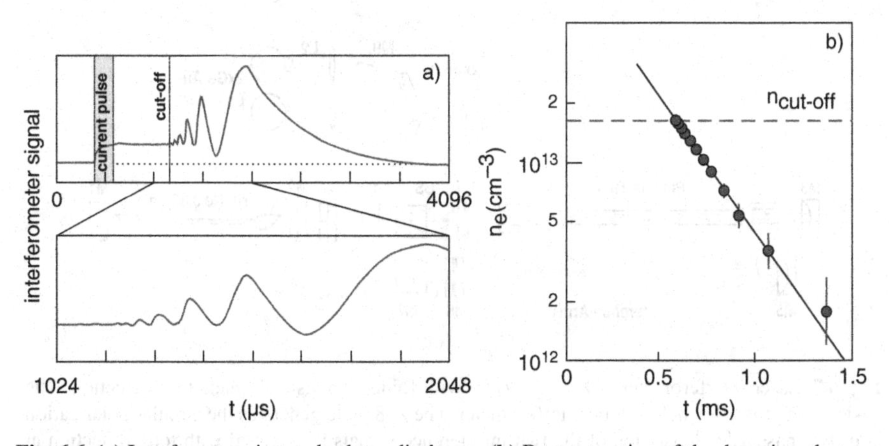

Fig. 6.6 **(a)** Interferogram in a pulsed gas discharge. **(b)** Reconstruction of the decaying electron density by counting interferometer *fringes*

During the current pulse the plasma density is much higher than the cut-off density. About 600 μs after the end of the current pulse, the interferometer signal leaves the cut-off regime and the signal oscillates about a mean value. One interferometer fringe corresponds to a phase shift of 2π. Reading the maxima, minima and zero crossings of the interferometer signal gives a resolution of a quarter fringe. These give the data points in Fig. 6.6b, which are plotted on a logarithmic scale to demonstrate the exponential decay of plasma density in the afterglow of this discharge. In this evaluation, the data points close to the cut-off were evaluated with the exact formula (6.48).

6.4.2 Folded Michelson Interferometer

The sensitivity of an interferometer can be improved in two ways: First, a different type of interferometer should be used, such as the Michelson type, which has already a two-fold passage of the beam. The sensitivity can be further enhanced by folding the beam to a z-shape, which gives a six-fold passage, however at the expense of spatial resolution. Such an improved laser interferometer [102] is shown in Fig. 6.7. The laser wavelength (3.39 μm) is still in the transmission band of quartz windows. A longer wavelength would require special materials for windows and optical components.

Second, counting interferometer fringes gives only $\approx \pm 45°$ reading accuracy. Therefore, the true phase angle should be measured with a more sensitive technique. This can be achieved with quadrature detection, in which two independent interferograms with a sine wave and cosine wave in the reference branch are made. This is technically realized by using a circularly polarized wave in the reference beam, which is generated by a $\lambda/8$ plate that is traversed twice, and by splitting the two interferometer signals by orthogonal polarizers.

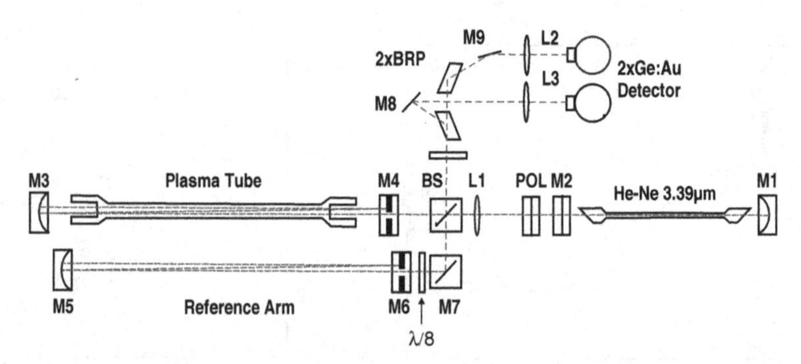

Fig. 6.7 Laser interferometer at $\lambda = 3.39\,\mu$m with folded beams and quadrature detection. The basic concept is a Michelson type interferometer. The $\lambda/8$ plate generates the circular polarization after two passages. Separation of the two interference signals is achieved with total reflection at the Brewster angle (from [102])

6.4.3 The Second-Harmonic Interferometer

A special kind of two-wavelength interferometer is the second-harmonic interferometer, which was introduced by Hopf and coworkers [103–106]. This technique is presently used, e.g., for the diagnostics of the Alcator C-mode tokamak [107, 108]. Second-harmonic interferometry has the advantage that probe and reference beam take the same path, which reduces the sensitivity of the interferometer to mechanical vibrations that affect conventional interferometers with a separate reference branch. The original signal from a Nd:YAG laser at 1064 nm wavelength is used as the probe beam (see Fig. 6.8). The frequency-doubled wave at 532 nm also traverses the plasma but experiences a different phase shift. Behind the plasma the probe wave is still strong and can be frequency doubled with a second crystal. The fundamental wave is then blocked by a filter. Both waves at 532 nm produce interference fringes on the detector.

Between the first and second doubler crystal, the fundamental laser wave at ω experiences a phase shift $\varphi_{\mathrm{p}}(\omega)$ by the plasma and $\varphi_{\mathrm{air}}(\omega)$ in the air gaps

$$\varphi_{\mathrm{p}}(\omega) = \frac{\omega}{c}\left(d - \frac{1}{2}\frac{\bar{n}_{\mathrm{e}} d e^2}{\omega^2 \varepsilon_0 m_e}\right) \tag{6.52}$$

$$\varphi_{\mathrm{air}}(\omega) = \frac{\omega}{c}(D-d)\,. \tag{6.53}$$

Behind the second doubler, the phase of the frequency doubled signal is

$$\varphi_1(2\omega) = 2\frac{\omega}{c}\left(D - \frac{1}{2}d\frac{\bar{n}_{\mathrm{e}}}{\omega^2}a\right), \tag{6.54}$$

where $a = e^2(\varepsilon_0 m_{\mathrm{e}})^{-1}$. The frequency-doubled signal has a total phase shift between the doublers of

$$\varphi_2(2\omega) = \frac{2\omega}{c}\left(D - \frac{1}{2}d\frac{\bar{n}_{\mathrm{e}}}{(2\omega)^2}a\right). \tag{6.55}$$

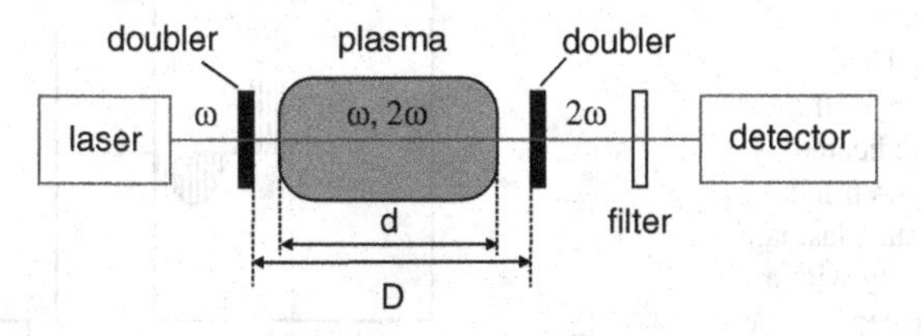

Fig. 6.8 Principle of a second-harmonic interferometer. The beam of a Nd:YAG laser at the fundamental wavelength 1064 nm is frequency doubled with a LiB_3O_5 crystal. Both the fundamental and frequency-doubled wave traverse the plasma. Behind the plasma the remaining fundamental wave is doubled and the two signals at 532 nm wavelength interfere at the detector

The phase difference of these two signals becomes

$$\varphi_2(2\omega) - \varphi_1(2\omega) = \frac{3}{2} d \frac{\bar{n}_e}{2\omega} a \,. \tag{6.56}$$

We see that the contribution from the air gap cancels when we neglect the different refractive indices of air at the two wavelengths. The second-harmonic interferometer is, by a factor of 1.5, more sensitive than a conventional interferometer operating at the fundamental frequency ω.

6.4.4 Plasma-Filled Microwave Cavities

The refractive index of a plasma can also be used to detune the resonance frequency of cylindrical microwave cavities. This is particularly useful at low electron densities because the resonance frequency is very sensitive to a change in electron density. A typical setup for this technique is shown in Fig. 6.9a. Suitable cavity modes for detecting the resonance are the TM_{0m0} modes, which have an electric field aligned with the z-axis of the cylinder. (See, e.g., [74] for microwave cavities and waveguides) This ensures that the electric field is homogeneous along the z-axis and perpendicular to the top and bottom of the cylinder. Therefore, the eigenfrequencies of the TM_{0m0} modes are independent of the resonator height. The radial boundary condition at the cylinder radius, $E_z(R) = 0$, defines the eigenfrequency of the cavity.

When the cavity is filled with a homogeneous dielectric material of dielectric constant ε, the eigenfunctions have the shape

$$E_z(r) = \hat{E} J_0 \left(\chi_{0m} \frac{r}{\sqrt{\varepsilon} R} \right) \,. \tag{6.57}$$

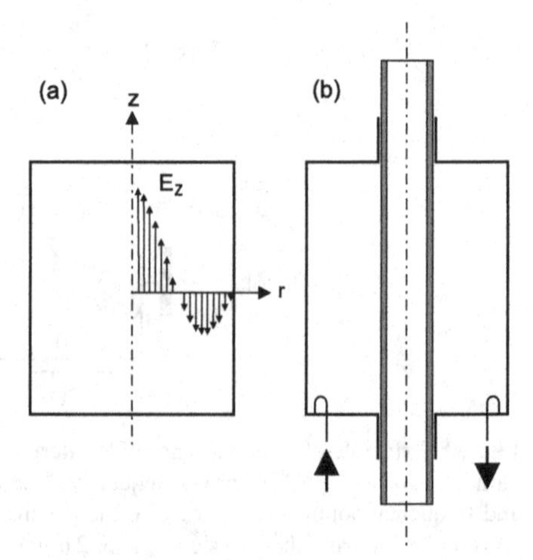

Fig. 6.9 (**a**) Cylindrical microwave cavity. The electric field vectors of the TM_{020} mode are indicated by arrows. The cavity is filled with a homogeneous plasma. (**b**) Cylindrical cavity with a plasma tube in the center. The microwave signal is coupled to the cavity with small loops that excite an azimuthal magnetic field

The coefficients χ_{0m} are the zeroes of the Bessel function J_0 and distinguish the various modes. The resonance frequencies are

$$f_m = \frac{1}{2\pi} \frac{c}{\sqrt{\varepsilon} R} \chi_{0m} \,, \tag{6.58}$$

which in the limit $n \ll n_{\mathrm{co}}$ gives a linear relation between resonance frequency and plasma density n

$$f_m \approx \frac{c \chi_{0m}}{2\pi R} \left(1 + \frac{1}{2} \frac{n}{n_{\mathrm{co}}} \right) . \tag{6.59}$$

In this approximation, the cut-off density is determined by the frequency of the empty cavity. The Q-factor $\omega/\Delta\omega$ of the cavity can be made very high, $Q_{010} > 5000$. This gives a density resolution $\Delta n \approx n_{\mathrm{co}}/Q$. An empty cavity with $R = 100\,\mathrm{mm}$ has a fundamental resonance $f_1 = 1.15\,\mathrm{GHz}$, which corresponds to a cut-off density $n_{\mathrm{co}} = 1.6 \times 10^{16}\,\mathrm{m}^{-3}$. Hence, densities as low as $1 \times 10^{14}\,\mathrm{m}^{-3}$ can be accurately measured. Cavity detuning was, for instance, used to study the effect of electron depletion in a dusty plasma [109]. The cavity method is also suitable in connection with discharge tubes of radius $a < R$ that fill only part of the cavity [Fig. 6.9b]. Consequently, the sensitivity to the plasma density is reduced by roughly a factor a/R.

6.5 Electrostatic Waves

After having discussed the transverse electromagnetic wave in an unmagnetized plasma, we now return to the longitudinal mode with $\mathbf{k}||\mathbf{E}$. Longitudinal waves are electrostatic as can be seen from Faraday's induction law (6.10) because $\mathbf{k} \times \mathbf{E} = 0$ and the wave magnetic field vanishes.

6.5.1 The Longitudinal Mode

We return here to the solutions of the wave (6.35). There, we had postponed the discussion of the third mode, which is defined by $\hat{E}_x \neq 0, \hat{E}_y = \hat{E}_z = 0$. The first row of the system of equations (6.35) reads

$$\frac{\omega^2}{c^2} \varepsilon(\omega) \hat{E}_x = 0 \tag{6.60}$$

with $\varepsilon(\omega)$ given by (6.33). This implies that

$$\varepsilon(\omega) = 0 \,, \tag{6.61}$$

which is the defining condition for the dispersion of an electrostatic wave. It further implies that, in a cold plasma, this electrostatic wave only exists for $\omega = \omega_{\rm pe}$. These are Langmuir's plasma oscillations, in which the electrons oscillate about their equilibrium at the electron plasma frequency. Although we have found a wave solution, the dispersion relation turns out to be independent of k. This means that the plasma oscillations cannot form propagating wave packets because the group velocity is zero.

6.5.2 Bohm-Gross Waves

When we consider a warm electron gas, in which pressure forces have a similar magnitude as the electric force, the Langmuir oscillations discussed above become dispersive. The dispersion relation can be derived as follows: We start with adding the pressure per particle to Newton's equation in one space dimension, because the electrostatic waves are one-dimensional

$$m\dot{v} = -q\frac{\mathrm{d}\phi}{\mathrm{d}x} - \frac{\gamma}{n}\frac{\mathrm{d}(nk_{\rm B}T)}{\mathrm{d}x}\,. \tag{6.62}$$

We have introduced the concept of adiabatic compression with an adiabatic exponent $\gamma = 3$ (for one-dimensional motion) to take into account that the pressure in the wave field changes on a rapid time scale. The velocity fluctuations can be transformed into density fluctuations by using the equation of continuity

$$\frac{\partial n}{\partial t} + \frac{\partial}{\partial x}(nv) = 0\,. \tag{6.63}$$

First, we split the density and velocity into equilibrium part and fluctuating part, $n = n_0 + \hat{n}\exp[\mathrm{i}(kx - \omega t)]$, $v = v_0 + \hat{v}\exp[\mathrm{i}(kx - \omega t)]$. We further assume that the electron gas is at rest, $v_0 = 0$. The wave amplitudes $\hat{n}$ and $\hat{v}$ and the potential fluctuation $\hat{\phi}$ are first-order quantities. Then, we replace the differential operators by frequency and wavenumber according to the substitution rules (6.9). This gives the equation of motion (6.62) as

$$-\mathrm{i}\omega m\hat{v} = -\mathrm{i}kq\hat{\phi} - \mathrm{i}k\gamma k_{\rm B}T\hat{n}\,. \tag{6.64}$$

Likewise, the continuity equation takes the form

$$-\mathrm{i}\omega\hat{n} + \mathrm{i}kn_0\hat{v} = 0\,, \tag{6.65}$$

which we use to substitute the velocity fluctuation by the corresponding density fluctuation

$$\hat{v} = \frac{\omega}{k}\frac{\hat{n}}{n_0}\,. \tag{6.66}$$

For eliminating the potential fluctuations, we use Poisson's equation $\partial^2\phi/\partial x^2 = (q/\varepsilon_0)(n_e - n_i)$ in Fourier notation, and insert the linearized electron density with the result

$$-k^2\hat{\phi} = \frac{q}{\varepsilon_0}\hat{n}\,. \tag{6.67}$$

Combining (6.64), (6.66) and (6.67), we obtain

$$\omega = \left(\omega_{pe}^2 + \frac{3}{2}k^2 v_{Te}^2\right)^{1/2} = \omega_{pe}\left(1 + 3k^2\lambda_{De}^2\right)^{1/2} \tag{6.68}$$

with the characteristic electron thermal speed $v_{Te} = (2k_B T_e/m_e)^{1/2}$. This is the dispersion relation for electron acoustic waves in a warm plasma, which were first described by David Bohm (1917–1992) and Eugene P. Gross (1926–) [110, 111].

We will see in Sect. 9.3.3 that the electron acoustic waves experience damping by kinetic effects (which are not contained in this fluid model) as soon as $k\lambda_{De} \approx 1$. Therefore, weakly damped waves are only found in the long wavelength limit. The dispersion relation is displayed in Fig. 9.8 of Sect. 9.3.2.

6.5.3 Ion-Acoustic Waves

When we allow that the ions can take part in the wave motion, there is a second electrostatic wave in a plasma with warm electrons. This is possible for wave frequencies much smaller than the electron plasma frequency. Note that the plasma cut-off was a feature of the transverse electromagnetic mode and does not affect the existence of low-frequency electrostatic modes.

When we consider low-frequency modes, the electron motion is only governed by pressure forces and inertial forces can be neglected. On the other hand, we can treat the ions as a fluid that is governed by the interplay of electric field force, ion inertia and ion pressure. It is wise to allow for different equilibrium densities of electrons and ions. While a two-component plasma of electrons and positive ions has $n_{e0} = n_{i0}$ because of quasineutrality, we will consider a more general case, where the difference of the densities is caused by the presence of a third negative species. These can either be negative ions or negatively charged dust.

The equation of motion for electrons and ions reads in Fourier notation

$$-\mathrm{i}\omega m_i \hat{v}_i = e\hat{E} - \frac{\mathrm{i}k}{n_{i0}}(\gamma_i k_B T_i)\hat{n}_i \tag{6.69}$$

$$0 = -e\hat{E} - \frac{\mathrm{i}k}{n_{e0}}(k_B T_e)\hat{n}_e\,. \tag{6.70}$$

Eliminating the ion velocity fluctuations by means of the continuity (6.66), we obtain the density fluctuations of electrons and ions for a given wave field $\hat{E}$ as

$$\hat{n}_\mathrm{i} = \frac{ek}{-\mathrm{i}\omega^2 m_\mathrm{i} + \mathrm{i}k^2\gamma_\mathrm{i} k_\mathrm{B} T_\mathrm{i}}\hat{E} \tag{6.71}$$

$$\hat{n}_\mathrm{e} = \frac{-e}{\mathrm{i}k k_\mathrm{B} T_\mathrm{e}}\hat{E}\,, \tag{6.72}$$

where we have assumed that the electron gas experiences an isothermal compression while the ion compression is adiabatic. This assumption is justified because the electrons move across many wavelengths during one cycle of this low-frequency wave, which justifies to consider the electron gas as a heat reservoir for the wave. The latter aspect also justifies to neglect temperature fluctuations of the electrons. The ions, on the other hand, are slow and do not move far from their starting position during one wave period.

At last, Poisson's equation becomes

$$\mathrm{i}k\hat{E} = \frac{e}{\varepsilon_0}(\hat{n}_\mathrm{i} - \hat{n}_\mathrm{e}) \tag{6.73}$$

and defines the condition for the consistency of the fluctuating field with the space charges. We then obtain

$$\mathrm{i}k\hat{E} = \left(\frac{n_{\mathrm{i}0}e^2}{\varepsilon_0 m_\mathrm{i}}\right)\frac{k}{-\mathrm{i}\omega^2 + \mathrm{i}k^2\gamma_\mathrm{i} k_\mathrm{B} T_\mathrm{i}/m_\mathrm{i}}\hat{E} + \left(\frac{n_{\mathrm{e}0}e^2}{\varepsilon_0 k_\mathrm{B} T_\mathrm{e}}\right)\frac{1}{\mathrm{i}k}\hat{E}\,. \tag{6.74}$$

Introducing the ion plasma frequency $\omega_\mathrm{pi} = (n_{\mathrm{i}0}e^2/\varepsilon_0 m_\mathrm{i})^{1/2}$ and the electron Debye length $\lambda_\mathrm{De} = (n_{\mathrm{e}0}e^2/\varepsilon_0 k_\mathrm{B} T_\mathrm{e})^{1/2}$, we find the following dielectric function

$$\varepsilon(k,\omega) = 1 - \frac{\omega_\mathrm{pi}^2}{\omega^2 - k^2\gamma_\mathrm{i} k_\mathrm{B} T_\mathrm{i}/m_\mathrm{i}} + \frac{1}{k^2\lambda_\mathrm{De}^2}\,. \tag{6.75}$$

The dispersion relation of the electrostatic wave is again given by $\varepsilon(k,\omega) = 0$ and can be solved for ω^2

$$\omega^2 = k^2\left(\frac{\gamma_\mathrm{i} k_\mathrm{B} T_\mathrm{i}}{m_\mathrm{i}} + \frac{\omega_\mathrm{pi}^2\lambda_\mathrm{De}^2}{1 + k^2\lambda_\mathrm{De}^2}\right)\,. \tag{6.76}$$

Here $C_\mathrm{s} = \omega_\mathrm{pi}\lambda_\mathrm{De}$ is the ion sound speed and we call this wave mode the *ion acoustic wave*.

In most gas discharge plasmas $T_\mathrm{e} \gg T_\mathrm{i}$. In that limit the first term in the parentheses can be dropped and we find

$$\omega \approx \frac{k\,C_\mathrm{s}}{\sqrt{1 + k^2\lambda_\mathrm{De}^2}}\,. \tag{6.77}$$

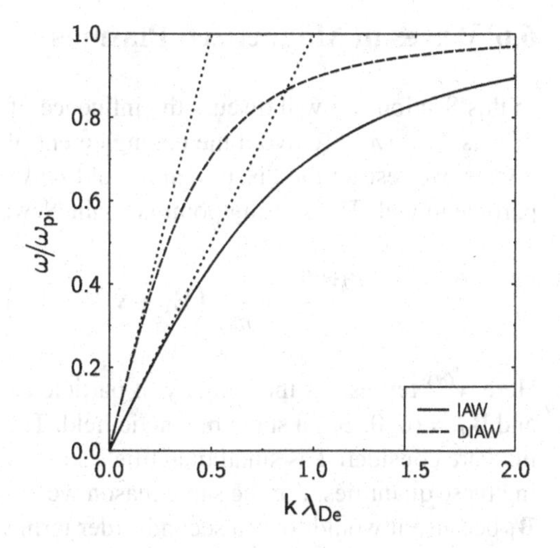

Fig. 6.10 Ion-acoustic wave (*solid line*) and dust-ion-acoustic wave (*dashed line*). The acoustic limits of the IAW and DIAW dispersion are indicated by dotted lines. The DIAW has an increased phase velocity

For small wavenumbers ($k^2\lambda_{\mathrm{De}}^2 \ll 1$) this wave has acoustic dispersion $\omega = kC_{\mathrm{s}}$ (see the asymptotes in Fig. 6.10). In the opposite case of large wavenumbers, the wave frequency approaches ω_{pi}.

In a plasma with $n_{\mathrm{e}0} \neq n_{\mathrm{i}0}$, the ion-sound speed can be rewritten as

$$C_{\mathrm{s}} = \left(\frac{n_{\mathrm{i}0} k_{\mathrm{B}} T_{\mathrm{e}}}{n_{\mathrm{e}0} m_{\mathrm{i}}} \right)^{1/2} . \tag{6.78}$$

When $n_{\mathrm{e}0} = n_{\mathrm{i}0}$, one is tempted to interpret the ion-acoustic wave as the interplay of a pressure force associated with the electrons and an inertia residing in the ions, as we have in ordinary sound waves in a neutral gas

$$c_{\mathrm{s}} = \left(\frac{\gamma p}{\rho} \right)^{1/2} . \tag{6.79}$$

This interpretation is obviously wrong, when we notice that the numerator in (6.78) is $n_{\mathrm{i}0} k_{\mathrm{B}} T_{\mathrm{e}}$ rather than $n_{\mathrm{e}0} k_{\mathrm{B}} T_{\mathrm{e}}$, as we would need for the electron pressure. The same problem arises in the denominator with the ion mass density. Hence, the picture of the mechanism behind the ion-acoustic wave must be revised. The apparent paradox can be resolved by considering the electrons not as a gas that exerts a pressure but rather as a fluid of the opposite charge that shields the electric repulsion between the ions. Therefore, the phase velocity increases, when the electron density is reduced, which means that the interaction between the ions is approaching their *naked* repulsion. This effect is well known from negative ion plasmas as can be read from the increase of the phase velocity with increasing ratio n_+/n_{e}, see Fig. 6.10b. Likewise, the ion-acoustic wave in a dusty plasma has a higher phase velocity than in the absence of dust, see Fig. 6.10a.

6.6 Waves in Magnetized Plasmas

In this Section, we will discuss the influence of a magnetic field on the propagation of plasma waves. To avoid the entanglement of magnetic field effects and pressure effects, we restrict the discussion to cold plasmas. This allows us to use the single particle model. The starting point is again Newton's equation of motion

$$\frac{\partial \mathbf{v}^{(\alpha)}}{\partial t} = \frac{q_\alpha}{m_\alpha}\left(\mathbf{E}_1 + \mathbf{v}^{(\alpha)} \times \mathbf{B}_0\right) \qquad \alpha = \mathrm{e}, \mathrm{i}\,. \tag{6.80}$$

Here, $\mathbf{v}^{(\alpha)}$ represents the velocity of particle oscillations, $\mathbf{E}_1$ the wave electric field and $\mathbf{B}_0 = (0, 0, B_0)$ a static magnetic field. The oscillation velocity and the electric field are considered as small quantities, so we will retain only linear terms containing these quantities. For the same reason we have neglected the wave magnetic field $\mathbf{B}_1$ because it would form a second-order term $\mathbf{v}^{(\alpha)} \times \mathbf{B}_1$ in the Lorentz force.

6.6.1 The Dielectric Tensor

To reduce the cluttering with subscripts and superscripts, we drop the symbol α for the particle species in the following and distinguish the particles by their q and m values. The interesting new effects in the dielectric tensor arise from the particle motion across the magnetic field

$$\hat{v}_x = \mathrm{i}\frac{q}{\omega m}(\hat{E}_x + \hat{v}_y B_0)\,, \quad \hat{v}_y = \mathrm{i}\frac{q}{\omega m}(\hat{E}_y - \hat{v}_x B_0)\,. \tag{6.81}$$

The ideal way to describe the gyromotion of the particles is using rotating vectors for the velocities and the electric field

$$\hat{v}^{\pm} = \hat{v}_x \pm \mathrm{i}\hat{v}_y\,, \quad \hat{E}^{\pm} = \hat{E}_x \pm \mathrm{i}\hat{E}_y\,. \tag{6.82}$$

In this way we can decouple the particle motion in (6.81)

$$\hat{v}^{\pm} = \mathrm{i}\frac{q}{\omega m}(\hat{E}^{\pm} \mp \mathrm{i}\hat{v}^{\pm} B_0)\,. \tag{6.83}$$

The cyclotron frequencies for electrons and ions are defined as

$$\omega_{\mathrm{ce}} = \frac{eB_0}{m_\mathrm{e}} \qquad \omega_{\mathrm{ci}} = \frac{|q|B_0}{m_\mathrm{i}}\,, \tag{6.84}$$

which results in

$$\hat{v}^{\pm} = \mathrm{i}\frac{q}{m}\hat{E}^{\pm}\frac{1}{\omega \mp s\omega_c}\,. \tag{6.85}$$

Here, $s = q/|q|$ is the sign of the particle charge. Transforming back to Cartesian coordinates

$$\hat{v}_x = \frac{1}{2}(\hat{v}^+ + \hat{v}^-)\,, \quad \hat{v}_y = \frac{1}{2\mathrm{i}}(\hat{v}^+ - \hat{v}^-)\,, \tag{6.86}$$

we obtain the matrix relation

$$\begin{pmatrix} \hat{v}_x \\ \hat{v}_y \\ \hat{v}_z \end{pmatrix} = \mathrm{i}\frac{q}{\omega m} \begin{pmatrix} \dfrac{\omega^2}{\omega^2 - \omega_\mathrm{c}^2} & \mathrm{i}\dfrac{s\omega\omega_\mathrm{c}}{\omega^2 - \omega_\mathrm{c}^2} & 0 \\ -\mathrm{i}\dfrac{s\omega\omega_\mathrm{c}}{\omega^2 - \omega_\mathrm{c}^2} & \dfrac{\omega^2}{\omega^2 - \omega_\mathrm{c}^2} & 0 \\ 0 & 0 & 1 \end{pmatrix} \cdot \begin{pmatrix} \hat{E}_x \\ \hat{E}_y \\ \hat{E}_z \end{pmatrix}. \tag{6.87}$$

In the last line of this matrix equation we have used the result from the unmagnetized plasma. Using the definition of the particle oscillating current $\hat{\mathbf{j}} = \sum_\alpha n_\alpha q_\alpha \hat{\mathbf{v}}^{(\alpha)}$ we obtain the conductivity tensor as

$$\boldsymbol{\sigma}(\omega) = \mathrm{i}\omega\varepsilon_0 \begin{pmatrix} \sum\limits_\alpha \dfrac{\omega_{\mathrm{p}\alpha}^2}{\omega^2 - \omega_{\mathrm{c}\alpha}^2} & \mathrm{i}\sum\limits_\alpha s_\alpha \dfrac{\omega_{\mathrm{p}\alpha}^2}{\omega^2 - \omega_{\mathrm{c}\alpha}^2}\dfrac{\omega_{\mathrm{c}\alpha}}{\omega} & 0 \\ -\mathrm{i}\sum\limits_\alpha s_\alpha \dfrac{\omega_{\mathrm{p}\alpha}^2}{\omega^2 - \omega_{\mathrm{c}\alpha}^2}\dfrac{\omega_{\mathrm{c}\alpha}}{\omega} & \sum\limits_\alpha \dfrac{\omega_{\mathrm{p}\alpha}^2}{\omega^2 - \omega_{\mathrm{c}\alpha}^2} & 0 \\ 0 & 0 & \sum\limits_\alpha \dfrac{\omega_{\mathrm{p}\alpha}^2}{\omega^2} \end{pmatrix} \tag{6.88}$$

and with the aid of (6.14) the dielectric tensor

$$\boldsymbol{\varepsilon}(\omega) = \begin{pmatrix} S & -\mathrm{i}D & 0 \\ \mathrm{i}D & S & 0 \\ 0 & 0 & P \end{pmatrix}, \tag{6.89}$$

in which we have used the parameters S, P, and D introduced by Thomas H. Stix [100, 112]

$$\begin{aligned} S &= 1 - \sum_\alpha \frac{\omega_{\mathrm{p}\alpha}^2}{\omega^2 - \omega_{\mathrm{c}\alpha}^2} \\ D &= \sum_\alpha s_\alpha \frac{\omega_{\mathrm{p}\alpha}^2}{\omega^2 - \omega_{\mathrm{c}\alpha}^2}\frac{\omega_{\mathrm{c}\alpha}}{\omega} \\ P &= 1 - \sum_\alpha \frac{\omega_{\mathrm{p}\alpha}^2}{\omega^2}\,. \end{aligned} \tag{6.90}$$

Introducing further the refractive index $\mathscr{N} = kc/\omega$ and the angle ψ between wave vector and magnetic field direction, the wave (6.35) takes the form

$$\begin{pmatrix} S - \mathscr{N}^2 \cos^2 \psi & -\mathrm{i}D & \mathscr{N}^2 \cos\psi \sin\psi \\ \mathrm{i}D & S - \mathscr{N}^2 & 0 \\ \mathscr{N}^2 \cos\psi \sin\psi & 0 & P - \mathscr{N}^2 \sin^2 \psi \end{pmatrix} \cdot \begin{pmatrix} \hat{E}_x \\ \hat{E}_y \\ \hat{E}_z \end{pmatrix} = 0 . \tag{6.91}$$

Because of the rotational symmetry of the problem about the direction of the magnetic field, we could arbitrarily choose the wave vector in the x-z plane, $\mathbf{k} = (k \sin\psi, 0, k\cos\psi)$. Equation (6.91) is now defining the refractive index $\mathscr{N}(\omega, k, \psi)$, which we will start discussing for the principal directions $\psi = 0$ and $\psi = \pi/2$.

6.6.2 Circularly Polarized Modes and the Faraday Effect

We begin with studying the wave propagation along the magnetic field ($\psi = 0$). Then the wave equation has the particular form

$$\begin{pmatrix} S - \mathscr{N}^2 & -\mathrm{i}D & 0 \\ \mathrm{i}D & S - \mathscr{N}^2 & 0 \\ 0 & 0 & P \end{pmatrix} \cdot \begin{pmatrix} \hat{E}_x \\ \hat{E}_y \\ \hat{E}_z \end{pmatrix} = 0 . \tag{6.92}$$

Here, we have to distinguish two cases:

1. $\hat{E}_x = \hat{E}_y = 0$ und $\hat{E}_z \neq 0$. This is a longitudinal wave that is described by the dispersion relation $P = 1 - (\omega_{\mathrm{pe}}^2 + \omega_{\mathrm{pi}}^2)/\omega^2 = 0$. In fact, we find the plasma oscillations again, which appeared in the unmagnetized case. Obviously, the magnetic field has no effect on the wave because the oscillations are aligned with the magnetic field and the Lorentz force vanishes.
2. $\hat{E}_x \neq 0 \neq \hat{E}_y$ und $\hat{E}_z = 0$. In this case we have transverse electromagnetic waves that are described by a 2×2 system of equations

$$\begin{pmatrix} S - \mathscr{N}^2 & -\mathrm{i}D \\ -\mathrm{i}D & S - \mathscr{N}^2 \end{pmatrix} \cdot \begin{pmatrix} \hat{E}_x \\ \hat{E}_y \end{pmatrix} = 0 . \tag{6.93}$$

Introducing again the rotating electric field $\hat{E}^{\pm}$ with (6.82)—this corresponds to a circular polarization of the wave—the two equations are decoupled:

$$(S - D - \mathscr{N}^2)\hat{E}^+ + (S + D - \mathscr{N}^2)\hat{E}^- = 0 . \tag{6.94}$$

When $\hat{E}^+ \neq 0$ und $\hat{E}^- = 0$, we have a *left-handed circularly polarized* wave (L-wave) with a refractive index $\mathscr{N}_L = \sqrt{S - D}$. In the other case, $\hat{E}^+ = 0$ und

$\hat{E}^- \neq 0$, the wave is a *right-handed circularly polarized* (R-mode), and the refractive index is $\mathscr{N}_R = \sqrt{S + D}$.

Using the definitions of the parameters S and D we obtain

$$\mathscr{N}_R = \left(1 - \frac{\omega_{pe}^2}{\omega(\omega - \omega_{ce})} - \frac{\omega_{pi}^2}{\omega(\omega + \omega_{ci})}\right)^{1/2} \tag{6.95}$$

$$\mathscr{N}_L = \left(1 - \frac{\omega_{pe}^2}{\omega(\omega + \omega_{ce})} - \frac{\omega_{pi}^2}{\omega(\omega - \omega_{ci})}\right)^{1/2} . \tag{6.96}$$

For $\omega = \omega_{ce}$ the refractive index of the R-mode approaches $\mathscr{N}_R \to \infty$. The R-mode is said to have a *resonance* at the electron cyclotron frequency. This resonance becomes immediately evident when we see that the sense of rotation of the wave vector and the electron are the same (Fig. 6.11). In the rotating frame of reference the electron experiences a DC electric field and can gain energy indefinitely. The same consideration applies to the L-mode, which has a resonance at the ion cyclotron frequency.

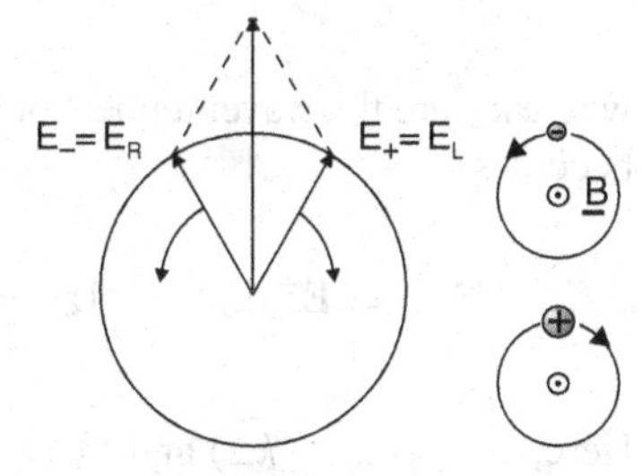

Fig. 6.11 The sense of rotation for the R-mode and L-mode compared to the gyromotion of electrons and positive ions

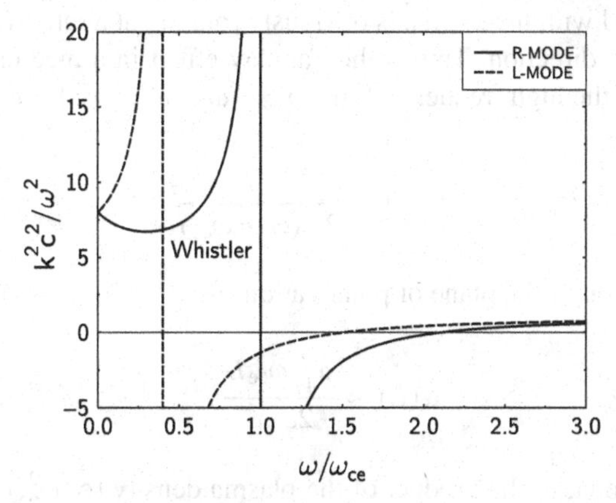

Fig. 6.12 The square of the refractive index for wave propagation along the magnetic field as a function of frequency. For clarity, an artificial mass ratio $m_e/m_i = 0.4$ was chosen. The R-mode has a resonance, $\mathscr{N}^2 \to \infty$, at the electron cyclotron frequency whereas the L-wave shows a resonance at (the lower) ion cyclotron frequency. In the high density limit $\omega_{pe}^2 \gg \omega_{ce}^2$ considered here, only the R-wave is propagating between ion and electron cyclotron frequency while the L-wave is in the cut-off, $\mathscr{N}^2 < 0$

The refractive index as a function of wave frequency determines the regimes where the R-mode and L-mode are either propagating or in the cut-off. In Fig. 6.12 the existence regimes are given for an artificial mass ratio of $m_e/m_i = 0.4$ to reduce the difference between ion and electron cyclotron frequency. Just above their respective cyclotron frequencies, the L-wave and the R-wave are in a cut-off band, $\mathcal{N}^2 < 0$, until they reach a propagating band beyond the cut-off frequency. At the highest frequencies, both refractive indices approach $\mathcal{N} = 1$.

6.6.2.1 Faraday Rotation

The small difference between $\mathcal{N}_R$ and $\mathcal{N}_L$ at high frequencies gives rise to the Faraday effect, namely the plane of polarization of a linearly polarized electromagnetic wave that propagates along the magnetic field line is rotated about the field.

A linearly-polarized transverse wave propagating along the magnetic field can be decomposed into a pair of R- and L-mode. Using circular coordinates and noting that $\mathrm{i} = \exp(\mathrm{i}\pi/2)$, we find for the electric field components $\mathbf{E}^{\pm}$

$$\mathbf{E}^{\pm} = \frac{\hat{E}}{2}\begin{pmatrix} \exp[\mathrm{i}(k_{\pm}z - \omega t)] \\ \exp[\mathrm{i}(k_{\pm}z \mp \frac{\pi}{2} - \omega t)] \end{pmatrix}, \tag{6.97}$$

where $k_{\pm}$ are the wavenumbers of the R- and L-mode. The electric field pattern then becomes

$$\mathbf{E}(z) = \mathbf{E}^{+}(z) + \mathbf{E}^{-}(z) = \hat{E}\exp[\mathrm{i}(\bar{k}z - \omega t)]\begin{pmatrix} \cos(\delta k\, z) \\ \sin(\delta k\, z) \end{pmatrix}. \tag{6.98}$$

Here, $\bar{k} = \frac{1}{2}(k_{+} + k_{-})$ and $\delta k = \frac{1}{2}(k_{+} - k_{-})$. The plane of polarization, which, at $z = 0$, was aligned with the x-axis, is obviously rotating at a rate $\alpha(z) = \delta k\, z$ about the magnetic field direction. This is the Faraday effect in a medium with circular bi-refringence. In the high frequency limit $\omega \gg (\omega_{pe}, \omega_{ce})$ we have

$$\mathcal{N}_{\pm} \approx 1 - \frac{\omega_{pe}^2}{2\omega(\omega \pm \omega_{ce})}. \tag{6.99}$$

This gives a rotation of the plane of polarization

$$\alpha(L) \approx \frac{\omega_{pe}^2 \omega_{ce} L}{2c\,\omega^2}, \tag{6.100}$$

which is proportional to the product of the plasma density ($\propto \omega_{pe}^2$), magnetic field ($\propto \omega_{ce}$) and path length L. In an inhomogeneous medium, the local product of density and magnetic field has to be integrated along the ray path.

Faraday rotation is a standard technique to study galactic magnetic fields (e.g., [113]). The magnetic structure of the solar corona was investigated with back-illumination by a satellite-borne transmitter [114] or by polarized radiation from natural radio-sources [115]. In the ionospheric plasma, modern techniques comprise Faraday rotation imaging with multiple satellites [116] or polarization analysis of coherent radar backscatter [117]. In fusion devices, polarimetry with many ray paths, after its demonstration in the TEXTOR device [118], is now a well established method, which is capable of measuring the poloidal component of the magnetic field. From the sensitivity point of view, see (6.51), long wavelengths in the far infrared are preferred [119]. Such far-infrared wavelengths were also applied in a reversed field pinch [120, 121] or in the Compact Helical System [122].

6.6.2.2 Example: Occultation of Radio Sources

A typical example for studying the magnetic field in the solar corona by means of Faraday rotation is shown in Fig. 6.13a. Here, the rotation measure $RM = \alpha(L)\lambda^{-2}$ for different radio sources near the ecliptic is determined with a radio telescope while the sun and its corona pass by different sources at various distances [115]. Because the rotation measure gives only the line-averaged product of electron density and parallel magnetic field, a model for the density distribution and the coronal magnetic field is used to predict the magnitude of the effect. The comparison of measurement and expectation is shown in Fig. 6.13b. A similar technique was applied with radio transmitters aboard solar orbiting satellites (e.g., [123]).

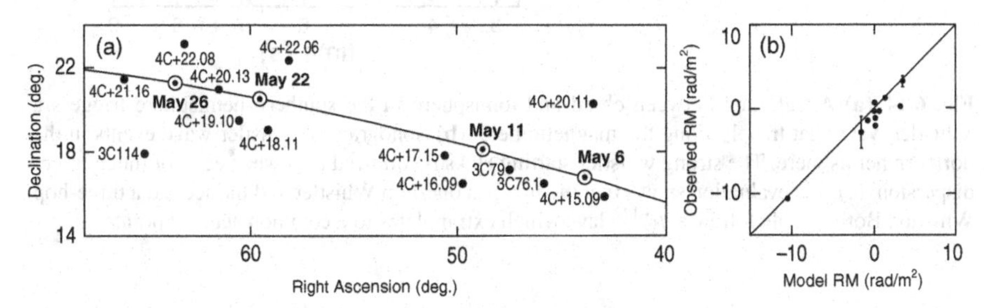

Fig. 6.13 Measuring the magnetic field in the solar solar corona by means of Faraday rotation. (**a**) Ecliptic with positions of the sun relative to radio sources. (**b**) Resulting rotation measure compared to model prediction. (Reproduced from [115] by permission of the AAS)

6.6.2.3 Whistler Waves

In most parts of the ionosphere and plasmasphere the plasma density is high enough to establish $\omega_{\mathrm{pe}}^2 \gg \omega_{\mathrm{ce}}^2$. Therefore, the R-mode is the only propagating mode in the frequency range between ion cyclotron frequency and electron frequency, as can be seen in Fig. 6.12.

A lightning event in the southern hemisphere triggers a wave pulse that is dispersed into a low frequency ($\omega^2 \ll \omega_{\mathrm{ce}}^2$) wave train while it propagates along a magnetic field line according to the refractive index $\mathcal{N}_{\rm R} \approx \omega_{\rm pe}/\sqrt{\omega\omega_{\rm ce}}$. The group delay time for these low-frequency wave packets becomes $T_{\rm g} \propto \omega^{-1/2}$. For an observer in the northern hemisphere, this gives an electric wave field in the audible range with a slowly decaying pitch, which explains the name *Whistler wave* for this phenomenon [124]. This effect can be visualized in a sonogram, in which the instantaneous frequency of the signal is plotted vs. time, see Fig. 6.14b. The analysis of the sonogram shows that the decaying pitch follows the $t \propto \omega^{-1/2}$ law and that the subsequent weaker echo has the same origin and, because its time scale is three times longer, represents a signal that has bounced three times between southern and northern hemisphere. The example is event c09m04 from Stephen Mc Greevy's VLF recordings.[1]

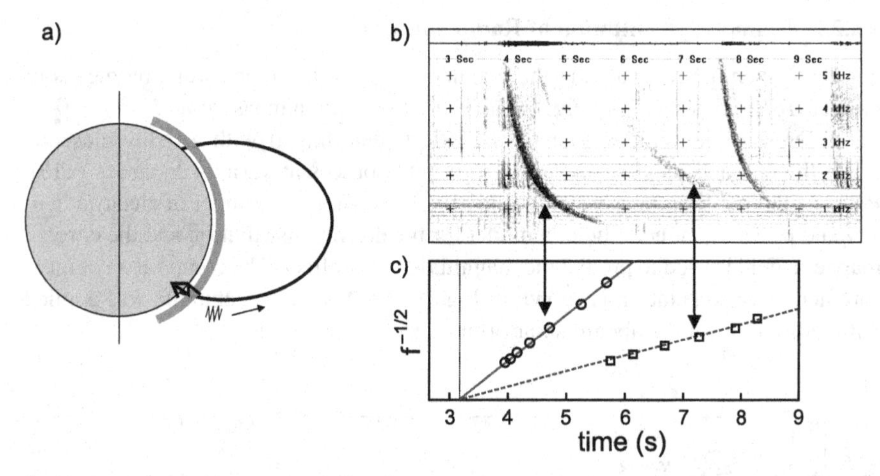

Fig. 6.14 (**a**) A lightning between cloud and ionosphere in the southern hemisphere triggers a Whistler wave that travels along the magnetic field. (**b**) Sonogram of whistler wave events in the northern hemisphere. The strong whistler starting at 4 s is followed by a weak echo of much larger dispersion. (**c**) The evaluation shows that the first is a one-hop Whistler and the second a three-hop Whistler. Both signals follow a $f^{-1/2}$ law, which extrapolates to a common starting point

6.6.3 Propagation Across the Magnetic Field

We now turn to wave propagation across the magnetic field ($\psi = \pi/2$). This still leaves the polarization of the wave open, which can be parallel to the magnetic field, or perpendicular, or at any angle in between. When the electric field vector is aligned with the static magnetic field, the wave is called the *ordinary mode* or O-mode. The refractive index for the O-mode is not affected by the magnetic field, because the ion and electron motion is purely along the magnetic field, and is given by (6.46) and (6.45). This is why the mode is called ordinary. The ordinary mode is used, e.g., for interferometry in magnetized plasmas, where it is a suitable density diagnostics.

[1] `http://www-pw.physics.uiowa.edu/mcgreevy/`

The *extraordinary mode* or X-mode has $\mathbf{E} \perp \mathbf{B}_0$ and is described by the 2×2 system of equations

$$\begin{pmatrix} S & -\mathrm{i}D \\ \mathrm{i}D & (S - \mathcal{N}^2) \end{pmatrix} \cdot \begin{pmatrix} \hat{E}_x \\ \hat{E}_y \end{pmatrix} = 0 . \tag{6.101}$$

Again, non-vanishing solutions for $\mathbf{E}$ are found when the determinant of the matrix becomes zero, yielding a refractive index given by

$$\mathcal{N}_\mathrm{X} = \left(\frac{S^2 - D^2}{S} \right)^{1/2} . \tag{6.102}$$

Resonances appear when the Stix parameter S vanishes ($S = 0$). In the case of very high frequencies, we can neglect the ion contributions in S, and find the so-called *upper-hybrid resonance* frequency

$$\omega_\mathrm{uh} = (\omega_\mathrm{ce}^2 + \omega_\mathrm{pe}^2)^{1/2} . \tag{6.103}$$

For intermediate frequencies, there is a second zero of S, which defines the *lower hybrid resonance* frequency

$$\omega_\mathrm{lh} = \left(\omega_\mathrm{ci}^2 + \frac{\omega_\mathrm{pi}^2 \omega_\mathrm{ce}^2}{\omega_\mathrm{pe}^2 + \omega_\mathrm{ce}^2} \right)^{1/2} . \tag{6.104}$$

In the limit of high electron density, $\omega_\mathrm{pe}^2 \gg \omega_\mathrm{ce}^2$, the lower hybrid frequency becomes $\omega_\mathrm{lh} \approx (\omega_\mathrm{ci}\omega_\mathrm{ce})^{1/2}$ The behavior of the refractive index for the X-mode and O-mode as a function of wave frequency is shown in Fig. 6.15.

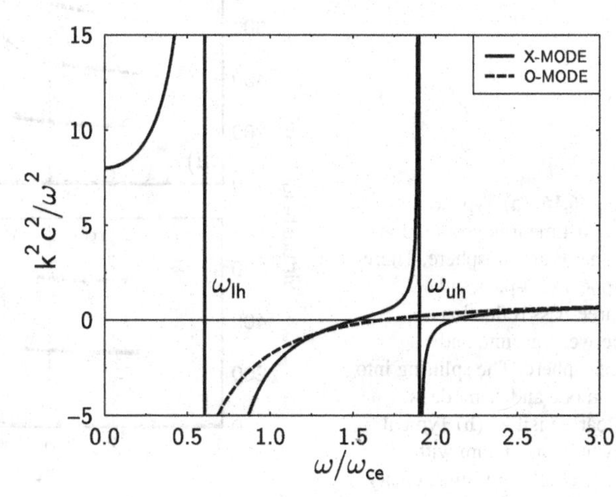

Fig. 6.15 The square of the refractive index for wave propagation perpendicular to the magnetic field as a function of frequency. An artificial mass ratio $m_\mathrm{e}/m_\mathrm{i} = 0.4$ is chosen. The X-mode has resonances at the lower hybrid frequency ω_lh and the upper hybrid frequency ω_uh

The X-mode is propagating at frequencies below the hybrid resonances. At the hybrid frequencies, $\mathcal{N}^2$ changes sign and the wave propagation is cut-off. The O-mode cut-off occurs at the electron plasma frequency. For very high frequencies the refractive index of both modes approaches unity.

6.6.3.1 Ionosondes

Vertical sounding of the ionosphere is a standard technique to study the electron density profile in the lower ionosphere, which is characterized by different layers.

Nowadays, ground-based (digital) ionosondes [125] are used, which emit wave bursts at various frequencies and determine the height of the reflective layer for each particular frequency from the echo delay time. The sonograms in Fig. 6.16 show the echoes from the O-mode and X-mode as well as multiple reflections between ionosphere and ground. These examples show times, where only the F-layer was present. After sunset, the E-layer rapidly disappears by recombination.

The night-time ionogram (taken on day 109 of 1993 at 04:00 h LT over Shriharikota, India), where the electron cyclotron frequency is $f_{\mathrm{ce}} = 980\,\mathrm{kHz}$, shows an O-mode cut-off at 4.5 MHz, which corresponds to an electron density of $2.5 \times 10^{11}\,\mathrm{m}^{-3}$ (Fig. 6.16a). The X-mode cut-off is found at 4.8 MHz (its expected value from (6.103) is 5.0 MHz). According to the higher electron density in the evening ionosphere, the O-mode and X-mode cut-offs shift to higher frequencies (Fig. 6.16b). The maximum electron density at 21:00 h local time for the same location reaches $1.6 \times 10^{12}\,\mathrm{m}^{-3}$.

The time delay for a particular reflection is commonly expressed in terms of the *virtual height* $h' = cT/2$, in which the refractive index of the plasma is not yet corrected. The density profile results from the reflection condition at the O-mode (X-mode) cut-off and the traversed part of the plasma up to the cut-off is used to convert the virtual height to the real height of the reflecting layer. There are

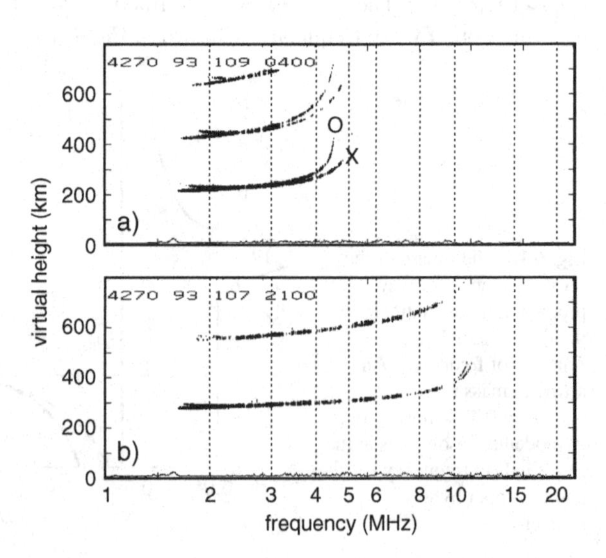

Fig. 6.16 (**a**) Typical night-time ionogram in the equatorial ionosphere. There appear two-pass and three-pass reflections between ground and ionosphere. The splitting into O-mode and X-mode is clearly visible. (**b**) Typical evening ionogram with a much higher plasma density

analytical methods [126] and computer programs [127, 128] available for this evaluation. A survey of this technique can be found in [129]. Ionosondes have recently been applied in the equatorial ionosphere [130] to study the *Equatorial Spread-F* phenomenon, at mid-latitude [131] for investigating sporadic E-events or to study plasma drift effects at the southern polar cap [132].

6.7 Resonance Cones

So far, the discussion was restricted to the principal wave modes, which propagate at $\theta = 0°$ or $\theta = 90°$ and possess resonances either at the cyclotron frequencies or at the hybrid frequencies. For oblique wave propagation, there is a so-called *lower oblique resonance*, which occurs for $\omega < \min(\omega_{pe}, \omega_{ce})$ in the Whistler wave regime. In a cold plasma, the resonance angle is given by [133, 134]

$$\sin^2(\theta_c) = \frac{\omega^2(\omega_{pe}^2 + \omega_{ce}^2 - \omega^2)}{\omega_{pe}^2\omega_{ce}^2} . \tag{6.105}$$

When the wave is excited with a small antenna (which can be the protruding inner conductor of a rigid coaxial cable), the resonance is found on the surface of a (double) cone of opening angle θ_c, which is aligned with the magnetic field direction and has its apex at the antenna, see Fig. 6.17a.

The resonance angle θ_c is a function of the electron density and can be used as a diagnostic method. The resonance angle is easily found by moving a receiver antenna around the transmitting antenna. For low plasma density ($\omega_{pe}^2 \ll \omega_{ce}^2$) and for $\omega^2 \ll \omega_{pe}^2$ (6.105) takes the limiting form

$$\sin(\theta_c) \approx \frac{\omega}{\omega_{pe}} . \tag{6.106}$$

A typical resonance curve with two pronounced maxima is shown in Fig. 6.17b. The dependence of the resonance angle on the transmitter frequency is shown in

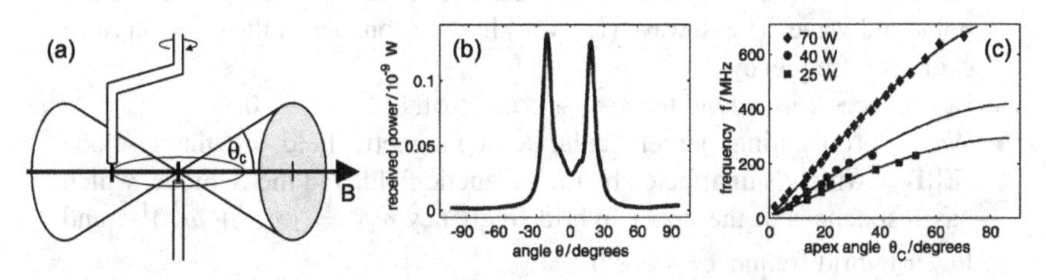

Fig. 6.17 (**a**) Geometry for recording a resonance cone in the laboratory. (**b**) Two resonance maxima at the intersection with the resonance cone. (**c**) Dependence of resonance angle on exciter frequency and plasma density (from [135])

Fig. 6.17c for three different values of the rf power that generates the plasma. Theoretical curves from (6.106), with ω_{pe} as fit-parameter, closely match the measured resonance angle. From this fit the electron density is obtained.

The resonance cone method was used for diagnostic purposes on sounding rockets in the ionosphere [136–139], for kinetic and non-thermal effects in laboratory plasmas [140–142], and in dusty plasmas [135].

The Basics in a Nutshell

- In Fourier notation, Maxwell's equations become:

Induction law	$\mathrm{i}\mathbf{k} \times \hat{\mathbf{E}} = \mathrm{i}\omega\hat{\mathbf{B}}$
Ampere's law	$\mathrm{i}\mathbf{k} \times \hat{\mathbf{B}} = -\mathrm{i}\omega\varepsilon_0\mu_0\hat{\mathbf{E}} + \mu_0\hat{\mathbf{j}}_0$
Poisson's law	$\mathrm{i}\mathbf{k} \cdot \hat{\mathbf{E}} = \hat{\rho}/\varepsilon_0$
no longitudinal $\hat{B}$	$\mathrm{i}\mathbf{k} \cdot \hat{\mathbf{B}} = 0\,.$

- The wave equation: $\{\mathbf{kk} - k^2 \boldsymbol{I} + \frac{\omega^2}{c^2}\boldsymbol{\varepsilon}_\omega\} \cdot \hat{\mathbf{E}} = 0$.
- The phase and group velocities are defined as $v_\varphi = \omega/k$, $v_g = \mathrm{d}\omega/\,\mathrm{d}k$.
- Transverse electromagnetic waves in an unmagnetized plasma have the refractive index $\mathscr{N} = \varepsilon(\omega) = (1 - \omega_{pe}^2/\omega^2)^{1/2}$. They exist only above a cut-off frequency, $\omega > \omega_{pe}$.
- The transverse mode is used for plasma interferometry to determine the plasma density. The phase shift of an interferometer is proportional to the product $n_e L\lambda$.
- The dispersion of an electrostatic wave in an unmagnetized plasma is determined by $\varepsilon(\omega) = 0$.
- Electrostatic waves have $\mathbf{k}||\hat{\mathbf{E}}$ and are found in two frequency regimes: Bohm-Gross modes for $\omega > \omega_{pe}$ and ion-acoustic waves for $\omega < \omega_{pi}$. The ion-acoustic speed is $C_s = (k_B T_e/m_i)^{1/2}$.
- In magnetized plasma, the fundamental modes for propagation along the magnetic field line have circular polarization. The refractive index of the R-wave and L-wave are different. This leads to Faraday rotation of a linearly polarized wave. The R-wave (L-wave) has a resonance at the electron (ion) cyclotron frequency.
- Resonances correspond to $\mathscr{N}^2 \to \infty$, cut-offs to $\mathscr{N}^2 \to 0$.
- Waves propagating perpendicular to a magnetic field are the O-mode ($\mathbf{E}||\mathbf{B}_0$), which is unaffected by the magnetic field, and the X-mode, which has resonances at the upper hybrid frequency $\omega_{uh} = (\omega_{pe}^2 + \omega_{ce}^2)^{1/2}$ and lower hybrid frequency $\omega_{lh} \approx (\omega_{ce}\omega_{ci})^{1/2}$.

Problems

6.1 In the limit $T_\mathrm{i} \ll T_\mathrm{e}$ the ion-acoustic wave has the dispersion relation

$$\omega(k) = \frac{\omega_{\mathrm{pi}}\lambda_{\mathrm{De}}\, k}{(1+k^2\lambda_{\mathrm{De}}^2)^{1/2}}$$

(a) Derive an expression for the phase velocity $v_\varphi(k)$ and group velocity $v_\mathrm{g}(k)$ as a function of the wave number k.
(b) Discuss the result with respect to "acoustic behavior" at $k\lambda_{\mathrm{De}} \ll 1$.

6.2 Assume that in a dielectric medium the relation $v_\varphi \cdot v_\mathrm{g} = c^2$ holds. What is the general shape of the dispersion relation $\omega(k)$ for this case?

6.3 (a) Show that for $\omega_{\mathrm{pe}}^2 \gg \omega_{\mathrm{ce}}^2 \gg \omega^2$ the refractive index for Whistler waves takes the limiting form

$$\mathcal{N} = \frac{\omega_{\mathrm{pe}}}{(\omega\omega_{\mathrm{ce}})^{1/2}}$$

(b) Calculate phase and group velocity and show that $v_{\mathrm{gr}} = 2v_\varphi$.

6.4 Determine the minimum plasma density at which a He-Ne Laser at $\lambda = 633\,\mathrm{nm}$ wavelength will be reflected.

6.5 Consider an electron-positron plasma with $n_\mathrm{e} = n_\mathrm{p}$. What is the cut-off frequency for electromagnetic waves in this system?

6.6 The plasma of the ionospheric F-layer has a density $n_\mathrm{e} \approx 2 \times 10^{12}\,\mathrm{m}^{-3}$. The typical magnetic field at mid-latitude is $B = 50\,\mu\mathrm{T}$. Calculate the electron plasma frequency f_{pe}, electron cyclotron frequency f_{ce} and the upper hybrid frequency f_{uh}.

6.7 Prove that $v_\varphi = v_{\mathrm{gr}}$ requires $\omega = v_\varphi\, k$.

Chapter 7
Plasma Boundaries

What I tell you three times is true.

Lewis Carroll, The Hunting of the Snark

A plasma separates itself from metallic or dielectric surfaces by forming a boundary layer, which appears darker than the bulk plasma itself. This is a first hint that the boundary layer is depleted of electrons that are needed to excite the neutral atoms producing the glow of an electric discharge. It was Langmuir who identified these dark spaces as regions that are not electrically neutral but are governed by a net (positive) space charge. The particle motion is determined by physical mechanisms that are different from those discussed for the quasineutral part of the plasma. The interaction of an ion with the electric field from the space charge of all the other ions is a new type of many-body interaction that is characteristic for the collective behavior of a plasma.

7.1 The Space-Charge Sheath

Let us consider the situation of a plasma that is in contact with a plane conducting wall located at $x = 0$ (Fig. 7.1) and fills the half-space $x < 0$. Because of the action of Debye shielding, we can expect that beyond a certain position $x = -d$ from the wall the plasma particles will have established quasineutrality. We call the region $-d < x < 0$, in which a significant deviation from quasineutrality is allowed, the *space charge sheath* or simply the *sheath*. The position $x = -d$ is named the *sheath edge*. It is further assumed that the ion motion in the sheath is collisionless, i.e., $\lambda_{\mathrm{mfp}} \gg d$.

Because the thermal velocity of the electrons is much higher than that of the ions, an initially uncharged wall will be hit more often by electrons than by ions, which accumulates a net negative surface charge on the wall and makes the wall potential negative with respect to the electric potential inside the quasineutral plasma. For the moment, we assume that the wall is electrically floating and that the charges are not flowing away in an external circuit. When the wall potential becomes negative, the number of electrons that can reach the wall diminishes until an equilibrium is reached, in which the residual electron flux to the wall equals the ion flux, and the net charge on the wall reaches an equilibrium value, the *floating potential*.

A. Piel, *Plasma Physics*, DOI 10.1007/978-3-642-10491-6_7,

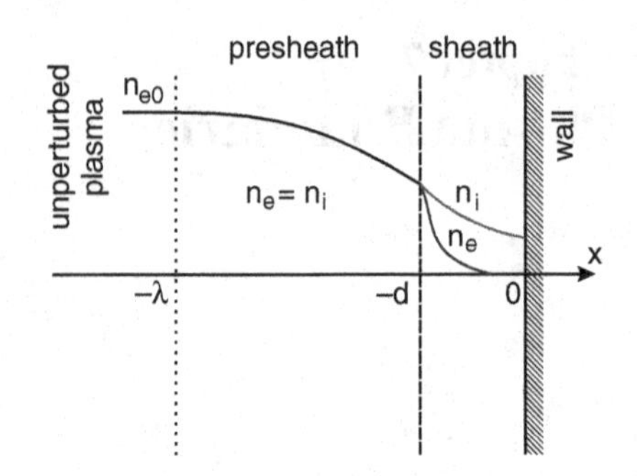

Fig. 7.1 Geometry of the plasma-wall boundary layer. A space charge sheath of thickness d with $n_e < n_i$ is formed at the wall. The matching between sheath and bulk plasma occurs in a quasi-neutral presheath of a size comparable to the ion mean free path λ

This simple picture gives us an impression why all isolated bodies inside a plasma charge up negatively. This applies to the fine metal wires which Langmuir introduced as *probes* into the plasma, to satellites in the Earth's plasmasphere, or to dust particles in a plasma. Of course, the real situation is more complex because we will see that the ion flux to a negative body (except for the very first moment) is not determined by the ion thermal velocity. Moreover, we may have neglected other processes that lead to charging. In the case of satellites, these processes are photoemission from solar UV-radiation or secondary emission by impact of energetic particles. We will discuss these effects in Sect. 10.1.

For completeness, we introduce a transition layer adjacent to the sheath, called the *presheath*, that matches the conditions between the space charge sheath and the unpertubed plasma. The presheath will be quasineutral, but the densities of electrons and ions will depend on position, and the ion drift velocity will be non-zero. This transition region has a thickness of roughly one ion mean free path.

7.2 The Child-Langmuir Law

Here, we consider a situation, where a potential difference between the wall at $\Phi(0)$ and the sheath edge at $\Phi(-d)$ is determined by an external voltage applied to the wall. We are mostly interested in cases where this potential difference creates a high potential barrier for thermal electrons $|\Phi(0) - \Phi(-d)| \gg k_B T_e/e$. Then the Boltzmann factor for the electron gas,

$$n_e(x) = n_e(-d) \exp\left\{\frac{e[\Phi(x) - \Phi(-d)]}{k_B T_e}\right\}, \tag{7.1}$$

ensures that only few electrons can overcome the barrier, and that a significant number of electrons is only found close to the sheath edge. In other words, for large negative voltages applied to the wall, most of the sheath will be a pure ion sheath. For simplicity of the calculation, we will completely ignore the electron space charge for the moment.

The ion motion inside the sheath will be discussed here only for the collisionless sheath, $d \ll \lambda_{\text{mfp}}$. Then, the ion velocity $u_i(x)$ is determined by energy conservation

$$\frac{1}{2}m_{\text{i}}u_{\text{i}}^2(x) + e\Phi(x) = \frac{1}{2}m_{\text{i}}u_{\text{i}}^2(-d) + e\Phi(-d) \tag{7.2}$$

assuming an initial velocity $u_i(-d) = u_0$. Then, setting $\Phi(-d) = 0$, we obtain

$$u_{\text{i}}(x) = \left[u_0^2 - \frac{2e\Phi(x)}{m_{\text{i}}}\right]^{1/2} . \tag{7.3}$$

In the following, we are interested to describe a steady-state solution for the ion flow towards the wall. In the absence of ionisation or recombination, the continuity equation reads

$$n_{\text{i}}(x)u_{\text{i}}(x) = n_{\text{i}}(-d)u_0. \tag{7.4}$$

Hence, the acceleration of ions leads to a reduction of ion density

$$n_{\text{i}}(x) = n_{\text{i}}(-d)\left[1 - \frac{2e\Phi(x)}{m_{\text{i}}u_0^2}\right]^{-1/2} . \tag{7.5}$$

This ion density must be used to determine the self-consistent electric potential distribution $\Phi(x)$. Therefore, potential and ion density must fulfill Poissons' equation

$$\Phi'' \approx -\frac{en_{\text{i}}(-d)}{\varepsilon_0}\left(-\frac{2e\Phi(x)}{m_{\text{i}}u_0^2}\right)^{-1/2}, \tag{7.6}$$

where we have used $e|\Phi(x)| \gg m_{\text{i}}u_0^2/2$, i.e., stating that the initial energy of the ion is small compared to the energy gained by free fall in the sheath potential. The classical solution of this problem according to Langmuir starts by multiplying both sides of Eq. (7.6) by Φ' and integrating from $x = -d$ to $x = 0$,

$$\frac{1}{2}\left[\Phi'^2(x) - \Phi'^2(-d)\right] = \frac{en_{\text{i}}(-d)u_0}{\varepsilon_0}\left(\frac{2m_{\text{i}}}{e}\right)^{1/2} \times \left\{[-\Phi(x)]^{1/2} - [-\Phi(-d)]^{1/2}\right\} . \tag{7.7}$$

We can neglect $\Phi'^2(-d)$ compared to $\Phi'^2(x)$, because the electric field at the sheath edge is small compared to that inside the sheath. By definition, $\Phi(-d) = 0$. Noting that $en_{\text{i}}(-d)u_0 = j_{\text{i}}$ is the (constant) ion current density inside the sheath, we have to perform a second integration of the equation

$$\Phi'(x) = 2\left(\frac{m_\mathrm{i}}{2e}\right)^{1/4}\left(\frac{j_\mathrm{i}}{\varepsilon_0}\right)^{1/2}[-\Phi(x)]^{1/4}\,, \tag{7.8}$$

which can be done by separation of the variables and leads to

$$\frac{4}{3}\Phi^{3/4} = 2\left(\frac{m_\mathrm{i}}{2e}\right)^{1/4}\left(\frac{j_\mathrm{i}}{\varepsilon_0}\right)^{1/2}(x+d)\,. \tag{7.9}$$

This result defines the potential distribution in a space charge sheath

$$\Phi(x) = \left(\frac{3}{2}\right)^{4/3}\left(\frac{m_\mathrm{i}}{2e}\right)^{1/3}\left(\frac{j_\mathrm{i}}{\varepsilon_0}\right)^{2/3}(x+d)^{4/3}\,, \tag{7.10}$$

and gives a relation between the total voltage drop $U = \Phi(-d) - \Phi(0)$, the ion current density j_i and the sheath thickness d:

$$U^{3/2} = \frac{9}{4}\left(\frac{m_\mathrm{i}}{2e}\right)^{1/2}\left(\frac{j_\mathrm{i}}{\varepsilon_0}\right)d^2\,. \tag{7.11}$$

Solving for the current density, we obtain the famous Child-Langmuir law, [143, 144]

$$j_\mathrm{i} = \frac{4}{9}\varepsilon_0\left(\frac{2e}{m_\mathrm{i}}\right)^{1/2}\frac{U^{3/2}}{d^2}\,, \tag{7.12}$$

which was originally formulated for the space-charge limited electron flow in a vacuum diode.

In a vacuum diode, the separation d between cathode and anode is fixed and the Child-Langmuir law defines the volt-ampere characteristic of the diode. In a plasma sheath, the voltage drop is fixed and we will see below that the ion current is also defined by the properties of the unperturbed plasma. Hence, the plasma sheath reacts by adjusting the sheath thickness d to fulfill the constraints by space-charge limited flow described by the Child-Langmuir law.

7.3 The Bohm Criterion

The matching of a space charge sheath with a plasma raises the question, why such a huge violation of quasi-neutrality does not set up a large-amplitude ion acoustic wave, by which the charge perturbation could propagate into the plasma bulk. So, what mechanism holds the space charge from spreading into the plasma? Obviously, we are asking for the stability of the plasma-sheath boundary. Under which conditions tends a neutral plasma to develop a charge imbalance? Such a question cannot be answered by the steady-state considerations of the previous Section. Rather, we must use more general concepts, e.g., those for mechanical stability.

7.3.1 Stability Analysis

Consider the equilibria of a point mass in the mechanical potentials shown in Fig. 7.2. In situation (a), the point mass attains a stable equilibrium at the minimum of the potential well. The potential well exerts a restoring force, when the mass is displaced from the minimum. The stable "trajectory" of the point mass is a dull function, $x(t) = 0$. In situation (b), the mass sits on top of a potential hill (think of an inverted pendulum). Any displacement from the maximum position leads to a force that drives the point mass further away from its initial equilibrium. Therefore, the equilibrium is unstable and the point mass follows a non-trivial trajectory $x(t)$. Obviously, the sign of the second derivative of the mechanical potential determines whether the equilibrium is stable or unstable.

The equation of motion for a point mass m in a mechanical potential $V(x)$

$$m\frac{\mathrm{d}^2x}{\mathrm{d}t^2} = -\frac{\mathrm{d}V}{\mathrm{d}x} \tag{7.13}$$

determines the trajectory $x(t)$ of the particle after it has experienced its first small displacement from the equilibrium position. What has this to do with our problem of the development of a space charge sheath? Consider the general shape of Poisson's equation

$$\frac{\mathrm{d}^2\Phi}{\mathrm{d}x^2} = f(\Phi) = -\frac{\mathrm{d}\Psi}{\mathrm{d}\Phi}, \tag{7.14}$$

in which the r.h.s. is a function of Φ that can be interpreted as being the derivative of a so-called *pseudopotential* Ψ (also known as *classical potential* or *Sagdeev potential*). This problem becomes mathematically equivalent to the mechanical problem when we make the identifications listed in Table 7.1.

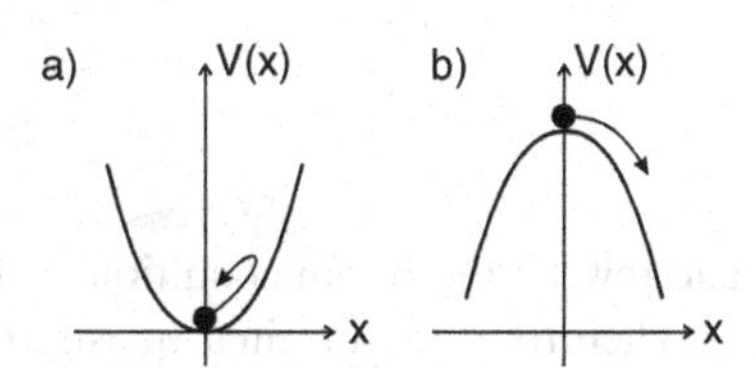

Fig. 7.2 (**a**) Stable mechanical equilibrium, $V''(0) > 0$. (**b**) Unstable mechanical equilibrium, $V''(0) < 0$

Table 7.1 Analogy between mechanical stability and sheath stability

Mechanical stability		Sheath stability	
Particle trajectory	$x(t)$	Electric potential distribution	$\Phi(x)$
Time	t	Space coordinate	x
Mechanical potential	$V(x)$	Pseudopotential	$\Psi(\Phi)$

7.3.2 The Bohm Criterion Imposed by the Sheath

Our remaining task is to calculate the pseudopotential and to determine its second derivative $d^2\Psi/d\Phi^2$, which has to be negative at the point of equilibrium to allow a plasma to develop a space charge sheath. When it is positive, the plasma remains neutral, which corresponds to the case in which the point mass rests in its stable minimum position.

Instead of calculating the second derivative of the pseudopotential, we can simply calculate the first derivate of the space-charge function, $-\mathrm{d}f(\Phi)/\mathrm{d}\Phi$. For this calculation it is essential to retain the electron space charge at the sheath edge given by (7.1). Hence, we have

$$f(\Phi) = \frac{en_\mathrm{e}(-d)}{\varepsilon_0}\left[\exp\left(\frac{e\Phi}{k_\mathrm{B}T_\mathrm{e}}\right) - \left(1 - \frac{2e\Phi}{m_\mathrm{i}u_0^2}\right)^{-1/2}\right] \tag{7.15}$$

and finally

$$-\left.\frac{\mathrm{d}f}{\mathrm{d}\Phi}\right|_{\Phi=0} = \frac{e}{k_\mathrm{B}T_\mathrm{e}} - \frac{e}{m_\mathrm{i}u_0^2} \le 0\,. \tag{7.16}$$

This gives the Bohm-criterion, named after the U.S.-born British physicist David Bohm (1917–1992), for the formation of a space charge sheath

$$u_0 \ge v_\mathrm{B} = \left(\frac{k_\mathrm{B}T_\mathrm{e}}{m_\mathrm{i}}\right)^{1/2}\,. \tag{7.17}$$

Hence, the speed of the ions at the sheath edge must be equal to or exceed the *Bohm velocity* v_B, which is obviously identical with the ion sound speed. We can also define a Mach number

$$M = \frac{u_0}{v_\mathrm{B}} \tag{7.18}$$

and rewrite the Bohm condition as $M \ge 1$, i.e., the ion flow has to be supersonic.

Therefore, the original question, why the space charge layer does not simply expand into a plasma by means of an ion acoustic wave can be answered as follows: The plasma in the presheath is not at rest. Rather, there is a mass motion with ion sound speed, or faster, into the sheath. An ion sound wave in this medium would be stationary in the laboratory frame of reference or would be swept back into the sheath. Hence, the Bohm criterion represents a *sound barrier* for the propagation of information from the sheath into the plasma. In this language of information, the plasma "does not know" about the presence of a space charge sheath.

7.3.3 The Bohm Criterion as Seen from the Presheath

The inequality in the Bohm criterion leaves a non-satisfying aspect. Why should Mother Nature allow for a wide range of possible entrance speeds u_0 into the sheath while she is choosing the most economic solutions elsewhere?

The answer is connected to the observation that a directed ion motion at the sheath edge requires an electric field in the presheath that accelerates the ions to the Bohm speed (or more). On the other hand, the presheath should be considered as quasineutral. Hence, the electron and ion density are nearly equal and therefore the ion density is determined by the electron Boltzmann factor

$$n_{\mathrm{i}}(x) = n_{\mathrm{e}}(x) = n_{\mathrm{e}0} \exp\left[\frac{e\Phi(x)}{k_{\mathrm{B}}T_{\mathrm{e}}}\right], \tag{7.19}$$

with $n_{\mathrm{e}0}$ being the electron density in the unperturbed plasma and $\Phi(x)$ the electric potential in the presheath. The ion motion in the presheath is affected by ion-neutral collisions at a collision frequency ν_{mi} for momentum loss. In steady state, the mean ion motion is therefore described by the equation

$$m_{\mathrm{i}} v_{\mathrm{i}} \frac{\mathrm{d}v_{\mathrm{i}}}{\mathrm{d}x} + m_{\mathrm{i}} \nu_{\mathrm{mi}} v_{\mathrm{i}} = -e\frac{\mathrm{d}\Phi}{\mathrm{d}x}, \tag{7.20}$$

in which the acceleration of the ion flow is described by the convective derivative $v(\mathrm{d}v_{\mathrm{i}}/\mathrm{d}x)$. We can eliminate the electric field $E = -\mathrm{d}\Phi/\mathrm{d}x$ from the ion equation of motion by rearranging (7.19) as

$$\Phi = \frac{k_{\mathrm{B}}T_{\mathrm{e}}}{e} \ln\left(\frac{n_{\mathrm{i}}}{n_{\mathrm{i}0}}\right). \tag{7.21}$$

Noting the continuity of the ion flow, $n_{\mathrm{i}} v_{\mathrm{i}} = \mathrm{const.}$, we easily obtain

$$\frac{\mathrm{d}\Phi}{\mathrm{d}x} = -\frac{k_{\mathrm{B}}T_{\mathrm{e}}}{e} \frac{1}{v_{\mathrm{i}}} \frac{\mathrm{d}v_{\mathrm{i}}}{\mathrm{d}x}. \tag{7.22}$$

Inserting this expression in (7.20) we have

$$\frac{\mathrm{d}v_{\mathrm{i}}}{\mathrm{d}x} = \frac{\nu_{\mathrm{mi}} v_{\mathrm{i}}^2}{v_{\mathrm{B}}^2 - v_{\mathrm{i}}^2} \tag{7.23}$$

For all subsonic velocities, $v_{\mathrm{i}} \leq v_{\mathrm{B}}$, the ion acceleration is positive. However, when the Bohm velocity is approached, the acceleration becomes singular. This singularity is also found in the electric field E. The appearance of a singularity in the electric field is the defining property for the position of the sheath edge because this singularity is a consequence of the constraint from assuming strict quasineutrality in the presheath.

In conclusion, the ion motion in a quasineutral presheath requires that $v_{\rm i} \le v_{\rm B}$. Hence, this is a second Bohm criterion, which follows from the conditions on the presheath side of the sheath edge, while the condition on the sheath side required $v_{\rm i} \ge v_{\rm B}$. Therefore, the complete Bohm criterion for the ion speed at the sheath edge can only be fulfilled by a unique velocity, the Bohm velocity,

$$v_{\rm i}(-d) = v_{\rm B}\,, \tag{7.24}$$

or, in other words, the Mach number has to be $M = 1$.

Does the singularity in the electric field mean that there is also a singularity in the electric potential? The answer is no. On its way from the plasma bulk through the presheath, an ion has gained the kinetic energy $\frac{1}{2} m_{\rm i} v_{\rm B}^2 = \frac{1}{2} k_{\rm B} T_{\rm e}$. Neglecting the energy dissipated in ion-neutral collisions, the potential at the sheath edge can be estimated from energy conservation as

$$\Phi(-d) \approx -\frac{1}{2}\frac{k_{\rm B} T_{\rm e}}{e}\,. \tag{7.25}$$

Accordingly, the plasma density at the sheath edge is reduced to

$$n_{\rm i}(-d) = n_{\rm e}(-d) = n_{\rm e0} \exp\left(-\frac{1}{2}\right) \approx 0.61\, n_{\rm e0}\,. \tag{7.26}$$

7.4 The Plane Langmuir Probe

In 1925, Mott-Smith and Langmuir [145] had introduced small additional electrodes into a plasma and studied its volt-ampere characteristic. These *Langmuir probes* are widely used in plasma physics because of their simple construction and versatility. We will show below, how the probe characteristic can be used to determine the electron density and electron temperature of a plasma.

The fundamental electric circuit of a Langmuir probe measurement is shown in Fig. 7.3a. A small plane electrode is inserted into a gas discharge. The discharge tube is typically operated from a high-voltage supply via a current-limiting series resistor $R_{\rm s}$. The probe is biased by an external voltage that is applied between the probe and a suitable electrode. For reasons of lab safety, this electrode must be properly grounded. Likewise, the power supply must be able to operate in a mode where the negative output is the "hot lead" and the positive output grounded. In this case, the anode (positive electrode) was chosen because the voltage drop in the anode layer is usually much smaller than that in the cathode (negative electrode) layer (see Chap. 11). A voltmeter gives the probe bias voltage $U_{\rm p}$ and a current meter the probe current $I_{\rm p}$.

A modern realisation of the circuit for recording probe characteristics with a computer is shown in Fig. 7.3b. The bias voltage is generated by a digital-to-analog

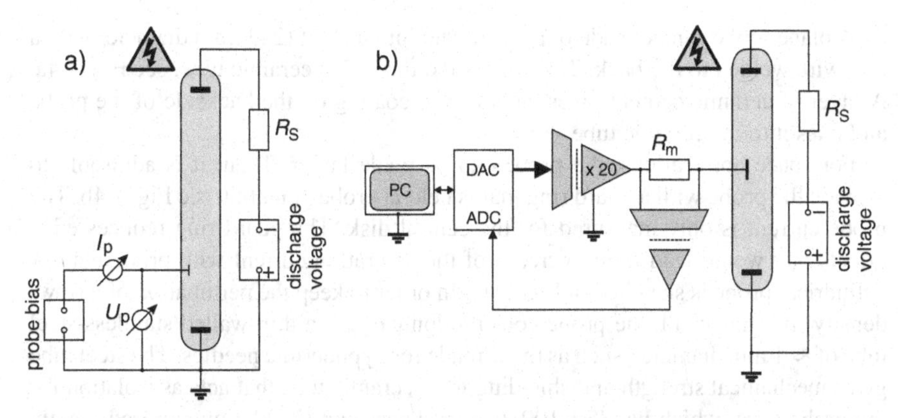

Fig. 7.3 (**a**) Arrangement for a plane Langmuir probe in a dc-discharge. The probe is biased with a voltage U_p with respect to a proper reference electrode. (**b**) Computer-controlled Langmuir probe circuit. A digital-to-analog converter (DAC) with subsequent amplifier provides a probe bias, between −100 V and +100 V. The probe current is measured with a series resistor R_m and an isolation amplifier, and finally A-D converted for numerical processing

converter (DAC), which delivers $(-5 \ldots +5)$ V and is amplified 20-times by a high-voltage operational amplifier. To protect the DAC and the computer from any unwanted plasma currents, an optically-isolated operational amplifier is used. The probe current is sensed as the voltage drop (< 1 V) across a small series resistor R_m by a second optically-isolated operational amplifier. The current signal is then read out by the computer via an analog-to-digital converter (ADC). Finally, the probe bias is corrected by the computer for the voltage drop across the series resistor, and the probe characteristic can be displayed and stored. Again, the probe circuit is closed by a connection between the *ground* terminal of the high-voltage opamp and the reference electrode that is connected to *protective ground* of the lab electrics. For your own safety, be sure that the computer is also properly grounded. The high voltage symbols are a reminder that all parts of the dc discharge and the probe circuit must be properly insulated and must not be touched during operation.

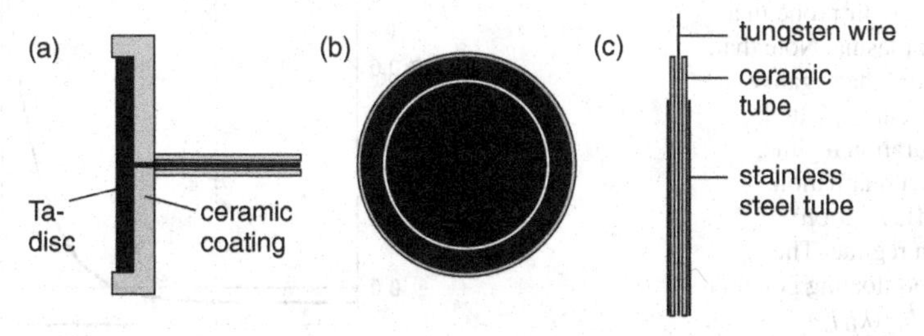

Fig. 7.4 (**a**) Design of a simple plane probe. (**b**) Plane probe with guard ring. (**c**) Construction of a cylindrical probe

A plane probe can be made of a small tantalum disk of (2–3) mm diameter with a fine wire welded to the back. The wire is fed through a ceramic tube, see Fig. 7.4a. A layer of ceramic cement gives an isolating coating on the backside of the probe and fixes it to the ceramic tube.

For space applications, the probe can be made larger. Then, it is advisable to provide the probe with a guard ring that is held at probe potential, see Fig. 7.4b. The probe current is only measured for the central disk. The guard ring reduces edge effects that would lead to an increase of the saturation current with probe bias. A cylindrical probe is sketched in Fig. 7.4c. In order to keep the pertubation of a (low-density) plasma small, the probe construction can use a thin-walled stainless-steel tube of $\approx$ 1 mm diameter, such as those made for hypodermic needles. The steel tube gives mechanical strength and shielding for a ceramic tube that acts as isolation for the probe wire, which has (50–100) μm thickness and (5–20) mm exposed length, depending on the application.

A typical I_p–U_p characteristic of a plane Langmuir probe is shown in Fig. 7.5. The characteristic can be subdivided into three regimes. At high negative bias (region I), no electrons reach the probe and a constant ion saturation current is extracted from the plasma. At high positive bias (region III), a constant electron saturation current is found. The magnitude of the electron saturation current is much higher than the ion saturation current, which will be explained below. In the intermediate region II, called the electron retardation regime, part of the electrons can overcome the energy barrier and reach the probe. According to the Boltzmann factor, the electron current increases exponentially with the bias voltage.

There are two points of specific physical significance on the characteristic. The probe potential, at which no net current flows in the probe circuit, is identical with that of a floating piece of metal in the plasma. This is the *floating potential* Φ_f of the probe. The boundary between electron retardation and electron saturation defines the *plasma potential* Φ_p, i.e., the potential inside the ambient plasma, which is usually as zero-reference, $\Phi_p = 0$.

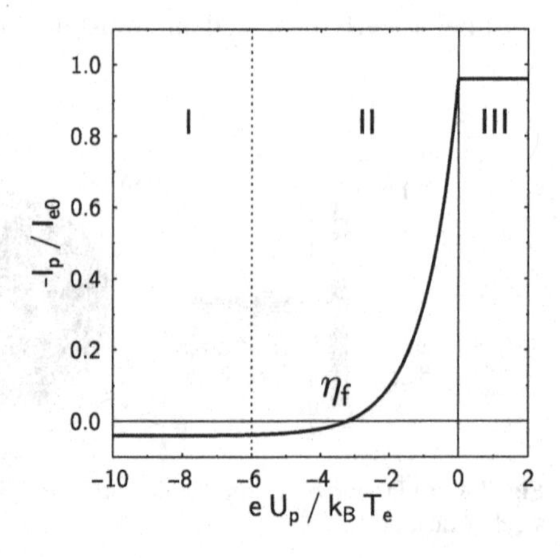

Fig. 7.5 Characteristic of a plane Langmuir probe in a hydrogen plasma. Note that, by tradition, the negative probe current is plotted. I: ion saturation regime, II: electron retardation regime, III: electron saturation regime. The normalized floating potential is $\eta_f = e\Phi_f/(k_B T_e) \approx -3.3$. The plasma potential is at $U_p = 0$

7.4.1 The Ion Saturation Current

According to the Bohm criterion, the ion current flowing into the sheath is defined by the conditions at the sheath edge and is independent of the voltage drop inside the sheath. The ion flow is strictly perpendicular to the probe surface. Therefore, the ion saturation current for a probe of surface area A reads

$$I_{\mathrm{i,sat}} = 0.61 n_{\mathrm{i}0} e v_{\mathrm{B}} A = 0.61 n_{\mathrm{i}0} e \sqrt{\frac{k_{\mathrm{B}} T_{\mathrm{e}}}{m_{\mathrm{i}}}} A \,. \tag{7.27}$$

Note that the ion saturation current depends on the ion density and the *electron* temperature When the electron temperature is known, this formula can be used to derive the ion density $n_{\mathrm{i}0}$ of the unperturbed plasma.

7.4.2 The Electron Saturation Current

The electron current at the plasma potential is the electron saturation current. This situation is different from the case of the ion saturation current. The ions in a discharge plasma with a temperature $T_{\mathrm{i}} \ll T_{\mathrm{e}}$ have gained energy in the presheath and enter the sheath as a nearly monoenergetic group with a directed velocity. At the plasma potential, however, there is no sheath formation. Rather, the Maxwellian electrons of the plasma are no longer hindered to reach the probe. Therefore, all electrons with a starting velocity directed towards the probe will reach the probe. However, each electron that has a velocity vector inclined by an angle θ from the normal to the probe surface, will only contribute with its perpendicular velocity, $v_{\perp} = v_{\mathrm{e}} \cos\theta$, to the probe current.

Because the Maxwell distribution of the electrons is isotropic, we can first integrate over the magnitude of the velocities, yielding the representative velocity as the mean thermal velocity $v_{\mathrm{th,e}}$. It remains to do the proper angular average. The fraction of electrons in the angular range between θ and $\theta + \mathrm{d}\theta$ is only determined by the geometry

$$\frac{\mathrm{d}n_{\mathrm{e}}}{n_{\mathrm{e}0}} = \frac{2\pi \sin\theta \, \mathrm{d}\theta}{4\pi} \,. \tag{7.28}$$

Hence, the angular integration of the current contributions over the halfspace of positive normal velocities becomes

$$\begin{aligned} I_{\mathrm{e,sat}} &= -A\, e \int_{\mathrm{halfspace}} v_{\mathrm{th,e}} \cos\theta \, \mathrm{d}n_{\mathrm{e}} = -\frac{1}{2} A\, e\, n_{\mathrm{e}0}\, v_{th,e} \int_0^{\pi/2} \cos\theta \, \sin\theta \, \mathrm{d}\theta \\ &= -\frac{1}{4} A\, e\, n_{\mathrm{e}0}\, v_{\mathrm{th,e}} = -\frac{1}{4} A\, e\, n_{\mathrm{e}0} \sqrt{\frac{8}{\pi} \frac{k_{\mathrm{B}} T_{\mathrm{e}}}{m_{\mathrm{e}}}} \,. \end{aligned} \tag{7.29}$$

We finally see that the electron saturation current is proportional to the product of the electron density and the square root of the electron temperature. Comparing electron and ion saturation curent, we find

$$\frac{|I_{\mathrm{e,sat}}|}{I_{\mathrm{i,sat}}} = \frac{0.25}{0.61}\sqrt{\frac{8m_{\mathrm{i}}}{\pi m_{\mathrm{e}}}} = 0.65\sqrt{\frac{m_{\mathrm{i}}}{m_{\mathrm{e}}}} \tag{7.30}$$

which explains why, in an argon plasma, the electron saturation is 177 times the ion saturation current.

It is often found in probe measurements that the results for the electron density derived from the electron current and ion density from the ion regime do not agree. This is not a hint at a violation of the quasineutrality. Rather, it shows that some assumptions made in probe theory, e.g., ion collisions or the effective probe geometry, are not properly taken care of.

7.4.3 The Electron Retardation Current

Without proof (which can be found in Sect. 7.5.1), we can state that the electron current in the electron retardation regime is determined by the saturation current multiplied by the Boltzmann factor

$$I_{\mathrm{e}}(U) = I_{\mathrm{e,sat}} \exp\left(\frac{e(U - \Phi_{\mathrm{p}})}{k_{\mathrm{B}} T_{\mathrm{e}}}\right) . \tag{7.31}$$

This exponential increase with U can be used for determining the electron temperature. The electron current can be retrieved by subtracting the ion saturation current from the probe current in the retardation region. When we now plot the logarithm of the electron current vs. the probe bias voltage,

$$\ln\left(\frac{|I_{\mathrm{e}}(U)|}{\mathrm{mA}}\right) = \ln\left(\frac{|I_{\mathrm{e,sat}}|}{\mathrm{mA}}\right) + \frac{e(U - \Phi_{\mathrm{p}})}{k_{\mathrm{B}} T_{\mathrm{e}}} , \tag{7.32}$$

we obtain a straight line with slope $e(k_{\mathrm{B}} T_{\mathrm{e}})^{-1}$.

An example for the determination of the electron temperature is shown in Fig. 7.6. This characteristic was obtained with a cylindrical probe, but the probe geometry does not affect the electron saturation regime. However, the different geometry is the reason why the currents in the electron and ion saturation regime increase with applied voltage, as will be discussed in Sect. 7.5.4. The plasma potential is found at the inflection point of the characteristic. The evaluation starts with fitting a model function $I_{\mathrm{i}}(U_{\mathrm{p}})$, such as given by (7.55), to the ion saturation current in the regime $\Phi_{\mathrm{p}} - U_{\mathrm{p}} > 5k_{\mathrm{B}} T_{\mathrm{e}}/e$, see Fig. 7.6a. Then, the electron retardation current is given as $I_{\mathrm{e}} = I_{\mathrm{p}} + I_{\mathrm{i}}$. This electron current $I_{\mathrm{e}}(U_{\mathrm{p}})$ is shown in a semi-log plot in Fig. 7.6b. Between -10 and 0 V, the logarithm of the electron retardation

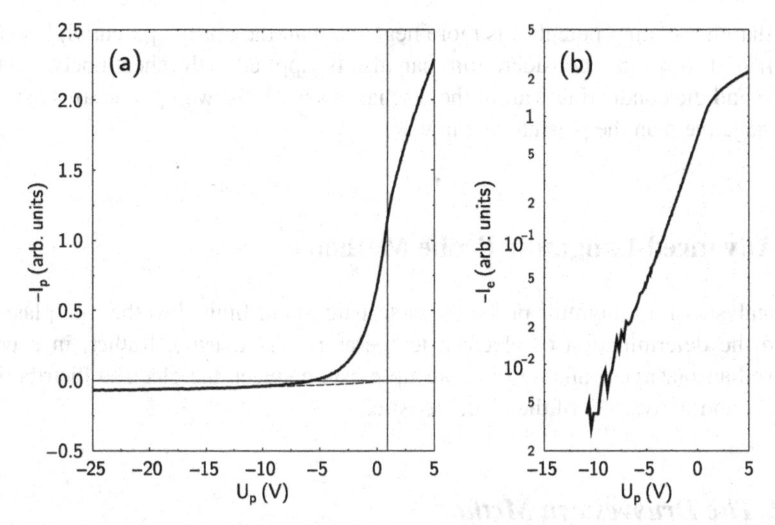

Fig. 7.6 (**a**) Linear plot of $-I_p(U_p)$. A fit to the ion saturation current is shown as *dashed line*. The vertical line marks the plasma potential. (**b**) A log-lin plot of the (negative) electron current vs. probe voltage shows that a Maxwellian results in a *straight line*, which can be used to determine the electron temperature

current shows the expected linear increase with probe bias. This Maxwellian part of the probe characteristic extends over more than two decades in probe current.

The electron temperature can be quickly determined from two points in this exponential part of the characteristic, e.g., (−3.42 V / 0.1 mA) and (−0.64 V / 1 mA). Using (7.31) and $\Delta U_p = (3.42\text{–}0.64)$ V, this gives the electron temperature in volts

$$\frac{k_B T_e}{e} = \frac{\Delta U_p}{\ln(10)} = 1.21\,\text{V}\,. \tag{7.33}$$

More accurate values can be obtained by fitting an exponential to the electron current. Note that a least-square fit to the logarithmic data can be misleading, because noisy (or distorted) data at low electron currents would have the same statistical weight as smooth data at higher currents.

7.4.4 The Floating Potential

The floating potential is defined by the vanishing of the probe current $I_p = 0$ and depends on the electron temperature and ion-to-electron mass ratio (cf. Problem 7.1)

$$\Phi_f - \Phi_p = \frac{k_B T_e}{e} \ln\left[0.61(2\pi)^{1/2}\left(\frac{m_e}{m_i}\right)^{1/2}\right]. \tag{7.34}$$

Note that the floating potential is more negative than the plasma potential, because $m_e/m_i \ll 1$. Since our considerations can also be applied to the sheath between the plasma and the conducting wall of the discharge vessel, the wall potential is usually more negative than the plasma potential.

7.5 Advanced Langmuir Probe Methods

The analysis of a Langmuir probe characteristic is not limited to thermal plasmas and to the determination of electron temperature and density. Rather, in a non-Maxwellian plasma, a unique relationship exists between the electron distribution function and derivatives of the characteristic.

7.5.1 The Druyvesteyn Method

In gas discharges, the real distribution function may significantly deviate from a Maxwellian. Druyvesteyn [146], in 1930, introduced a method to derive the electron distribution function from the second derivative $\mathrm{d}^2 I_p/\mathrm{d}U_p^2$ of the probe characteristic. His method can be derived from our previous considerations, as follows: Let us define the z-direction as the normal to the probe surface. In the case of electron retardation, only those electrons can reach the probe surface that have a sufficiently large z-component to overcome the potential barrier

$$\frac{m}{2} v_z^2 > eU_p \,, \tag{7.35}$$

where $-U_p$ is the probe potential with respect to the plasma. For a given magnitude of the velocity v of an electron, this defines a restriction for the maximum angle θ with respect to the probe normal given by

$$\frac{m}{2} (v \cos\theta)^2 > eU_p \,. \tag{7.36}$$

The electron retardation current to a plane probe for a given probe bias U_p can be obtained by summing up all contributions in spherical coordinates and taking care of the restrictions for a minimum velocity $v_{\min}$ and the maximum angle $\theta(v)$

$$j_e = -e \int_0^{2\pi} \mathrm{d}\varphi \int_{v_{\min}}^{\infty} v^2 \mathrm{d}v \int_0^{\theta(v)} [f(v,\theta)\, v \,\cos\theta] \sin\theta \,\mathrm{d}\theta \,. \tag{7.37}$$

Here, $v_{\min} = (2eU_p/m)^{1/2}$ is the minimum velocity and $\theta(v) = \arccos(v_{\min}/v)$ the maximum angle in the integration. The Druyvesteyn method requires that the electron distribution function is *isotropic*. For this case, it is useful to replace the velocity unit by equivalent kinetic energies expressed in volt units

$$\tilde{f}(U) = f\left(\sqrt{2eU/m}\right) . \tag{7.38}$$

Substituting $U = mv^2/2e$ and $\mathrm{d}U = (m/e)v\mathrm{d}v$, (7.37) reads

$$\begin{aligned} j_\mathrm{e} &= -2\pi e \int_{v_\mathrm{min}}^{\infty} v^3 \mathrm{d}v f(v) \int_0^{\theta(v)} \sin\theta\cos\theta\,\mathrm{d}\theta \\ &= -\frac{4\pi e^3}{m^2} \int_{U_\mathrm{p}}^{\infty} \mathrm{d}U U \tilde{f}(U) \frac{1}{2}\left(1 - \frac{U_\mathrm{p}}{U}\right) \\ &= -2\pi \frac{e^3}{m^2} \int_{U_\mathrm{p}}^{\infty} (U - U_\mathrm{p}) \tilde{f}(U)\,dU . \end{aligned} \tag{7.39}$$

The electron distribution function, which appears in the integrand, can be recovered by differentiating the probe current twice with respect to the applied voltage. For this purpose, we have to apply the following rule in the first step

$$\frac{\mathrm{d}}{\mathrm{d}y} \int_{\alpha(y)}^{\beta(y)} f(x, y)\mathrm{d}x = \int_{\alpha(y)}^{\beta(y)} \frac{\partial f(x, y)}{\partial y} \mathrm{d}x + \beta'(y) f[\beta(y), y] - \alpha'(y) f[\alpha(y), y] \tag{7.40}$$

and obtain

$$\frac{\mathrm{d}j_e}{\mathrm{d}U_\mathrm{p}} = -2\pi \frac{e^3}{m^2} \left\{ \int_{U_\mathrm{p}}^{\infty} [-\tilde{f}(U)]\mathrm{d}U - \underbrace{(U_\mathrm{p} - U_\mathrm{p})}_{=0} \tilde{f}(U_\mathrm{p}) \right\} , \tag{7.41}$$

in which the contribution from the lower integration boundary cancels. In the second derivative we obtain

$$\frac{\mathrm{d}^2 j_e}{\mathrm{d}U_\mathrm{p}^2} = -2\pi \frac{e^3}{m^2} \tilde{f}(U_\mathrm{p}) . \tag{7.42}$$

In this way, the second derivative of the probe characteristic represents the electron probability distribution on the volt scale. However, this is not yet the velocity distribution, because the individual intervals in volt units cover quite different velocity segments. Therefore, we must remove the nonlinear distortion originating from the volt scale. This can be done by observing that

$$f(v) = \tilde{f}(U)\frac{\mathrm{d}U}{\mathrm{d}v} = \tilde{f}(U)\frac{mv}{e} = \tilde{f}(U)\left(2\frac{m}{e}U\right)^{1/2} . \tag{7.43}$$

Therefore, the second derivative of the characteristic must be multiplied by the factor $(2mU_\mathrm{p}/e)^{1/2}$ to obtain the proper velocity distribution.

In a magnetized plasma, in which the electron flow is one-dimensional along the magnetic field line, the distribution function can be derived from the first derivative of the probe characteristic (cf. Problem 7.2).

7.5.2 A Practical Realization of the Druyvesteyn Technique

The straightforward application of the Druyvesteyn method would employ calculating two numerical derivatives. For data superimposed by noise, this would dramatically enhance the noise. This is why many plasma scientists apply smoothing of the data before taking the derivatives.

A different method of noise reduction is based on modulation techniques and narrow-band detection of Fourier components. The idea of the *second harmonic probe* or the *two frequency method* is to use the curvature of the probe characteristic as a nonlinear element that produces harmonics or combination frequencies. Let us assume that the probe voltage U_dc is superimposed by two sine voltages of frequencies ω_1 and ω_2 and small amplitudes U_1 and U_2. Then the probe current can be expanded into a Taylor series about U_dc, and the addition theorems for sines and cosines give the resulting spectral components

$$\begin{aligned} I_\mathrm{e} &= I_\mathrm{e}(U_\mathrm{dc} + U_1 \sin\omega_1 t + \sin\omega_2 t) \\ &= I_\mathrm{e}(U_\mathrm{dc}) + \left.\frac{\mathrm{d}I_e}{\mathrm{d}U_\mathrm{p}}\right|_{U_\mathrm{dc}} \Big[U_1 \sin\omega_1 t + U_2 \sin\omega_2 t\Big] \\ &\quad + \left.\frac{1}{2}\frac{\mathrm{d}^2 I_\mathrm{e}}{\mathrm{d}U_\mathrm{p}^2}\right|_{U_\mathrm{dc}} \left[\frac{1}{2}\Big(U_1^2 + U_2^2\Big) - \frac{1}{2}\Big(U_1^2 \cos 2\omega_1 t + U_2^2 \cos 2\omega_2 t\Big)\right. \\ &\quad \left. + U_1 U_2\Big(\cos(\omega_1 - \omega_2)t - \cos(\omega_1 + \omega_2)t\Big)\right] . \end{aligned} \tag{7.44}$$

The probe current contains Fourier components at $2\omega_1$ and $2\omega_2$ as well as combination frequencies $\omega_1 - \omega_2$ and $\omega_1 + \omega_2$ that are proportional to the second derivative of the probe characteristic. When only a single modulation voltage is applied, detecting the second harmonic at $2\omega_1$ gives the desired distribution function. The drawback of the method is that any other nonlinearity in the electric circuit, which applies the modulation voltage or detects the current, also produces frequency components at $2\omega_1$, and limits the dynamic range of the method. This is often a problem when the applied signal at ω_1 leaks through the narrow-band filter centered at $2\omega_1$.

7.5.3 Double Probes

The Langmuir probe method, as described above, requires that the electric circuit is closed through a suitably large reference electrode, often one of the discharge electrodes. There are cases, where such a reference electrode is missing, as in inductively generated rf discharges, or on a satellite, where it may be forbidden to use the satellite body as a reference electrode. For these purposes, the double-probe method was introduced [147] in which two identical Langmuir probes are operated in series. The principle of the probe circuit is shown in Fig. 7.7a. The probes 1 and 2 are connected by a (battery operated) floating voltage source U_p and a current measuring instrument that yields I_p.

Both probes are operating near the floating potential Φ_f, as can be seen from Fig. 7.7b. Note that the ratio of electron saturation current and ion saturation current was set to an artificial value, for better readability. The maximum current, which can flow in the probe circuit is the ion-saturation current. The other probe is then operating in the electron retardation region. Let us write the two probe characteristics as

$$I_p = I_{i0} + I_{e0} \exp\left(\frac{e(U_1 - \Phi_p)}{k_B T_e}\right) \tag{7.45}$$

$$-I_p = I_{i0} + I_{e0} \exp\left(\frac{e(U_2 - \Phi_p)}{k_B T_e}\right) . \tag{7.46}$$

Here, we have used the current continuity, which requires that the current drawn by probe 2 must be supplied by probe 1. These symmetrical current values $\pm I_p$ are marked as horizontal dotted lines in Fig. 7.7b. Further, we require $U_p = U_2 - U_1$. By eliminating U_1 and U_2 from (7.45) and (7.46), one easily obtains (cf. problem 7.3) the double probe characteristic

$$I_p = I_{i0} \tanh\left(\frac{eU_p}{2k_B T_e}\right) , \tag{7.47}$$

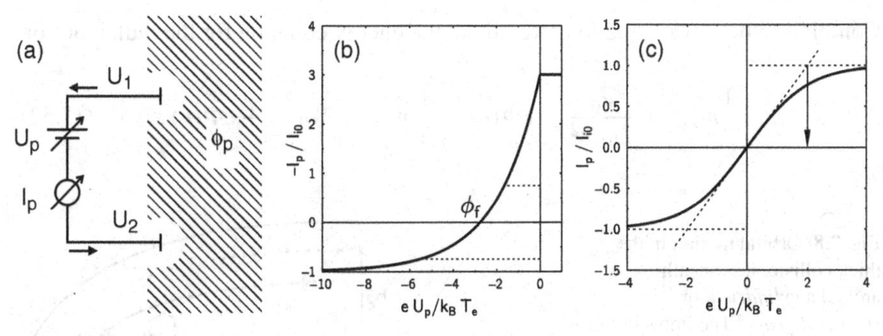

Fig. 7.7 (**a**) Circuit of the double probe method. (**b**) Points of operation of the individual probes. (**c**) Characteristic of a double probe. The asymptotes intersect at $eU_p = 2k_B T_e$

which is shown in Fig. 7.7c. The probe current reaches the ion-saturation current I_{i0} for large positive or negative bias. Therefore, the double probe can measure plasma density, when the electron temperature is known. The electron temperature can be derived from the slope of the characteristic at the origin (cf. Problem 7.4).

7.5.4 Orbital Motion about Cylindrical and Spherical Probes

The ion motion in a thick sheath, $\lambda_{De} \gg a$, around a small sphere or thin wire of diameter a with negative bias leads to an increase of the effective probe area that exceeds the geometrical probe area. The geometry of this problem is sketched in Fig. 7.8. Far away from the probe, the ion has a velocity v_0. When collisions in the sheath are rare, the ion motion can be described by the conservation of energy and angular momentum, as in celestial mechanics. Therefore, this model is named the *Orbital Motion Limit* (OML) of probe theory [148]. There are orbits with $b < b_c$, which hit the probe and contribute to the probe current. b_c is the critical impact parameter for grazing collisions. Orbits with $b > b_c$ do not contribute to the ion current but provide space charge for shielding. In Chap. 10 we will see that all trajectories contribute to momentum transfer to a small spherical dust grain.

The ion energy and angular momentum at large distance are

$$W_0 = \frac{1}{2} m_i v_0^2 \tag{7.48}$$

$$J_0 = m_i v_0 b \,, \tag{7.49}$$

where b is the impact parameter. Then the conservation of energy and angular momentum can be written as

$$W_0 = \frac{1}{2} m_i (v_r^2 + r^2 \dot{\theta}^2) + e\Phi(r) \tag{7.50}$$

$$J_0 = m_i r^2 \dot{\theta} \,. \tag{7.51}$$

Combining these two equations we obtain the energy equation for the radial motion

$$W_0 = \frac{1}{2} m_i v_r^2 + \frac{J_0^2}{2 m_i r^2} + e\Phi(r) = \frac{1}{2} m_i v_r^2 + W_0 \frac{b^2}{r^2} + e\Phi(r) \,, \tag{7.52}$$

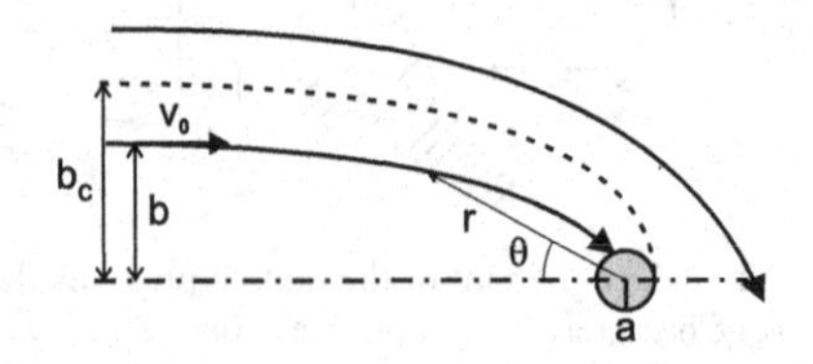

Fig. 7.8 Orbital motion in the thick collisionless sheath around a *cylindrical* or *spherical* probe. The impact parameter b_c determines the effective probe cross section

which can be rearranged into an expression for the impact parameter

$$b = r\left[1 - \frac{e\Phi(r)}{W_0} - \frac{m_i v_r^2}{2W_0}\right]^{1/2} . \tag{7.53}$$

The critical impact parameter b_c is defined by a grazing collision, $v_r = 0$ at $r = a$

$$b_c = a\left[1 - \frac{e\Phi(a)}{W_0}\right]^{1/2} \tag{7.54}$$

and depends only on the potential energy at the surface $r = a$. Because the potential is attractive, we have $b_c > a$. This means that the probe area becomes effectively larger than the geometrical cross section. Let us denominate b_c/a as the *OML factor*. Then this factor has to be applied to the ion current of a cylindrical probe. For a spherical probe, the effective cross section is a circle πb_c^2 that replaces the geometrical cross section πa^2 and the square of the OML factor must be applied.

Replacing $W_0 \approx k_B T_i$, we can estimate the shape of the ion saturation current of a cylindrical and a spherical probe, as shown schematically in Fig. 7.9. The arguments for the electron saturation current are similar and give the same functional dependence with $W_0 \approx k_B T_e$.

In cylindrical (or spherical) probe geometry, the attractive regime gives no longer a constant current. We define the electron saturation current $j_{e,sat}$ as the value of the electron current at the plasma potential ($\Phi_p = 0$). Then the electron current in the attractive region of a cylindrical or spherical probe has the shape

$$j_{e,cyl}(U) = j_{e,sat}\left(1 + \frac{eU}{k_B T_e}\right)^{1/2} \tag{7.55}$$

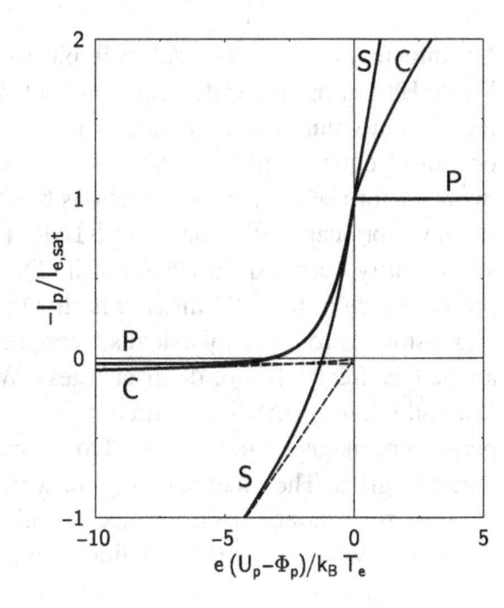

Fig. 7.9 Characteristics of *cylindrical* (C) and *spherical* probes (S) from the OML model compared to a *plane* probe (P) for a hydrogen plasma with $T_e/T_i = 100$. The dashed lines give the ion saturation currents for cylindrical and spherical probes

$$j_{\mathrm{e,sph}}(U) = j_{\mathrm{e,sat}}\left(1 + \frac{eU}{k_{\mathrm{B}}T_{\mathrm{e}}}\right) . \tag{7.56}$$

For a cylindrical probe, the plasma potential is defined by the inflection point of the characteristic ($\mathrm{d}^2 I_{\mathrm{p}}/\mathrm{d}U_{\mathrm{p}}^2 = 0$). For a spherical probe, there is no inflection point, and the plasma potential, which is needed to read the electron saturation current, can only be estimated by calculating the distance between plasma potential and floating potential and fixing the plasma potential in this way in the measured curve.

There is a first caveat with the simple OML model presented here. The reader will certainly be puzzled that we have not used a Bohm criterion for the ion current, which leads to a scaling of the ion current with the electron temperature. Rather, the ion OML-factor is based on the ion temperature. This is a side effect from using the thick sheath approximation, which is the limit $\lambda_{\mathrm{De}} \to \infty$. Hence, the sheath edge is at infinity and the whole volume of interest is a sheath region. In practice, the limit of validity of the OML model is already reached at a distance from the probe of the order of the ion mean free path. Therefore, the OML model describes the orbital motion during the last mean free path. Moreover, the OML model is consistent because elastic or charge-exchange collisions with the neutral gas would "cool" the ions from Bohm energy to gas (ion) temperature. The observed increase of the ion collection current with applied probe bias is the typical feature of orbital motion.

There is a second caveat. Cylindrical probes of finite length behave like spherical probes as soon as the sheath diameter becomes comparable with the probe length. For a deeper understanding of Langmuir probes, the reader should inspect more specialized literature, e.g., [148–151].

7.6 Application: Ion Extraction From Plasmas

Ion thrusters are the classical example for efficient ion extraction from a plasma. While Hall thrusters, as described in Sect. 4.3.5, perform the ion acceleration inside the plasma volume, other concepts use a pair of grids that are in contact with a dc or radio frequency plasma. NASA has developed a xenon ion thruster with 30 cm diameter for use in planetary missions by the NASA Solar electric propulsion Technology Application Readiness (NSTAR) program [152]. The NSTAR engine was successfully operated, in 1998, on the Deep Space 1 (DS1) mission [153]. A simplified sketch of the DS1 thruster is shown in Fig. 7.10.

Plasma is produced by a dc discharge between a hollow cathode and a large-area anode (see Sect. 11.1 for dc discharges). When operated at low gas pressure, a hollow cathode is an efficient source of electrons. Plasma confinement is enhanced by permanent magnets (not shown). On the exhaust side, the plasma is in contact with a pair of grids. The inner screen grid, which has the same potential as the cathode, confines the primary electrons electrostatically and terminates the plasma. Between the anode, which essentially defines the plasma potential, and the electron source

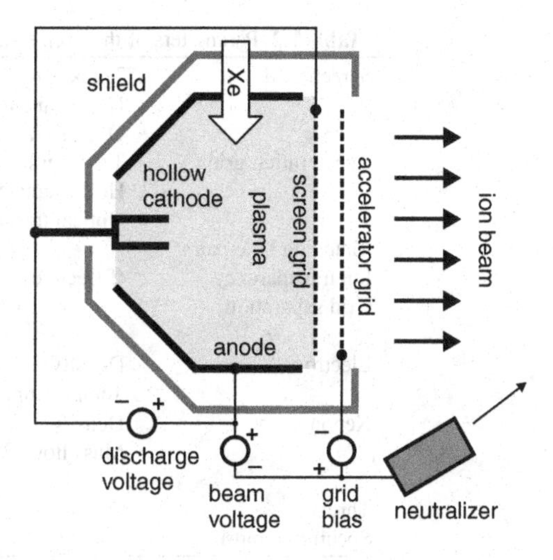

Fig. 7.10 Sketch of the 30 cm diameter ion thruster for the Deep Space 1 mission (cross section). Ion acceleration occurs between the screen grid and accelerator grid. The device reaches ≈ 90 mN thrust and a specific impulse of 3,200 s. (Reprinted with permission from [153]. © 2002, American Institute of Physics)

that acts as neutralizer, a high acceleration voltage of typically 1,000 V is applied. The accelerator grid is held at about −200 V w.r.t. to the neutralizer to prevent the electron beam to enter the discharge region. The total voltage difference of 1,200 V between the two grids is used for ion acceleration. An outer shield prevents that electrons are accelerated towards the anode.

A few words on the language of rocketry: Thrust is the reaction force from the ion beam leaving the thruster with exhaust speed v_{ex} at a mass-flow rate dm/dt,

$$F_{\mathrm{t}} = \frac{\mathrm{d}m}{\mathrm{d}t} v_{\mathrm{ex}} \,. \tag{7.57}$$

Specific impulse is a measure for the fuel efficiency, and is defined by the change in momentum per unit weight on Earth, i.e.,

$$I_{\mathrm{sp}} = \frac{F_{\mathrm{t}}}{g\,\mathrm{d}m/\mathrm{d}t} = \frac{v_{\mathrm{ex}}}{g} \,, \tag{7.58}$$

where g is the gravitational acceleration at the Earth surface. Chemical propellants reach a specific impulse of (250–450) s, typically.

The parameters of the DS1 thruster are compiled in Table 7.2. The holes in the acceleration grid are much smaller than those in the screen grid to reduce the loss of neutral propellant gas. The hole combinations act as ion lenses and yield a transparency for ions of up to 83%.

To apply our basic knowledge of space charge limited ion-flow, we can try some reverse engineering of the DS1 engine. From the Child-Langmuir law (7.11) and setting $m_{\mathrm{i}} \approx 131 m_{\mathrm{p}}$, $U = 1200\,\mathrm{V}$ and $d = 0.66\,\mathrm{mm}$, we obtain a maximum space-charge limited ion current density of $460\,\mathrm{A\,m^{-2}}$. Taking 15,000 holes of $2.8\,\mathrm{mm}^2$ each, yields a maximum current of 20 A. The actual maximum ion beam current of

Table 7.2 Parameters of the deep space 1 thruster [152, 153]

Screen grid	Thickness	0.38 mm
	Hole diameter	1.91 mm
	Optical transparency	67%
Acceleration grid	Thickness	0.51 mm
	Hole diameter	1.14 mm
	Optical transparency	24%
Matching hole pairs		15,000
Ion transparency	(Effective)	83%
Grid separation		0.66 mm
Electron	Density	$10^{17}\,\mathrm{m}^{-3}$
	Temperature	5 eV
Xenon	Density	$10^{18}\,\mathrm{m}^{-3}$
	Mass flow	$\leq$ 23 sccm
Thrust		$\leq$ 92 mN
Specific impulse		$\approx$ 3,200 s

the DS1 thruster was 1.8 A. This is a hint that the ion current is not limited by space charge in the grid region but is determined by the ionization rate in the plasma. The theoretical maximum current that can leave a plasma is the Bohm current $j_{\mathrm{B}} = 0.61 n_{\mathrm{i}} e (k_{\mathrm{B}} T_{\mathrm{e}}/m_{\mathrm{i}})^{1/2} \approx 190\,\mathrm{A\,m}^{-2}$, which would correspond to a total current of 8.4 A entering the grid openings. This is a further hint at a reduction of the electron density in the region before the screen grid that results from the limited ionization rate.

7.7 Double Layers

We have learnt in Sect. 7.3 that a collisionless plasma can only deliver a maximum ion current, the Bohm current, which initiates the formation of a space charge sheath. On the other hand, this plasma can carry a much higher electron current. Langmuir [154] had found that sudden potential jumps form inside the plasma volume, mostly close to constrictions of the diameter of his discharge tubes. He called these structures *double layers* (DLs).

Double layers are found in laboratory and astrophysical plasmas. Raadu had [155] pointed out that, because particles are accelerated by the net potential difference Φ_0 of a DL, the DL acts as an electric load dissipating energy at a rate $I\,\Phi_0$, where I is the total current through the DL. In this way, a DL exhibits an internal resistance. The nature of this resistance is quite different from an ohmic resistance, which transforms electric energy to random motion. Rather, a DL resembles an old-fashioned television tube, in which an electron beam is generated that produces the moving luminous dot on the screen. Since the DL acts as a load, there has to be an external source that maintains the potential difference. In the laboratory, this is an electrical power supply, whereas in space it may be the magnetic energy stored in an extended current system, which responds to a change in current with an inductive voltage.

7.7.1 Langmuir's Strong Double Layer

The geometry of a double layer (DL) is sketched in Fig. 7.11. The potential step is caused by a positive space charge, which causes the negative curvature and a negative space charge responsible for the positive curvature of the electric potential. The system is assumed one-dimensional with variation in x-direction only.

Let us assume that ions enter the system at $x = 0$ and at an initial potential $\Phi = \Phi_0$ with a current density $j_i = en_{i0}v_{i0}$ and are accelerated by the DL potential $\Phi(x)$. Electrons enter at the low-potential side $x = L$ with a negative velocity (but positive current density) $j_e = -en_{e0}v_{e0}$. By continuity, each of these currents is conserved. From energy conservation, the instantaneous velocities of electrons and ions during the transit are

$$v_i = \sqrt{v_{i0}^2 + 2e(\Phi_0 - \Phi)/m_i}$$

$$v_e = -\sqrt{v_{e0}^2 + 2e\Phi/m_e}\,, \tag{7.59}$$

in which the initial velocities have been introduced to avoid singularities at $x = 0$ and $x = L$. These initial velocities can be neglected later for sufficiently high values of the potential step $\Phi_0 \gg k_B T_e/e$. This limiting case defines a *strong DL*. Weak DLs, which require a description by kinetic theory, are discussed in [155, 156]. DL experiments were reviewed in [157, 158]. Using the conservation of the currents, we can replace the densities in Poisson's equation by velocities

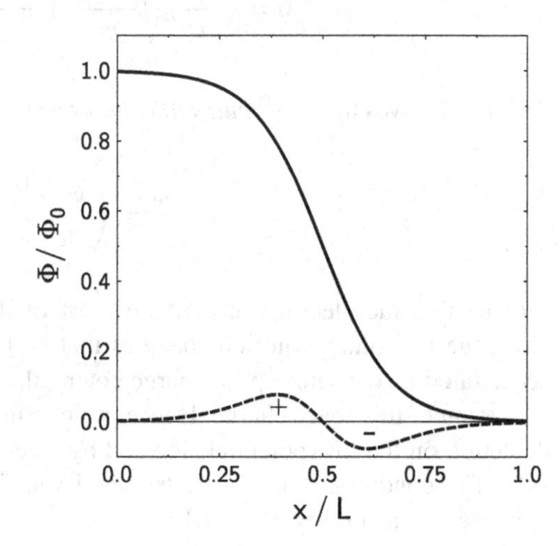

Fig. 7.11 The potential distribution in a strong double layer (*full line*) and the associated space charge in arbitrary units (*dashed line*)

$$\varepsilon_0 \Phi'' = -\frac{j_\mathrm{i}}{\left(v_{\mathrm{i}0}^2 + 2e(\Phi_0 - \Phi)/m_\mathrm{i}\right)^{1/2}} + \frac{j_\mathrm{e}}{\left(v_{\mathrm{e}0}^2 + 2e\Phi/m_\mathrm{e}\right)^{1/2}} . \quad (7.60)$$

Multiplying by Φ' and integrating from 0 to x we obtain

$$\frac{\varepsilon_0}{2}\left[\Phi'(x)^2 - \Phi'(0)^2\right] = \frac{m_\mathrm{i}}{e} j_\mathrm{i} \left[\left(v_{\mathrm{i}0}^2 + \frac{2e(\Phi_0 - \Phi)}{m_\mathrm{i}}\right)^{1/2} - \left(v_{\mathrm{i}0}^2 + \frac{2e\Phi_0}{m_\mathrm{i}}\right)^{1/2}\right]$$
$$+ \frac{m_\mathrm{e}}{e} j_\mathrm{e} \left[\left(v_{\mathrm{e}0}^2 + \frac{2e\Phi}{m_\mathrm{e}}\right)^{1/2} - |v_{\mathrm{e}0}|\right] . \quad (7.61)$$

The electric field must vanish at $x = 0$ and $x = L$, because we assume that the plasma is quasineutral for $x < 0$ and $x > L$. However, this quasi-neutrality cannot be established by the free streaming electrons and ions alone [156]. This becomes evident, when we require quasineutrality on the high potential side and notice that the ions become diluted by acceleration in the sheath. Using the same argument for the electrons, their density must be higher on the low-potential side and quasineutrality would be violated on the low-potential side. Therefore, additional populations of trapped particles must exist, electrons on the high-potential side and ions on the low-potential side, which establish the quasineutrality. However, these trapped (thermal) particles cannot penetrate into the sheath because of the high potential step in a strong double layer. This is why we can approximately set $\Phi'(0)^2 \approx 0$ near the edges of the strong double layer, because the square of the electric field becomes small compared to the values inside the DL. Therefore, we can rewrite (7.61) in the limit of vanishing initial velocities as

$$0 = -\frac{m_\mathrm{i}}{e} j_\mathrm{i} \sqrt{\frac{2e\Phi_0}{m_\mathrm{i}}} + \frac{m_\mathrm{e}}{e} j_\mathrm{e} \sqrt{\frac{2e\Phi_0}{m_\mathrm{e}}} . \quad (7.62)$$

This result gives the *Langmuir citerion* for a strong DL

$$\frac{j_\mathrm{e}}{j_\mathrm{i}} = \left(\frac{m_\mathrm{i}}{m_\mathrm{e}}\right)^{1/2} . \quad (7.63)$$

We see that the electrons contribute most of the current in a DL. On the other hand, the Langmuir criterion specifies that both charge carrier species, which are accelerated by the same space charge potential, contribute to the same degree in the formation of this space charge. Hence, we obtain a symmetrical space charge by ion depletion on the low-potential side and by electron depletion on the high-potential side. The condition that the electric field vanishes at $x = L$ is equivalent to the macroscopic neutrality of the DL

$$E(L) = E(0) + \frac{e}{\varepsilon_0} \int_0^L (n_i - n_e) \mathrm{d}x \,. \tag{7.64}$$

The physical meaning of the Langmuir criterion can be understood as follows: Remember that j_i/e is the (constant) ion flux density and $m_i(2e\Phi_0/m_i)^{1/2}$ is the momentum, which the ion has gained in traversing the DL. Then (7.53) states that the ion momentum flux, which leaves the DL at $x = L$, is the same as the electron momentum flux leaving at $x = 0$ [159]. This relation can be interpreted as a force equilibrium, which ensures that the DL stays at rest. On the other hand, when this condition is violated, the DL will move at a speed v_{DL}, which ensures that, in the moving frame of reference, the Langmuir criterion is again fulfilled.

We can go even further and apply the argument of momentum fluxes to the interior of the DL. Rearranging (7.61) in the limit of vanishing initial velocities yields

$$-\frac{\varepsilon_0}{2}\Phi'(x)^2 + \frac{j_i}{e} m_i \left(\frac{2e[\Phi_0 - \Phi(x)]}{m_i}\right)^{1/2} + \frac{j_e}{e} m_e \left(\frac{2e\Phi(x)}{m_e}\right)^{1/2} = \mathrm{const}\,. \tag{7.65}$$

This means that, at any place inside the DL, the sum of the negative electric Maxwell stress (which represents the tension of the electric field lines) and the particle momentum fluxes is constant. In other words, the potential is shaped by the ion and electron ram pressure like a rubber membrane.

At the end of this paragraph, a word of caution is necessary: Langmuir's strong DL is still a simplified toy model which neglects that there is a kinetic pressure of the electron and ion gas on both sides of the DL, and that part of the electron population on the high potential side as well as an ion population on the low-potential side will be transmitted by the DL.

7.7.2 Experimental Evidence of Double Layers

A detailed comparison of DL potential structure and the electron distribution function was made by Coakley and Hershkowitz [160]. The strong double layer with $e\Phi/k_B T_e = 14$ was generated between two grids that separate the three sections of a triple-plasma device, as shown in Fig. 7.12.

The voltage difference of the grids defined the DL potential. The electron distribution function was measured with a Langmuir probe. In the present situation, a directed electron beam is generated. Therefore, the beam appears in the first derivative of the probe characteristic (see Sect. 7.5.1 and Problem 7.2). The distribution functions are plotted for a sequence of positions that are approximately 2 cm apart. The dashed horizontal line marks the energy zero on the low-potential side. The solid line indicates the variation of the plasma potential.

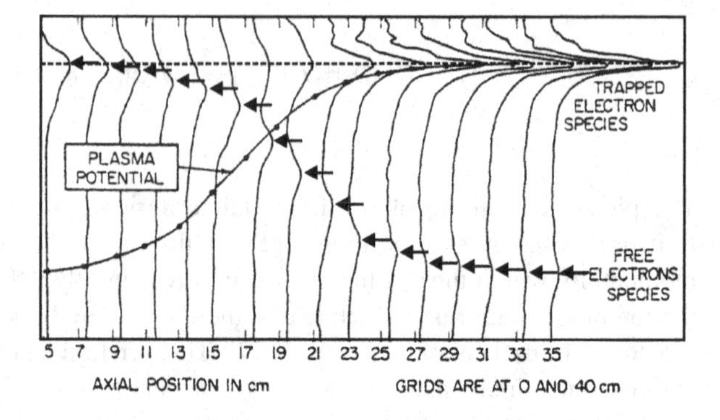

Fig. 7.12 Comparison of the potential shape in a DL and the electron energy distribution. The accelerated population is marked with *arrows*, the trapped electrons are marked with a *dashed line*. (Reprinted with permission from [160]. © 1979, American Institute of Physics)

The arrows mark a group of free electrons that becomes accelerated when going from the low-potential side (left) to the high-potential side (right) of the DL. A second group of trapped electrons is found on the high-potential side. Its position is marked by the dashed line. Note, how the height of the latter peak decreases when approaching the repulsive potential of the DL. These experimental results confirm the general description of strong double layers given in Sect. 7.7.

The Basics in a Nutshell

- The Child-Langmuir Law

$$j = \frac{4}{9}\left(\frac{2e}{m}\right)^{1/2}\frac{U^{3/2}}{d^2}$$

describes the maximum, space-charge limited current in a single-species system of length d for an applied voltage U.
- Space-charge limited currents appear in plasma sheaths and in grid regions for ion extraction.
- The Bohm criterion for a sheath, $v_\mathrm{i} = v_\mathrm{B} = (k_\mathrm{B}T_\mathrm{e}/m_\mathrm{i})^{1/2}$ states that ions must enter the sheath with ion-sound speed.
- The current–voltage characteristic of a plane Langmuir probe has the parts: ion saturation regime, electron retardation regime and electron saturation regime. The floating potential is defined by $I = 0$, the plasma potential is the transition point from electron retardation to electron saturation current.
- The ion saturation curent of a plane probe is $I_{\mathrm{i,sat}} = \exp(-1/2)env_\mathrm{B}$. The electron saturation current is $I_{\mathrm{e,sat}} = -(1/4)env_{\mathrm{th,e}}$. Both currents can be used to determine the plasma density n, when the electron temperature is known.
- The electron temperature is obtained from a semi-log plot of the electron retardation current vs. the probe voltage.
- A current-carrying collisionless plasma can spontaneously form a localized internal potential drop, called a double layer.

Problems

7.1 Derive (7.34) for the floating potential. Which values takes the floating potential for a plasma of $k_B T = 3\,\text{eV}$ containing hydrogen ions or argon ions?

7.2 In a magnetized plasma, electrons can only move freely along the magnetic field lines. Therefore, the Druyvesteyn method for a magnetized plasma can be established from a one-dimensional distribution function $f(v_z)$. Then the electron current to the probe is given by the integral

$$j_e = -e \int_{v_{min}}^{\infty} v_z f(v_z) \mathrm{d}v_z \,.$$

Show that the distribution function for a one-dimensional situation can be recovered from the *first* derivative of the probe characteristic.

7.3 Derive the double probe characteristic (7.47) by eliminating U_1 and U_2 from (7.45) and (7.46).

7.4 Show that the slope of the double probe characteristic (7.47) at the origin can be used for determining the electron temperature. Explain, why the asymptotes in Fig. 7.7c intersect at $U_p = 2k_B T_e/e$.

Chapter 8
Instabilities

"And if you take one from three hundred and sixty-five, what remains?"
"Three hundred and sixty four, of course."
Humpty-Dumpty looked doubtful. "I'd rather see that worked out on paper."

Lewis Carroll, Through the Looking-Glass

The stability of a plasma system can be analyzed by different methods. For a simple mechanical system, such as the pendulum consisting of a massless rod and a bob shown in Fig. 8.1a, stability is defined by the property that a deflection from the equilibrium position (shown in grey) leads to a restoring force $F_{\rm rest}$, which drives the pendulum back to its original position. The interplay of a restoring force, which is proportional to the deflection, and the inertia of the pendulum bob leads to harmonic oscillations.

The inverted pendulum in Fig. 8.1b has a quite different behavior. A deflection from the unstable equilibrium leads to a force $F_{\rm defl}$ that tends to increase the initial deflection. Because the deflecting force is again proportional to the deflection angle, an initial perturbation will grow exponentially in time.

Stability of a system can also be studied by analyzing its potential energy $W_{\rm pot}$ as shown in Fig. 8.1c,d. The system is stable, when the potential energy increases for any possible perturbation. Therefore, the minimum of the potential energy is such a

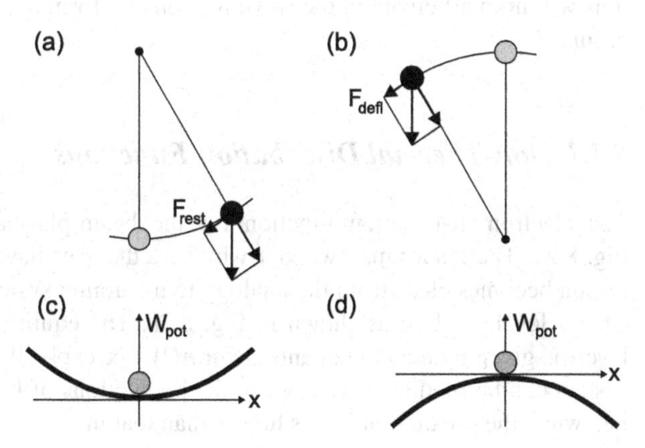

Fig. 8.1 (**a**) A classical pendulum experiences a restoring force after an initial deflection. (**b**) The inverted pendulum develops a growing deflection because the deflecting force increases with the deflection angle. (**c**) A system is stable when the potential energy takes a minimum at the equilibrium. (**d**) Instability occurs, when there is a neighboring state of lower potential energy

A. Piel, *Plasma Physics*, DOI 10.1007/978-3-642-10491-6_8,

stable point. The original equilibrium is unstable if there is any neighboring point of lower potential energy. While the calculation of restoring or deflecting forces allowed an immediate quantitative prediction of the pendulum's motion, the energy analysis does not tell immediately how the system reaches the lower energy state.

A third way of studying stability is to decompose a small initial perturbation into Fourier components, called *modes*, $\propto \exp(-i\omega t)$. If the frequencies of all modes are real, the system is stable. But if any of these Fourier modes has a complex frequency ω with a positive imaginary part, this mode will grow in time. This technique is called normal mode analysis.

In this Chapter, we will study two different classes of instabilities. The first class, named *microinstabilities*, describes homogeneous plasmas, which have a distribution function that deviates substantially from a Maxwellian. Typical members of this class are situations with a beam (electron or ion) traversing a population of plasma particles at rest. Because these systems can be treated by simple mathematical methods, we will use a beam-system to study the influence of finite length on the instability, which is of high practical interest for laboratory experiments.

The second class of *macroinstabilties* is characterized by inhomogeneity in real space. Here, we are interested in the stability of current-carrying pinch plasmas and in situations that resemble the situation of a heavy fluid resting on a lighter fluid under the influence of gravity.

8.1 Beam-Plasma Instability

It was Langmuir who noticed that oscillations at the electron plasma frequency (*Langmuir oscillations*) can spontaneously grow in a non-equilibrium plasma [154]. Such a non-equilibrium distribution function can consist of a background plasma and a group of fast electrons travelling in the same direction (*beam*). When the velocity of the beam is large compared to the thermal velocity of the background electrons, the plasma bulk can be simply described by a cold plasma model. Since we are interested in high-frequency waves near the electron plasma frequency, the ions will not participate in the wave motion and form a uniform neutralizing background.

8.1.1 Non-Thermal Distribution Functions

The electron distribution function for the beam-plasma system is sketched in Fig. 8.2a. The reader may wonder why here the axes have been interchanged. The reason becomes clear from the analogy to an atomic system with (non-degenerate) energy levels a, b, c as shown in Fig. 8.2b. The equilibrium population of these levels is given by the Boltzmann factor $n(W) \propto \exp[-W/(k_{\mathrm{B}}T)]$. Such an atomic system can be used as a laser, wenn the populations of levels b and c are inverted, i.e., when the population in c is higher than that in b.

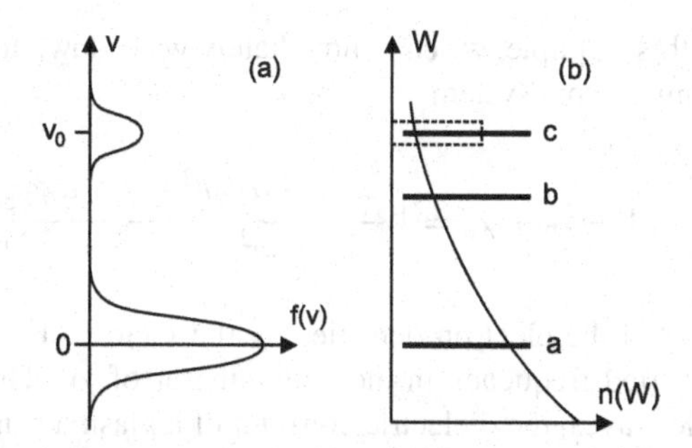

Fig. 8.2 (**a**) Electron velocity distribution $f(v)$ for the beam-plasma system. (**b**) Analogy to population inversion in an atomic laser system. The solid line gives the thermal population according to a Boltzmann factor. The dashed box indicates the overpopulation of level c w.r.t. the thermal population in level b

Consider the production rate of photons, which is given by the (negative) rate at which the upper laser niveau c is net depopulated by the competition of stimulated emission $B_{bc}n_{\mathrm{ph}}n_c$ and absorption $B_{cb}n_{\mathrm{ph}}n_b$,

$$\frac{\mathrm{d}n_{\mathrm{ph}}}{\mathrm{d}t} = -\frac{\mathrm{d}n_c}{\mathrm{d}t} = B_{bc}n_{\mathrm{ph}}n_c - B_{cb}n_{\mathrm{ph}}n_b\,. \tag{8.1}$$

Spontaneous emission processes between levels c and b can be neglected at high photon density. For non-degenerate levels, the Einstein coefficients are identical, $B_{cb} = B_{bc}$. The number of photons grows exponentially in time when $n_c > n_b$. The associated exponential growth in the photon density is the laser process. This analogy shows that instability in terms of exponentially growing waves is a consequence of the strong deviation from thermal equilibrium.

For the beam-plasma situation, let us denote the total electron density by n_{e0} with a correponding electron plasma frequency ω_{pe}. The beam population represents a fraction $n_{\mathrm{b}} = \alpha_{\mathrm{b}}n_{\mathrm{e0}}$ of the total electron population, and we will assume $\alpha_{\mathrm{b}} \ll 1$. The beam velocity is v_0. The problem is considered as one-dimensional with the beam propagating in x-direction.

8.1.2 Dispersion of the Beam-Plasma Modes

In Sect. 6.4 we had used first-order perturbation theory to derive the dielectric function of a cold plasma as

$$\varepsilon(\omega) = 1 - \frac{\omega_{\mathrm{pe}}^2}{\omega^2} = 1 + \chi_{\mathrm{p}}\,. \tag{8.2}$$

The dielectric function is the sum of the permittivity of the vacuum ("1") and the susceptibility $\chi_{\mathrm{p}}(\omega)$ of the plasma electrons. χ_{p} can also be interpreted as the ratio of the electron conduction current to the vacuum displacement current, at a given frequency ω. From this example, we can immediately write down the total dielectric function for the beam-plasma system

$$\varepsilon(\omega, k) = 1 + \chi_p + \chi_b = 1 - \frac{(1-\alpha_b)\omega_{pe}^2}{\omega^2} - \frac{\alpha_b \omega_{pe}^2}{(\omega - kv_0)^2}. \tag{8.3}$$

Here, we have adjusted the electron densities by the factors $(1 - \alpha_b)$ and α_b and used the Doppler-shifted frequency in the denominator of χ_b. The reader can see that, for electrostatic waves, the dielectric constant of a plasma can be simply composed by adding up the susceptibilities of the participating plasma constituents. For completeness, a derivation of the Doppler shift in the beam susceptibility will be given below in Sect. 8.1.4.

Electrostatic (longitudinal) waves are characterized by $\varepsilon = 0$, which defines the dispersion relation $\omega(k)$. Inspecting (8.3), we see that determining the zeroes of this equation is mathematically equivalent to solving a fourth-order polynomial with real coefficients. As stated by the *fundamental theorem of algebra*, such polynomials have either real roots or pairs of complex-conjugate roots. Therefore, we expect four independent branches of wave dispersion $\omega(k)$. If a pair of complex-conjugate roots appears, we are finished because one of these complex roots will be exponentially growing in time and defines instability.

Let us first inspect the functional dependence of the dielectric function $\varepsilon(\omega, k)$ (8.3) on the wave frequency ω, which is plotted in Fig. 8.3 for various values of α_b. The dielectric function has (negative) singularities for $\omega = 0$ and $\omega = kv_0$, as expected from the vanishing denominators of (8.3). There is one real root, $\omega/\omega_{pe} \approx -1$ that is nearly unaffected by the presence of the beam. The roots with $\omega/\omega_{pe} > 0$ show a different pattern. For the weakest beam fraction ($\alpha_b = 0.001$), there is one root close to $\omega/\omega_{pe} = +1$. We identify these two roots as the *plasma modes*. A second pair of roots is found symmetric about $\omega/\omega_{pe} = kV_0/\omega_{pe}$, which we call *beam modes*. With increasing α_b, the separation between the beam roots increases until the left beam root merges with the right plasma root. For even

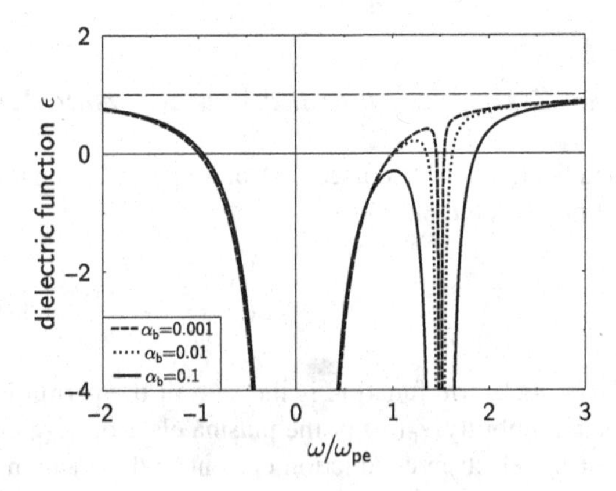

Fig. 8.3 The dielectric function of the beam-plasma system (8.3) for $kv_0/\omega_{pe} = 1.5$. For $\alpha_b = 0.001$ and $\alpha_b = 0.01$ there are four real roots. A pair of complex roots is expected for $\alpha_b = 0.01$

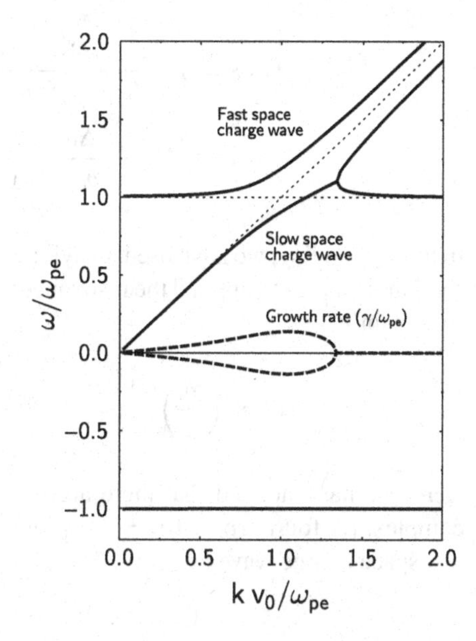

Fig. 8.4 The dispersion relation for the beam-plasma modes at $\alpha_b = 0.01$. The dotted lines mark the asymptotes $\omega = \omega_{pe}$ and $\omega = kv_0$. The plasma mode develops into the *fast space-charge wave* which then approaches the fast beam mode. For $kv_0/\omega_{pe} < 1.3$ the beam mode is a complex conjugate *slow space-charge wave*. At the triple point it splits into stable modes, a slow beam mode and a plasma mode. The second plasma mode with negative ω remains unaffected by the beam

higher values of α_b, the dielectric function stays negative in the entire interval $0 < \omega < kv_0$. However, if ω is taken as a complex quantity, there will be a pair of conjugate complex roots with the real part of ω lying in this interval.

The dispersion relation for the beam-plasma modes consists of different branches $\omega(k)$, see Fig. 8.4. According to our wave perturbation $\propto \exp[i(kx - \omega t)]$ and allowing for a complex $\omega = \omega_r + \mathrm{i}\omega_i$, there will be growing waves $\propto \exp[\mathrm{i}(kx - \omega_r t)]\exp[\omega_i t]$, when $\omega_i > 0$. In the limit $\alpha_b \ll 1$, the plasma modes at $\omega = \pm\omega_{pe}$ and the (degenerate) beam mode $\omega = kv_0$ are uncoupled (see dotted lines in Fig. 8.4). For non-vanishing α_b, the positive plasma mode connects to the beam mode and becomes the *fast space-charge wave*, which has $\omega/k > v_0$. For $kv_0/\omega_{pe} < 1.3$ the beam modes form a conjugate pair. These waves are propagating more slowly than the beam and are called *slow space-charge waves*. The one with $\omega_I > 0$ is exponentially growing in time. The growth rate takes a maximum value near the intersection $\omega_{pe} = kv_0$. At the triple point, the slow space-charge waves become real and form the slow beam mode and the plasma mode. The second plasma mode with negative ω remains unaffected by the beam.

8.1.3 Growth Rate for a Weak Beam

For small values of α_b, the slow space-charge wave that has the maximum growth rate $\gamma = \omega_I$, is found close to $kv_0/\omega_{pe} = 1$. This means that the phase velocity of the wave is nearly resonant with the electron beam, $v_\varphi \approx v_0$. Therefore, it is reasonable to seek an approximate solution for $\varepsilon(\omega, k) = 0$ in the vicinity of the resonance point $\omega_{pe} = kv_0$. Introducing $\omega = \omega_{pe} + \Delta\omega$, we can rewrite the dielectric function in this regime as

$$0 = \varepsilon = 1 - \frac{\omega_{\text{pe}}^2}{(\omega_{\text{pe}} + \Delta\omega)^2} - \frac{\alpha_{\text{b}}\omega_{\text{pe}}^2}{(\omega_{\text{pe}} + \Delta\omega - kv_0)^2} \tag{8.4}$$

$$\approx 1 - \frac{\omega_{\text{pe}}^2}{\omega_{\text{pe}}^2} + \frac{2\Delta\omega}{\omega_{\text{pe}}^3} - \frac{\alpha_{\text{b}}\omega_{\text{pe}}^2}{(\Delta\omega)^2} \,. \tag{8.5}$$

In the second line, we have used a Taylor expansion of the plasma dielectric function for $\Delta\omega \ll \omega_{\text{pe}}$, and applied the resonance condition in the last term. Solving for $\Delta\omega$, we obtain

$$\Delta\omega = \left(\frac{\alpha_{\text{b}}}{2}\right)^{1/3} \omega_{\text{pe}}\, \mathrm{e}^{n2\pi \mathrm{i}/3} \qquad \text{with :} \quad n = 0, 1, 2\,. \tag{8.6}$$

Here, we have noticed that there are three roots in this region, two of which are complex. (A fourth root at $\omega = -\omega_{\text{pe}}$ is non-resonant). This gives, for $n = 0$, the fast space charge wave as

$$\omega = \omega_{\text{pe}} \left[1 + \left(\frac{\alpha_{\text{b}}}{2}\right)^{1/3}\right]. \tag{8.7}$$

The slow space-charge wave has

$$\omega_{\text{R}} = \omega_{\text{pe}} \left[1 - \frac{1}{2}\left(\frac{\alpha_{\text{b}}}{2}\right)^{1/3}\right] \tag{8.8}$$

$$\omega_{\text{I}} = \pm\frac{3^{1/2}}{2}\left(\frac{\alpha_{\text{b}}}{2}\right)^{1/3} \omega_{\text{pe}}\,. \tag{8.9}$$

The most spectacular result for the unstable mode is the fact, that the growth rate depends on the third root of the beam fraction α_{b}. Therefore, a beam fraction of $\alpha = 0.002$ generates a wave that has an e-folding after only ten wave periods ($\omega_{\text{R}}/\omega_{\text{I}} \approx 10$). The real part of the frequency is close to ω_{pe} and this wave can therefore be identified as the Langmuir wave. The high growth rate explains why self-excited Langmuir oscillations are so ubiquitous in dc discharges.

8.1.4 Why is the Slow Space-Charge Wave Unstable?

Let us first consider the motion of a single electron in an oscillating electric field, which is described by (6.30) and results in an an oscillatory velocity of amplitude

$$\hat{v}_{\text{e}} = \frac{e}{\mathrm{i}\omega m_{\text{e}}}\hat{E}\,. \tag{8.10}$$

Because of the inertia of the electron, the oscillation velocity $\hat{v}_e$ lags behind the electric force $-e\hat{E}$ by a phase angle of 90°. In the language of electronics, the electric "current" $-e\hat{v}_e$ lags behind the "voltage" $\hat{E}$. Hence, an electron behaves like an inductor.

Consider now an electron with velocity v_0 in a wave field. This beam electron obeys the linearized equation of motion

$$\frac{\partial v_b}{\partial t} + v_0 \frac{\partial v_b}{\partial x} = -\frac{e}{m_e} \hat{E} \exp[i(kx - \omega t)] . \tag{8.11}$$

In Fourier notation, and solving for the oscillation velocity, this becomes

$$\hat{v}_b = \frac{e}{i(\omega - kv_0)m_e} \hat{E} . \tag{8.12}$$

Again, the electron behaves as an inductor, as long as $\omega - kv_0 > 0$, which is realized for the fast space-charge wave. But, for the slow space-charge wave, the opposite case is realized with $\omega - kv_0 < 0$. Therefore, beam electrons show a strange behavior when they interact with the slow wave. Their motion in the wave field is such as if they had a "negative mass". In the language of electronics, the electron now behaves like a capacitor for which the current leads the voltage by a phase shift of 90°.

We can easily see that the concept of a "negative mass" has a physical meaning. We will discuss this effect in terms of the average kinetic energy of the wave. For this purpose, we must calculate the density fluctuations in the beam, which are related to the velocity modulation by the continuity of the flow

$$\frac{\partial n_b}{\partial t} + \frac{\partial (n_b v_b)}{\partial x} = 0 . \tag{8.13}$$

Linearizing $n_b = n_{b0} + n_{b1}$, $v_b = v_0 + v_{b1}$, and setting the first-order quantities $\propto \exp[i(kx - \omega t)]$, we obtain for the perturbed quantities

$$(-i\omega + kv_0)\hat{n} + ikn_{b0}\hat{v}_b = 0 \tag{8.14}$$

and

$$\hat{n}_b = \frac{n_{b0}k}{\omega - kv_0} \hat{v}_b . \tag{8.15}$$

The density fluctuations are in phase with the velocity fluctuations for the fast wave $\omega - kv_0 > 0$ but opposite to the velocity fluctuations for the slow wave. Using the velocity fluctuations from (8.12), we find the density fluctuations as

$$\hat{n}_b = \frac{en_{b0}k}{i(\omega - kv_0)^2} \hat{E} . \tag{8.16}$$

The mean kinetic energy of the beam can be written as

$$\begin{aligned}\langle W_{\rm kin}\rangle &= \frac{1}{2}m_{\rm e}\big\langle(n_{\rm b0}+\hat{n}_{\rm b})(v_0+\hat{v}_{\rm b})^2\big\rangle \\ &= \frac{1}{2}m_{\rm e}\big\langle n_{\rm b0}v_0^2+2n_{\rm b0}v_0\hat{v}_{\rm b}+n_{\rm b0}\hat{v}_{\rm b}^2+\hat{n}_{\rm b}v_0^2+2\hat{n}_{\rm b}v_0\hat{v}_{\rm b}+\hat{n}_{\rm b}\hat{v}_{\rm b}^2\big\rangle\,, \quad (8.17)\end{aligned}$$

where $\langle\cdots\rangle$ denotes the average over one wavelength. In the sum on the r.h.s., all odd powers of fluctuating quantities vanish in the average. Therefore, the only remaining terms are the zero-order beam energy and the second-order corrections

$$\langle W_{\rm kin}\rangle = \frac{1}{2}n_{\rm b0}m_{\rm e}v_0^2+\frac{1}{2}m_{\rm e}\big\langle n_{\rm b0}\hat{v}_{\rm b}^2+2\hat{n}_{\rm b}v_0\hat{v}_{\rm b}\big\rangle\,. \quad (8.18)$$

Using (8.15), we find that for nearly resonant particles, $|\omega/k-v_0|\ll v_0$, the second term in the angle brackets is much larger than the first. Moreover, this contribution of the wave to the beam energy has different signs for the slow wave and the fast wave.

For deriving the energy relations, Fig. 8.5 shows the electric field, the velocity fluctuation, the density fluctuation of the beam, and the resulting beam energy from the second-order contribution (8.18). On average, the beam energy is lowered by the presence of the wave. This justifies the terminology of a *negative energy wave* [161].

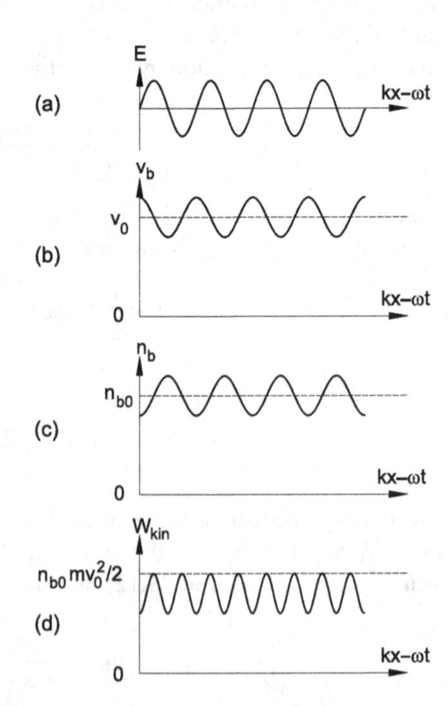

Fig. 8.5 The slow space-charge wave. (**a**) Electric field, (**b**) beam velocity modulation, (**c**) beam density modulation, (**d**) resulting beam kinetic energy. The mean beam energy is reduced by the presence of the wave

The coupling of this negative energy wave with the plasma fluctuations (which represent a positive energy wave) is the reason for wave growth. The loss in kinetic energy appears as gain for the wave potential energy. In this way, the wave potential grows in time at the expense of the beam velocity. In Sect. 9.4 we will trace this evolution into the non-linear regime by means of computer simulations.

Returning to the concept of "negative mass", for the slow wave, the density clumps are growing by decelerating the original beam whereas, for the fast wave, the clumps are filled up by accelerating slower electrons. In the first case, energy is transfered from the beam to the wave, and the wave grows, in the second case. the wave accelerates electrons, thereby losing energy.

8.1.5 Temporal or Spatial Growth

In the last paragraph it was assumed that the wave is characterized by an imaginary part ω_{i} of the wave frequency. This means that everywhere the wave amplitude grows at the same rate. In the beam-plasma system, one can also imagine an unmodulated beam that enters a finite plasma. While the beam propagates through the plasma, any small wave-like perturbation will grow in space. For this case, we have to solve the dispersion relation $\varepsilon(\omega, k) = 0$ with the dielectric function from (8.3) for real frequency ω and complex wavenumber k. Simple algebraic manipulation gives

$$k v_0 = \omega + \mathrm{i} \left(\frac{\alpha_{\mathrm{b}} \omega_{\mathrm{pe}}^2 \omega^2}{\omega_{\mathrm{pe}}^2 - \omega^2} \right)^{1/2}, \tag{8.19}$$

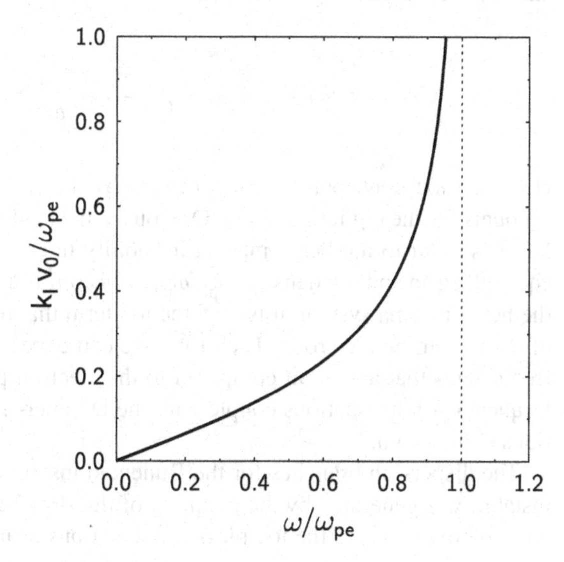

Fig. 8.6 The spatial growth rate in the beam-plasma system for $\alpha_{\mathrm{b}} = 0.1$ as a function of the real wave frequency ω. Note that the growth rate even becomes infinite at $\omega = \omega_{\mathrm{pe}}$

which yields the imaginary part of the wavenumber (as shown in Fig. 8.6)

$$k_I = \frac{\omega_{pe}}{v_0} \frac{\alpha_b^{1/2} \omega}{(\omega_{pe}^2 - \omega^2)^{1/2}} . \tag{8.20}$$

The subtle differences between spatial and temporal growth are discussed in [162].

8.2 Buneman Instability

A second, related example for the instability of counter-streaming charged particles is found in a current carrying plasma, in which a dc electric field leads to a flow of all plasma electrons relative to the ions. This instability was first discussed by Oscar Buneman (1913–1993) [163]. Again we neglect collisions and assume that the drift velocity of the plasma ions is much smaller than the electron beam velocity. Therefore, we describe the instability in the rest frame of the ions. Further, we assume that the electron beam velocity v_0 is much higher than the thermal spread of the electron distribution (cold beam approximation).

8.2.1 Dielectric Function

The dielectric function for this system contains the susceptibilities of stationary ions and beam electrons

$$\varepsilon(\omega, k) = 1 + \chi_i + \chi_e = 1 - \frac{\omega_{pi}^2}{\omega^2} - \frac{\omega_{pe}^2}{(\omega - kv_0)^2} . \tag{8.21}$$

Here, the ion contribution is determined by the ion plasma frequency ω_{pi}, which accounts for the higher ion mass. Obviously, the mathematical structure of the problem is similar to the beam-plasma instability in (8.3) when we recognize that, for equal electron and ion density, $\omega_{pi}^2/\omega_{pe}^2 = m_e/m_i$ is a small quantity. Different from the beam-plasma system, it is now the ion term that represents a small perturbation of the streaming electrons. Therefore, we can expect that the unstable waves have frequencies that are small compared to the electron plasma frequency. These low-frequency ion fluctuations couple with the Doppler-shifted electron plasma oscillations in the beam.

The dispersion branches for the Buneman instability are shown in Fig. 8.7. The instability is generated by the coupling of the slow beam mode ω_4, which is a negative energy wave, to the ion plasma fluctuations near ω_{pi}, resulting in the unstable branch $\omega_{1,2}$.

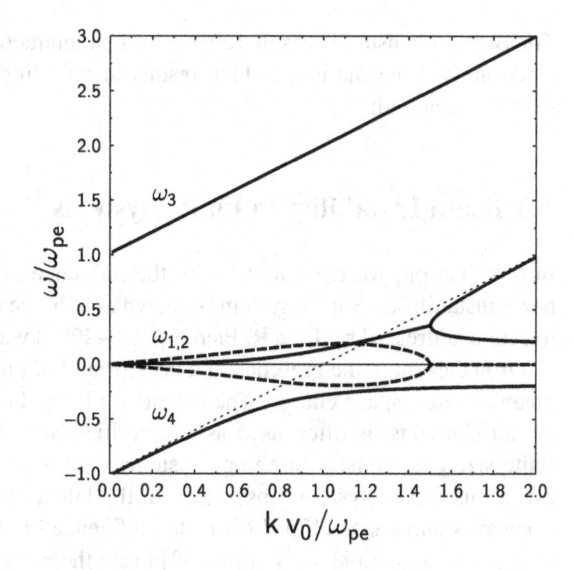

Fig. 8.7 Dispersion branches $\varepsilon = 0$ for the Buneman instability. The dotted line gives the (uncoupled) slow beam mode. The real part (full lines) and imaginary part (dashed lines) of ω are shown as function of normalized wavenumber. For clarity, an artificial mass ratio $m_e/m_i = 0.3$ was chosen

8.2.2 Instability Analysis

The dispersion relation $\varepsilon(\omega, k) = 0$ can be rewritten as

$$kv_0 - \omega = \pm \frac{\omega_{pe}}{(1 - \omega_{pi}^2/\omega^2)^{1/2}} \approx \pm\omega_{pe}\left(1 + \frac{\omega_{pi}^2}{2\omega^2}\right). \tag{8.22}$$

In the second step, we have made the approximation $\omega_{pi}^2 \ll \omega^2$, which justifies a Taylor expansion, This assumption has to be confirmed later. The minus sign gives the real solution while the plus sign leads to complex ω. Setting $\omega = |\omega|e^{i\theta}$ and requiring that the imaginary parts cancel in (8.22) yields

$$|\omega|^3 = \omega_{pi}^2\omega_{pe}\cos(\theta). \tag{8.23}$$

The maximum unstable mode can be found by setting $d\omega_I/d\theta = d[|\omega|\sin(\theta)]/d\theta = 0$, which gives $\theta = \pi/3$. Hence, the maximum unstable mode has

$$\omega_R = \frac{1}{2}\left(\frac{m_e}{2m_i}\right)^{1/3}\omega_{pe}$$
$$\omega_I = \frac{\sqrt{3}}{2}\left(\frac{m_e}{2m_i}\right)^{1/3}\omega_{pe}. \tag{8.24}$$

In the end, we see that $\omega_R^2/\omega_{pi}^2 = 0.157(m_i/m_e)^{1/3} \approx 1.9\ldots6.3$ (for ion masses from hydrogen to argon), which justifies the Taylor expansion made above. Note that, for this instability, the growth rate is slightly larger than the wave frequency.

The Buneman instability is therefore a violent interaction mechanism between beam electrons and plasma ions, which results in e-folding of the amplitude within less than a wavelength.

8.3 Beam Instability in Finite Systems

In this Section, we are interested in the influence of finite length on the electron beam instabilities. Such a system is known as a Pierce diode [164–169]. The original question addressed by John R. Pierce (1910–2002) was finding the maximum stable electron current in the presence of a background of positive ions that neutralizes the mean electron space charge. The extended Pierce diode, which allows for external circuit elements, is often used as a simplified model for studying the stability of finite-size collisionless discharge systems with an electron flow [170–174]. Interest in this model system arose again in the late 1980s with respect to the nonlinear waves and chaos [175–177] or the influence of ion dynamics [169, 178, 179]. Computer simulations [177, 180–182] made the nonlinear and chaotic states accessible. The system was also used as a model for chaos control in plasma systems [183–186].

8.3.1 Geometry of the Pierce Diode

The Pierce diode consists of two conducting planes at $x = 0$ and $x = L$. The electrode at $x = 0$ can be considered as a transparent grid through which an unmodulated electron beam with a velocity v_0 can enter the system. The diode is filled with a homogeneous neutralizing background of immobile ions. The name Pierce diode alludes to vacuum diodes, in which the left electrode (cathode) is a thermal emitter of electrons, which are accelerated by a positive voltage on the right electrode (anode). The electron flow in diodes will be discussed in detail in Sect. 9.2.

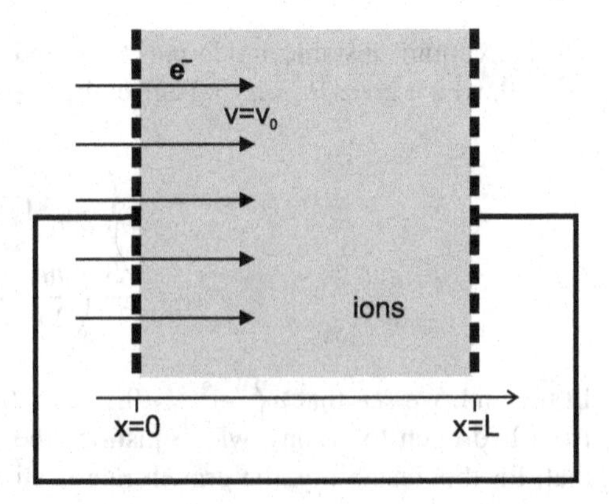

Fig. 8.8 Schematic of the classical Pierce diode. An unmodulated electron beam with velocity v_0 enters the diode through the grid at $x = 0$. The diode is neutralized by a homogeneous background of immobile ions

Different from the vacuum diode, both electrodes in the classical Pierce diode are connected to each other and held at ground potential $\Phi = 0$ (Fig. 8.8). The classical Pierce diode is completely characterized by a single parameter $\alpha_P = \omega_{pe} L / v_0$, as we will see below.

8.3.2 The Dispersion Relation for a Free Electron Beam

The dielectric function for an electron beam traversing an immobile ion background is the same as for the Buneman instability (8.21), except for $\omega_{pi} = 0$ due to the assumed immobility of the ions that can be achieved by assigning an infinite ion mass. Then, the oscillations in the electron beam follow from

$$0 = \varepsilon(\omega, k) = 1 - \frac{\omega_{pe}^2}{(\omega - k v_0)^2} . \quad (8.25)$$

These are waves travelling with the beam or against the beam with wavenumbers

$$k_+ = \frac{\omega + \omega_{pe}}{v_0} \quad \text{and} \quad k_- = \frac{\omega - \omega_{pe}}{v_0} . \quad (8.26)$$

8.3.3 The Influence of the Boundaries

Intuitively, we could assume that the influence of the boundaries is the formation of standing waves. However, the situation is more complex, because the forward and backward propagating waves have different wavelength. Moreover, we must account for electric charges on the surface of the metallic boundaries, which also produce an electric field inside the diode. This is why Pierce composed the oscillating electric wave potential inside the diode from four ingredients:

$$\tilde{\Phi}(x, t) = \left(Ax + B \mathrm{e}^{-\mathrm{i}k_+ x} + C \mathrm{e}^{-\mathrm{i}k_- x} + D \right) \mathrm{e}^{-\mathrm{i}\omega t} . \quad (8.27)$$

$Ax + D$ is the solution for the potential in a parallel plate capacitor in vacuum. The other terms represent the two counterpropagating waves. The following boundary conditions must be met by the resulting full wave field:

$$0 = \tilde{\Phi}(0, t) = \tilde{\Phi}(L, t) = \tilde{n}_e(0, t) = \tilde{v}_e(0, t) . \quad (8.28)$$

The first two conditions require the vanishing of the electric potential at the grounded electrodes, the other two conditions state that the electron beam is not modulated when it enters the diode. The potential condition at $x = 0$ then gives

$$0 = \tilde{\Phi}(0, t) = B + C + D . \quad (8.29)$$

Noting that $\tilde{n}_e \propto \partial^2\tilde{\Phi}/\partial x^2$, we have

$$\tilde{n}_e(0,t) = 0 = -k_+^2 B - k_-^2 C\,. \tag{8.30}$$

The vanishing of the longitudinal oscillations in the electron beam requires some detailed considerations. Besides the wave-like terms there must also be a component $\bar{v}$ corresponding to the oscillation in the electric field of the surface charges:

$$\tilde{v} = \left[v_+ \mathrm{e}^{\mathrm{i}k_+ x} + v_- \mathrm{e}^{\mathrm{i}k_x} + \bar{v}\right]\mathrm{e}^{-\mathrm{i}\omega t}\,. \tag{8.31}$$

The entire oscillating velocity $\tilde{v}$ must obey the linearized equation of motion

$$\frac{\partial \tilde{v}}{\partial t} + v_0\frac{\partial \tilde{v}}{\partial x} = \frac{e}{m}\frac{\partial \tilde{\Phi}}{\partial x}\,, \tag{8.32}$$

which gives the relations between the independent coefficients v_+, v_- and $\bar{v}$

$$v_+ = -\frac{e}{m}\frac{k_+}{\omega - k_+ v_0}B\,,\quad v_- = -\frac{e}{m}\frac{k_-}{\omega - k_- v_0}C\,,\quad \bar{v} = \frac{e}{m}\frac{1}{\mathrm{i}\omega}A\,. \tag{8.33}$$

The vanishing of $\tilde{v}(0,t)$ then results in

$$\tilde{v}(0,t) = 0 = \frac{e}{m}\left[\frac{k_+}{\omega - k_+ v_0}B + \frac{k_-}{\omega - k_- v_0}C - \frac{1}{\mathrm{i}\omega}A\right]. \tag{8.34}$$

Using the beam dispersion relation (8.26), we obtain a further relation between the coefficients A, B, and C

$$0 = k_+ B - k_- C + \frac{\omega_{\mathrm{pe}}}{\mathrm{i}\omega}A\,. \tag{8.35}$$

We can now use the three equations (8.29), (8.30), and (8.35) to express all other coefficients by, say, coefficient B, which remains undetermined and describes the amplitude of the wave. In this way, we obtain the shape of the wave potential inside the Pierce diode as

$$\tilde{\Phi}(x) = B\left[-\frac{\mathrm{i}\omega}{\omega_p}\frac{k_+}{k_-}(k_- + k_+)x + \left(\mathrm{e}^{-\mathrm{i}k_+ x} - 1\right) - \left(\frac{k_+}{k_-}\right)^2\left(\mathrm{e}^{-\mathrm{i}k_- x} - 1\right)\right\}. \tag{8.36}$$

This solution will fulfill the remaining boundary condition $\tilde{\Phi}(L,t) = 0$ only for certain combinations of wave frequency, plasma frequency and beam velocity, which we will derive in the next step. Introducing the abbreviations $\theta = \omega L/v_0$ and $\alpha_{\mathrm{P}} = \omega_{\mathrm{pe}}L/v_0$ we obtain

$$0 = \omega\left\{\theta + \frac{\mathrm{i}\,\alpha_{\mathrm{P}}}{2\,\theta}\left[\frac{\theta - \alpha_{\mathrm{P}}}{\theta + \alpha_{\mathrm{P}}}\left(\mathrm{e}^{-\mathrm{i}(\theta+\alpha_{\mathrm{P}})} - 1\right) - \frac{\theta + \alpha_{\mathrm{P}}}{\theta - \alpha_{\mathrm{P}}}\left(\mathrm{e}^{-\mathrm{i}(\theta-\alpha_{\mathrm{P}})} - 1\right)\right]\right\}. \tag{8.37}$$

This is an implicit relation between θ, which contains the wave frequency ω, and the Pierce parameter α_P. In general, the solutions for θ will be complex and there will be infinitely many solutions, because of the transcendental nature of the equation.

8.3.4 The Pierce Modes

The existence diagram for the various modes of the Pierce instability is shown in Fig. 8.9. A stable mode with frequency $\omega = 0$, i.e., a homogeneous dc current flow of the beam, is found for $\alpha_P < \pi$. This electron flow becomes unstable when the imaginary part of θ becomes positive. This happens for $\pi < \alpha_P < 2\pi$ and leads to exponential but non-oscillatory growth of any initial perturbation. A first oscillatory

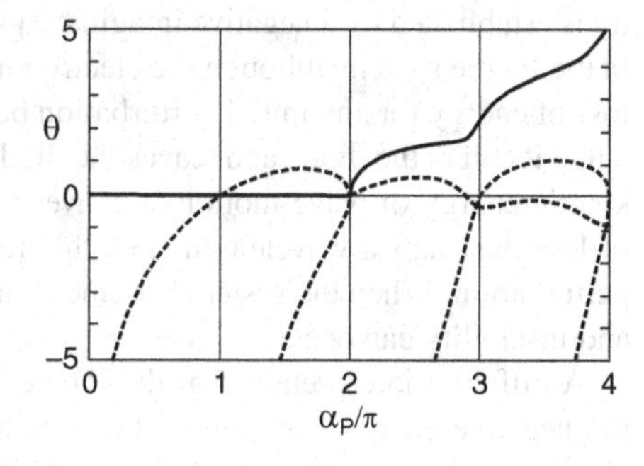

Fig. 8.9 Real part (*solid line*) and imaginary part (*dashed line*) of the normalized frequency θ as a function of the Pierce parameter α_P

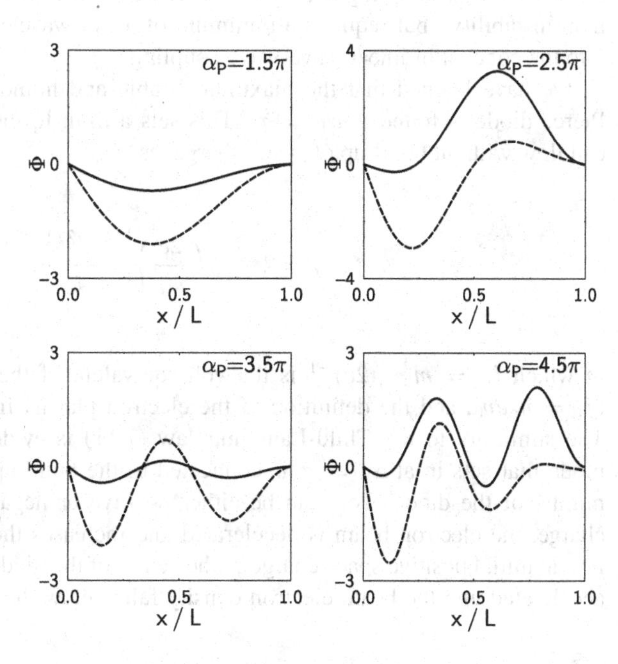

Fig. 8.10 The real (*solid line*) and imaginary part (*dashed line*) of the wave potential for the Pierce modes at $\alpha_P = 1.5$, 2.5, 3.5, and 4.5

unstable Pierce mode is found in the interval $2\pi < \alpha_P < 3\pi - \Delta\alpha_P$, which becomes stable again for higher α_P. Then a second oscillatory Pierce mode of even higher frequency takes over at $\alpha_P = 3\pi - \Delta\alpha_P$ and becomes unstable for $\alpha_P = 3\pi$ and vanishes again for $\alpha_P \approx 4\pi$. This pattern repeats with ever higher frequencies.

Although the Pierce modes are not simply standing waves between the end plates, there is still a distinguishing feature of the modes in terms of the number of maxima and minima of the wave potential in the interval $0 < x < L$. Figure 8.10 shows that the number of half wavelengths inside the diode increases stepwise as one ($\alpha_P = 1.5$), two ($\alpha_P = 2.5$), three ($\alpha_P = 3.5$) and four ($\alpha_P = 4.5$).

8.3.5 Discussion of the Pierce Model

The Pierce model gives a stable homogeneous electron flow for $\alpha_P < \pi$. The stability is established by a negative imaginary part of $\omega = \theta v_0/L$. Why is there damping in the Pierce system although the electron motion is collisionless? There is indeed a loss of energy for any initial perturbation because the electron beam is unmodulated when it enters the diode and leaves the diode with some density modulation. Hence, kinetic energy of wave motion is convected out of the system. For a short system of less than half a wavelength size, this removal is efficient and damps the initial perturbation. When the system becomes longer, this mechanism becomes inefficient and instability can occur.

A different interpretation of the Pierce instability is to consider the coupling of the negative energy wave in the beam with the oscillating surface charges on the electrodes. This coupling leads to instability as in the beam-plasma or in the Buneman instability, but requires a minimum of a full wavelength in the system to fit into the concept of linear wave-wave coupling.

We have learned that the maximum stable and homogeneous solution of the Pierce diode is found at $\alpha_P = \pi$. This sets a limit to the current density, which can flow without build-up of space charge, as

$$j = 2\pi^2 \varepsilon_0 \left(\frac{2e}{m_e} \right)^{1/2} \frac{U^{3/2}}{L^2} , \tag{8.38}$$

in which $U = m_e v_0^2 (2e)^{-1}$ is the volt-equivalent of the beam injection energy, $|j| = n_e e v_0$, and the definition of the electron plasma frequency has been used. The similarity to the Child-Langmuir law (7.11) is evident. The non-oscillatory mode that sets in at $\alpha_P = \pi$ is connected to the build-up of space charge in the middle of the diode. This can be either positive or negative. For negative space charge, the electron beam is decelerated and increases the negative space charge. For an initial positive space charge in the center of the diode, the electron beam gets accelerated and the beam electron density falls below the ion density, thus adding

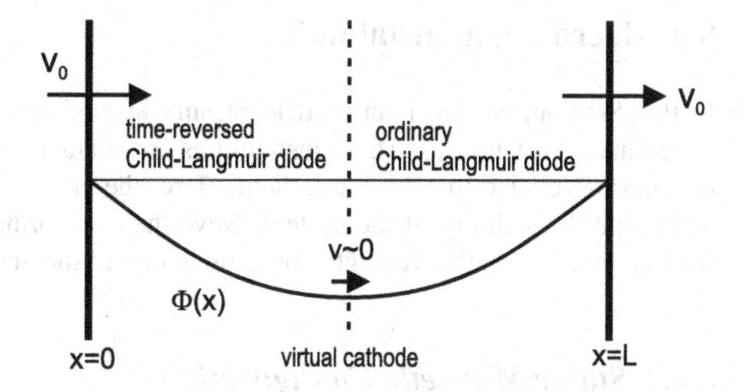

Fig. 8.11 The injection of an electron beam into a vacuum gap between grounded electrodes can be considered as operating two ordinary diodes back to back with a reversed current flow in the left diode

net positive space charge. Both cases are obviously unstable and create a growing, non-oscillatory space charge.

For comparing the maximum stable current in a Pierce diode with the Child-Langmuir law of a vacuum diode, we must bear in mind that, in a common vacuum diode, the electrons start with $v = 0$ at the cathode and exit with $v = v_0$ at the anode. Let us consider the injection of beam electrons with $v = v_0$ into the gap between two grounded electrodes, as shown in Fig. 8.11. At the maximum possible current, the electrons will nearly come to rest in the middle of the gap ($x = L/2$) because they are decelerated by their own space-charge electric field. This leads to a deep potential minimum $\Phi \approx -mv_0^2/(2e)$ in the center. In the right half of the diode, the electrons are accelerated again by the space-charge field and exit with $v = v_0$ at $x = L$. The left half of the gap can be considered as a time-reversed Child-Langmuir situation. It is no surprise that the same law applies, when v is replaced by $-v$, because our derivation of the Child-Langmuir law was based on kinetic energy, which remains the same under time reversal for a stationary flow. Hence, the beam injection into a vacuum gap is the same as operating two Child-Langmuir diodes back to back, each of length $L/2$. The minimum of the potential in the center of the gap takes the role of a *virtual cathode* and the grounded electrodes can be considered as the anodes. The stability of electron flow injected into a gap between grounded electrodes was studied e.g. in [187, 188]. The formation of virtual cathodes in front of thermal emitters will be discussed in Sect. 9.2.2.

Hence, an honest comparison of the maximum current in the neutralized Pierce diode with a non-neutralized electron flow must bear in mind that in the latter case the Child-Langmuir law must be written for a diode of length $L/2$. Hence the ratio of the maximum currents becomes

$$\frac{j_{\text{max,Pierce}}}{j_{\text{max,vacuum}}} = \frac{2\pi^2}{16/9} \approx 11\,. \tag{8.39}$$

This is a respectable gain by one order of magnitude for the maximum stable current in a neutralized diode over a vacuum diode.

8.4 Macroscopic Instabilities

In this Section, we are interested in plasma instabilities occurring in real space, called macroinstabilities. These instabilities are characterized by a displacement of the plasma relative to a magnetic field. Here, the energy principle can be used to determine the stability of the system. Nevertheless, normal mode analysis will be the tool to detect the wavelength and growth rate of the unstable modes.

8.4.1 Stable Magnetic Configurations

Consider the magnetic field topologies for a magnetic mirror and a magnetic cusp in Fig. 8.12. We had seen in (3.27) that the gradient of the magnetic field intensity points always towards the center of field line curvature. In the center of the mirror field the gradient points inwards whereas, near the magnets, the gradient points outwards.

Let us now consider the total energy of a small volume of plasma, which consists of kinetic and magnetic energy, which represents the potential energy for this case,

$$W_{\mathrm{tot}} = W_{\mathrm{kin}} + \frac{B^2}{2\mu_0}\,. \tag{8.40}$$

When this plasma volume is shifted into a region of weaker magnetic field, the magnetic energy will decrease accordingly. The existence of such a state of lower potential energy makes the situation unstable. However, we cannot give the detailed mechanism, how the plasma manages to get to this lower energy state. We can only say that the plasma in the center of a mirror field has no stable confinement against radial displacements. Consider now the field line topology of a magnetic cusp in Fig. 8.12b. There, the magnetic field increases in all directions and the plasma is in a stable confinement. Such a situation is called a *minimum B* configuration.

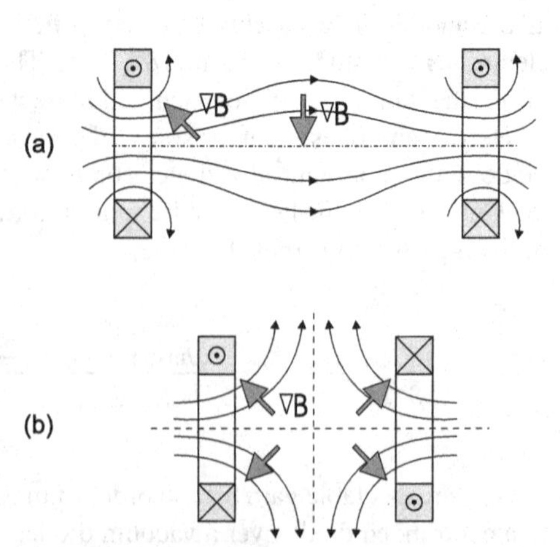

Fig. 8.12 (**a**) A magnetic mirror field is generated by currents of the same polarity in the magnets, (**b**) A magnetic cusp forms when the current in one magnet is reverted. Note that the direction of the gradient in magnetic field strength always points towards the center of the local field line curvature

8.4.2 Pinch Instabilities

The pinch effect was already introduced in Sect. 5.3.4. The pinch effect is not necessarily a homogeneous mechanism. When we assume that the plasma cross section is reduced at some point, the magnetic pressure at the plasma surface will increase, because $B_\varphi = \mu_0 I\,(2\pi a)^{-1}$, as shown in Fig. 8.13a. This increased magnetic pressure further reduces the plasma radius at this point, and the plasma column develops a *sausage instability*.

The magnetic pressure can also deviate from its equilibrium value, when the plasma column is curved, see Fig. 8.13b. Because the magnetic field lines are perpendicular to the local current direction, the field line density, and the associated magnetic pressure, is higher on the inner side and lower on the outer side of the curved plasma column. Hence, the imbalance of magnetic pressure will further displace the column forming a kink.

The sausage and the kink instability can be stabilized by a superimposed longitudinal magnetic field, which is frozen in the plasma. The magnetic field lines have a tension $\mathcal{T} = B^2/\mu_0$ that tends to straighten the field lines, see Sect. 5.2.2. This gives a net restoring force that counteracts the instablity from the magnetic pressure imbalance of the azimuthal magnetic field component, as shown in Fig. 8.13c.

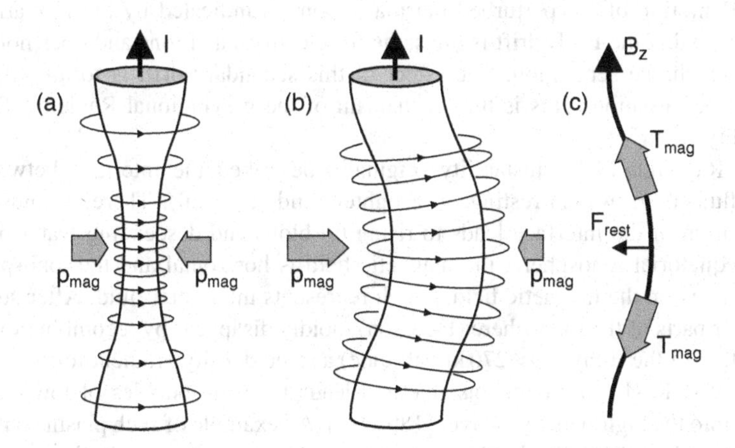

Fig. 8.13 (**a**) Sausage instability, (**b**) kink instability of a pinch plasma. The magnetic pressure increases when the cross-section shrinks or becomes asymmetric when the plasma column is curved. (**c**) The magnetic tension of a superimposed longitudinal magnetic field counteracts the instability

8.4.3 Rayleigh–Taylor Instability

A cartoon of the Rayleigh-Taylor instability for the equatorial ionosphere is shown in Fig. 8.14. The unpurturbed plasma boundary is shown by the horizontal dashed line. The plasma fills the upper halfspace. The magnetic field is perpendicular to

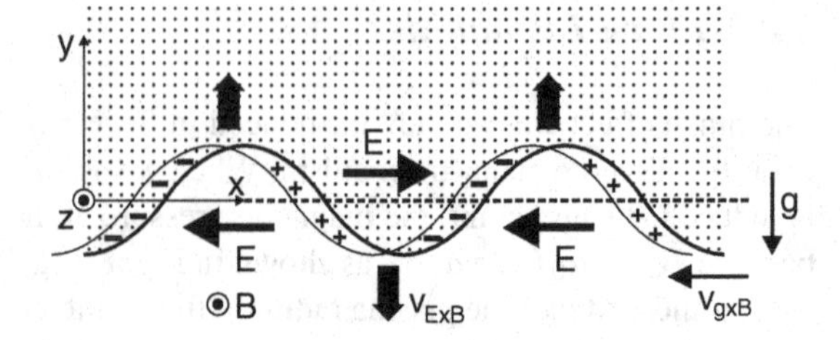

Fig. 8.14 A cartoon of the Rayleigh-Taylor instability in the equatorial ionosphere. The plasma is the heavy fluid that rests on the horizontal magnetic field

the $x - y$ plane. Under the action of the gravitational force, the ions experience a g×B drift with a velocity given by (3.14) when we can neglect ion-neutral collisions. The (opposing) drift velocity of the electrons is smaller by a factor m_e/m_i and will be omitted here. The initial homogeneous equilibrium of the boundary can be understood from the force balance $\mathbf{j}_i \times \mathbf{B} + n_i m_i \mathbf{g} = 0$, in which $j_i = e n_i v_g$.

For understanding the instability mechanism, we consider an initial sinusoidal perturbation of the boundary, as indicated by the heavy line. The effect of the g×B drift is to shift the ions slightly in $-x$ direction, as indicated by the light line. This generates positive surplus-charges at the surface by an overshoot of ions on the leading edge and a lack of ions on the trailing edge. These surface charges generate an E×B motion of the perturbed plasma region, as indicated by the box arrows. Remember that the E×B drift is the same for electrons and ions and does not lead to further charge separation. The effect of this secondary drift is to amplify the original perturbation. This is the mechanism of the gravitational Rayleigh-Taylor instability.

The Rayleigh-Taylor instability originally described the interface between a heavy fluid (e.g., water) resting on a lighter fluid (e.g., oil). There, a sinusoidal perturbation of the interface leads to rising oil blobs and descending water blobs. In the equatorial ionosphere, the magnetic field is horizontal and the ionospheric plasma rests on the magnetic field, which represents the lighter fluid. After sunset, the lower parts of the ionosphere (E-region) rapidly disappear by recombination. At the bottom of the F-layer (≈ 270 km altitude) a steep density gradient forms, which can become Rayleigh-Taylor unstable and leads to rising bubbles of low-density plasma into the high-density F-layer [189–193]. An example of such plasma bubbles is shown in Fig. 8.15. The bubbles appear as reduced plasma density in a comparison of the density profile during the upleg and downleg of the rocket trajectory. The upleg intersected the bubble region whereas the downleg traversed unperturbed plasma. This result was obtained during the DEOS (Dynamics of the Equatorial Ionosphere Over Shar) rocket campaign [193].

Magnetized plasmas are generally susceptible to Rayleigh-Taylor-like instabilities. Here, the role of gravity can be taken over by the internal kinetic pressure of the plasma particles, as shown in Fig. 8.16a,b. The plasma surface develops a periodic perturbation $\propto \exp(im\varphi)$. The azimuthal mode number m gives the number of grooves in the plasma column. This pattern resembles the fluted columns in ancient Greece, as shown in Fig. 8.16c and explains the name *flute instability*.

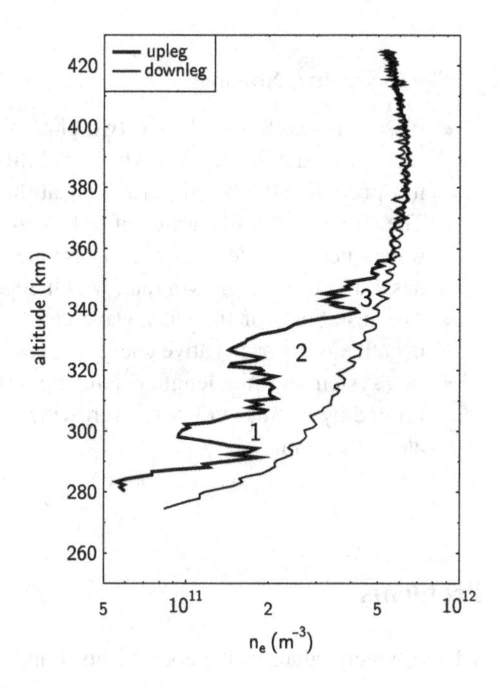

Fig. 8.15 Observation of plasma bubbles in the equatorial ionosphere with Langmuir probes aboard a sounding rocket (from [193]). The upleg of the rocket trajectory intersected three bubbles (marked 1–3) whereas the downleg went through unperturbed plasma. The fine wiggle seen on the downleg results from the wake of the spinning rocket

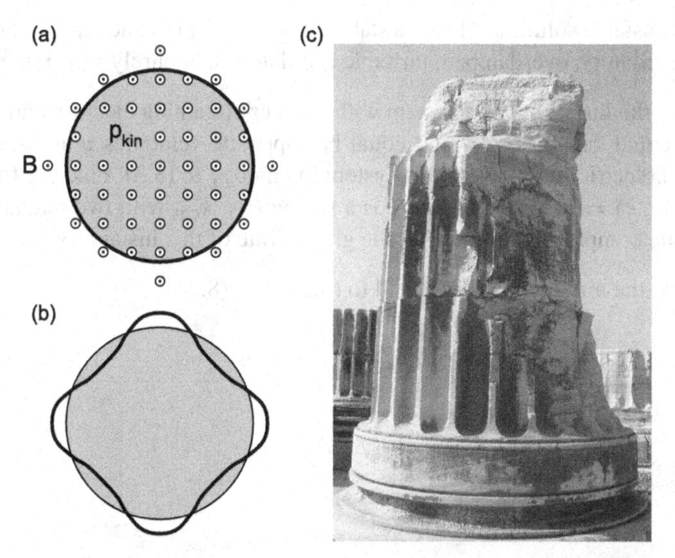

Fig. 8.16 Flute instability of a magnetized plasma column as a generalized Rayleigh-Taylor mechanism. (**a**) Unperturbed plasma column, (**b**) $m = 4$ mode, (**c**) fluted Greek columns (Photo: J. Piel)

The Basics in a Nutshell

- Plasma instabilities fall into two classes, macroscopic instabilities in real space, like the Rayleigh-Taylor instability, and microinstabilities in velocity space, like the beam-plasma instability.
- The directed flow of a group of fast electrons (beam) can excite electrostatic waves near the electron plasma frequency. This beam-plasma instability has a tremendous growth rate, which depends on $(n_b/n_p)^{1/3}$.
- The instability of the slow wave can be understood from the concept of negative mass or negative energy waves.
- In a system of finite length (Pierce diode) the maximum electron current is limited by the onset of purely-growing, non-oscillating disturbances of the electron beam.

Problems

8.1 For which values of the coefficients a and b has the differential equation

$$\ddot{x} + a\dot{x} + bx = 0 .$$

stable and unstable solutions? Draw a stability map $b = f(a)$ and mark regions with damped oscillatory, overdamped, unstable oscillatory and purely growing modes.

8.2 Discuss the instability of a system with counter-streaming electron and positron beams of equal density $\propto \omega_b^2$ and equal but opposite velocities v and $-v$. Write down the dielectric function for this system in analogy to (8.3). Find the four solutions of $\varepsilon(\omega, k) = 0$. Show that there is a region $k < k_{crit}$ with two real and a pair of conjugate complex solutions. Plot the growth rate of this instability vs. kv/ω_b.

8.3 Perform the missing steps that lead to (8.23) and (8.24).

Chapter 9
Kinetic Description of Plasmas

"All right", said the Cat; and this time it vanished quite slowly, beginning with the end of the tail, and ending with the grin, which remained some time after the rest of it had gone.

Lewis Carroll, Alice in Wonderland

In the previous chapters, the description of the plasma state was refined step by step. In the single-particle model, we were interested in the motion of individual particles in typical magnetic field configurations, but the interaction between the particles and the modification of the fields by the presence and motion of charged particles was neglected. In the fluid model, we had considered the average behavior of particles filling a small volume of space. In this approximation, only moments of a shifted Maxwell distribution, like mean flow velocity or gas temperature, were retained, but, by combining with Maxwell's equations, the model became self-consistent. The fluid model goes beyond the single-particle model in that pressure effects are now included. This fluid model, and its formulation in terms of MHD-equations, became capable to describe the combined macroscopic motion of plasma and magnetic field lines. A first attempt to deal with non-Maxwellian velocity distributions was the introduction of a beam-plasma system, which generates self-excited electrostatic waves near the electron plasma frequency.

When we go to high-temperature plasmas, the thermal effects are incompletely described by the concept of pressure. Rather, individual groups of particles in the distribution function have quite different interactions with a wave. In the present Chapter, we will give a brief introduction to the kinetic description of a plasma with an arbitrary velocity distribution by means of the Vlasov equation. This is the third stage of refinement in the description of the plasma state, as sketched in Fig. 9.1. Here, the emphasis is on velocity-space effects like the collisionless Landau damping of waves. As a second example, we will study the relationship between single-particle motion and kinetic theory for space-charge-limited electron flow in diodes. At last, particle simulation as a means of kinetic plasma description will be briefly discussed.

Besides this hierarchy of models, which can be sorted according to plasma temperature and collisionality, there are additional ways of plasma description, such as treating the plasma as a dielectric. We have seen that plasmas support various types of waves, among them light waves, plasma oscillations, ion sound waves or Alfvén waves. Depending on the necessary refinement, we can combine the idea

A. Piel, *Plasma Physics*, DOI 10.1007/978-3-642-10491-6_9,

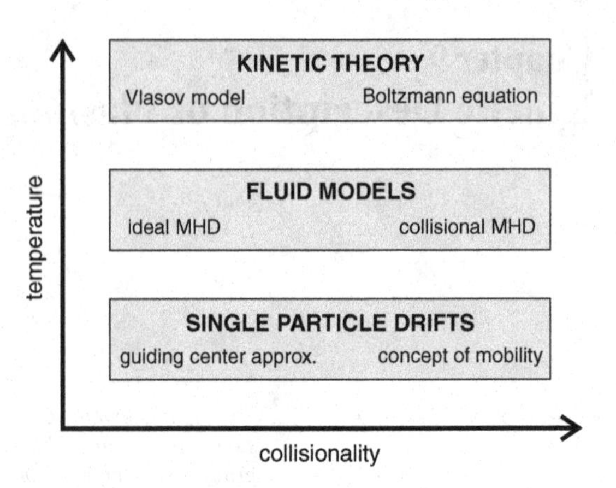

Fig. 9.1 The hierarchy of plasma models

of a dielectric with any of the three levels of plasma description. In particular, we will see in kinetic theory, what the concepts of *cold plasma* and *warm plasma* really mean.

9.1 The Vlasov Model

A complete description of a plasma must on the one hand include fluid aspects and self-consistent fields, and on the other hand the velocity distributions of the particle species. Such a concept is developed in kinetic theory. In this Section, we will abandon the true particle positions, but use the probability distribution in real space and in velocity space. For collisionless plasmas this can be done in terms of the *Vlasov model* that was introduced, in 1938, by Anatoly Vlasov (1908–1975).

9.1.1 Heuristic Derivation of the Vlasov Equation

In the fluid model we became acquainted with the concept of replacing particle trajectories by a statistical description of the mean properties of the plasma particles within small fluid elements. There, we had defined the mass density $\rho_m(\mathbf{r}, t)$ and the flow velocity $\mathbf{u}(\mathbf{r}, t)$, which are connected by the conservation of mass

$$\frac{\partial}{\partial t}\rho_m(\mathbf{r}, t) + \nabla \cdot [\rho_m(\mathbf{r}, t)\mathbf{u}(\mathbf{r}, t)] = 0\,. \tag{9.1}$$

In kinetic theory, it is no longer sufficient to consider a mean flow velocity, but the evolution of the number of particles in a certain velocity interval $\mathrm{d}^3 v$ about a velocity vector $\mathbf{v}$ has to be explicitely described. The mass Δm inside a small volume $\Delta x \Delta y \Delta z$ of real space was defined by

$$\Delta m = \rho_m(\mathbf{r}, t)\Delta x \Delta y \Delta z\,. \tag{9.2}$$

In analogy, we now subdivide velocity space into small bins, $\Delta v_x \Delta v_y \Delta v_z$, and consider the number of particles $\Delta N^{(\alpha)}$ of species α inside an element of a six-dimensional phase space that is spanned by three spatial coordinates and three velocity coordinates

$$\Delta N^{(\alpha)} = f^{(\alpha)}(\mathbf{r}, \mathbf{v}, t)\Delta x \Delta y \Delta z\, \Delta v_x \Delta v_y \Delta v_z\,. \tag{9.3}$$

Taking the limit of infinitesimal size, $\mathrm{d}^3 r\ \mathrm{d}^3 v$, needs a short discussion. When phase space is subdivided into ever finer bins, the problem arises that, in the end, we will find one or no plasma particle inside such a bin. The distribution function $f^{(\alpha)}$ would then become a sum of δ-functions

$$f^{(\alpha)}(\mathbf{r}.\mathbf{v}, t) = \sum_k \delta(\mathbf{r} - \mathbf{r}_k(t))\delta(\mathbf{v} - \mathbf{v}_k(t))\,, \tag{9.4}$$

which represents the exact particle positions and velocities. However, then we had recovered the problem of solving the equations of motion for a many-particle system, of say 10^{20} particles; instead, we are searching for a mathematically simpler description by statistical methods.

For this purpose, we start with finite bins, $\Delta x \Delta y \Delta z\, \Delta v_x \Delta v_y \Delta v_z$, of macroscopic size, which contain a sufficient number of particles to justify statistical techniques. Then we define a continuous distribution $f^{(j)}$ on this intermediate scale and require that $f^{(\alpha)}$ remains continuous in taking the limit. One could imagine that this is equivalent to grind the real particles into a much finer "Vlasov sand", where each grain of sand has the same value of q/m (which is the only property of the particle in the equation of motion) as the real plasma particles, and is distributed such as to preserve the continuity of $f^{(\alpha)}$. This approach is called the *Vlasov picture*. This subdivision comes at a price, because we loose the information of the arrangement of neighboring particles, i.e., correlated motion or collisions. Hence, the Vlasov model does only apply to weakly coupled plasmas with $\Gamma \ll 1$.

A different way to give a kinetic description will be introduced below in Sect. 9.4 by combining the particles inside a mesoscopic bin into a *superparticle* of the same q/m. Then we may end up with only 10^4–10^5 superparticles for which the equations of motion can be solved on a computer. However, forming superparticles enhances the grainyness of the system and the particles inside a superparticle are artificially correlated.

The function $f^{(\alpha)}$ has the following normalisation,

$$N^{(\alpha)} = \iint f^{(\alpha)}(\mathbf{r}, \mathbf{v}, t)\, \mathrm{d}^3 r\, \mathrm{d}^3 v\,, \tag{9.5}$$

where $N^{(\alpha)}$ is the total number of particles of species α. The particle density in real space, the mass density, and the charge density then become

$$n^{(\alpha)}(\mathbf{r}, t) = \int f^{(\alpha)}(\mathbf{r}, \mathbf{v}, t)\mathrm{d}^3 v \tag{9.6}$$

$$\rho_m(\mathbf{r},t) = \sum_\alpha m^{(\alpha)} n^{(\alpha)}(\mathbf{r},t) \tag{9.7}$$

$$\rho(\mathbf{r},t) = \sum_\alpha q^{(\alpha)} n^{(\alpha)}(\mathbf{r},t)\,. \tag{9.8}$$

9.1.2 The Vlasov Equation

We now seek an equation of motion for the distribution function $f(\mathbf{r},\mathbf{v},t)$ that generalizes the continuity equation (9.1). Let us first recall that, in the fluid model, $\mathbf{u}(\mathbf{r},t)$ represents a physically measurable variable. Now, the velocity $\mathbf{v}$ becomes a coordinate of velocity space. The difference lies in the fact that, in the fluid model, the mean flow velocity is attached to a group of particles whereas in the kinetic model the particles have this velocity because they happen to be in a bin with the label $\mathbf{v}$. However, when we arbitrarily select a small volume of phase space $\mathrm{d}^3r\,\mathrm{d}^3v$ about the vector $(\mathbf{r},\mathbf{v})$, the particles in this bin form a group that behaves like a fluid with the streaming velocity $\mathbf{v}$.

To simplify our arguments, we consider the phase space of a one-dimensional system, which has only the coordinates (x, v_x). The particle balance within a phase space volume $\Delta x \Delta v_x$ is determined by the difference of inflow and outflow in real space and, in addition, by acceleration and deceleration (see Fig. 9.2). For the moment, we drop the superscript (α) and consider only one of the plasma species, e.g., the electrons. Since $f \Delta x \Delta v_x$ is the number of particles in that small phase-space element, we can write in analogy to the continuity equation (9.1)

$$\frac{\partial f}{\partial t} = -\frac{\partial}{\partial x}(f v_x) - \frac{\partial}{\partial v_x}(f a)\,, \tag{9.9}$$

in which $f v_x$ is the flux in real space and $f a$ the flux in v_x direction caused by an acceleration a, as indicated by the arrows in Fig. 9.2. Here, we have neglected creation and annihilation of charge carriers by ionization and recombination, as well

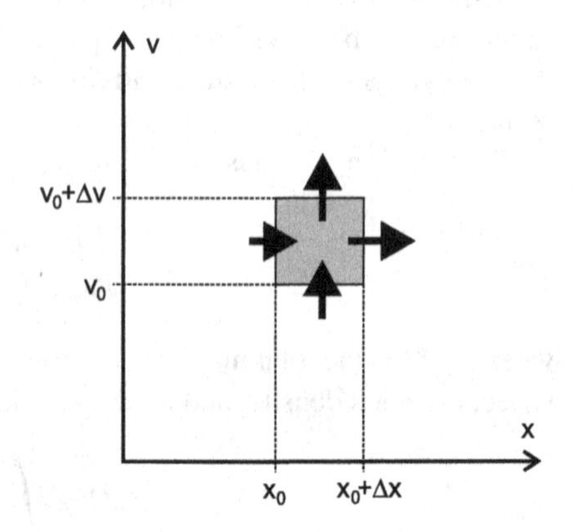

Fig. 9.2 The Vlasov equation describes the flow of a probability fluid in phase space. A gain within the shaded phase-space volume $\Delta V = \Delta x \Delta v$ can be achieved by a flux imbalance (horizontal arrows) in real space (x) or by a difference of acceleration (vertical arrows) in velocity space (v)

as collisions that kick particles from one phase-space cell to another cell at far distance. Noting that the phase-space coordinate v_x is independent of x and that the x-component of the Lorentz force is independent of v_x, we have

$$\frac{\partial f}{\partial t} + v_x \frac{\partial f}{\partial x} + a \frac{\partial f}{\partial v_x} = 0 . \tag{9.10}$$

Generalizing to three space coordinates and three velocities, we obtain

$$\frac{\partial f}{\partial t} + \mathbf{v} \cdot \nabla_r f + \mathbf{a} \cdot \nabla_v f = 0 . \tag{9.11}$$

Here, we have introduced the short-hand notations $\nabla_r = (\partial/\partial x, \partial/\partial y, \partial/\partial z)$ and $\nabla_v = (\partial/\partial v_x, \partial/\partial v_y, \partial/\partial v_z)$. The particle acceleration a is determined by the electric and magnetic fields, which are the sum of external fields and internal fields from the particle currents

$$\mathbf{a} = \frac{q}{m} (\mathbf{E} + \mathbf{v} \times \mathbf{B}) . \tag{9.12}$$

It must be emphasized here that the internal electric and magnetic fields result from average quantities like the space charge distribution $\rho = \sum_\alpha q_\alpha \int f_\alpha \mathrm{d}^3 v$ and the current distribution $\mathbf{j} = \sum_\alpha q_\alpha \int \mathbf{v}_\alpha f_\alpha \mathrm{d}^3 v$, which are both defined as integrals over the distribution function. In this sense, the fields are average quantities of the Vlasov system and any memory of the pair interaction of individual particles is lost. This is equivalent to assuming weak coupling between the plasma particles and neglecting collisions.

Combining (9.11) and (9.12) we obtain the *Vlasov equation*

$$\frac{\partial f}{\partial t} + \mathbf{v} \cdot \nabla_r f + \frac{q}{m} (\mathbf{E} + \mathbf{v} \times \mathbf{B}) \cdot \nabla_v f = 0 . \tag{9.13}$$

There are individual Vlasov equations for electrons and ions.

9.1.3 Properties of the Vlasov Equation

Before discussing applications of the Vlasov model, we consider general properties of the Vlasov equation:

1. The Vlasov equation conserves the total number of particles N of a species, which can be proven, for the one-dimensional case, as follows:

$$\frac{\partial N}{\partial t} = \frac{\partial}{\partial t} \iint f \, \mathrm{d}x \mathrm{d}v = - \iint v \frac{\partial f}{\partial x} \mathrm{d}x \mathrm{d}v - \iint a \frac{\partial f}{\partial v} \mathrm{d}x \mathrm{d}v$$

$$= -\int_{-\infty}^{\infty} \mathrm{d}v \left\{ \left[vf \right]_{x=-\infty}^{x=\infty} - \int_{-\infty}^{\infty} f \frac{\mathrm{d}v}{\mathrm{d}x} \mathrm{d}x \right\}$$

$$- \int_{-\infty}^{\infty} \mathrm{d}x \left\{ \left[af \right]_{v=-\infty}^{v=\infty} - \int_{-\infty}^{\infty} f \frac{\mathrm{d}a}{\mathrm{d}v} \mathrm{d}v \right\} = 0 \,. \tag{9.14}$$

Here we have used that the expressions in square brackets vanish, because f decays faster than x^{-2} for $x \to \pm\infty$, otherwise the total number of particles would be infinite. Similarly, f decays faster as v^{-2} for $v \to \pm\infty$, otherwise the total kinetic energy would become infinite. Further, $\mathrm{d}v/\mathrm{d}x = 0$, because v and x are independent variables, and $\mathrm{d}a/\mathrm{d}v = 0$ because the x component of the Lorentz force does not depend on v_x.

2. Any function, $g[\frac{1}{2}mv^2 + q\Phi(x)]$, which can be written in terms of the total energy of the particle, is a solution of the Vlasov equation (cf. Problem 9.1).
3. The Vlasov equation has the property that the phase-space density f is constant along the trajectory of a test particle that moves in the electromagnetic fields $\mathbf{E}$ and $\mathbf{B}$. Let $[\mathbf{x}(t), \mathbf{v}(t)]$ be the trajectory that follows from the equation of motion $m\dot{\mathbf{v}} = q(\mathbf{E} + \mathbf{v} \times \mathbf{B})$ and $\dot{\mathbf{x}} = \mathbf{v}$, then

$$\begin{aligned} \frac{\mathrm{d}f(\mathbf{x}(t), \mathbf{v}(t), t)}{\mathrm{d}t} &= \frac{\partial f}{\partial t} + \frac{\partial f}{\partial \mathbf{x}} \cdot \frac{\mathrm{d}\mathbf{x}}{\mathrm{d}t} + \frac{\partial f}{\partial \mathbf{v}} \cdot \frac{\mathrm{d}\mathbf{v}}{\mathrm{d}t} \\ &= \frac{\partial f}{\partial t} + \frac{\partial f}{\partial \mathbf{x}} \cdot \mathbf{v} + \frac{\partial f}{\partial \mathbf{v}} \cdot \frac{q}{m}(\mathbf{E} + \mathbf{v} \times \mathbf{B}) = 0 \,. \end{aligned} \tag{9.15}$$

4. The Vlasov equation is invariant under time reversal, $(t \to -t)$, $(\mathbf{v} \to -\mathbf{v})$. This means that there is no change in entropy for a Vlasov system.

9.1.4 Relation Between the Vlasov Equation and Fluid Models

Obviously, the Vlasov model is more sophisticated than the fluid models in that now arbitrary distribution functions can be treated correctly. The fluid models did only catch the first three moments of the distribution function: density, drift velocity and effective temperature. Does this mean that the Vlasov model is just another model that competes with the fluid models in accuracy?

The answer is that the collisionless fluid model is a special case of the Vlasov model. The fluid equations can be exactly derived from the Vlasov equation by taking the appropriate velocity moments for the terms of the Vlasov equation. We give here two examples for this procedure and restrict the discussion to the simple 1-dimensional case.

Integrating the individual terms of the Vlasov equation over all velocities gives

$$0 = \frac{\partial}{\partial t} \int f \,\mathrm{d}v + \frac{\partial}{\partial x} \int v\, f \,\mathrm{d}v + a\big[f\big]_{-\infty}^{\infty} = \frac{\partial n}{\partial t} + \frac{\partial}{\partial x}(nu) \,, \tag{9.16}$$

which is just the continuity equation (5.8). Here, $u = (1/n)\int vf\,\mathrm{d}v$ is again the fluid velocity. Likewise, we can multiply all terms by mv and perform the integration to obtain

$$\begin{aligned}
0 &= \frac{\partial}{\partial t}\int mvf\,\mathrm{d}v + \frac{\partial}{\partial x}\int v^2 f\,\mathrm{d}v + a\int v\frac{\partial f}{\partial v}\,\mathrm{d}v \\
&= \frac{\partial}{\partial t}\int mvf\,\mathrm{d}v + \frac{\partial}{\partial x}\left[\int m(v-u)^2 f\,\mathrm{d}v + nmu^2\right] \\
&\quad + a\left(\left[vf\right]_{-\infty}^{\infty} - \int f\frac{\mathrm{d}v}{\mathrm{d}v}\,\mathrm{d}v\right) \\
&= \frac{\partial}{\partial t}(nmu) + \frac{\partial p}{\partial x} + u\frac{\partial}{\partial t}(nmu) + (nmu)\frac{\partial u}{\partial x} - nma \\
&= nm\left(\frac{\partial u}{\partial t} + u\frac{\partial u}{\partial x}\right) + \frac{\partial p}{\partial x} - nma\,,
\end{aligned} \tag{9.17}$$

which is the momentum transport equation (5.28). In the second line, we have used Steiner's theorem for second moments of a distribution, and in the last line, we have used the continuity equation, which cancels two terms. $p = \int m(v-u)^2 f\,\mathrm{d}v$ is the kinetic pressure.

By multiplying with v^n and integrating the terms in the Vlasov equation, we can define an infinite hierarchy of moment equations. Note that each of these equations is linked to the next higher member in the hierarchy: The continuity equation links the change in density to the divergence of the particle flux. The momentum equation describing the particle flux invokes the pressure gradient, which is defined in the equation for the third moments, and so on. Hence, the fluid model must be terminated by truncation. Instead of using a third moment equation that describes the heat transport, one is often content with using an equation of state, $p = nk_\mathrm{B}T$, to truncate the momentum equation.

9.2 Application to Current Flow in Diodes

As a first example, we use the Vlasov equation to study the steady-state current flow in electron diodes under the influence of space charge. The difference from the treatment of the Child-Langmuir law in Sect. 7.2 is that we now allow for a thermal velocity distribution of the electrons at the entrance point of a vacuum diode.

Before starting with the calculation, we summarize our expectations. The electrons are in thermal contact with a heated cathode at $x = 0$, and only electrons with a positive velocity leave the cathode. An anode with a positive bias voltage is assumed at some distance $x = L$. Close to the cathode, the velocity distribution function will be a half-Maxwellian with a temperature determined by the cathode temperature. The limiting current from the Child-Langmuir law corresponds to the situation that the electric field at the cathode vanishes. When the emitted current is lower than the limiting current, the electric field force on an electron is positive and all electrons can flow to the anode. However, when the emitted current is higher than the limiting current, the electric field at the cathode is reversed because a significant amount of negative space charge is formed in front of the cathode. Such a situation with a potential minimum is shown in Fig. 9.3.

Now, only those electrons can overcome the potential barrier that have a sufficiently high initial velocity. Electrons with lower starting velocity will be reflected back to the cathode. Some sample trajectories in $(x - v)$ phase space are shown for transmitted and reflected populations. The velocity distribution can be considered as being partitioned into intervalls of equal velocity, which propagate through the system like the test particles. The separatrix (dotted line in Fig. 9.3) between the populations of *free* and *trapped* electrons is defined by $v = 0$ at the potential minimum.

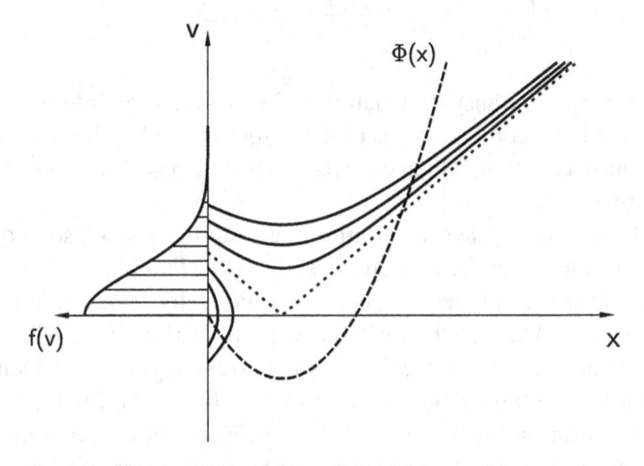

Fig. 9.3 A combination of the half-Maxwellian of the electrons at the cathode of a vacuum diode with the trajectories in phase space (x,v). The potential distribution $\Phi(x)$ is shown as an overlay to the phase space. Only part of the electrons can overcome the potential minimum, the others are reflected back to the cathode

9.2.1 Construction of the Distribution Function

With these prerequisites, we can now state the problem of a stationary flow in terms of the Vlasov and Poisson equations, which we write down for a one-dimensional system

$$v\frac{\partial f(x,v)}{\partial x} + \frac{e}{m_e}\frac{\partial \Phi}{\partial x}\frac{\partial f(x,v)}{\partial v} = 0 \tag{9.18}$$

$$\frac{\partial^2 \Phi}{\partial x^2} = \frac{e}{\varepsilon_0}\int_{-\infty}^{\infty} f(x,v)\,\mathrm{d}v\,. \tag{9.19}$$

The phase space trajectories of test particles form the *characteristic curves* of the Vlasov equation and result from integrating the equation of motion for

$$\frac{dx}{d\tau} = v \quad \text{and} \quad \frac{dv}{d\tau} = \frac{e}{m}\frac{d\Phi}{dx} . \tag{9.20}$$

Here we have introduced the *transit time* τ, which must be distinguished from the absolute time. The considered problem of a stationary flow is independent of absolute time. However, for each electron an individual time τ elapses after injection at the cathode. This time τ can be considered as a series of tick marks along the characteristic curve. The trajectory $v(x)$ follows by eliminating the parameter τ from the solution of (9.20).

Our initial remarks on the properties of the Vlasov equation are now very helpful. Since the value of the distribution function is constant along a phase-space trajectory, the construction of the distribution function at any place x inside the diode is reduced to a mapping problem. This mapping is accomplished by the conservation of total energy for a test electron

$$\frac{1}{2}m_e v^2 - e\Phi = \frac{1}{2}m_e v_0^2 - e\Phi_0 , \tag{9.21}$$

with v_0 the initial velocity at the cathode and Φ_0 the cathode potential. We can set $\Phi_0 = 0$ for convenience. Then the mapping of velocities reads

$$v(\Phi, v_0) = \pm\left(v_0^2 + \frac{2e\Phi}{m_e}\right)^{1/2} . \tag{9.22}$$

This means, that for a given electric potential $\Phi(x)$, we can immediately give the starting velocity v_0 and read the corresponding value of the Maxwellian distribution that we have postulated for a position immediately before the cathode. The two signs of the velocity in (9.22) represent the forward (+) and backward (−) flows of electrons.

We define the velocity distribution at the cathode as the half-Maxwellian

$$f(0, v_0) = A \exp\left(-\frac{m_e v_0^2}{2k_B T_e}\right) . \tag{9.23}$$

The normalization $A = n_e m_e^{1/2}(2\pi k_B T_e)^{-1/2}$ is that of a full Maxwellian. This choice ensures that n_e approximately represents the density of trapped electrons, when the potential minimum is very deep and most of the emitted electrons are reflected.

Those electrons that have a nearly-vanishing positive velocity at the potential minimum, will gain energy from the electric field. This group of electrons represents the lowest velocity in the transmitted electron distribution and defines a cut-off velocity v_c for the distribution

$$v_\mathrm{c} = \left\{ \frac{2e}{m_\mathrm{e}} [\Phi(x) - \Phi_\mathrm{min}] \right\}^{1/2} . \tag{9.24}$$

Then, we find the positive half of the distribution function as

$$f_+(x,v) = A \exp\left(\frac{e\Phi(x)}{k_\mathrm{B}T_\mathrm{e}}\right) \exp\left(-\frac{m_\mathrm{e}v^2}{2k_\mathrm{B}T_\mathrm{e}}\right) \begin{cases} \Theta(v - v_\mathrm{c}) \; ; \; x > x_\mathrm{min} \\ \Theta(v) \qquad ; \; 0 < x < x_\mathrm{min} \end{cases} . \tag{9.25}$$

Θ is the Heaviside step function

$$\Theta(x) = \begin{cases} 0 : x \leq 0 \\ 1 : x > 0 \end{cases} . \tag{9.26}$$

For $x > x_\mathrm{min}$, the distribution function is a cut-off Maxwellian with a density modified by the Boltzmann factor $\exp[e\Phi/(k_\mathrm{B}T_\mathrm{e})]$. All positive velocities are found in the region between cathode and potential minimum.

Negative velocities are only found for $0 < x < x_\mathrm{min}$. However, the distribution only extends up to the negative cut-off velocity because all electrons with a higher velocity have escaped towards the anode. Therefore, the distribution of negative velocities reads in this region

$$f_-(x,v) = A \exp\left(\frac{e\Phi(x)}{k_\mathrm{B}T_\mathrm{e}}\right) \exp\left(-\frac{m_\mathrm{e}v^2}{2k_\mathrm{B}T_\mathrm{e}}\right) \Theta(v + v_\mathrm{c})[1 - \Theta(v)] . \tag{9.27}$$

The distribution functions before and behind the potential minimum are shown in Fig. 9.4.

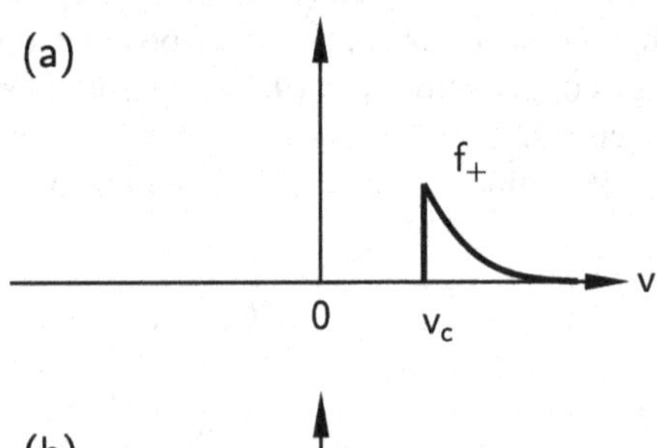

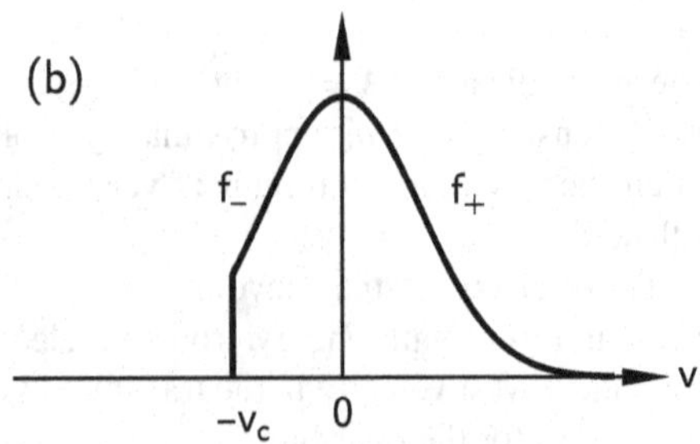

Fig. 9.4 (**a**) Electron distribution function f_+ beyond the potential minimum. (**b**) Distribution function $f_+ + f_-$ between cathode and the potential minimum. The cut-off velocity v_c is determined by the condition to overcome the potential barrier

9.2.2 Virtual Cathode and Current Continuity

The current density at the potential minimum is defined by the integral over positive velocities, $j = -e \int v f_+(v)\, dv$. A simple calculation shows that

$$j = -e n_e \exp\left(\frac{e\Phi_{min}}{k_B T_e}\right)\left(\frac{k_B T_e}{2\pi m_e}\right)^{1/2} . \tag{9.28}$$

This expression contains the density of the full Maxwellian at the cathode multiplied by the Boltzmann factor (which gives the density reduction at the potential minimum) and the mean velocity of a half-Maxwellian distribution, see Problem 9.2. The reader may notice that the result for $\Phi_{min} = 0$ is identical to the electron saturation current (7.29).

Hence, the starting distribution at the potential minimum is again a half-Maxwellian and the potential minimum acts as a *virtual cathode* that feeds the r.h.s. of the diode. It can be easily shown (see Problem 9.3) that the electron current density is conserved in this region. Remembering that the phase space density is also conserved for the characteristic that has $v = 0$ at the virtual cathode, this seems puzzling at first glance because the electron velocity increases on the way towards the anode. Inspecting Fig. 9.5 shows that the distribution function narrows on the velocity scale with increasing mean velocity and this effect compensates for the acceleration.

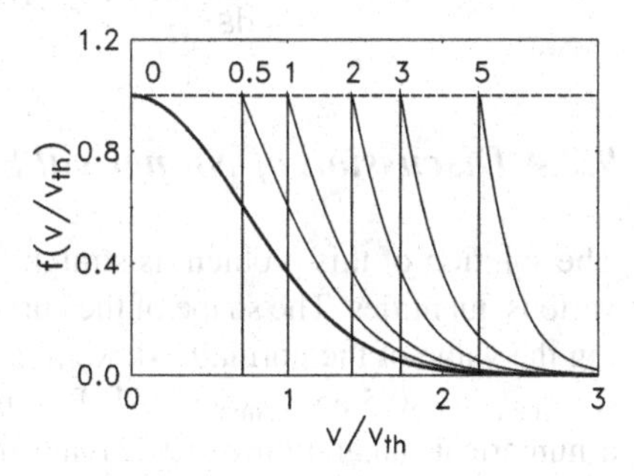

Fig. 9.5 Electron distribution functions between the virtual cathode (potential minimum) and the anode at various normalized potential values $e(\Phi - \Phi_{min})/(k_B T_e)$

9.2.3 Finding a Self-Consistent Solution

Up to now, we have assumed a kind of prescribed potential distribution that possesses a single minimum of negative potential and reaches positive values on the anode side of the diode. However, we were neither able to give the position of the potential minimum nor its depth. The true potential shape results from solving Poisson's equation (9.19) with the distribution functions given by (9.25) and (9.27), which makes the potential self-consistent with the distribution function. This gives the set of equations for the regions I ($0 \le x \le x_{min}$) and II ($x > x_{min}$)

$$\left.\frac{\mathrm{d}^2\Phi}{\mathrm{d}x^2}\right|_I = \frac{e}{\varepsilon_0}\int_{-\infty}^{\infty}(f_+ + f_-)\,\mathrm{d}v \quad ; \quad \left.\frac{\mathrm{d}^2\Phi}{\mathrm{d}x^2}\right|_{II} = \frac{e}{\varepsilon_0}\int_{-\infty}^{\infty} f_+\,\mathrm{d}v \tag{9.29}$$

It is customary to introduce normalized quantities for a later numerical solution of the problem, $\eta = e\Phi/(k_\mathrm{B}T_\mathrm{e})$, $\xi = x/\lambda_\mathrm{De}$, $\lambda_\mathrm{De} = [\varepsilon_0 k_\mathrm{B} T_\mathrm{e}/(n_\mathrm{e}e^2)]^{1/2}$, which gives

$$\left.\frac{\mathrm{d}^2\eta}{\mathrm{d}\xi^2}\right|_I = \frac{\mathrm{e}^\eta}{\sqrt{\pi}}\int_{-\sqrt{\eta-\eta_\mathrm{min}}}^{\infty} \mathrm{e}^{-t^2}\mathrm{d}t \quad ; \quad \left.\frac{\mathrm{d}^2\eta}{\mathrm{d}\xi^2}\right|_{II} = \frac{\mathrm{e}^\eta}{\sqrt{\pi}}\int_{+\sqrt{\eta-\eta_\mathrm{min}}}^{\infty} \mathrm{e}^{-t^2}\mathrm{d}t\,. \tag{9.30}$$

The integral on the r.h.s. can be expressed in terms of the error functions

$$\mathrm{erf}(x) = \frac{2}{\sqrt{\pi}}\int_0^\infty \mathrm{e}^{-t^2}\mathrm{d}t \quad \text{and} \quad \mathrm{erfc}(x) = 1 - \mathrm{erf}(x)\,, \tag{9.31}$$

which results in the compact system

$$\left.\frac{\mathrm{d}^2\eta}{\mathrm{d}\xi^2}\right|_I = \frac{\mathrm{e}^\eta}{2}\left[1 + \mathrm{erf}\left(\sqrt{\eta - \eta_\mathrm{min}}\right)\right] \tag{9.32}$$

$$\left.\frac{\mathrm{d}^2\eta}{\mathrm{d}\xi^2}\right|_{II} = \frac{\mathrm{e}^\eta}{2}\mathrm{erfc}\left(\sqrt{\eta - \eta_\mathrm{min}}\right)\,. \tag{9.33}$$

9.2.4 Discussion of Numerical Solutions

The solution of this problem is straightforward, but involves implicit definitions of various quantities. The shape of the curves $\eta(\xi)$ is completely determined by choosing the value of the normalized minimum potential η_min. Fig. 9.6a shows examples for $\eta_\mathrm{min} = -0.5$ and $\eta_\mathrm{min} = -1$. For the latter starting point, we can first perform a numerical integration of (9.32) and find the position of the cathode ($\eta = 0$) at ξ_1. The integration of (9.33) gives an increasing potential that intersects the anode potential, say at $\eta_a = 3$. This defines the normalized anode position ξ_2. Note that we would obtain different values of ξ_1 and ξ_2 for the integration from $\eta_\mathrm{min} = -0.5$.

How do we find η_min for a given problem? Let us assume that the electron temperature T_e is given by the cathode temperature. Then the emission current of the cathode follows from Richardson's law (see Sect. 11.1.5)

$$|j_\mathrm{em}| = A_\mathrm{R}T^2 \exp\left(\frac{W_\mathrm{R}}{k_\mathrm{B}T}\right)\,, \tag{9.34}$$

in which W_R is the effective work function of the emitter. We can use (9.28), with $\Phi_\mathrm{min} = 0$, to define the electron density n_e and the Debye length λ_De.

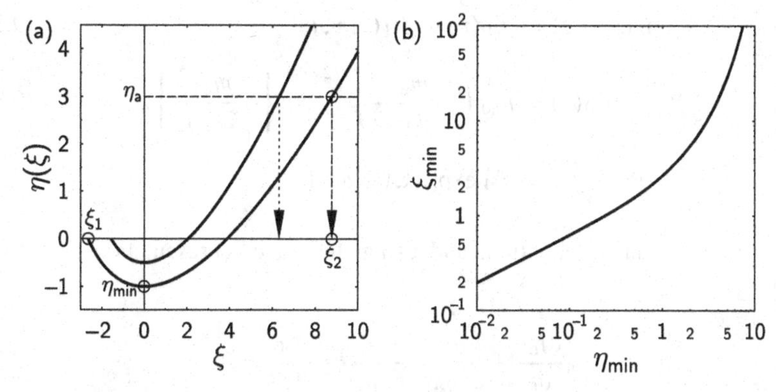

Fig. 9.6 (**a**) Normalized potential shape in a diode with thermal emitter for two values of the minimum potential, $\eta_{min} = -0.5$ and $\eta_{min} = -1$. (**b**) Dependence of the minimum position ξ_{min} on the minimum potential η_{min}

Equation (9.28) then gives the minimum potential as $\eta_{min} = \ln(j_{em}/j)$. The remaining steps are to specify the normalized anode position from the absolute diode length, $L = (\xi_2 - \xi_1)\lambda_{De}$, and to read the anode bias from a curve similar to Fig. 9.6a.

The method described above can be applied to many systems with steady collisionless electron flows, such as low-pressure arc discharges, Q-machines, thermionic converters, or neutralized diodes. Particular emphasis was laid on the simplicity of constructing solutions by characteristics, which are the trajectories of test particles.

9.3 Kinetic Effects in Electrostatic Waves

The second example for kinetic phenomena in plasmas is the discussion of small-amplitude electrostatic waves in unmagnetized plasmas. These problems are one-dimensional and the mathematical apparatus will not overshadow the physical content. Kinetic effects can be expected from the temperature of the plasma constitutents. One of these effects is Landau damping. Other kinetic effects may arise from a more general shape of the distribution function and as a result the waves can become unstable.

9.3.1 Electrostatic Electron Waves

In this Section we search for electron waves near the electron plasma frequency. In this frequency regime, the ions do not participate in the wave motion and form a neutralizing background only. The normal mode analysis starts from splitting the electron distribution function into a homogeneous ($\partial/\partial x = 0$) and stationary ($\partial/\partial t = 0$) distribution $f_{e0}(v)$, which we assume to be a Maxwellian, and a small superimposed wave-like pertubation $f_{e1}(x, v, t)$

$$f_e(x, v, t) = f_{e0}(v) + f_{e1}(x, v, t) \tag{9.35}$$

$$f_{e0}(v) = n_{e0}\left(\frac{m_e}{2\pi k_B T_e}\right)^{1/2} \exp\left\{-\frac{m_e v^2}{2k_B T_e}\right\} \tag{9.36}$$

$$f_{e1} = \hat{f}_{e1} \exp[i(kx - \omega t)] . \tag{9.37}$$

Linearizing the Vlasov equation, and using the wave representation (9.36), we obtain

$$\frac{\partial f_{e1}}{\partial t} + v\frac{\partial f_{e1}}{\partial x} - \frac{e}{m_e}E_1\frac{\partial f_{e0}}{\partial v} = 0 \tag{9.38}$$

$$-i\omega \hat{f}_{e1} + ikv\hat{f}_{e1} - \frac{e}{m_e}\hat{E}_1\frac{\partial f_{e0}}{\partial v} = 0 , \tag{9.39}$$

which yields the perturbed electron distribution function as

$$\hat{f}_{e1} = i\frac{e}{m_e}\frac{\partial f_{e0}/\partial v}{\omega - kv}\hat{E}_1 . \tag{9.40}$$

The vanishing of the denominator $(\omega - kv)$ causes a singularity in the perturbed distribution function, which we will have to address carefully. The electrons with $v \approx \omega/k$ will be called *resonant particles*. In Sect. 8.1.2 we had already seen the particular role of resonant particles for beam-plasma interaction.

The perturbed electron distribution function represents a space charge

$$\rho = e\left(n_i - \int_{-\infty}^{\infty} f_e \, dv\right) = -e\int_{-\infty}^{+\infty} f_{e1} \, dv , \tag{9.41}$$

in which the unperturbed Maxwellian of the electrons is just neutralized by the ion background. Only the fluctuating part of the electron distribution contributes to the space charge. The relationship between the wave electric field and the perturbed distribution function is established by Poisson's equation, which takes the form

$$ik\hat{E}_1 = \frac{\rho}{\varepsilon_0} = \frac{1}{ik}\hat{E}_1\frac{\omega_{pe}^2}{n_{e0}}\int_{-\infty}^{+\infty}\frac{\partial f_{e0}/\partial v}{\omega/k - v}\, dv . \tag{9.42}$$

This equation can be rewritten in terms of the dielectric function $\varepsilon(\omega, k)$ with the result $ik\hat{E}_1\,\varepsilon(\omega, k) = 0$, which requires that $\varepsilon(\omega, k) = 0$ for non-vanishing wave fields. This is the dispersion relation for electrostatic electron waves. It now contains the dielectric function from kinetic theory

$$\varepsilon(\omega, k) = 1 + \frac{\omega_{\rm pe}^2}{k^2} \int_{-\infty}^{+\infty} \frac{1}{n_{\rm e0}} \frac{\partial f_{\rm e0}/\partial v}{\omega/k - v} \, {\rm d}v \tag{9.43}$$

with the derivative of the Maxwellian

$$\frac{\partial f_{\rm e0}}{\partial v} = -n_{\rm e0} \frac{2v}{\sqrt{\pi} v_{\rm Te}^3} \exp\left(-\frac{v^2}{v_{\rm Te}^2}\right) . \tag{9.44}$$

9.3.2 The Meaning of Cold, Warm and Hot Plasma

When the mean thermal speed of the electrons is sufficiently small compared to the phase velocity of the wave (see Fig. 9.7), the contribution from resonant particles in (9.43) is attenuated by the exponentially small factor in the numerator. Then, the main contributions to the integral in (9.43) originate from the interval $[-v_{\rm Te}, v_{\rm Te}]$, where we can expand the function $(\omega/k - v)^{-1}$ into a Taylor series

$$\frac{1}{\omega/k - v} = \frac{k}{\omega} + \frac{k^2}{\omega^2} v + \frac{k^3}{\omega^3} v^2 + \frac{k^4}{\omega^4} v^3 + \cdots . \tag{9.45}$$

The integral (9.43) can be solved analytically using the relations

$$\int_{-\infty}^{+\infty} x^{2n} {\rm e}^{-ax^2} = \frac{1 \times 3 \times \cdots \times (2n-1)}{(2a)^n} \left(\frac{\pi}{a}\right)^{1/2} \tag{9.46}$$

$$\int_{-\infty}^{+\infty} x^{2n+1} {\rm e}^{-ax^2} = 0 . \tag{9.47}$$

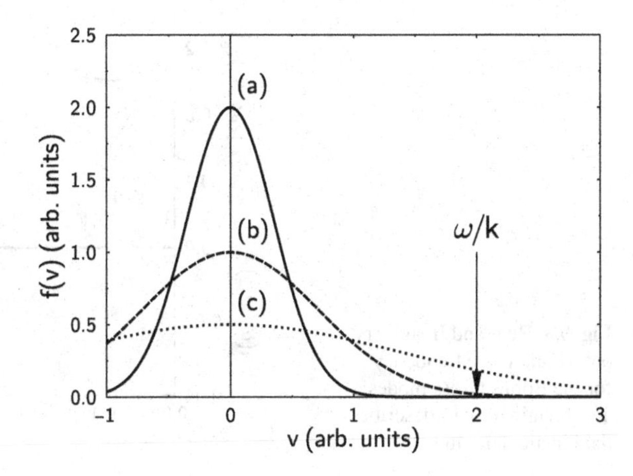

Fig. 9.7 Relation between phase velocity and width of the electron distribution function for a (**a**) cold plasma, (**b**) warm plasma, and (**c**) hot plasma

Using terms up to fourth order in the phase velocity, we obtain

$$\varepsilon(\omega, k) = 1 - \frac{\omega_{\text{pe}}^2}{\omega^2} - \frac{3}{2}\frac{\omega_{\text{pe}}^2}{\omega^4} k^2 v_{\text{Te}}^2 = 0\,. \tag{9.48}$$

The first two terms represent the cold-plasma result (6.45), which we had obtained from the single-particle model. The third term gives a thermal correction that leads to the dispersion relation of Bohm-Gross waves (6.68)

$$\omega^2 = \omega_{\text{pe}}^2 + \gamma_{\text{e}} k^2 \frac{k_{\text{B}} T_{\text{e}}}{m_{\text{e}}}\,. \tag{9.49}$$

Note that we did not have to specify the coefficient $\gamma_{\text{e}} = 3$ for a one-dimensional adiabatic compression. Rather, the adiabaticity of the process followed from the limit $v_{T,\text{e}} \ll \omega/k$ and was obtained from the coefficient for the lowest-order thermal correction in (9.46).

Summarizing, the cold-plasma approximation uses the lowest (non-vanishing, i.e., second) order in the expansion of the dielectric function $\varepsilon(\omega, k)$ in powers of $k v_{\text{Te}}/\omega$. A *warm plasma* description retains the next-higher non-vanishing terms, which are fourth order. Our Taylor expansion breaks down for hot plasmas, which are characerized by $\omega/k \le v_{\text{e}}$. Then, contributions from resonant particles will play a significant role. For the Bohm-Gross modes in Fig. 9.8, the resonant particles lead to wave damping, which we will discuss in the next paragraph.

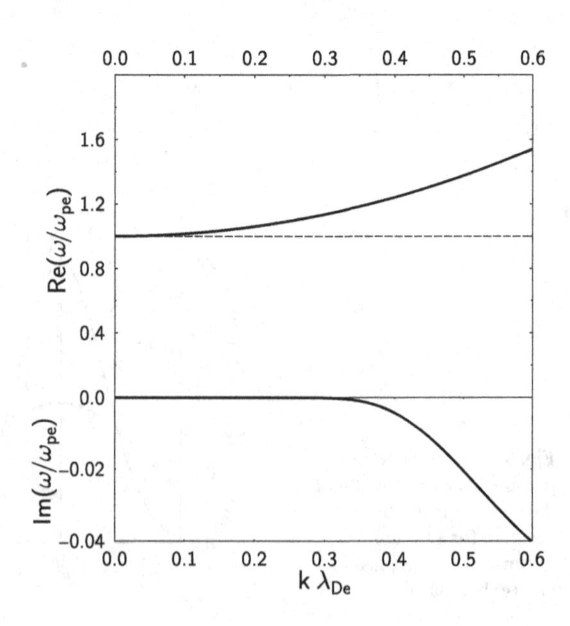

Fig. 9.8 Real and Imaginary part of the wave frequency for the Bohm-Gross modes. The imaginary part describes the kinetic damping

9.3.3 Landau Damping

Let us now allow for phase velocities in the vicinity of the thermal velocity and have a closer look at resonant particles. Up to now, we have only considered the Cauchy principal value of the integral (denoted by the symbol P)

$$\frac{\omega_{\mathrm{pe}}^2}{k^2} P \int_{-\infty}^{+\infty} \frac{\partial f_{\mathrm{e}0}/\partial v}{\omega/k - v} \mathrm{d}v \approx \frac{\omega_{\mathrm{pe}}^2}{\omega^2} + \frac{3}{2}\frac{\omega_{\mathrm{pe}}^2}{\omega^4} k^2 v_{\mathrm{Te}}^2 + \cdots \tag{9.50}$$

Integrals of the type

$$\int_{-\infty}^{\infty} \frac{F(u)}{v - u} \mathrm{d}v \tag{9.51}$$

require a treatment in the complex v-plane. In our case, $u = \omega/k$, will become a complex phase velocity and ω a complex frequency. The Soviet physicist Lev Davidovich Landau (1908–1968) has shown [194] that the proper analytic continuation of the integral (9.51) is found by deforming the integration path in such a way that it passes under the singularity at $v = u$. This integration path is called the Landau-contour and is shown in Fig. 9.9 for the cases of a growing wave ($\mathrm{Im}(u) > 0$), an undamped wave ($\mathrm{Im}(u) = 0$) and a damped wave ($\mathrm{Im}(u) < 0$).

In the following, we assume that the imaginary part of u is small compared to the real part. Therefore, in evaluating the integral in (9.43) we have to use the Cauchy principal value but can use the contribution from the semi-circle in the Landau contour, as shown in Fig. 9.9b. The latter is one half of the residue at the pole. We then obtain

$$0 = 1 - \frac{\omega_{\mathrm{pe}}^2}{k^2}\left(P \int_{-\infty}^{\infty} \frac{1}{n_{\mathrm{e}0}} \frac{\partial f_{\mathrm{e}0}/\partial v}{v - \omega/k} \mathrm{d}v + \mathrm{i}\pi \frac{1}{n_{\mathrm{e}0}} \left.\frac{\partial f_{\mathrm{e}0}}{\partial v}\right|_{v=\omega/k} \right) \tag{9.52}$$

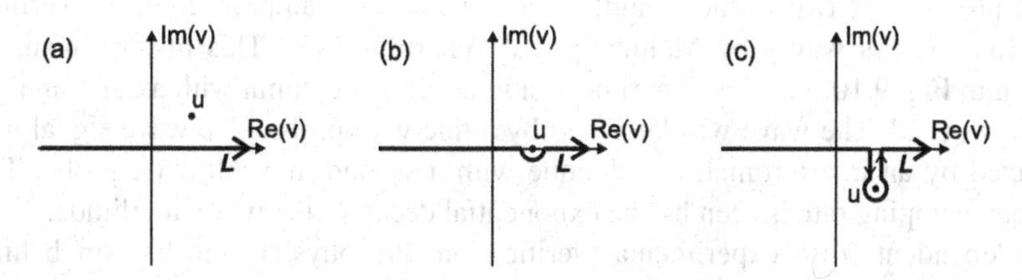

Fig. 9.9 (**a**) The Landau contour L for $\mathrm{Im}(u) > 0$ follows the $\mathrm{Re}(v)$ axis. (**b**) The Landau contour passes with a semi-circle below the pole $\mathrm{Im}(u) = 0$. (**c**) The Landau contour encircles the pole for $\mathrm{Im}(u) < 0$

The contribution from the residue makes the dielectric function complex and we can expect that the solution ω will also be complex. Because of the assumed smallness of the imaginary part of ω, we can use perturbation methods. For the Cauchy principal value, we can use the cold plasma result and obtain

$$0 = 1 - \frac{\omega_{\rm pe}^2}{\omega^2} - {\rm i}\pi \frac{\omega_{\rm pe}^2}{k^2} \frac{1}{n_{\rm e0}} \left. \frac{\partial f_{\rm e0}}{\partial v} \right|_{v=\omega/k} . \tag{9.53}$$

Solving for ω and expanding the square root yields

$$\omega = \omega_{\rm pe} \left[1 + {\rm i}\frac{\pi}{2} \frac{\omega_{\rm pe}^2}{k^2} \frac{1}{n_{\rm e0}} \left. \frac{\partial f_{\rm e0}}{\partial v} \right|_{v=\omega/k} \right] . \tag{9.54}$$

For calculating the derivative of the Maxwellian, however, we have to use the full phase velocity of the Bohm-Gross wave, $\omega_{\rm R} \approx \omega_{\rm pe}[1 + \frac{3}{2}k^2\lambda_{\rm De}^2]$, in the exponent, but it is sufficient to use the cold plasma result in the forefactor. This gives Landau's result for the imaginary part of the wave frequency,

$${\rm Im}(\omega) = -\sqrt{\frac{\pi}{8}} \frac{\omega_{\rm pe}}{k^3\lambda_{\rm De}^3} \exp\left(-\frac{1}{2k^2\lambda_{\rm De}^2} - \frac{3}{2} \right) , \tag{9.55}$$

which is shown in Fig. 9.8. The electrostatic electron waves are damped for short wavelength, $k\lambda_{\rm De} > 0.4$, and the reason for this damping is the resonant interaction with a part of the Maxwell distribution.

This damping mechanism is called *Landau damping* or collisionless damping. Landau's arguments were mostly mathematical in nature in that he used a well-posed initial-value-problem that he treated by means of Laplace transform. Vlasov's normal mode analysis, which we have used above to find the principal value of the dielectric function, cannot predict this kind of wave damping. Only the contour deformation according to the rules of Laplace transform yields damped waves. In the calculation above, we have, for simplicity, incorporated the Landau contour into a normal mode analysis.

At the time of Landau's discovery, experiments on plasma waves were hampered by collisional damping, which masked the predicted effect. It took the technical progress of two decades until, in 1966, Landau damping could be verified for Bohm-Gross waves by Malmberg and Wharton [195]. This historic result is shown in Fig. 9.10. For this experiment, a long plasma column with axial magnetic field was used. The wave was launched by a fine wire probe. The wave signal was detected by an interferometric technique with a second movable wire probe. The Landau damping rate is seen by the exponential decay of the wave amplitude.

Independent from experimental verification, the physical mechanism behind Landau damping was puzzling. In 1961, John Dawson (1930–2001), presented an analysis for the energy exchange between resonant particles and the wave [161]. Many attempts were made during the 60 years after Landau's seminal paper, to

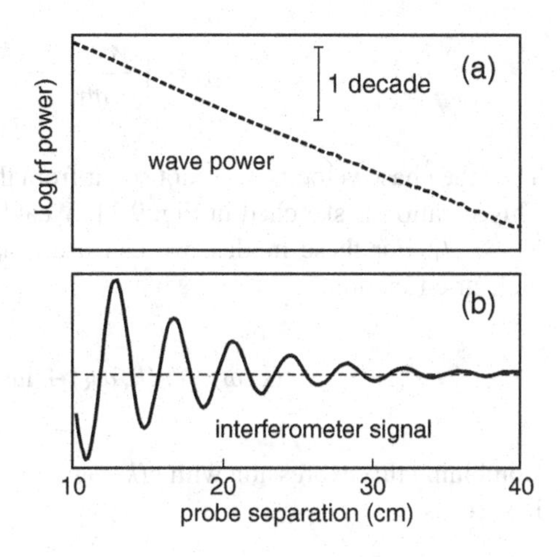

Fig. 9.10 A sketch of the experimental result of Malmberg and Wharton on Landau damping of Bohm-Gross waves. (**a**) power of wave signal received by an axially movable probe on a logarithmic scale. (**b**) interferometer signal on a linear scale

clarify the Landau mechanism, e.g., [196–203]. Other authors reexamined the Landau problem for simplified tutorial purposes [204–207].

The appearance of wave damping in a collisionless plasma is an astonishing result, since we have stated before that the Vlasov equation conserves entropy. This apparent contradiction had puzzled many researchers and it was only in the late 1960s, that the reversibility of the Landau process was demonstrated in terms of plasma wave echoes [208–210], which we will discuss in Sect. 9.3.6. Before doing so, we will give a second example for Landau damping.

9.3.4 Damping of Ion-Acoustic Waves

Our treatment of the ion-acoustic wave in Sect. 6.5.3 was based on a fluid model, which retained the influence of electron and ion temperature by appropriate pressure gradients. In the preceding kinetic treatment of the Bohm-Gross modes, we have learned that the real part of the dielectric function is identical in the fluid model that includes pressure forces, and in the kinetic treatment. Therefore, we can use the dispersion relation (6.76) for calculating the Landau damping of ion-acoustic waves.

The Landau method can be applied when the damping rate ω_{I} of the wave is much smaller than the wave frequency ω_{R}. A small value of Landau damping can be found when the phase velocity of the wave avoids the region of the thermal velocity where the distribution function has its steepest gradient in velocity space. This condition can always be met by the electrons, because $v_\varphi \approx (k_{\mathrm{B}}T_{\mathrm{e}}/m_{\mathrm{i}})^{1/2} \ll (k_{\mathrm{B}}T_{\mathrm{e}}/m_{\mathrm{e}})^{1/2}$ is ensured by the electron-to-ion mass ratio. Landau damping by ions becomes important, when $T_{\mathrm{e}} \approx T_{\mathrm{i}}$, as in Q-machine plasmas, where many investigations of this kind were made. From (6.76) the proper phase velocity of the ion-acoustic wave is

$$v_\varphi = \left(\frac{\gamma_i k_B T_i}{m_i} + \frac{k_B T_e}{m_i} \right)^{1/2} . \tag{9.56}$$

Then the phase velocity v_φ is not so far from the ion thermal speed $(8k_B T_i/\pi m_i)^{1/2}$. This situation is sketched in Fig. 9.11. Weakly damped ion-acoustic waves require $T_e \gg T_i$. For these modes, we can use Landau's treatment and can expand the dielectric function

$$\varepsilon(k, \omega) \approx \varepsilon(k, \omega_R) + i\omega_I \frac{\partial \varepsilon(k, \omega_R)}{\partial \omega_R} = 0 . \tag{9.57}$$

Combining this expression with $\varepsilon(k, \omega_R) = \varepsilon_R(k, \omega_R) + i\varepsilon_I(k, \omega_R)$, gives the damping rate as

$$\omega_I = -\frac{\varepsilon_I(k, \omega_R)}{\partial \varepsilon(k, \omega_R)/\partial \omega_R} . \tag{9.58}$$

The imaginary part of the dielectric function is determined by the ion distribution function

$$\varepsilon_I(k, \omega_R) = -\pi i \frac{\omega_{pi}^2}{k^2} \frac{1}{n_{i0}} \left. \frac{\partial f_{i0}}{\partial v} \right|_{v=\omega_R/k} . \tag{9.59}$$

Inserting the Maxwellian for f_{i0} and using the phase velocity of the ion-acoustic wave for $T_e \gg T_i$, $v_\varphi = \omega_{pi}\lambda_{De}/(1 + k^2\lambda_{De}^2)^{1/2}$, we obtain, after some simple algebra, the imaginary part of the dielectric function as

$$\varepsilon_I(k, \omega_R) = \sqrt{\frac{\pi}{8}} \frac{2\omega_{pi}^2/\omega_R^2}{(1 + k^2\lambda_{De}^2)^{3/2}} \left(\frac{T_e}{T_i} \right)^{3/2} \exp\left[-\frac{T_e/T_i}{2(1 + k^2\lambda_{De}^2)} \right] . \tag{9.60}$$

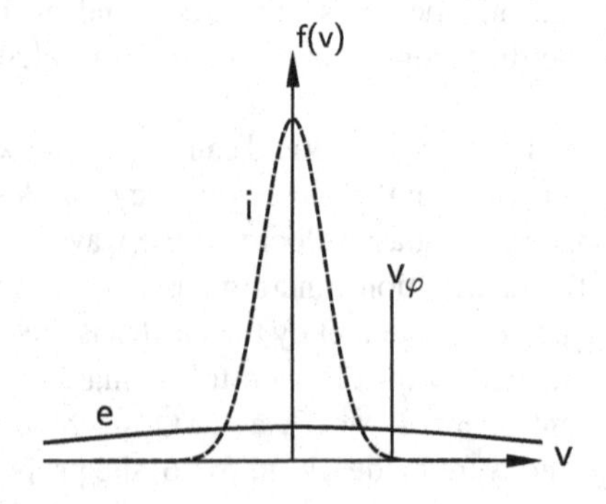

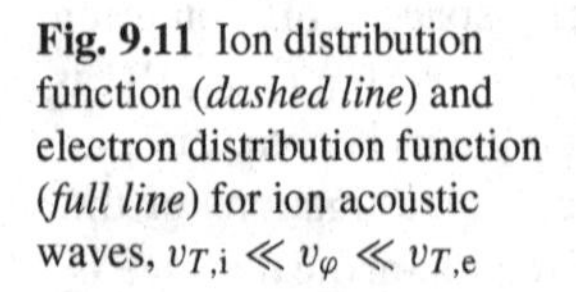
Fig. 9.11 Ion distribution function (*dashed line*) and electron distribution function (*full line*) for ion acoustic waves, $v_{T,i} \ll v_\varphi \ll v_{T,e}$

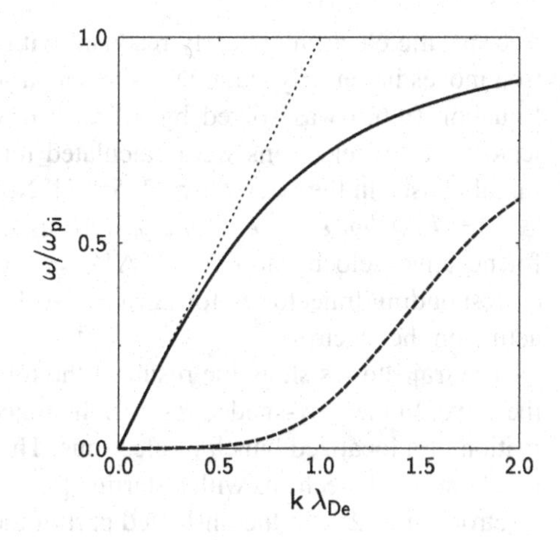

Fig. 9.12 Real part (*full line*) and magnitude of imaginary part (*dashed line*) of the ion-acoustic wave dispersion for $T_e/T_i = 20$. The acoustic dispersion at small $k\lambda_{De}$ is indicated by the dotted line. Landau damping becomes important for $k\lambda_{De} > 1$

Estimating $\partial\varepsilon(k,\omega_R)/\partial\omega_R \approx 2\omega_{pi}^2/\omega_R^3$, the Landau damping rate by ions for the ion-acoustic wave becomes

$$\omega_I = -\sqrt{\frac{\pi}{8}}\frac{\omega_R}{(1+k^2\lambda_{De}^2)^{3/2}}\left(\frac{T_e}{T_i}\right)^{3/2}\exp\left[-\frac{T_e/T_i}{2(1+k^2\lambda_{De}^2)}\right]. \tag{9.61}$$

The damping of the wave becomes significant for $k\lambda_{De} > 1$, as shown in Fig. 9.12 for $T_e/T_i = 20$.

9.3.5 A Physical Picture of Landau Damping

This Section gives a brief summary of the different mechanisms that shed light on Landau damping.

9.3.5.1 Bunching

Let us start with studying the motion of nearly resonant electrons in the potential of the wave $E(x,t) = \hat{E}\cos[kx(t) - \omega t]$. Here, we take care of the fact that the correct force on the electron is not only determined by the change in time of the electric field, but also by the change of position of the electron within the spatial wave pattern. Acceleration or deceleration by the wave field therefore leads to a phase modulation of the electron. When the electron is nearly resonant with the wave, its position can be described as $x(t) = x_0 + v_\varphi t + \Delta v(t)\,t$, with $v_\varphi = \omega/k$ being the phase velocity of the wave and $\Delta v(t)$ the small instantaneous speed of the electron in the wave frame. Then, the motion of the electron is given by

$$\frac{d\Delta v}{dt} = -\frac{e}{m}\cos[kx_0 + k\Delta v(t)t]. \tag{9.62}$$

Because the electron is nearly resonant with the wave, we can argue that the electron moves in a nearly stationary sinusoidal potential well represented by the wave. Equation (9.62) was solved by a fourth order Runge-Kutta integration. The trajectories of 16 electrons were calculated for same initial velocity and equidistant initial phases in the interval $[\pi/2, 5\pi/2]$. Normalized quantities $X = kx$, $T = \omega t$, $\Delta V = k\Delta v/\omega$, $\varepsilon = ek/(m\omega^2)$ were used. Figure 9.13a shows the trajectories for negative velocity at $T = 0$, $\Delta V_0 = -0.1$ and $\varepsilon = 0.01$. Panel c shows the corresponding trajectories for $\Delta V_0 = +0.1$. The panels b,d give the electric force acting on the electron.

The trajectories show the result of the relative motion between the electron and the wave. In the grey-shaded regime, the trajectories are focused and lead to the formation of a localized bunch of electrons. The bunching in panel a involves positive acceleration of electrons with a starting phase of $\approx \pi$ and a negative acceleration for electrons at $\approx 2\pi$. In the unshaded part of the wave phase, the electron trajectories are divergent and lead to a reduction of electron density. This process is known from klystron microwave tubes, where electron bunching in a time-varying electric field is used to amplify microwaves. The same mechanism is acting for positive injection velocity in panel c. Now the electrons starting at $\approx \pi$ overtake those starting at $\approx 2\pi$. The difference is that the bunching point is convected with the mean flow to a more negative position in a and to a more positive position in c.

Bunching is a non-linear process from its very beginning. After being injected with equal speed and homogeneous distribution over all phase angles in $[\pi/2, 5\pi/2]$, it becomes immediately evident that, after a short time, the focusing and defocusing of trajectories leads to an uneven distribution over the wave phase. An analytical

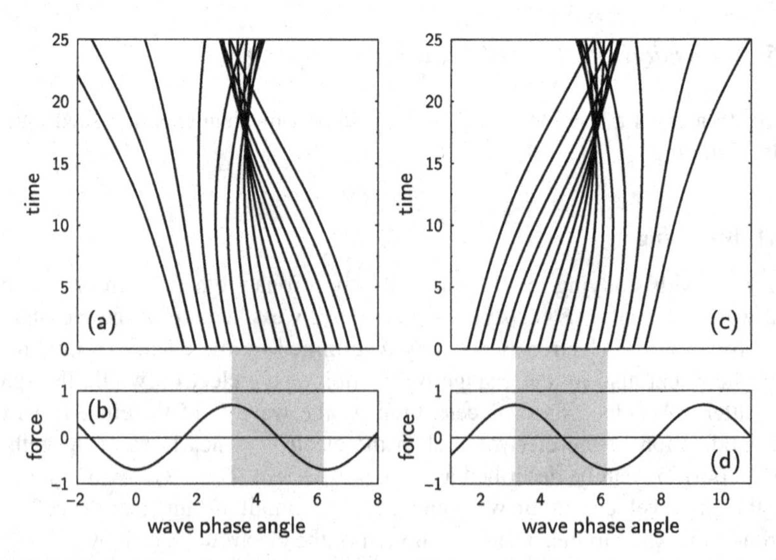

Fig. 9.13 (**a**) Trajectories of test electrons in a sinusoidal wave field for various phases at $t = 0$ and negative initial velocity. (**c**) Same as (**a**) but positive initial velocity. (**b, d**) Corresponding electric force exerted by the wave. The shaded areas indicate bunching of trajectories

expression for the evolution of the phase-averaged velocity $\langle \Delta v \rangle_{x_0}$ can be obtained by the following iterative method described by Nicholson [211]

$$\Delta v^{(2)}(t) = -\frac{e}{m}\hat{E} \int_0^t \cos[kx(t')]\,\mathrm{d}t' \tag{9.63}$$

$$x(t') = x_0 + \int_0^{t'} \Delta v^{(1)}(t'')\,\mathrm{d}t'' \tag{9.64}$$

$$\Delta v^{(1)}(t'') = \Delta v_0 - \frac{e}{m}\hat{E} \int_0^{t''} \cos[k(x_0 + \Delta v_0\, t''')]\,\mathrm{d}t' , \tag{9.65}$$

which starts with calculating the velocity change $\Delta v^{(1)}$ along the unperturbed orbit in (9.65), and uses the improved position (9.64) to obtain an improved velocity $\Delta v^{(2)}$. Using the result $\Delta v^{(1)}$ from (9.65) would give a zero net result on averaging over the initial phases. Therefore, an expression is needed, which is correct to second order. The individual steps of this calculation and averaging the result over the initial phase kx_0 are simple but lengthy. We give here only the final result,

$$\langle \Delta v \rangle_{x_0} = -\frac{1}{24}\left(\frac{e\hat{E}}{m_\mathrm{e}}\right)^2 k^2 \Delta v_0\, t^4 . \tag{9.66}$$

The dependence on the sign of Δv_0 shows that for the fast wave, which has $\Delta v_0 < 0$, the electrons, on average, gain momentum from the wave, whereas they loose momentum in the slow wave. The analytical result (9.66) compares very well with the numerical result $\langle \Delta v \rangle = \frac{1}{16}\sum_k \Delta v_k(t) - \Delta v_0$ obtained from the 16 trajectories in Fig. 9.13, as shown in Fig. 9.14.

A final caveat is necessary here. For clarity, a rather large value $\varepsilon = 0.01$ was chosen in Fig. 9.13. This leads to trapping of some particle orbits, which are reflected inside the potential well of the wave. Trapping becomes evident from the reversal of the slope of a trajectory at $T > 10$ in Fig. 9.13a,c. Linear Landau damping corresponds to the regime, where trapping has not yet occurred.

9.3.5.2 The Net Effect in a Velocity Distribution

The preceding arguments have shown that resonant particles that move slower than the wave extract energy from the wave while those moving faster than the wave give energy back to the wave. Landau's formula (9.55) says that the damping rate is proportional to the slope $\partial f_0/\partial v|_{v_\varphi}$ of the unperturbed velocity distribution at the phase velocity of the wave, as shown in Fig. 9.15a. When the slope is negative, this is equivalent to saying that there are more slower than faster particles in the resonance regime. Hence, the net effect for the wave is damping.

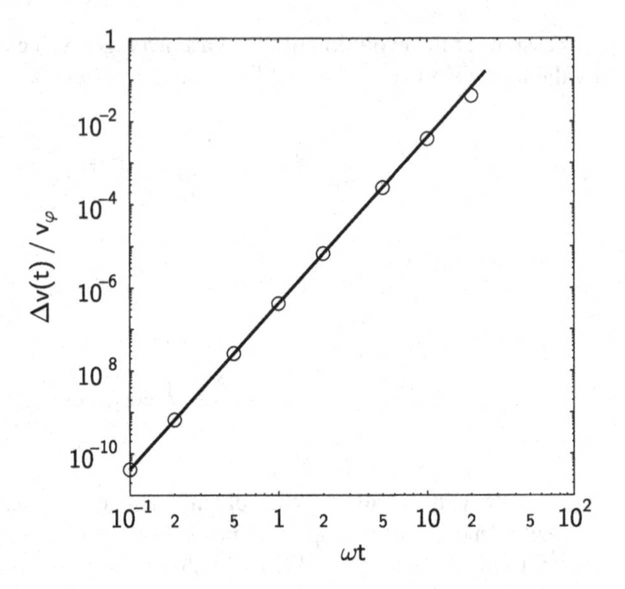

Fig. 9.14 Evolution of the phase-averaged velocity in the wave frame showing the t^4 dependence in (9.66). Circles: numerical result from trajectories in Fig. 9.10

The same arguments lead to growing waves, i.e., an instability, when the velocity distribution has a positive slope. This can happen, e.g., in situations with a warm electron population at rest and a second drifting electron group as shown in Fig. 9.15b. The latter effect is known as *inverse Landau damping*. When both populations are sufficiently narrow and well separated, we recover the beam-plasma instability discussed in Sect. 8.1.2.

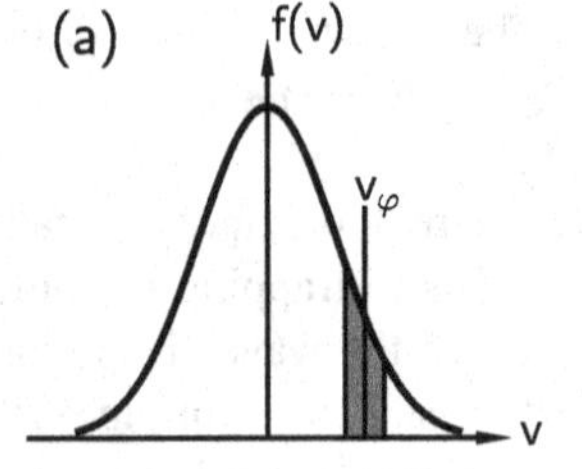

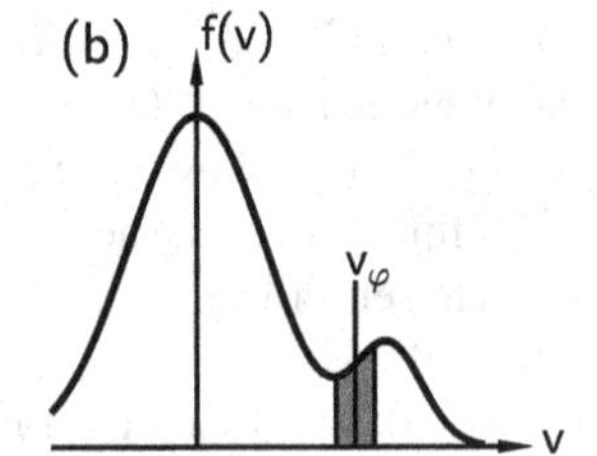

Fig. 9.15 **(a)** Maxwellian distribution with an intervall of resonant particles centered about v_φ. Landau damping is a consequence of $\mathrm{d}f/\mathrm{d}v < 0$. **(b)** A non-Maxwellian distribution with $\mathrm{d}f/\mathrm{d}v > 0$ for resonant particles

9.3.5.3 The Effect of Landau Damping on the Particles

Summarizing our understanding of longitudinal waves in a hot plasma, we can state that the wave dispersion is determined by the non-resonant particles. Loosely speaking, the dispersion is determined by the elastic properties of the plasma medium. Further, we have learned that the wave amplitude $\hat{E}$ decreases in time by a group of resonant electrons that experience a net acceleration at the expense of the wave energy. So far, so good. Still unsolved is the question how this wave damping is compatible with the reversibility of the Vlasov equation and the conservation of entropy.

Our discussion of Landau damping was so far restricted to the analysis of the decay of the wave electric field, which according to (9.42) is an integral over the perturbed distribution function. Therefore, the decay of the wave only reflects the average response of all particles. But what happens to individual groups of particles? Will any initial perturbation of this particle group also decay at the Landau damping rate? This question can be answered by seeking a solution of the linearized Vlasov equation (9.38) for the particle distribution.

For this purpose, we assume that the distribution function has been perturbed by a short wave flash, $E(x,t) \propto \exp(ikx)\delta(t)$, that is periodic in space but localized around $t = 0$. Then the perturbed distribution $f_1(x, v, t)$ obeys the linearized equation

$$\frac{\partial f_1}{\partial t} + v\frac{\partial f_1}{\partial x} - \frac{e}{m_\mathrm{e}}\hat{E}_1 \mathrm{e}^{ikx}\frac{\partial f_0}{\partial v}\delta(t) = 0\,. \tag{9.67}$$

Decomposing $f_1(x, v, t) = \bar{f}_1(v, t)\exp(ikx)$ we obtain

$$\frac{\partial \bar{f}_1}{\partial t} + ikv\bar{f}_1 = \frac{e}{m_\mathrm{e}}\hat{E}_1\frac{\partial f_0}{\partial v}\delta(t)\,. \tag{9.68}$$

Remembering that $\delta(t) = \int_{-\infty}^{\infty} \mathrm{e}^{-\mathrm{i}\omega t}$ we can seek the response $\bar{f}_1$ for each individual frequency of the wave packet that makes up the delta function

$$\bar{f}_1 = \frac{1}{2\pi}\int_{-\infty}^{\infty} \tilde{f}_1(\omega, t)\mathrm{d}\omega \tag{9.69}$$

Note, that $\tilde{f}_1$ is not the Fourier transform of $\bar{f}_1$. Then $\tilde{f}_1$ is obtained from

$$\frac{\partial \tilde{f}_1}{\partial t} + ikv\tilde{f}_1 = \frac{e}{m_\mathrm{e}}\hat{E}_1\frac{\partial f_0}{\partial v}\mathrm{e}^{-\mathrm{i}\omega t} \tag{9.70}$$

and must fulfill the initial condition $\tilde{f}_1(t = 0) = 0$. The solution of the differential equation consists of the solution of the homogeneous equation

$$\tilde{f}_1^{\,\mathrm{hom}} = \lambda \mathrm{e}^{-ikvt} \tag{9.71}$$

and any particular solution of (9.70), which can be assumed as $\propto \exp(-\mathrm{i}\omega t)$. Inserting in (9.70) gives

$$\tilde{f}_1^{\,\mathrm{part}} = \mathrm{i}\frac{e}{m_\mathrm{e}}\hat{E}_1\frac{\partial f_0}{\partial v}\frac{\mathrm{e}^{-\mathrm{i}\omega t}}{\omega - kv}\,, \tag{9.72}$$

which is our previous result (9.40) for the wave-like perturbation of the distribution function. However, the total solution for $\tilde{f}_1$

$$\tilde{f}_1 = \mathrm{i}\frac{e}{m_\mathrm{e}}\hat{E}_1\frac{\partial f_0}{\partial v}\,\frac{\mathrm{e}^{-\mathrm{i}\omega t}-\mathrm{e}^{-\mathrm{i}kvt}}{\omega - kv}\,, \tag{9.73}$$

now contains an additional term $\propto \mathrm{e}^{\mathrm{i}kvt}$, which we call the *ballistic response* of the distribution function to the initial wave flash. Note that the denominator $\omega - kv$ ensures that the main contribution to the wave-like response and the ballistic response is concentrated in the resonant particles. Again, a proper treatment of the pole in integrating $\tilde{f}_1$ over ω will give a complex ω. Therefore, the wave-like response of the distribution function will decay by Landau damping. However, the ballistic term contains real values for k and v and will not disappear for $t \to \infty$.

The ballistic response represents a superimposed corrugation of the distribution function as shown in Fig. 9.16. In the course of time, this corrugation becomes ever finer. Therefore, the macroscopic electric field associated with this perturbed distribution, which results from an integral over velocity, will vanish by *phase mixing*. This solves the paradox that a macroscopic quantity dies out while the information is still hidden in the subtleties of the distribution function. It is no surprise that Coulomb collisions, which generate slight velocity changes, will be an efficient mechanism to erase this memory of the initial perturbation.

In summary, any disturbance observed at (x, t) has two sources. One is the plasma waves that have propagated in space and time and reach x at time t. This is mainly the contribution of the non-resonant particles. The other source are particles, which reach (x, t) and carry with them the memory of their original perturbation. Here, the resonant particles are most important in draining energy from the wave. Their velocity spread is the reason for the phase mixing of the ballistic response.

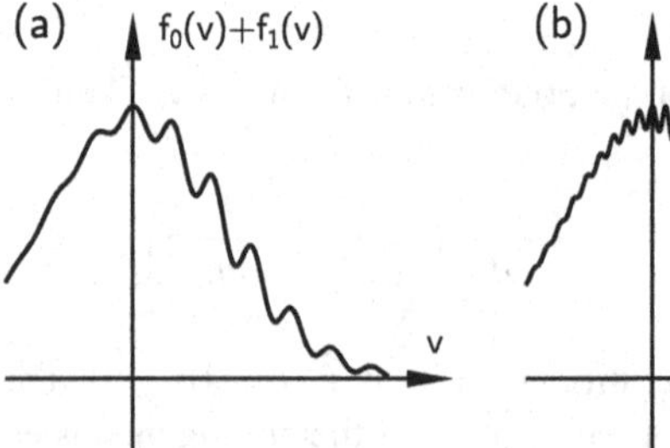

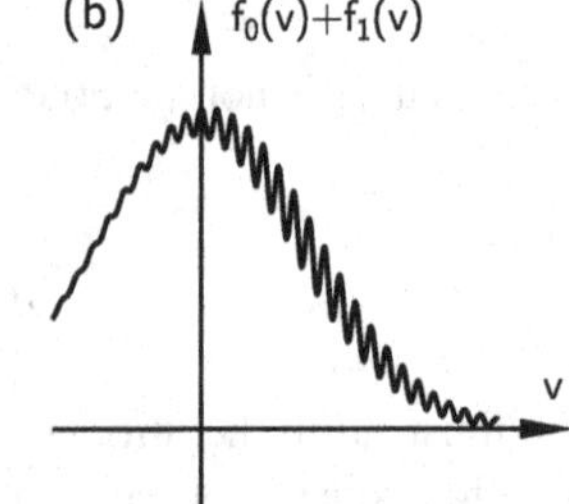

Fig. 9.16 A Maxwellian with superimposed ballistic response to an initial perturbation. (**a**) $t = t_0$, (**b**) $t = 3t_0$

9.3.6 Plasma Wave Echoes

The existence of this ballistic response, which is at the heart of Landau damping, could be proved experimentally by the generation of plasma echoes for electron waves [208, 209, 212] and for ion waves [213]. In the preceding Section, we had derived the ballistic response to a wave flash localized in time. Although it is possible to produce temporal echoes, it is experimentally much easier to study spatial echoes, which are produced by a localized source that imposes a continuous wave pattern in time $\propto e^{-i\omega t}\delta(x)$.

The principle of a wave echo experiment with ion waves [210] is shown in Fig. 9.17. A first wave with frequency ω_1 is excited by a grid at $x = 0$ and decays by Landau damping. The perturbed distribution function in the region $x > 0$ then takes the limiting form

$$\tilde{f}_1 \propto \cos\left(\omega_1 t - \frac{\omega_1 x}{v}\right). \tag{9.74}$$

When a voltage with a frequency $\omega_2 > \omega_1$ is applied to a grid at position $x = d$, where the macroscopic signal of the first wave has disappeared by phase mixing, two processes will be observed. A second wave with frequency ω_2 will propagate away from the grid and decay by Landau damping. Further, the perturbed distribution function will now contain a second-order ballistic memory from this modulation, which reads

$$\tilde{f}_2 \propto \cos\left(\omega_1 t - \frac{\omega_1 x}{v}\right)\cos\left(\omega_2 t - \frac{\omega_2(x-d)}{v}\right) \tag{9.75}$$

$$= \frac{1}{2}\left\{\cos\left[(\omega_1+\omega_2)t - \frac{\omega_1 x + \omega_2(x-d)}{v}\right]\right.$$
$$\left. + \cos\left[(\omega_2-\omega_1)t - \frac{\omega_2(x-d) - \omega_1 x}{v}\right]\right\}. \tag{9.76}$$

The first term in (9.76) is again rapidly oscillating and will give no macroscopic signal because of phase mixing. The second term, however, becomes stationary for $\omega_2(x-d) - \omega_1 x = 0$ and represents the echo signal, which will attain its maximum amplitude at

$$x_{\text{echo}} = \frac{\omega_2}{\omega_2 - \omega_1} d. \tag{9.77}$$

In this way, plasma wave echoes demonstrate the reversibility of the phase mixing process. This is also a proof of the entropy conservation of the Vlasov model.

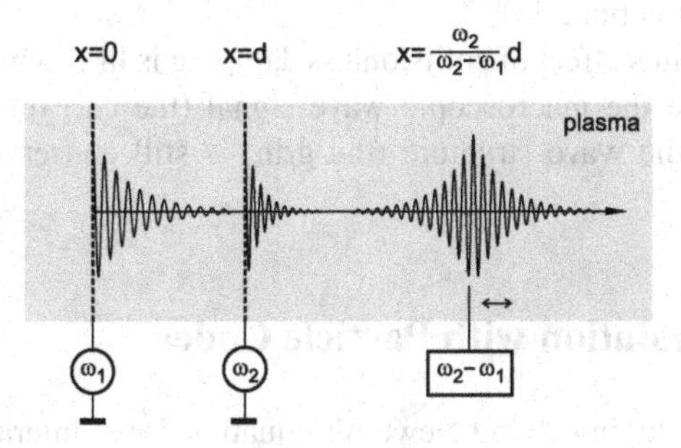

Fig. 9.17 Schematic of a wave echo experiment. The first wave is launched by a grid at $x = 0$, the second at $x = d$. The echo signal (magnified amplitude) is detected with a movable probe and a receiver tuned to $\omega_2 - \omega_1$. The echo maximum is found at $x = \omega_2 d/(\omega_2 - \omega_1)$

9.3.7 No Simple Route to Landau Damping

The discussion in the previous paragraphs was an attempt to grasp some constituents behind the physical mechanism of Landau damping. Each individual view, however, must be incomplete—otherwise Landau damping could be "explained" by fluid models or even single-particle motion whereas it is at the heart of kinetic plasma description (Fig. 9.18).

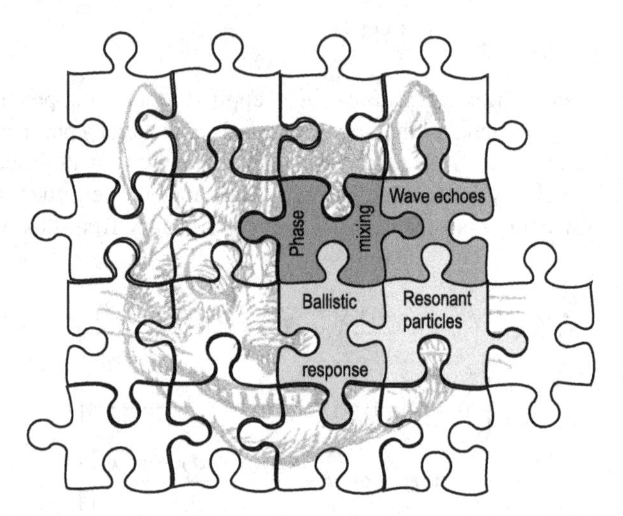

Fig. 9.18 The Landau damping puzzle showing the ingredients of a simplified description of the physical processes behind Landau's famous formula

We have learned that concepts like "the wave" must be considered carefully. The wave electric field is mostly due to the oscillatory motion of the non-resonant particles, which we had seen to be responsible for the wave dispersion. Landau damping describes the interaction between the wave and the resonant particles, in other words, the collective scattering of a subgroup of particles by the majority of particles, which gives rise to wave damping. Our simplification was the introduction of an artificial boundary that separates the resonant from the non-resonant particles, which allowed some useful estimates. Landau's treatment, however, does not make such an explicit distinction.

In the end, Landau's effect of collisionless damping is like Lewis Carrol's famous Cheshire cat, where the macroscopic wave signal (the cat) disappears while the information about the wave structure (the grin) is still conserved in the ballistic response.

9.4 Plasma Simulation with Particle Codes

The principal difficulty in solving Newton's equation for N interacting particles lay in the sheer number of $N = 10^{10} - 10^{20}$ particles in a typical plasma. Moreover, calculating the interaction force between N particles involves $\approx N^2$ evaluations of the shielded Coulomb force. We had overcome this difficulty in the previous section by grinding the particles into ever finer "Vlasov sand" that has the same q/m for each grain, and therefore preserves the interaction forces between volume elements of finite size. This concept allowed a statistical treatment in terms of the Vlasov equation.

In this Section, we go into the other direction and merge all particles within a volume element into a *superparticle*. Again this superparticle has the same q/m as the individual particles it consists of. Typical numbers of particles within a superparticle can be $N_s = 10^4 - 10^6$. A further improvement for the numerical simulations of electrostatic problems with superparticles is the assignment of the charge distribution, the resulting electric field and potential to a fixed grid with N_g grid points. This reduces the calculation effort for a one-dimensional system to $N N_g \log_2 N_g$ instead of N^2 steps, which can be a substantial reduction, if $N = 10^5$ and $N_g = 100$, typically.

Plasma physics by computer simulation is now an established branch of our field. The fundamental methods are described in textbooks, e.g., [214, 215]. In the following, the particle-in-cell (PIC) method will be described, which is implemented in many codes. Some of these codes are available for free.[1] Have fun playing yourself with the codes. It will give you the impression that you can master the plasma. The experimental plasma physicists often experience that the plasma masters the experimenter.

9.4.1 The Particle-in-Cell Algorithm

The discussion of plasma simulation will be restricted to one-dimensional (1-D) electrostatic problems, which we had studied before by analytical methods. The PIC method assumes that the particle can be found with the same probability at any place within a cell of the computational grid. This is equivalent to assigning a box-shaped profile of width Δx for the particle. When the superparticle moves over the grid, there is a continuous change of its contribution to a cell p and its neighboring cell $p+1$, as shown in Fig. 9.19.

The charge assignment to grid point x_p is made by

$$\rho_p = \frac{q N_s}{\Delta x} \sum_{i=1}^{N_p} W(x_i - x_p) \tag{9.78}$$

with the linear weighting function

$$W(x) = \begin{cases} 1 - |x| & : \quad |x| < 1 \\ 0 & : \quad |x| \geq 1 \,. \end{cases} \tag{9.79}$$

[1] http://ptsg.eecs.berkeley.edu/

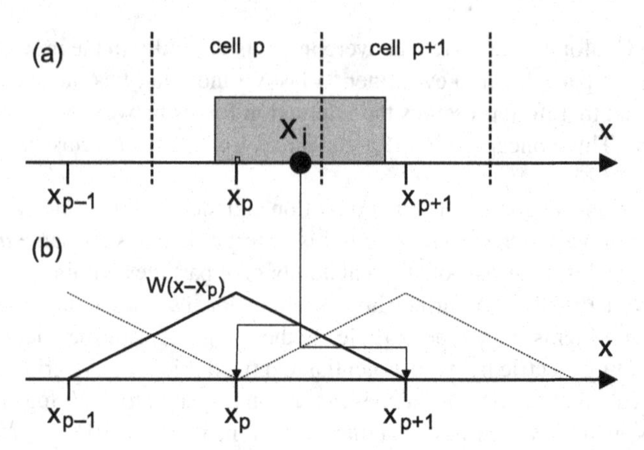

Fig. 9.19 (**a**) The particle at x_i is represented by a box-like charge cloud of width Δx. When it moves over the calculation grid, charge is assigned to cells p and $p+1$ according to the overlap of the cloud with the cell. (**b**) This charge assignment is described by the weighting function $W(x - x_p)$ for the cell p

The advantage of such extended charge clouds lies in the smooth variation of the interaction force between two such clouds. If the particle was represented by a thin charge sheet, then the interaction force between two such sheets, which is independent of the distance between the sheets, would suddenly switch sign when the sheets penetrate each other.

The electric field results from solving Poisson's equation on this grid. First the second derivative is replaced by a second difference

$$\frac{\Phi_{p-1} - 2\Phi_p + \Phi_{p+1}}{(\Delta x)^2} = -\frac{\rho_p}{\varepsilon_0}\,. \tag{9.80}$$

Then, the electric field results from

$$E_p = \frac{\phi_{p-1} - \phi_{p+1}}{2\Delta x}\,. \tag{9.81}$$

Poisson's equation can be readily solved by diagonalization of the matrix, see e.g., [214]. For periodic boundary conditions, methods based on fast Fourier transform may be even superior. The interpolation of the field force at the position of the particle is made with the same weighting function (9.79) as used for the charge assignment on the grid

$$F_i = qN_s \sum_{p=0}^{N_g-1} W(x_i - x_p)E_p\,. \tag{9.82}$$

The particle position is advanced by a discrete representation of Newton's equation in terms of a *leap-frog* scheme

$$\frac{x_i^{n+1} - x_i^n}{\Delta t} = v_i^{n+1/2}$$
$$\frac{v_i^{n+1/2} - v_i^{n-1/2}}{\Delta t} = \frac{F(x_i)\Delta t}{m_i}, \tag{9.83}$$

in which the superscript labels the number of the time step. The advancement of the velocity is made at half timesteps. A full cycle of the PIC time step is shown in Fig. 9.20.

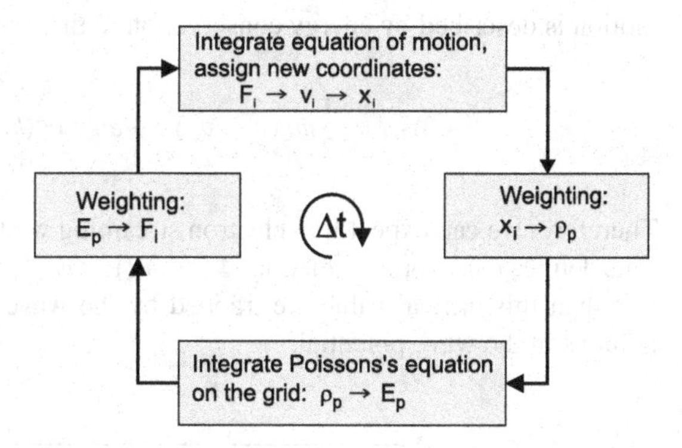

Fig. 9.20 Time step of the particle-in-cell technique

9.4.2 Phase-Space Representation

Before discussing the interaction of electrons with wave fields, let us shortly recall the description of a dynamical system in phase space. A simple one-dimensional system, the pendulum, is described by the potential energy

$$W_{\text{pot}} = -W_0 \cos(\varphi). \tag{9.84}$$

For small excitation energies, the pendulum performs harmonic oscillations about the equilibrium position at $\varphi = 0$. The potential well and the phase space φ–$(d\varphi/dt)$ of this pendulum are shown in Fig. 9.21. The phase space contours in Fig. 9.21b correspond to various values of total energy

$$W_{\text{tot}} = \frac{1}{2}\mathsf{I}\left(\frac{d\varphi}{dt}\right)^2 - W_0 \cos(\varphi), \tag{9.85}$$

I being the moment of inertia for this pendulum.

The phase space representation has the following properties:

- For small total energy, the energy contour is an ellipse.
- There are bound oscillating states for $W_{\rm tot} < 2W_0$ and free rotating states for $W_{\rm tot} > 2W_0$, separated by a separatrix, which is shown dashed line in Fig. 9.21b.
- The motion of a phase space point is always clockwise, as indicated by the arrows in Fig. 9.21b.
- The oscillation period becomes longer when the oscillation amplitudes is increased. It becomes infinite at the separatrix.

We will use this phase space picture to study the motion of nearly resonant electrons in a wave field. The resonance condition $v \approx v_\varphi$ ensures that the electron "sees" a nearly constant potential well of the wave. Therefore, in a first approximation, its motion is described by energy conservation in the moving frame of reference:

$$W_{\rm tot} = \frac{1}{2} m_{\rm e}(v - v_\varphi)^2 + e\hat{\Phi}\cos(kx) = {\rm const}\,. \tag{9.86}$$

Therefore, we can expect free electron streaming w.r.t. the wave when $W_{\rm tot} > 2e\hat{\Phi}$. This defines the trapping potential $\Phi_{\rm t} = m(v - v_\varphi)^2/(4e)$. Electrons with an energy less than this critical value are trapped by the wave and perform bouncing oscillatiuons in the wave potential.

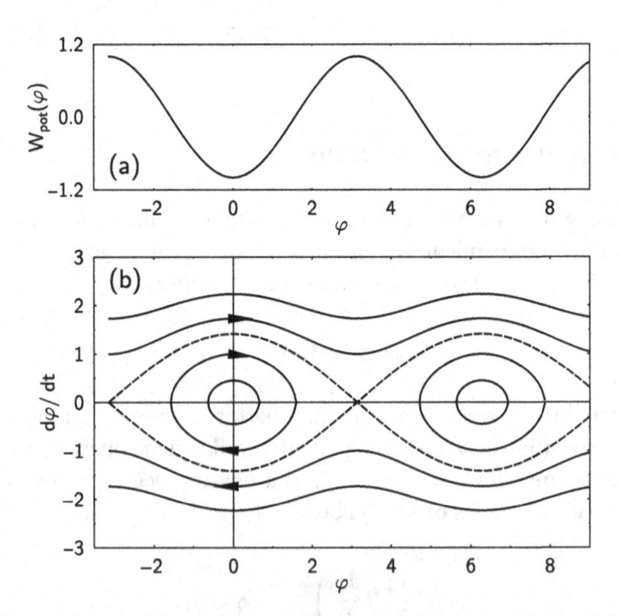

Fig. 9.21 (**a**) Potential energy of a pendulum. (**b**) Phase space contours of the pendulum for various values of total energy. The dashed line separates bound oscillating states inside from free rotating states

9.4.3 Instability Saturation by Trapping

For comparison with the analytical treatment, we will study here a PIC-simulation of the beam-plasma instability. The simulation is made with the ES1 code[2] described in [215].

The system consists of a majority of electrons at rest and an electron beam with a beam fraction $\alpha_b = 0.01$. The system is neutralized by immobile ions. The number of plasma particles is 256 and there are 512 beam particles with q/m reduced according to the beam fraction. The number of grid points is 64 and the time step is chosen as $\omega_{pe}\Delta t = 0.05$. For those readers, who want to run their own simulations, it should be mentioned that the code uses normalized quantities

$$t_{norm} = \omega_{pe} t_{phys}\,, \quad x_{norm} = k x_{phys}\,, \quad v_{norm} = \frac{v_{x,phys}}{v_0}\,, \; W_{norm} = \frac{2W_{phys}}{m_e v^2}\,.$$

In the text below we will refer to the real physical quantities.

The result of this simulation is shown in Fig. 9.22b–f as a series of electron phase space plots combined with the electric potential of the wave. Note that the potential energy $W_{pot} = -e\Phi$ has the opposite sign. The ES1 code uses periodic boundary conditions and the system length is just one wavelength of the unstable mode. The x coordinate is therefore given as a wave phase angle between 0 and 2π. The system of reference rests in the moving frame of the unperturbed beam. Therefore, the unstable slow space-charge wave, which is propagating nearly at the beam speed, is moving slowly to the left in this representation.

The exponential growth of the instability and its eventual saturation can be seen in a semi-log plot of the electric field energy and the beam kinetic energy in Fig. 9.22a. For $\omega_{pe}t < 72$, an exponential growth of the beam-plasma instability is found, which becomes a straight line in the semi-log representation. The growth rate of the wave energy $\varepsilon_0 \hat{E}^2/2$, is twice the growth rate (8.9) of the wave amplitude. From the slope of the straight line we obtain the growth rate as $2\gamma/\omega_{pe} = 0.272$, which compares well with the value from linear instability analysis, $2\gamma/\omega_{pe} = 3^{1/2}(\alpha_b/2)^{1/3} = 0.296$.

The time $\omega_{pe}t = 72$ marks the end of the exponential growth phase. Figure 9.21b shows that the beam electrons are still free streaming, but experience a considerable velocity modulation. The initial beam velocity $v_0 = 1$ is indicated by a fine horizontal line. The non-resonant plasma electrons show a much smaller velocity modulation. The field energy in Fig. 9.21a begins to saturate and then performs oscillations, in which field energy and beam kinetic energy are exchanged. At the first maximum of the field energy, the corresponding phase space in Fig. 9.21c shows that the beam particles have been trapped by the wave field and begin to perform bouncing oscillations in the potential well represented by the positive half-wave. In panel d, the wave energy has reached a minimum, which is associated with a large

---

[2] The code is available from `http://ptsg.eecs.berkeley.edu/#Software`

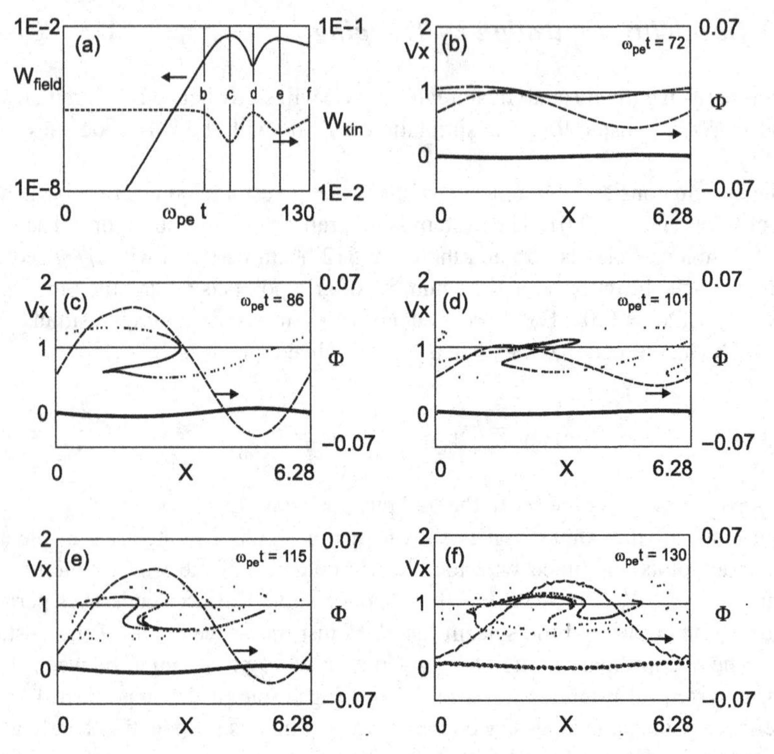

Fig. 9.22 Particle-in-cell simulation of the beam-plasma instability for $\alpha_b = 0.01$ (**a**) Semi-log plot of the electric field energy (*full line*) and beam kinetic energy (*dashed line*). (**b–f**) x–v_x phase space showing plasma electrons at $v \approx 0$ and beam electrons at $v \approx 1$. For comparison, the wave potential (*long-dashed line*) is superimposed. The selected examples correspond to the times marked by vertical lines in panel (**a**)

group of beam electrons performing again a forward motion in the potential well. These electrons have gained energy from the wave. At this point, the beam kinetic energy has nearly reached its original unpertubed value. Panels e and f show that the beam electrons begin to spread over all phases of the trapping motion but are mostly confined in the potential well of the wave. The spread of the electrons over the entire potential well is a consequence of the dependence of the bouncing frequency in a sinusoidal potential well on the electron energy.

The bounce frequency can be calculated from the curvature at the minimum of the potential well; the wave field energy $\frac{1}{2}\varepsilon_0 E^2$ at trapping can be obtained from the trapping potential Φ_t, see problems 9.5 and 9.6.

9.4.4 Current Flow in Bounded Plasmas

At last, let us revisit the questions of current flow in diodes. The first example is intended to illustrate the similarity of the analytic treatment of the virtual cathode problem in diodes with thermal emitters with PIC-simulations. The second example introduces an instability of the electron beam in a diode, which shows that an inhomogeneous equilibrium flow can switch to an oscillating state of a different topology in phase space.

9.4.4.1 The Virtual Cathode of a Thermal Emitter

When the emission current of a cathode exceeds the limiting value given by the Child-Langmuir law (7.11) a potential minimum forms that reflects slow electrons back to the cathode. The current beyond the potential minimum is just the Child-Langmuir current. Therefore, the potential minimum was named *virtual cathode*. An analytical treatment of this problem by the Vlasov-theory was given in Sect. 9.2.

Here, we re-examine the problem by PIC-simulation with the PDP1 code [216]. The simulation parameters for the diode with thermal emitter in Fig. 9.23 are given in Table 9.1.

The thermal velocity of the electrons is here defined as $v_T = (k_B T_e/m_e)^{1/2}$ and corresponds to a cathode temperature of 650 K. The Child-Langmuir current for an empty diode with this set of parameters is $j_{CL} = 0.0235\,\mathrm{A\,m^{-2}}$. The injection current is therefore about 40 times larger, which leads to the formation of a potential minimum of $-0.16\,\mathrm{V}$ depth. The actual anode current density in this diode is $0.059\,\mathrm{A\,m^{-2}}$.

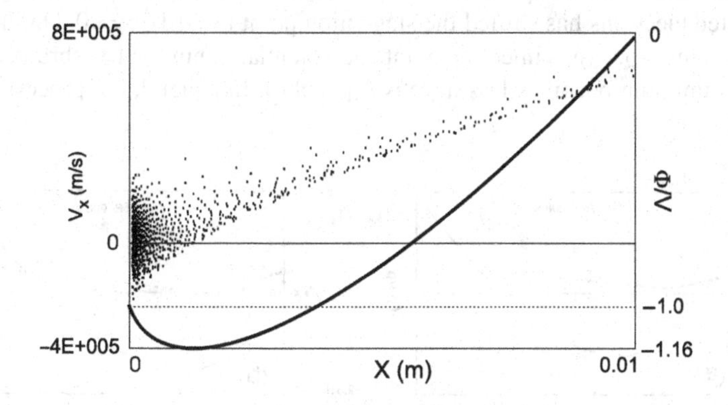

Fig. 9.23 Formation of a potential minimum (virtual cathode) in front of a thermal emitter. The electron phase space is superimposed. Note the reflected electrons that return to the cathode and the narrowing of the electron distribution by acceleration beyond the virtual cathode

Table 9.1 Simulation parameters for virtual cathode formation

Length of diode	0.01 m
Applied voltage	1 V
Injection current density	1 $\mathrm{Am^{-2}}$
Electron thermal velocity	1×10^5 m/s
Number of cells	400
Particles per superparticle	5×10^5
Time step	5×10^{-11} s

We can now check the consistency of the simulation results by comparing with the analytical model presented in Sect. 9.2. The depth of the virtual cathode $\eta_{\min} = \ln(j/j_{\mathrm{em}})$ gives

$$\Phi_{\min} = \frac{k_{\mathrm{B}} T_{\mathrm{e}}}{e} \ln\left(\frac{j}{j_{\mathrm{em}}}\right) = 0.056\,\mathrm{V} \ln(0.059) = -0.16\,\mathrm{V}\,. \tag{9.87}$$

This is just the observed depth of the potential minimum in the simulation.

9.4.4.2 Blocking Oscillations of an Electron Beam

Let us now investigate the injection of a monoenergetic electron beam into a vacuum diode without any applied external voltage. In Sect. 8.3.5 we had argued that an electron beam in a vacuum diode is slowed down by its own space charge and a virtual cathode forms in the center of the diode. The maximum stable current in this diode is reached, when the electrons are slowed down to practically zero velocity at the potential minimum. This was the situation that can be described by back-to-back operation of a time-reversed and a normal Child-Langmuir diode.

When a much larger current is injected into the diode, the virtual cathode will form much closer to the injection point. The simulation in Fig. 9.24a–d shows how the injected electron beam comes to a stagnation point (panel b) and is reflected back towards the cathode (panel c). At this time, the additional space charge from the reflected electrons has shifted the stagnation point towards $x = 0$. On the other hand, the volume between injection point and potential minimum has shrunk and the potential minimum becomes less negative (panel c). In panel d, the process begins to repeat.

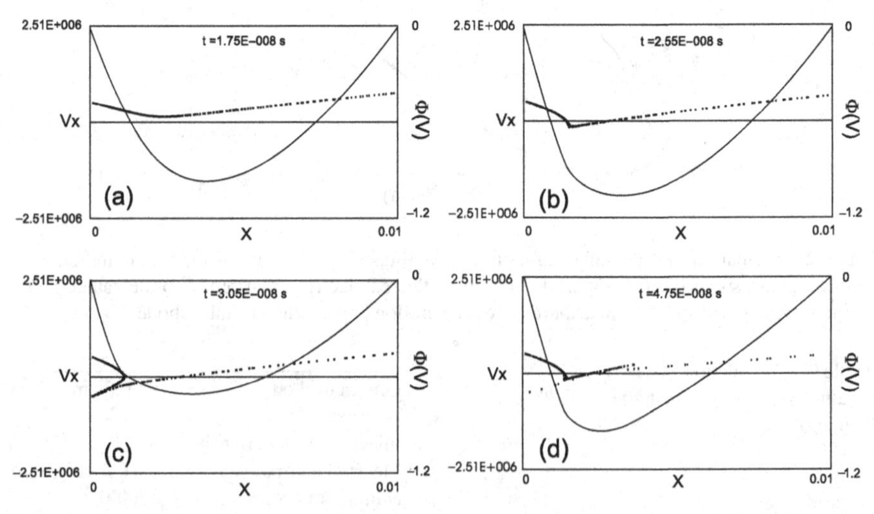

Fig. 9.24 Virtual cathode oscillations from a monoenergetic electron beam, which is stopped by its own space charge

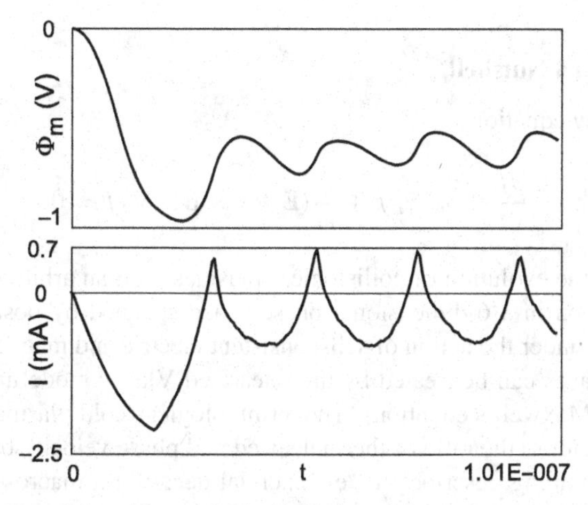

Fig. 9.25 Oscillations of the mid-potential in the diode and the total current

The final state of the system is non-stationary and shows strongly-nonlinear current oscillations as shown in Fig. 9.25. For a short time, the current is even disrupted. Note that the total current, which is the sum of the electron current and the displacement current by motion of the virtual cathode, becomes even positive for a short time. This phenomenon is called *blocking oscillations* and is typical for a relaxation oscillator. A typical feature of relaxation oscillators is the appearance of two different time scales in the oscillation period, a fast evolution near the point of current zero, and a gradual evolution in between. The spiked waveform of the current contains several harmonics.

Virtual cathode oscillations are an efficient means to produce high-power microwaves [217, 218]. By injecting a relativistic electron beam with currents from 1 to 200 kA, and energies from 0.1 to 10 MeV, microwaves with frequencies of 1–100 GHz and pulse durations from 1 to 500 ns are produced. For example, Davis et al. [219] reported output powers of 1.4 GW at 3.9 GHz with several hundred megawatts in harmonic radiation. Other authors [220] report 4 GW at 6.5 GHz. A one-dimensional analytical model of virtual cathode oscillations [221] predicts microwave frequencies and power. Under optimum conditions, up to 30% conversion efficiency (microwave energy to beam energy) can be expected. In most experiments, the conversion efficiency was close to 3%.

Summarizing, the two examples presented above, of a diode with a thermal emitter or with an injected electron beam, both being operated at supercritical injection current, reveal similar principles as we had found in the analysis of Maxwellian distributions and electron beams in kinetic theory. Electron beams can efficiently pile up space charge by bunching. In a Maxwellian distribution, bunching is counteracted by phase mixing of the contributions from many beams that make up the Maxwellian. Hence, the apparent stability of the virtual cathode for a thermal emitter can be understood as the Landau damping of those modes that are responsible for blocking oscillations in the case of monoenergetic injection.

The Basics in a Nutshell

- The Vlasov equation

$$\frac{\partial f}{\partial t} + \mathbf{v} \cdot \nabla_r f + \frac{q}{m}(\mathbf{E} + \mathbf{v} \times \mathbf{B}) \cdot \nabla_v f = 0$$

  describes the evolution of collisionless plasmas with an arbitrary distribution function in a 6-dimensional phase space spanned by position $\mathbf{x}$ and velocity $\mathbf{v}$ under the action of self-consistent electric and magnetic fields.
- Plasma waves can be treated by the linearized Vlasov model in combination with Maxwell's equations. The terminologies "cold plasma" and "hot plasma" refer to the ratio of thermal speed and phase velocity of the wave.
- Landau damping describes the exponential decay of a macroscopic wave electric field while the information is retained in the perturbed distribution function. The information can be partly recovered in echo experiments.
- The rate of Landau damping is determined by the slope of the unperturbed distribution function at the wave's phase velocity. For an unshifted Maxwellian, this is always negative. Velocity distributions having an additional shifted Maxwellian can produce a velocity interval, where the slope becomes positive leading to inverse Landau damping or instability.
- A physical picture of the mechanisms behind Landau damping involves charge bunching, ballistic response of particles and phase mixing.
- Plasma simulation with particle codes is complementary to the Vlasov approach. It describes the motion of superparticles that represent clumps of some thousand real particles. The particle-in-cell technique overcomes the limitation of N^2 scaling of the computation time for particle-particle codes.
- Plasma simulations make the nonlinear evolution of plasma processes accessible. Examples are: the trapping of electrons in beam-plasma interaction or the onset of blocking oscillations in diodes above the critical current.

Problems

9.1 Show that any function $g(\frac{1}{2}mv^2 + q\Phi)$, which only depends on the total energy of a particle, solves the Vlasov equation

$$\frac{\partial f}{\partial t} + v\frac{\partial f}{\partial x} - \frac{q}{m}\frac{\partial \Phi}{\partial x}\frac{\partial f}{\partial v} = 0\,.$$

9.2 Verify that the mean velocity of a one-dimensional half-Maxwellian electron distribution is given by

$$v_{\text{mean}} = \frac{1}{n_{\text{e}0}} \int_0^\infty v f_{\text{M}}^{(1)} \, \text{d}v = \left(\frac{k_{\text{B}} T_{\text{e}}}{2\pi m_{\text{e}}} \right)^{1/2} .$$

9.3 Start with the velocity distribution (9.25) and prove that the current density in the right half of the diode is the constant (9.28).

9.4 Derive the Landau damping rate for the Bohm-Gross mode (9.55) from (9.44) and (9.54).

9.5 Using (8.8) and (9.86), show that the trapping potential is given by

$$\Phi_{\text{t}} = \frac{m_{\text{e}}}{4e} v_0^2 \left[\frac{1}{2} \left(\frac{\alpha_{\text{b}}}{2} \right)^{1/3} \right]^2 .$$

9.6 The mean energy density of the electric wave field is given by $W_E = \frac{1}{2}\varepsilon_0 \langle E^2 \rangle = \frac{1}{4}\varepsilon_0 \hat{E}^2$. Consider the fastest growing mode of the beam-plasma instability (8.24) at the onset of trapping. Use $|\hat{E}| = k\hat{\Phi}$ and $k = \omega_{\text{pe}}/v_0$ and show that the mean field energy is given by

$$W_E = \left(\frac{1}{2} n_{\text{b}0} \, m v_0^2 \right) 2^{-31/3} \alpha_{\text{b}}^{1/3} .$$

Chapter 10
Dusty Plasmas

You boil it in sawdusts: you salt it in glue:
You condense it with locusts and tape:
Still keeping one principle object in view—
to preserve its symmetrical shape.

Lewis Carroll, The Hunting of the Snark

Since the 1980s, a new branch of plasma physics has emerged—the study of dusty plasmas. Because of similarities with complex fluids, dusty plasmas are also known as *complex plasmas*. This field has roots in astrophysics [222] and became interesting for laboratory plasma research when powder formation in plasma-enhanced chemical vapour deposition was identified to limit the deposition rate [223, 224], and when dust formation and dust trapping was observed during plasma etching of silicon wafers [225].

Dusty plasmas contain, among electrons and atomic or molecular ions, microscopic particles with sizes ranging from some ten nanometers to several ten micrometers. The dust particles become electrically charged and interact with the other plasma constituents. When the density of dust particles is sufficiently high, the electrostatic interparticle forces become important. The dust subsystem can develop collective behavior, which manifests itself as wave phenomena or, for micrometer particles carrying several thousand elementary charges, by the formation of liquid or solid phases. The discovery of plasma crystallization, in 1994, [32–34] gave a strong boost to the field of dusty plasmas. Much of the attractivity of this field of research is due to the fact that the high mass of dust particles—a dust particle of 1 μm diameter has a typical mass of 3×10^{11} proton masses—leads to a long dynamic response time of milliseconds or longer. For dusty plasmas with micrometer sized particles, the motion of all individual particles can be followed by fast video-cameras. This is a unique opportunity to study the collective behavior of an ensemble of charged particles at the kinetic level.

The field of dusty plasmas has now reached maturity and various aspects have been summarized in a host of review articles in the past decade [226–236] or in monographs [237–241]. The topics addressed in this chapter are therefore not aimed at giving a balanced overview of the many existing observations. Rather, we will focus on the following questions

A. Piel, *Plasma Physics*, DOI 10.1007/978-3-642-10491-6_10,

- What is the new physics introduced by dust particles carrying thousands or hundred-thousands of elementary charges?
- What are the specific methods to study dusty plasmas?
- What does classical plasma physics learn from dusty plasmas?

The striking difference between a dusty plasma and a three-component plasma, containing electrons, positive ions and an additional population of negative ions, is the tremendous charge, $q_\mathrm{d} = -(10^3 \ldots 10^5)\, e$, on a dust grain of several micrometer size. Whereas in gas discharges Coulomb collisions of electrons with ions are negligible compared to collisions with atoms, negative dust particles lead to efficient scattering of positive ions. Therefore, orbital motion of ions becomes a central concept in dusty plasmas and turns out to be of equal importance as gyromotion in magnetized plasmas. Orbital motion determines the nonlinear shielding effects of these highly charged dust grains and momentum exchange between ions and dust leads to the new phenomenon of ion wind forces.

The huge dust charge is also the reason for the high value of the coupling parameter between the dust grains that can lead to liquid and solid phases of the dust system at room temperature. Plasma crystals [32–34] and Yukawa balls [36, 242] are suitable systems to study structural properties of solids, phase transitions, or phonon dynamics with "atomic resolution".

Dust particles differ from negative ions in that their charge is not a fixed quantity. Charge, being the integral of the charging currents over time, depends on the changing environment along the the particle's past trajectory. This makes the Coulomb force non-conservative and can be the source of instability. For nanometer sized dust, fluctuations of the charge due to the discrete steps of collecting an ion or electron become important. Coagulation of small particles becomes possible when one of the partners is either neutral or attains the opposite charge for a short time.

For simplicity, we will discuss dusty plasma effects in the following only for spherical particles. In situations with many particles, all the particles are assumed to have the same size.

10.1 Charging of Dust Particles

In most laboratory or industrial plasmas, the charging of dust grains occurs primarily by collecting electrons and positive ions. Dust grains behave like small isolated probes at floating potential. Because of the higher thermal speed of the electrons compared to the ions, dust grains are negatively charged. In addition to collecting thermal plasma particles, dust particles in space are subjected to fluxes of energetic photons or particles, e.g., solar wind particles, which release electrons by photoemission or secondary emission. These processes can even be more important than charge collection and make the dust positively charged. A survey of charging processes in space can be found in [243].

In this Section, we will first analyze the elementary mechanisms and then turn to the questions of charge variability: How long is the relaxation time to a new equilibrium charge? What is the statistics of charge fluctuations? How do charges compete for the charging resources?

10.1.1 Secondary Emission

Secondary emission is a process in which a primary energetic particle penetrates the surface of a solid and creates free electrons along its path by ionization of atoms in the solid until it is stopped. The secondary electrons can reach the surface by diffusion and leave the solid. The yield of secondary electrons is defined as the ratio of the emitted electron current to the current of primary particles. For electron impact, it is described by $\delta = I_s/I_e$, i.e., the average number of released electrons per incident electron.

Secondary electron emission by ion impact is described by a coefficient γ, the average number of released electrons per incident ion. For most materials, we have $\gamma \ll 1$, and γ depends only weakly on ion energy. The release of electrons from metal cathodes by ion impact, which is essential for plasma production in glow discharges, will be discussed in Sect. 11.1.3.

The shape of the secondary emission yield by electron impact as a function of primary electron energy is given by the formula [244],

$$\delta(W) = \delta_m \frac{W}{W_m} \exp\left[-2\left(\frac{W}{W_m}\right)^{1/2} + 2\right], \qquad (10.1)$$

which is an accepted model for metal surfaces at normal incidence. Here, W_m is the energy where the secondary yield takes its maximum of height δ_m. In [245] a slightly different formula is used to describe materials of astrophysical interest,

$$\delta(W) = 4\delta_m \frac{W/W_m}{[1 + (W/W_m)]^2}. \qquad (10.2)$$

Both curves are compared in Fig. 10.1. It turns out that both models are practically identical for $W/W_m < 2$ but have different asymptotics at high energies. The coefficients δ_m and W_m for the secondary electron yield of different materials are compiled in Table 10.1 and can be used for both models.

The secondary emission from bulk material and small grains can be substantially different. In a small spherical grain, a diffusing electron finds a surface in any direction rather than only in one direction for bulk matter. This effect increases the secondary yield. On the other hand, small dust grains become transparent for energetic projectiles, which deposit only part of their energy thus lowering the yield.

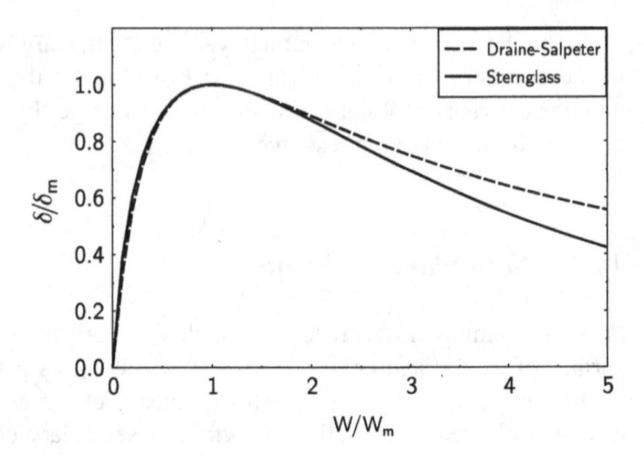

Fig. 10.1 Secondary emission yield for electron impact on a solid as a function of primary electron energy W. *Solid line*: Sternglass [244], *dashed line*: Draine & Salpeter [245]

Table 10.1 Secondary emission yield by electron impact for materials of astrophysical interest

Material	δ_m	E_m (eV)	Source
Graphite	1	250	[245]
	1	300	[246]
SiO_2	2.9	420	[245]
Mica	2.4	340	[245]
Fe	1.3	400	[245]
Al	0.95	300	[245]
Al_2O_3	2.6	300	[246]
MgO	23	1200	[245]
Lunar dust	≈ 1.5	≈ 500	[245]
	1.4	275	[246]
	1.7	340	[246]

10.1.2 Photoemission

Photoelectric emission is considered to be the dominant mechanism of cosmic grain charging in many astrophysical environments. The elementary process involves a photon of energy $W_{ph} = h\nu$ which liberates an electron, which is bound to the grain with a binding energy W_b, called the work function of that material. The electron leaves the grain surface with an excess energy $W_{ex} = h\nu - W_b$, and the grain charge increases by one (positive) elementary charge.

In the laboratory, where the source of energetic photons may be a UV laser or a mercury lamp with a strong emission line in the UV, this concept leads to a final state, in which the grain attains a high positive potential W_{pot}. The potential is just high enough that the next liberated photoelectron cannot excape from the attractive potential well of the dust grain and returns to the surface of the dust grain. This gives the depth of the potential well $W_{pot} = -e\Phi_{pot} = -(h\nu - W_b)$ and determines the dust charge.

In reality, the situation in space is more complex. When photoemission becomes competitive with electron and ion collection, two different equilibria with positive and negative charge can coexist (named the *flip-flop effect*) [247]. This is an important point for the formation of dust agglomerates—the first step in the formation of planets around protostars. Like-charged dust grains, by their mutual repulsion, would be prevented from coagulation.

Moreover, dust charging by photoemission becomes more complex than the estimates above, because the source of UV radiation is usually not monochromatic. Rather, the spectrum of energetic photons from the Sun extends from the infrared to the extreme ultraviolet regime. Figure 10.2a gives the Solar Irradiance Reference Spectrum measured during the solar Carrington rotation 2068 (20 March to 16 April 2008) [248].

To gain insight into the photoionization of typical elements in cosmic dust, the cross section for photoionization of neutral carbon and silicon atoms is shown in Fig. 10.2b [249]. The thresholds for ionization lie at 110 nm (C) and 152 nm (Si). Therefore, the Lyman-α line of atomic hydrogen at 121.6 nm, which marks the boundary between the UV and EUV region, contributes to the ionization of silicon but not of carbon atoms. For both materials, the cross sections decay fast towards shorter wavelength. Therefore, the main contribution for photoionization comes from the region near the threshold.

For an extended spectrum, the photoelectron current released from the surface of a dust grain of radius a is given by

$$I_{\mathrm{ph}} = \pi a^2 e F_{\mathrm{ph}} \quad \text{with} \quad F_{\mathrm{ph}} = \int \xi(W_\lambda) S(W_\lambda) \mathrm{d}W_\lambda \,. \tag{10.3}$$

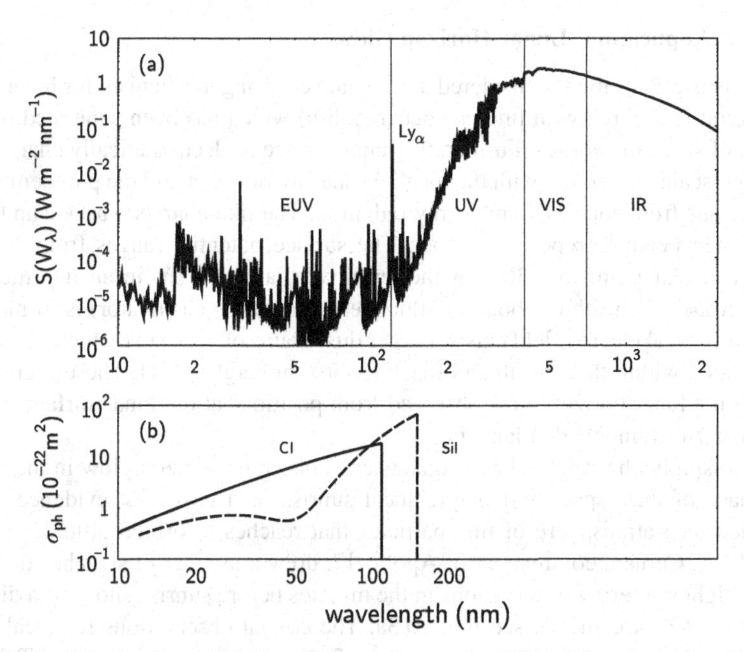

Fig. 10.2 (**a**) Solar Irradiance Reference Spectrum from exteme UV to infrared wavelengths (from [248]). (**b**) Cross section for photoionization of carbon and silicon atoms

Here, we have to integrate the product of the photoelectric efficiency $\xi(W_\lambda)$ (i.e., the number of electrons per incident photon) and the spectral energy density $S(W_\lambda)$ over the entire solar spectrum. The photoelectric efficiency of a dust particle cannot simply be derived from atomic data but must be determined experimentally. For Lunar dust, the photoelectric yield was measured in [250].

The spectral extent of the solar photon flux results in a distribution of excess energies. The photoelectrons can then be considered to have an effective temperature T_{ph}. This means, that there will be a fraction of photoelectrons in the tail of the distribution function, which can overcome the potential well of a positively charged grain. This fraction can be described by an appropriate Boltzmann factor. In [246] the following expression for F_{ph} is used:

$$F_{ph} = 3 \times 10^{10} \chi \frac{1}{[r(\mathrm{AU})]^2} \exp\left(-\frac{e\Phi}{k_B T_{ph}}\right) . \tag{10.4}$$

The factor r^{-2} takes care of the decay of the solar photon flux with heliocentric distance. χ is an empirical constant, which is 0.9 for metallic or graphitic grains, and 0.1 for icy grains. A typical value for the effective electron temperature is $k_B T_{ph} = 1.3\,\mathrm{eV}$. The exponential represents the Boltzmann factor for a (positive) grain potential Φ.

10.1.2.1 Application: Lunar Horizon Glow

Photoelectric charging is considered as the main charging mechanism for lunar dust. The Moon is covered with fine powder (regolith) which has been generated by the impact of small meteorites. Further, the lunar surface is electrostatically charged by the large-scale interaction with the local plasma environment and the photoemission of electrons from solar UV and X-ray radiation. The like-charged surface and dust grains then begin to repel each other. The surface potential ranges from +4.1 V at the subsolar point to −36 V at the terminator, and +3.1 V in an intermediate range. A Debye sheath of about 1 m thickness (8 m at the terminator) is formed, in which a vertical electric field exists. Larger dust grains of $\approx 5\,\mu\mathrm{m}$ diameter can thus be levitated within this sheath and reach (3–30) cm height [251]. These grains can explain the lunar horizon glow observed from positions at the lunar surface by the Surveyor-7 or Lunochod-II landers.

Surprisingly, the Apollo-17 astronauts even observed horizon glow in the orbiting phase of their spacecraft before local sunrise, and gave first evidence of an extended dust atmosphere of fine particles that reaches to orbital altitudes [252]. Capt. E. A. Cernan, commander of Apollo-17, drew a number of sketches describing the light scattering observations in the minutes before sunrise showing a diffuse "corona" and "streamers", see Fig. 10.3a. The crucial observations for local light scattering are found in the sketches made at T-2 minutes, T-1 minute, and T-5 seconds. While the corona has been visible for more than 4 min, the streamers were

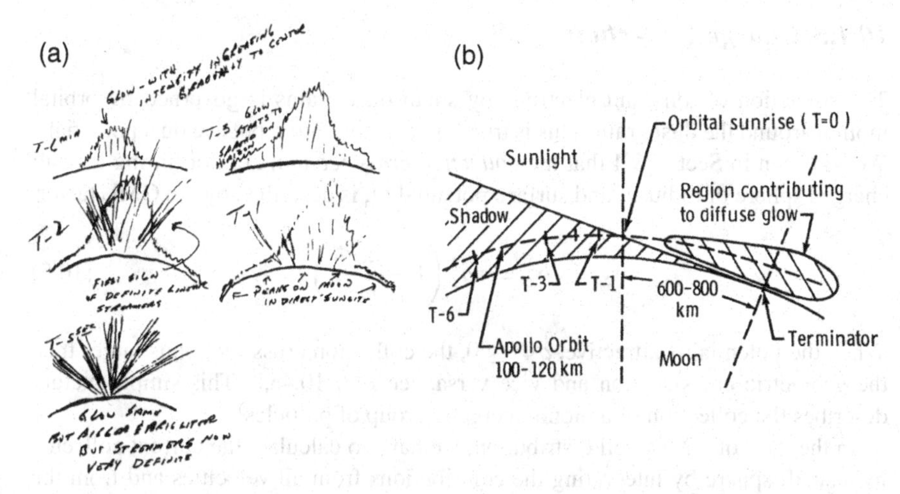

Fig. 10.3 Lunar horizon glow before sunrise as observed by Apollo-17 astronauts. (**a**) "Corona" and rays sketched by Captain E. A. Cernan. The handwritten comments read: "T-6^{min} Glow with intensity increasing gradually to center", "T-3 Glow starts to spread on horizon", "T-2 First sign of definite linear streamers", "T-1 Peaks on Moon in direct sunlight", "T-5^{sec} Glow same but bigger & brighter, but streamers now very definite". (**b**) The observation geometry and a tentative dust distribution. (Reproduced with permission of the author from [252])

observed only from T-2 minutes and intensified at a progressively increasing rate compared to the corona. The increase during the final 5 s exceeds that during the previous 2 min.

Such a phenomenon can only be explained by light scattering from submicron particles, which, according to Mie-scattering theory, is strongly enhanced in a narrow cone in forward and backward direction [253]. Therefore, forward scattered light becomes visible only seconds before crossing the point of orbital sunrise ($T = 0$), see Fig. 10.3b. Another example for effective light scattering by small particles is the Zodiakal light—backscattered sunlight from interplanetary dust in the ecliptic.

The lunar dust atmosphere is still a field of controversal debate because of a lack of new observational results. One of the current models of lunar dust *lofting* involves an electrostatic fountain mechanism [254]. It is argued that a dust grain of 10 nm radius carrying (20–200) elementary charges has initially an electrostatic potential energy between (60–6000) eV. This energy compares to the difference of gravitational potential energy between the lunar surface and an altitude $h = 100\,\mathrm{km}$ for that grain, $mg_{\mathrm{L}}h = 1250\,\mathrm{eV}$, which makes such a fountain process energetically possible. Larger grains will reach only lower altitudes. However, there are many open questions left, e.g., the number density of these particles in the dust atmosphere, the size distribution, the efficiency of light scattering, the transport of submicron dust by horizontal electric fields or by radiation pressure etc., which require more experimental data before the lunar dust atmosphere can be completely understood.

10.1.3 Charge Collection

The collection of ions and electrons by small dust grains is governed by orbital motion around the dust grain. This is true for attractive and repulsive dust potentials. We had seen in Sect. 7.5.4 that the *collection cross section*, i.e., for hitting a small charged sphere of radius a and surface potential Φ, is described by the OML factor

$$\sigma_{\mathrm{c}} = \pi b_{\mathrm{c}}^2 = \pi a^2 \left(1 - \frac{2q\Phi}{mv^2}\right) . \tag{10.5}$$

When the potential is attractive, $q\Phi \leq 0$, the collection cross section is larger than the geometric cross section and vice versa, see Fig. 10.4a,b. This simple picture describes the collection of a mono-energetic group of particles.

In the case of a Maxwell distribution, we have to calculate the current collected by a small sphere by integrating the contributions from all velocities and from the full solid angle 4π, as shown in Fig. 10.4c. In the case of a repulsive potential, $q\Phi > 0$, a minimum velocity $v_0 = (2q\Phi/m)^{1/2}$ ensures that $\sigma_{\mathrm{c}} \geq 0$. The correct distribution function for this case is the Maxwell distribution of speeds, which results in

$$\begin{aligned} I &= q \int_{v_0}^{\infty} v\sigma_{\mathrm{c}}(v) f_{\mathrm{M}}(v) \mathrm{d}v \\ &= qn\pi a^2 \left(\frac{m}{2\pi k_{\mathrm{B}}T}\right)^{3/2} \int_{v_0}^{\infty} 4\pi v \left(v^2 - v_0^2\right) \exp\left(-\frac{mv^2}{2k_{\mathrm{B}}T}\right) \mathrm{d}v \\ &= qn\pi a^2 \left(\frac{k_{\mathrm{B}}T}{2\pi m}\right)^{1/2} \exp\left(-\frac{q\Phi}{k_{\mathrm{B}}T}\right) . \end{aligned} \tag{10.6}$$

The exponential is the familiar Boltzmann factor that describes the current reduction by the potential barrier. This was illustrated in Fig. 10.4b in terms of a collection radius $b_{\mathrm{c}} < a$.

For an attractive potential, we can set the lower integration limit to zero and the integral to be solved reads

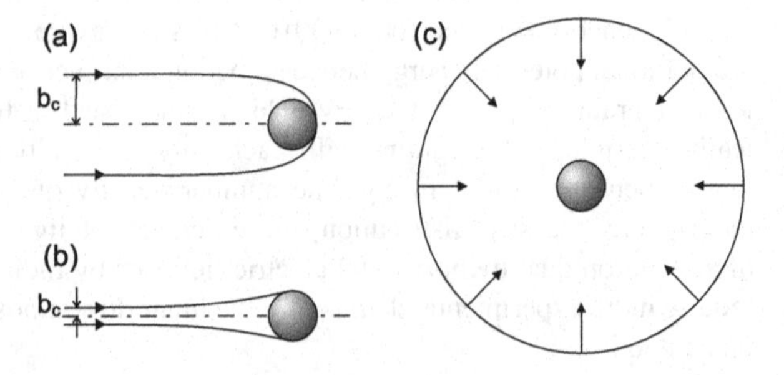

Fig. 10.4 (**a**) Increase of collection cross section by OML factor for an attracting potential, (**b**) Decrease of collection cross section for a repulsive potential, (**c**) Summing up contributions from all directions

$$I = qn\pi a^2 \left(\frac{m}{2\pi k_B T}\right)^{3/2} \int_0^\infty 4\pi v \left(v^2 + v_0^2\right) \exp\left(-\frac{mv^2}{2k_B T}\right) dv$$
$$= qn\pi a^2 \left(\frac{k_B T}{2\pi m}\right)^{1/2} \left[1 - \frac{q\Phi}{k_B T}\right] . \quad (10.7)$$

The expression in the bracket is the OML-factor that describes the focusing of ion orbits shown in Fig. 10.4a. This is the result, which we had already used in the discussion of the electron current to a spherical probe in Sect. 7.5.4.

10.1.3.1 The Floating Potential of Dust Grains

In the absence of other charging mechanisms, we can conjecture that, because of the much higher electron thermal velocity, the charge on a dust grain, and hence its surface potential, must be negative. Then we can equate the electron retardation current from (10.6) with the ion current from (10.7) to obtain an implicit equation for the floating potential

$$\left(\frac{k_B T_e}{2\pi m_e}\right)^{1/2} \exp\left(\frac{e\Phi_f}{k_B T_e}\right) = \left(\frac{k_B T_i}{2\pi m_i}\right)^{1/2} \left[1 - \frac{e\Phi_f}{k_B T_i}\right] , \quad (10.8)$$

which has to be solved numerically. Now, setting $\tau = T_e/T_i$, $\mu = m_i/m_e$, the normalized floating potential $\eta_f = -e\Phi_f/k_B T_e$ is found to be only a function of the mass ratio μ and the temperature ratio τ, but the floating potential turns out to be independent of the particle size a. Then, the equation for η reads

$$e^{-\eta} = (\mu\tau)^{-1/2}(1 + \tau\eta) . \quad (10.9)$$

Figure 10.5 shows that, over a wide temperature range, the floating potential is only a weak function of the temperature ratio. The hydrogen ions were chosen as typical for space plasmas whereas argon is a common gas in laboratory investigations. Note that the floating potential Φ_f was assumed negative.

10.1.3.2 The Charge on a Dust Grain

The relationship between potential and charge on a dust grain can be established when we consider the dust grain of radius a being a small spherical capacitor, which has the capacitance

$$C = 4\pi\varepsilon_0 a \quad (10.10)$$

and gives the dust charge

$$q_d = C\Phi_f = 4\pi\varepsilon_0 a\Phi_f . \quad (10.11)$$

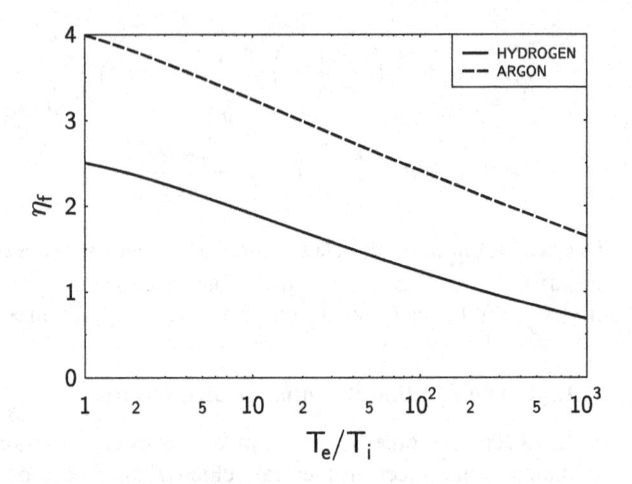

Fig. 10.5 The normalized floating potential, $\eta_f = -e\Phi_f/k_B T_e$, of a dust grain in a plasma with H^+ or Ar^+ ions

In cases where the condition $a \ll \lambda_D$ is not fulfilled, the opposing charges can be considered as being localized near $r = \lambda_D$, and the capacitance is then enlarged to

$$C = 4\pi\varepsilon_0 a\left(1 + \frac{a}{\lambda_D}\right) . \tag{10.12}$$

A good estimate for the number of elementary charges Z_d on a dust grain in a laboratory argon plasma with $T_e/T_i \approx 100$ can be obtained using $\Phi_f = -2.41\,k_B T_e$. Then

$$Z_d = 1675\,a\,(\mu m)\,T_e\,(eV) . \tag{10.13}$$

Nearly the same factor applies for an isothermal hydrogen plasma, which has the classical value [222], $\Phi_f \approx -2.5\,k_B T_e$.

10.1.4 Charging Time

The charging of a dust grain is governed by the charging equation

$$\frac{dq_d}{dt} = I_i + I_e , \tag{10.14}$$

which can used to describe three different cases:

1. $dq_d/dt = 0$ defines the floating potential Φ_f.
2. When $q(t = 0) = 0$, (10.14) yields the full nonlinear charging process of a dust grain.
3. When q_d deviates only slightly from its equilibrium value at the floating potential, (10.14) yields the linear relaxation time τ.

Case 2 is of little practical value because it describes the dust-charge evolution during the switch-on period of the plasma. Here we are mostly interested in the linear charge relaxation time of case 3. Using $\Phi = q/C$, and expanding the charging currents about the floating potential, we obtain a relaxation-type equation for the grain potential

$$\frac{\mathrm{d}\Phi}{\mathrm{d}t} = \frac{1}{C}\underbrace{\left[\left.\frac{\mathrm{d}I_\mathrm{i}}{\mathrm{d}\Phi}\right|_{\Phi_\mathrm{f}} + \left.\frac{\mathrm{d}I_\mathrm{e}}{\mathrm{d}\Phi}\right|_{\Phi_\mathrm{f}}\right]}_{-1/R}(\Phi - \Phi_\mathrm{f}) = -\frac{1}{RC}(\Phi - \Phi_\mathrm{f})\,. \tag{10.15}$$

This is the well-known equation for charging a capacitor of capacitance C through a resistor R, which gives a solution of the type

$$\Phi(t) = \Phi(t=0)\mathrm{e}^{-t/\tau} + \Phi_\mathrm{f}\left(1 - \mathrm{e}^{-t/\tau}\right) \tag{10.16}$$

with a relaxation time $\tau = RC$. In our case, the resistor R is related to the slope of the characteristic of a spherical probe at floating potential, $R = -\mathrm{d}U/\mathrm{d}I$, as shown in Fig. 10.6. The dependence of the relaxation time on the size of the dust grain follows from $C \propto a$ (from the capacitance model) and $R \propto a^{-2}$ (from the particle cross section), which makes $\tau = RC \propto a^{-1}$. Therefore, large grains have a much shorter relaxation time than small grains, which is at first counter-intuitive from considering only the increase of grain capacitance with particle size.

This relaxation model for small deviations from the floating potential yields a relaxation time that (for $T_\mathrm{e}/T_\mathrm{i} = 100$) scales as

$$\tau = 1.04 \times 10^{10}\,\mathrm{s}\,\frac{\sqrt{T_\mathrm{e}\,(\mathrm{eV})}}{n\,(\mathrm{m}^{-3})\,a\,(\mu\mathrm{m})}\,. \tag{10.17}$$

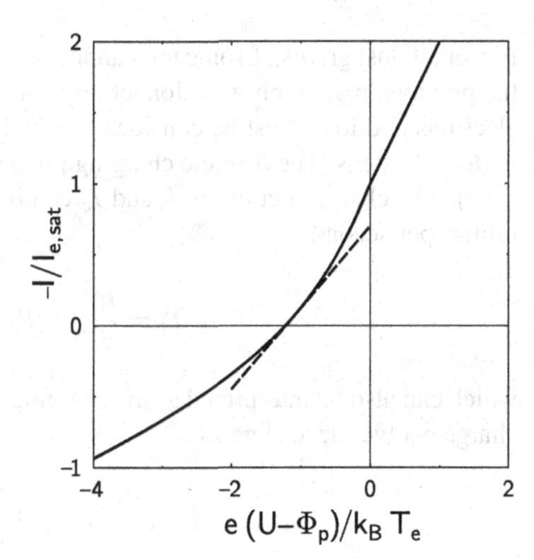

Fig. 10.6 The effective resistor R for the linear charge relaxation time is given by the slope of the current-voltage characteristic for a spherical grain at the floating potential (*dashed line*), $R = -\mathrm{d}U/\mathrm{d}I$

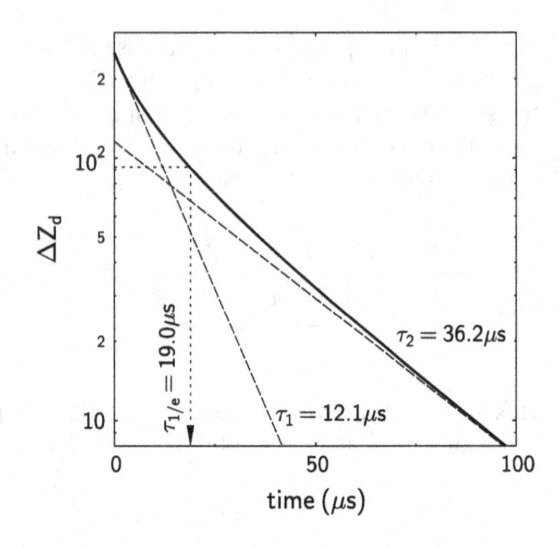

Fig. 10.7 Nonlinear relaxation of the dust charge deviation from its equilibrium value for a dust grain with $a = 500\,\text{nm}$ and plasma conditions $T_e = 3\,\text{eV}$, $T_i = 0.03\,\text{eV}$, $n = 10^{15}\,\text{m}^{-3}$. τ_1 and τ_2 are the time constants from the early and late relaxation. The relaxation time in [255] was defined as the $1/e$ point of the charge deviation

For example, a dust grain of $a = 1\,\mu\text{m}$ radius in a typical laboratory plasma with electron temperature $T_e = 3\,\text{eV}$ and density $n = 10^{15}\,\text{m}^{-3}$ has a charging time of $\tau = 18\,\mu\text{s}$.

Solving the charging equation (10.14) by direct numerical integration shows the non-exponential approach to the equilibrium value. The charge deviation ΔZ_d from the asymptotic value at large t is shown in Fig. 10.7 in a semilog plot. The final approach is indeed exponential with a time constant $\tau_2 = 36.2\,\mu\text{s}$, as predicted by (10.17). The initial slope, which has a shorter time constant $\tau_1 \approx \tau_2/3$, characterizes the nonlinear dust charging regime. In [255] the dust charging time was defined as the point where ΔZ_d decayed by $1/e$.

10.1.5 Charge Fluctuations

For small dust grains of some ten nanometers size, the continuum charging model of the previous paragraph is no longer applicable. Rather, the collection of individual electrons and ions must be considered, which introduces fluctuations of the charge in discrete steps. The discrete charging process by a random process was studied in [255]. The charging currents I_i and I_e can be replaced by charge collection probabilities per second

$$P_i = \frac{I_i}{e} \qquad P_e = -\frac{I_e}{e}\,, \tag{10.18}$$

which can also be interpreted as the charging frequencies by ions and electrons. The charge evolves according to

$$q_d(t_{k+1}) = q_d(t_k) + eH\left[x_1 - \frac{P_e}{P_e + P_i}\right]\,, \tag{10.19}$$

where x_1 is a random variable that is uniformly distributed in the interval [0, 1], and $H(y)$ is a step function that is -1 for $y < 0$ and $+1$ for $y \geq 0$. Since P_e and P_i depend on the instantaneous value of $q_d(t_k)$, the statistical process is nonlinear. The length of the time step $\Delta t_k = t_{k+1} - t_k$ is given by

$$\Delta t_k = \frac{\ln(x_2)}{P_e + P_i}, \tag{10.20}$$

where x_2 is a second random variable. When we replace x_2 by the probability $p(t)$ to "survive" a short interval t without suffering a change in charge, we find $p(t) = \exp[-(P_e + P_i)t]$. This result matches our earlier reasoning of mean free path in Sect. 4.2.2 and mean free time in Sect. 4.3.4.

A comparison of the discrete charging model with the continuous charging process (10.14) is shown in Fig. 10.8. The discrete model leads to charge fluctuations about the equilibrium value. Furthermore, deviations from the equilibrium value decay at the relaxation time (10.16), as becomes evident from comparing the behavior of 10 nm and 50 nm radius particles in Fig. 10.8.

Extensive numerical studies in [255] have shown that the fluctuations of the dust charge about its equilibrium value Z_d are described by a standard deviation $\delta Z_d = 0.5 Z_d^{1/2}$. In other words, the relative fluctuations of the dust charge $\delta Z_d / Z_d$ decrease with increasing size of the dust particle.

Charge fluctuations can be responsible for the coexistence of positive and negative particles. This mechanism is efficient for small grains carrying only a few elementary charges. The presence of oppositely charged grains then facilitates the coagulation of small dust grains e.g., in plasma reactors for nano-powder production [256]. However, as soon as the particles have grown to more than 25 nm radius, and their charge exceeds about 40 elementary charges, charge fluctuations can no longer produce grains of the opposite charge, and coagulation between particles of same size is stopped. Opposite to powder production in plasma reactors, which yields a

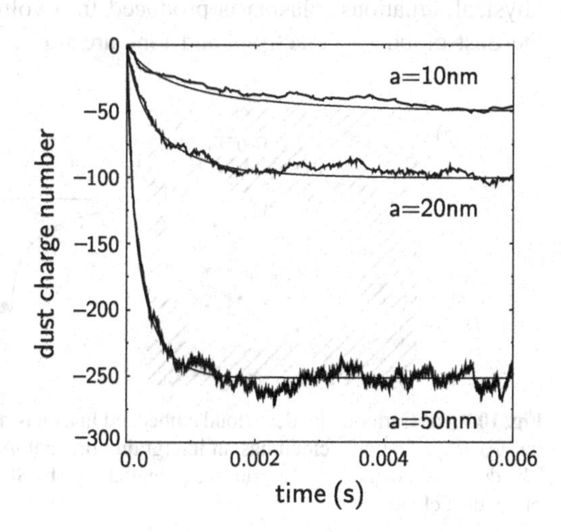

Fig. 10.8 Discrete charging of submicrometer particles according to the model in [255]. For comparison, the continuum model is shown by the *light lines*

narrow size distribution, there is a wide distribution of particle sizes in astrophysical situations, and agglomeration of small particles with larger particles is important as long as the small particles can attain the opposite charge [257] by fluctuations or by the flip-flop effect described in Sect. 10.1.2.

10.1.6 Influence of Dust Density on Dust Charge

Up to now, we have considered the charging of an isolated dust grain in a plasma environment that provides inexhaustible charging currents. The situation is different when a dust cloud contains many dust particles and the total dust charge becomes comparable with the total positive ion charge. The quasineutrality of plasmas then allows only for a small number n_e of free electrons

$$n_e = n_i - Z_d n_d \,. \tag{10.21}$$

The other electrons are bound to the dust particles.

When the dust particles are densely packed, the plasma potential Φ_c inside the cloud between the dust grains may be different from the ambient plasma, where we have chosen the potential as $\Phi_0 = 0$, as sketched in Fig. 10.9.

Let us assume that the electrons and ions in the dust cloud are in thermal equilibrium with the ambient plasma. Then the electron and ion concentrations will be governed by Boltzmann factors

$$n_e = n_\infty \exp\left(\frac{e\Phi_c}{k_B T_e}\right) \qquad n_i = n_\infty \exp\left(-\frac{e\Phi_c}{k_B T_i}\right) \,. \tag{10.22}$$

The Boltzmann response of the electrons is a proven concepts for all plasmas. However, using the Boltzmann factor for the ions needs additional justification. In astrophysical situations, plasma is produced in a volume, which is large compared to the dust cloud and electrons and ions are transported over large distances before

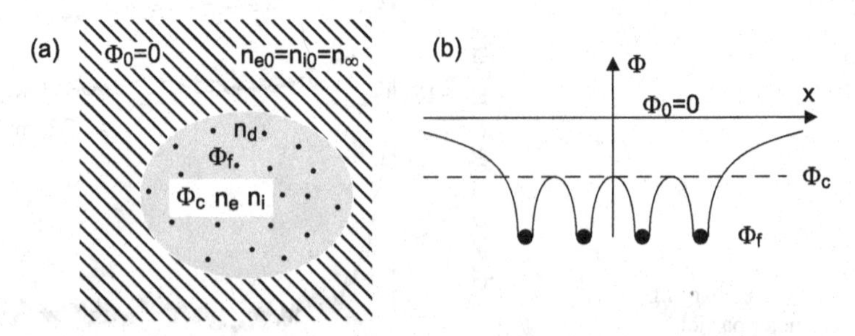

Fig. 10.9 (**a**) Cartoon of a dust cloud embedded in a quasineutral plasma of density n_∞ and potential $\Phi_0 = 0$. The dust cloud has an intergrain potential Φ_c, electron and ion densities n_e and n_i. The dust has a density n_d and surface potential Φ_f. (**b**) Sketch of the potential profile in a section of the dust cloud

they are finally lost by recombination at a surface. The ambient plasma acts as an inexhaustible reservoir for the dust charging process. Moreover, in astrophysical situations, except for the energetic particles in the Solar wind, the thermal plasma often has $T_e \approx T_i$ and the density adjustments by the Boltzmann factors are moderate.

In laboratory plasmas, however, it is a better approximation to assume a fixed ion density. The ion density is determined by the balance by ionization and losses and some ionization occurs inside the dust cloud. Losses can be ambipolar diffusion towards the walls or, in dense dust clouds, the charging currents of the dust particles, which act as an "internal wall".

10.1.6.1 Isothermal Plasma with Boltzmann Ions

Let us first consider the original model of an isothermal hydrogen plasma, which was conceived for astrophysical dust clouds [258]. The values for electron and ion density from (10.22) have to be used for calculating the charging currents for the dust grains, assuming that the grain potential is negative

$$I_i = n_i e \pi a^2 \left(\frac{k_B T_i}{2\pi m_i}\right)^{1/2} \left[1 - \frac{e(\Phi_f - \Phi_c)}{k_B T_i}\right] \quad (10.23)$$

$$I_e = -n_e e \pi a^2 \left(\frac{k_B T_e}{2\pi m_e}\right)^{1/2} \exp\left[\frac{e(\Phi_f - \Phi_c)}{k_B T_e}\right] . \quad (10.24)$$

Again, the expression in the bracket of (10.23) is the OML-factor that describes the increase of the particle cross-section for ion attraction, see Fig. 10.4a, and the exponential in (10.24) is the Boltzmann factor that represents the fraction of electrons that overcomes the potential barrier for a repulsive potential, see Fig. 10.4b. Note that in this model each dust particle receives its charging currents from its immediate neighborhood, where the particle densities n_e and n_i, as well as the plasma potential Φ_c differ from the values in the plasma encompassing the dust cloud. Then, the dust grains will attain a floating potential $\Phi_f - \Phi_c$ relative to the dust cloud potential and it is this potential difference that determines the dust charge. $\Phi_f - \Phi_c$ can considerably deviate from the value given by (10.8) that an isolated dust grain attains in a plasma with $n_e = n_i$.

Using again the normalized potential $\eta = -e\Phi/k_B T_e$, the mass ratio $\mu = m_i/m_e$, the temperature ratio $\tau = T_e/T_i$, and replacing the dust charge $q_d = 4\pi\varepsilon_0 a(\Phi_f - \Phi_c)$, we obtain a coupled set of equations that describes the quasineutrality condition within the dust cloud (10.21) and the floating condition, $I_i + I_e = 0$, for the dust grains

$$0 = e^{-\eta_c} - e^{\tau\eta_c} + P(\eta_f - \eta_c) \quad (10.25)$$

$$0 = (\mu\tau)^{-1/2}[1 + \tau(\eta_f - \eta_c)] - e^{-\eta_f - \tau\eta_c} . \quad (10.26)$$

$P = A\,a\,(\mu\mathrm{m})\,n_\mathrm{d}/n_\infty$ is the dimensionless dust density parameter introduced in [258, 259], and $A = 1\,\mu\mathrm{m}\,(4\pi\varepsilon_0 k_\mathrm{B} T_\mathrm{e}/e^2) = 695\,T_\mathrm{e}\,(\mathrm{eV})$. A different interpretation of the parameter P is

$$P = 3\frac{a}{\lambda_\mathrm{De}}\left(\frac{4\pi}{3}n_\mathrm{d}\lambda_\mathrm{De}^3\right), \tag{10.27}$$

which relates the parameter P to the number of dust particles in an electron Debye sphere. (10.25) and (10.26) are again solved numerically and give the dependence of the cloud potential and grain floating potential as a function of the parameter P, as shown in Fig. 10.10 for typical conditions in space (hydrogen ion, $T_\mathrm{e}/T_\mathrm{i} = 1$).

For $P \ll 1$, we recover the normalized floating potential $\eta_\mathrm{f} \approx 2.5$ of isolated grains, which we had already discussed in Sect. 10.1.3.1. Remember that *floating* means that a dust particle attains a sufficiently negative potential to reduce the electron flux to the same value as the ion flux to the particle, which makes the net electric current zero. In this limit, the dust particles are well separated and hence there is no distinction between the cloud potential Φ_c and the ambient plasma potential Φ_0. When the parameter P approaches unity, the normalized grain potential shows a substantial reduction and the normalized cloud potential is found to increase. It is not the reduction in the magnitude of the floating potential η_f, but the potential difference $\eta_\mathrm{f} - \eta_\mathrm{c}$ that demonstrates a reduction of the negative grain charge. For $P > 100$, the grain potential reaches a final equilibrium value, $\eta_\mathrm{f} = \eta_\mathrm{c}$, in which the charge per particle gets less and less because the total amount of charge in the dust cloud is finite and has to be shared by ever more particles.

This limiting case follows immediately from the quasineutrality condition (10.25) in the limit $P \to \infty$. From (10.26) we obtain $\eta_\mathrm{f} = \frac{1}{4}\ln(\mu\tau) = \frac{1}{4}\ln(m_\mathrm{i}/m_\mathrm{e}) = 1.88$. The Boltzmann equilibrium of the dust cloud with the ambient plasma then leads to a reduction of the electron density by a factor of $(m_\mathrm{i}/m_\mathrm{e})^{1/4} = 6.55$ and an increase of the ion density by the same factor. In this final state, the net current from the ambient plasma into the dust cloud is zero and the entire dust cloud is floating in the ambient plasma.

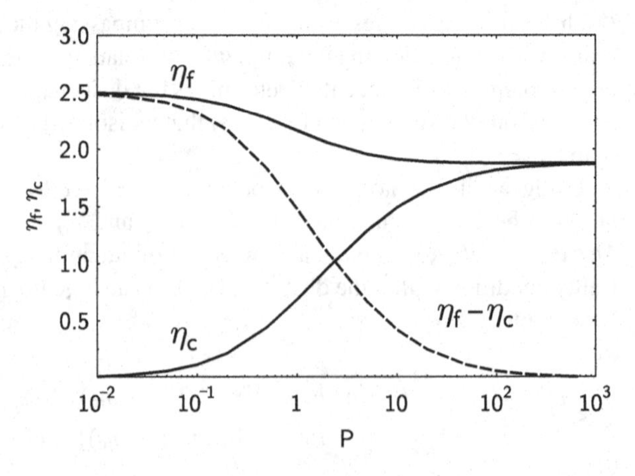

Fig. 10.10 Normalized cloud potential η_c and grain potential η_f (*solid lines*) for an isothermal hydrogen plasma with Boltzmann ion response as function of the parameter P. The potential difference $\eta_\mathrm{f} - \eta_\mathrm{c}$ determines the dust charge q_d

10.1.6.2 Non-isothermal Plasma with Fixed Ion Density

In a laboratory plasma with $T_e \gg T_i$, the ions show no Boltzmann response. Rather we assume that the ion density in the dust cloud is fixed at the same value n_∞ as in the surrounding plasma. Then the corresponding model equations become

$$0 = e^{-\eta_c} - 1 + P(\eta_f - \eta_c) \quad (10.28)$$

$$0 = (\mu\tau)^{-1/2}[1 + \tau(\eta_f - \eta_c)] - e^{-\eta_f} . \quad (10.29)$$

This system behaves basically in the same manner (see Fig. 10.11) as the original model with Boltzmann ions described above. When the parameter P reaches $P \approx 1$, the normalized cloud potential rises until it asymptotically reaches a final value of $\eta_c = (1/2)\ln(\mu\tau) = 7.905$. At this value, the entire dust cloud floats in the ambient plasma. At the same time the dust charge is becoming smaller and smaller.

At first glance, the increase of the floating potential of the cloud beyond the value for a floating sphere is surprising. In Sect. 10.1.3.1, we had emphasized that the floating potential of a small sphere is independent of its size. This puzzle is resolved when we inspect (10.29), in which the OML factor $[1+\tau(\eta_f-\eta_c)]$ becomes unity for vanishing dust charge. Therefore, the dust cloud as a whole does not benefit from orbital motion. As a consequence, the electron current must be suppressed much stronger, because it competes with a smaller ion current that is not enhanced by an OML factor.

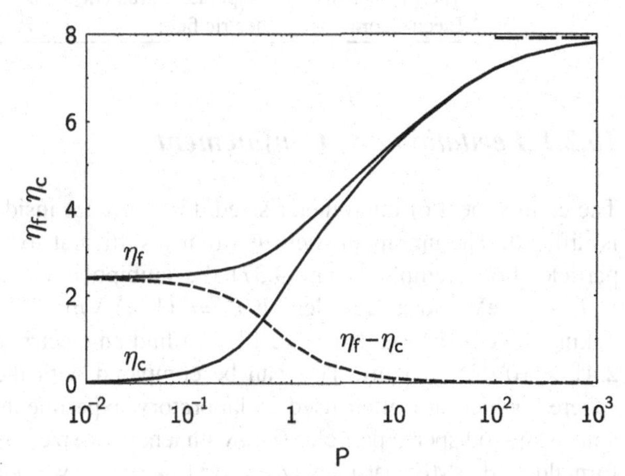

Fig. 10.11 Normalized cloud potential η_c and grain potential η_f for an argon plasma with fixed ion density and $T_e/T_i = 100$ as a function of the Havnes parameter P. The dust charge is determined by the difference $\eta_f - \eta_c$

10.1.6.3 Charge Sharing

In the end, when the free electron density in the dust cloud becomes much smaller than the ion density, we have a quasineutrality condition $n_d|q_d| \approx n_i e$ and, for fixed ion density, the dust charge decreases as

$$|q_d| = \frac{n_i}{n_d} e . \quad (10.30)$$

Therefore, the dust grains must share among them the maximum allowed negative charge. Any additional dust grain entering the cloud must steal its charge from the neighbors.

10.2 Forces on Dust Particles

Dust particles experience various external forces that depend in different ways on the size of the dust particle, as compiled in Table 10.2 [260]. Particles of several micrometer size are dominated by gravity whereas submicron particles are more susceptible to drag forces.

The interaction forces between dust particles are in most cases repulsive and can be described by shielded Coulomb potentials. There are, however, situations with streaming ions, where charging of the *wake* behind a dust particle can lead to net attractive force between dust particles.

Table 10.2 External forces on dust particles

Name	Origin	Size dependence
Weight force	gravity	a^3
Neutral drag	streaming neutrals	a^2
Ion drag force	streaming ions	a^2
Thermophoresis	temperature gradient	a^2
Electric force	electric field	a^1

10.2.1 Levitation and Confinement

The confinement of micrometer-sized dust particles inside the quasineutral plasma is difficult. The ambipolar field is often insufficient to balance the weight of the particle. For example, from (4.37) the ambipolar electric field in a plasma of $k_B T_e = 3\,\mathrm{eV}$ and a scale length $L = (1/n)|\nabla n| \approx 1\,\mathrm{cm}$ is $E \approx 300\,\mathrm{V\,m^{-1}}$. Taking the dust charge from (10.13), we find an electric field force $F_E = q_d\,E = 2.41 \times 10^{-13}\mathrm{N}\ a(\mu\mathrm{m})$. This can be compared with the weight force of plastic spheres, which are often used in laboratory experiments because they are available as monodisperse particles (i.e., with a narrow size distribution). For melamine-formaldehyde (MF) particles ($\rho_d = 1514\,\mathrm{kg\,m^{-3}}$) we obtain a weight force $F_g = 6.22 \times 10^{-14}\,[a\,(\mu\mathrm{m})]^3$. Figure 10.12 shows a comparison of these forces. The intersections of the curves give the particle size that can be levitated by the assumed ambipolar field. Therefore, experiments with particles confined in the bulk plasma, where the ambipolar field is low, are restricted to particles of $a < 1\,\mu\mathrm{m}$, typically. Larger particles will sediment to the bottom of the plasma, where the particles are finally trapped by stronger electric fields in the sheath region between plasma and wall (or electrode). Plasmas with steeper density gradients can levitate larger particles.

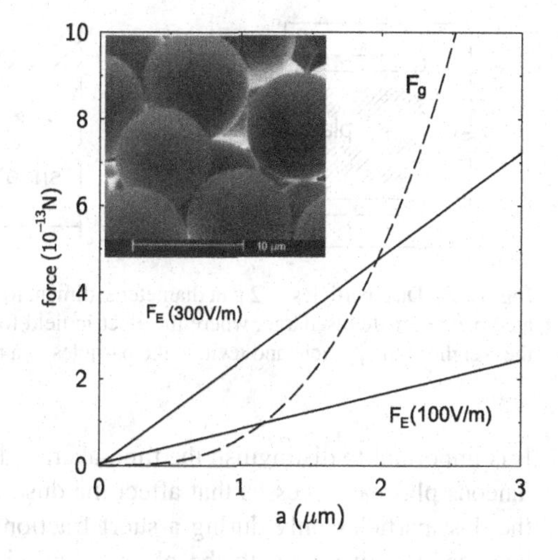

Fig. 10.12 Weight force F_g and electric field force F_E from the ambipolar field of $E = 100\,\mathrm{V\,m^{-1}}$ and $E = 300\,\mathrm{V\,m^{-1}}$. The inset shows an electron micrograph of monodisperse spherical melamine-formaldehyde particles

10.2.1.1 Levitation in the Radio-Frequency Sheath

For many years, the workhorse of dusty plasma research in the laboratory was a parallel plate discharge driven by a radio-frequency (rf) voltage at 13.56 MHz (Fig. 10.13).

The discharge forms a central quasineutral bulk plasma and two sheath regions that separate the plasma from the electrodes (Fig. 10.14). Details of this discharge type are discussed in Sect. 11.2. Here, we need only know that the sheath region has (on time-average) a net positive space charge.

The dust particles are too heavy to react to the rf electric field. Therefore, the levitation of the dust particles is determined by the *time-averaged electric field* $\langle E \rangle$, which is determined by Poisson's equation $\mathrm{d}\langle E \rangle/\mathrm{d}z = \langle \rho \rangle/\varepsilon_0$. Experiments have shown that the charge density $\langle \rho \rangle$ is nearly constant [261], which results in a linear increase

$$\langle E \rangle \approx \frac{\langle \rho \rangle}{\varepsilon_0} z \,. \tag{10.31}$$

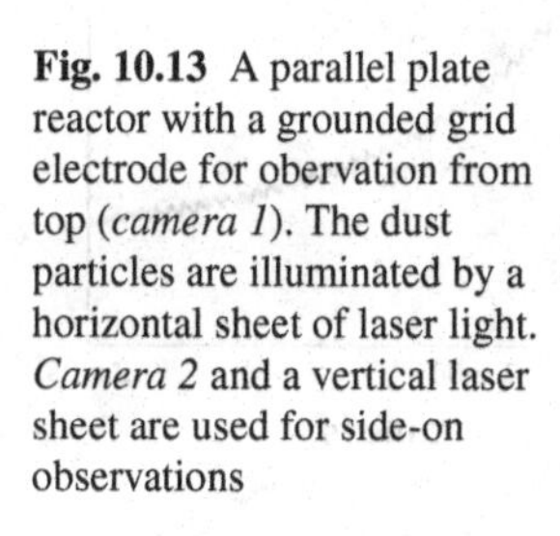

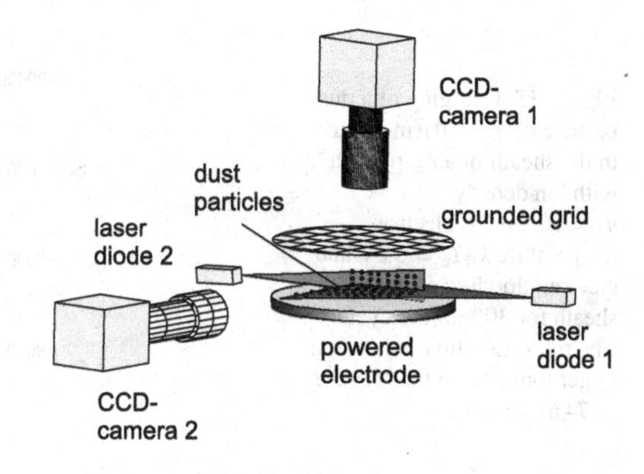

Fig. 10.13 A parallel plate reactor with a grounded grid electrode for obervation from top (*camera 1*). The dust particles are illuminated by a horizontal sheet of laser light. *Camera 2* and a vertical laser sheet are used for side-on observations

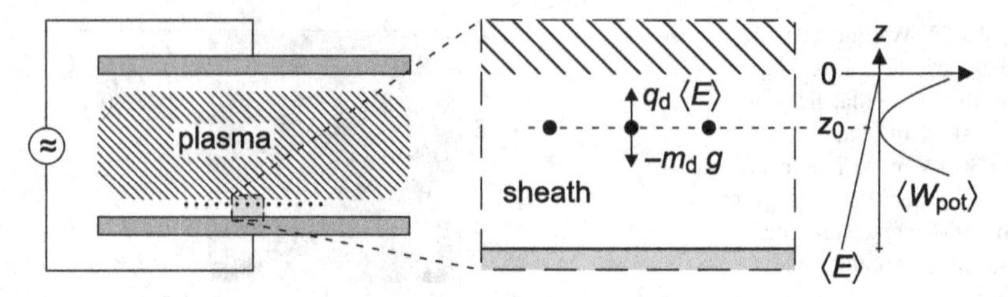

Fig. 10.14 Dust particles > 2 μm diameter sediment to the lower sheath of a radio-frequency operated parallel plate discharge, where the electric field force becomes sufficiently strong to balance the weight of the particle and levitate the particles in a thin layer

It is important to distinguish the time-averaged behavior of the dust and the instantaneous plasma processes that affect the dust. A negative charging current reaches the dust particles only during a short fraction of the rf cycle, when the electrode becomes positive w.r.t. to the plasma, and electrons are able to flood the sheath (see Sect. 11.2). Positive argon ions are also too heavy to follow the 13.56 MHz excitation. Therefore, the ion density profile is stationary and the positive ion flow to the dust particles is continuous. How does this affect the particle charge? Since the charge relaxation time from (10.17) is shortest for large particles, we illustrate the charging process for a large particle of $a = 10\,\mu\text{m}$ radius (Fig. 10.15).

For $k_B T_e = 3\,\text{eV}$, $n = 10^{15}\,\text{m}^{-3}$ and electron flooding of the sheath for 30% of the cycle, the relaxation time (in the quasineutral bulk plasma) is $\tau = 1.8\,\mu\text{s}$. Therefore, during the much shorter rf period of 74 ns (13.56 MHz) the dust grain acts like an RC-integrator and the charging curve becomes fairly smooth with a minor ripple from the charging—discharging during each rf cycle. For smaller particles, this ripple becomes even smaller. In particular, all dust grains in the range of interest, $a = (1 - 10)\,\mu\text{m}$ attain a fixed electric charge. The value of this charge, however, differs from the equilibrium charge (10.13) in a quasineutral environment.

Let us now return to the question of levitation. In the sheath, the (time averaged) electric field is one or two orders of magnitude stronger than the ambipolar field.

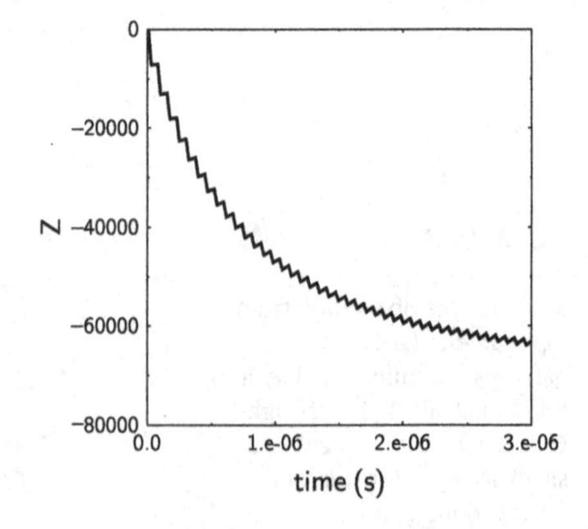

Fig. 10.15 Charging of a dust particle of $a = 10\,\mu\text{m}$ radius in the sheath of a r.f. plasma with ion density $n_i = 10^{15}\,\text{m}^{-3}$, electron temperature $k_B T_e = 3\,\text{eV}$ and electron flooding of the sheath for 30% of the cycle. The relaxation time τ is much larger longer than the rf cycle of 74 ns

It has a minimum value at the plasma boundary and a maximum at the electrode (Fig. 10.14). Dust particles are levitated at a height z_0 inside the sheath, where the weight force and the electric force are balanced

$$-m_\mathrm{d}g + q_\mathrm{d}\langle E(z_0)\rangle = 0\,. \tag{10.32}$$

This condition is similar to the levitation condition for an oil drop in the famous experiments of Robert Millikan (1868–1953) that proved the quantization of the electric charge, which won him the 1923 nobel prize. However, in Millikan's experiment, the electric field was independent of position while it is inhomogeneous in the space charge sheath, with a linear increase from the plasma edge to the electrode.

10.2.1.2 The Vertical Resonance

The equation of motion for a dust particle in the sheath involves the position dependence of the electric force

$$m_\mathrm{d}\ddot{z} + m_\mathrm{d}\nu_\mathrm{d}\dot{z} - q_\mathrm{d}E(z) = F_\mathrm{ext}(t)\,. \tag{10.33}$$

Here, ν_d is the dust-neutral collision frequency, which describes the friction of dust particles with the neutral gas. Such a friction coefficient was introduced by Epstein [262] to describe the motion of the oil drops in Millikan's experiment. Neutral drag will be discussed below in Sect. 10.2.2.

When we make a linear approximation for the variation of the electric field

$$E(z) \approx E(z_0) + \frac{\mathrm{d}E}{\mathrm{d}z}(z-z_0) = E(z_0) + \frac{\langle\rho\rangle}{\varepsilon_0}(z-z_0) \tag{10.34}$$

and assume that the dust charge q_d is a fixed quantity, the equation of motion (10.33) represents a damped harmonic oscillator with a fundamental frequency ω_0 given by

$$\omega_0^2 = \frac{|q_\mathrm{d}|\,\langle\rho\rangle}{\varepsilon_0 m_\mathrm{d}}\,. \tag{10.35}$$

The parabolic shape of the effective potential well $\langle W_\mathrm{pot}\rangle$ (Fig. 10.14) results from integrating the linear variation of the electric force $q_\mathrm{d}E(z)$ and adding the potential energy in the gravitational field $m_\mathrm{d}gz$. The energy minimum is at the equilibrium position defined by (10.32).

The vertical resonance depends on the ratio $q_\mathrm{d}/m_\mathrm{d}$ and can be used to determine the dust charge q_d when the dust mass m_d is well known, as in the case of monodisperse MF particles. Since the resonance frequency depends on the net charge density in the sheath, a measurement of the ion density in the plasma is necessary, which then must be extrapolated into the sheath.

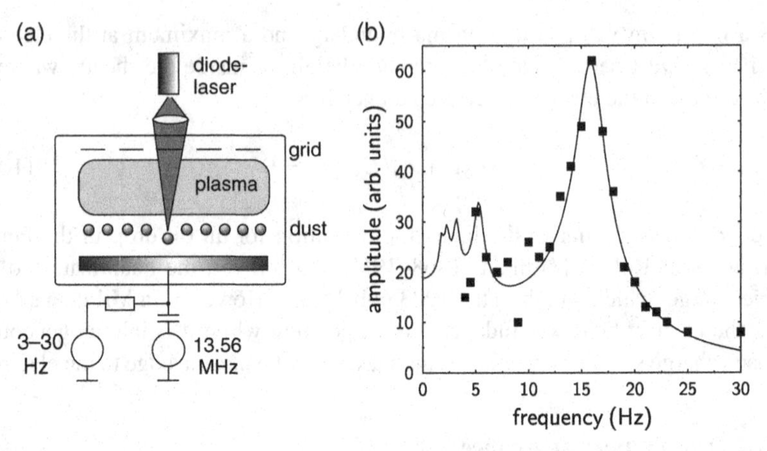

Fig. 10.16 Excitation of the vertical resonance by radiation pressure from a laser diode. A spurious resonances from the harmonics of the square-wave excitation appears at 1/3 of the resonance frequency (from [265])

The resonance curve can be recorded by applying an external force of frequency f to the dust particles and recording the oscillation amplitude, as shown in Fig. 10.16a. There are two ways to apply a force to the dust particles. In the original version of the resonance method [263, 264], an additional low-frequency bias voltage of (3–30) Hz frequency was added to the lower electrode, which leads to a periodic shift of the boundary between plasma and sheath. The technique was later refined in [265], where a focused diode laser was used that was chopped at a frequency f and exerts a periodic force on a single dust particle by radiation pressure.

The resonance curves obtained by the laser method is shown in Fig. 10.16b. The squares are the experimental points, which are compared with the resonance curve of a damped oscillator. The resonance frequency is $f_{\mathrm{res}} \approx 16\,\mathrm{Hz}$. There is also a spurious resonance at $f_{\mathrm{res}}/3$, which is caused by exciting the main resonance with the harmonic of a square wave at $3f$. The theoretical curve comprises also spurious resonances at $f_{\mathrm{res}}/5$ and $f_{\mathrm{res}}/7$, which are not resolved in the experiment.

The resulting dust charge is $Z_{\mathrm{d}} = (8320 \pm 2000)$ compared to $Z_{\mathrm{d}} = 16000$ from (10.13) for a particle of $4.8\,\mu\mathrm{m}$ radius in a quasineutral plasma environment. The difference can be attributed to the net electron depletion of the sheath [266].

10.2.1.3 Self-Excited Vertical Oscillations

Let us now consider the case of vertical oscillations with a variable dust charge. At different positions in the sheath, the dust charge may attain an equilibrium charge, which in linear approximation is given by

$$q_{\mathrm{d}}^{\mathrm{eq}}(z) \approx q_{\mathrm{d0}} + q_{\mathrm{d}}' z\,. \tag{10.36}$$

q_{d0} is the equilibrium charge at the levitation position defined by $q_{d0}E(z_0) = m_d g$. When the dust particle oscillates in the sheath, the dust charge lags behind due to the finite charge relaxation time. The evolution of the dust charge is described by

$$\dot{q}_d = -\frac{1}{\tau}\left[q_d(z) - q_d^{eq}(z)\right] = -\frac{1}{\tau}\left[q_d(z) - q_{d0} - q_d' z\right] . \quad (10.37)$$

Solving this charging equation together with the equation of motion (10.33) by a normal mode analysis $\propto \exp(-i\omega t)$ [267], gives unstable solutions with a growth rate

$$\gamma = \mathrm{Im}(\omega) = -\frac{1}{2}\left(\beta + \frac{q_d' E(z_0)\tau}{m_d}\right) , \quad (10.38)$$

when the term in parentheses on the r.h.s. becomes positive. This happens for $q_d' E(z_0) < 0$ and $\beta < |q_d' E(z_0)|\tau/m_d$.

Avoiding the tedious mathematics that leads to (10.38), we can grasp the basic physical mechanism of the instability by considering a dust particle during its vertical oscillation when it crosses z_0 with velocity v. To estimate the order of magnitude of the surplus charge, we go back in time by one charging time τ, when the particle was at $z_1 = z_0 - v\tau$. There, the equilibrium charge deviates from that at z_0 by $\Delta q_d = -q_d' v\tau$. Superimposed to its periodic motion in the potential well, the particle experiences an extra force from the surplus charge

$$\Delta F \approx \Delta q_d E(z_0) = -q_d' v\tau E(z_0) . \quad (10.39)$$

For instability, this extra force must overcome the frictional force $\beta m_d v$, which defines a critical friction coefficient

$$\beta_c = \frac{|q_d' E(z_0)|\,\tau}{m_d} \quad (10.40)$$

in accordance with the discussion of (10.38).

Unstable vertical oscillations were first reported in [268]. The instability is favoured by high values of q_d', which are found in the sheath of dc-discharges. There, electrons can only penetrate the sheath in terms of a Boltzmann factor, which prevents the presence of electrons deep inside the sheath. Contrariwise, in rf discharges, electrons even reach the electrode for a fraction of each rf cycle, and q_d' takes more moderate values. This is why this instability was only observed in dc discharges. Self-excited oscillations of finite amplitude were also found in [269], as shown in Fig. 10.17. There, a constant amplitude of the oscillations results from higher order corrections in the Taylor expansion of the electric field and the charge as functions of vertical position.

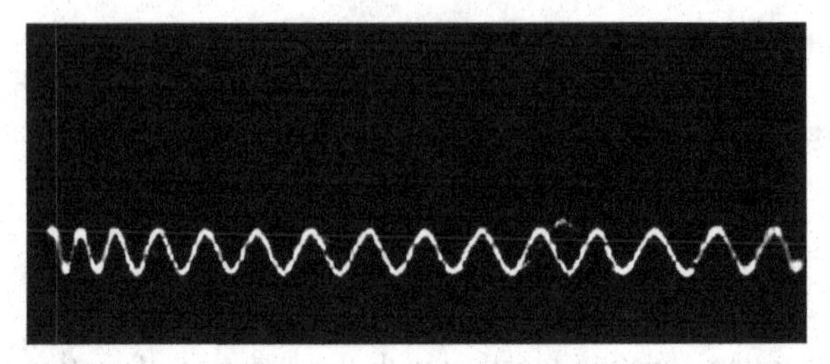

Fig. 10.17 Self-excited vertical oscillations of a particle in the sheath of a dc discharge, which is moving horizontally from left to right. This photo has an exposure time of 0.88 s. (Reproduced with permission from [269]. ©1999 by the American Physical Society)

10.2.1.4 Confinement Geometries

Because of the mutual repulsion of like-charged dust grains, a dust cloud can only exist in a particle trap which provides confinement of the dust particles from all sides. The vertical confinement in the space-charge sheath by parabolic potential wells was discussed in Sect. 10.2.1.2. Horizontal confinement can be achieved by bending the equipotential upwards, on which the particle is levitated.

The thickness of the space-charge sheath is basically determined by the Child-Langmuir law (7.12). Therefore, a radial decay of the plasma density profile results in a wider space charge sheath at lower electron density, which leads to an effective upward bending of the equipotential, see Fig. 10.18a. Metallic barriers arranged on the electrode have a similar effect and lead to radial electric field forces near the barrier, as shown in Fig. 10.18b. When a parabolic confining potential is needed, a shallow spherical depression in the electrode can shape the equipotentials, see Fig. 10.18c. The levitation and confinement of Yukawa balls (Sect. 10.3.3) is achieved by a combination of the thermophoretic force from a temperature gradient in the neutral gas (due to heating of the electrode) with radial electric forces originating from a sheath around the confining glass walls of a small box, as shown in Fig. 10.18d.

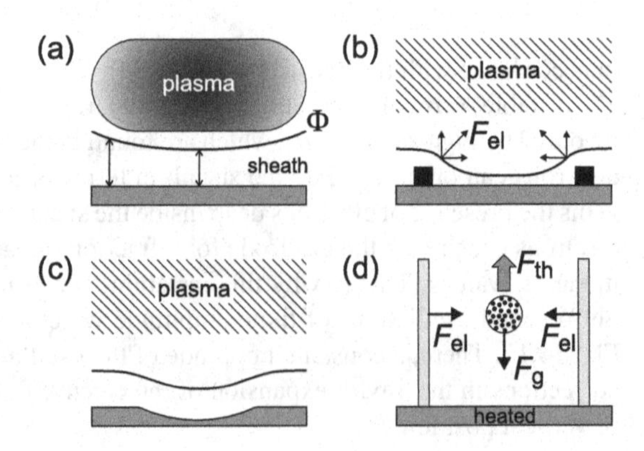

Fig. 10.18 (**a**) Curved equipotential due to sheath expansion caused by a radial density gradient. (**b**) Bending of equipotentials by metal barriers. (**c**) Parabolic depression in electrode. (**d**) Levitation by thermophoretic forces and confinement in a glass box by electric forces

10.2.2 Neutral Drag Force

The neutral drag force played an important historical role in Millikan's famous oil drop experiments that demonstrated the quantization of the electric charge. Epstein [262] had analyzed the drag force for various kinds of collision types between neutral gas atoms and oil drops, for which he introduced an *accomodation coefficient* δ. The neutral drag force for a spherical dust grain of radius a is given by the expression

$$\mathbf{F}_\mathrm{n} = -\frac{4}{3}\delta\pi a^2 n_\mathrm{n} v_{\mathrm{th,n}}(\mathbf{v}_\mathrm{d} - \mathbf{v}_\mathrm{n}) \,. \tag{10.41}$$

Here, $\mathbf{v}_\mathrm{d}$ ist the dust velocity, $v_{\mathrm{th,n}}$ the mean thermal velocity of the neutral gas, and $\mathbf{v}_\mathrm{n} \ll v_{\mathrm{th,n}}$ the drift velocity of the neutral gas. The coefficient δ has the values

$$\delta = \begin{cases} 1 & : \text{specular reflection} \\ 1+\pi/8 & : \text{perfect diffuse reflection} \,, \end{cases} \tag{10.42}$$

which can be considered as the limiting cases for the reflection process. Refinements for the drag force were discussed in [237].

Instead of calculating the force that a neutral wind of velocity $\mathbf{v}_\mathrm{n}$ exerts on a dust particle at rest, we can also ask for the friction force on a dust particle that moves with velocity $\mathbf{v}_\mathrm{d}$ through a neutral gas at rest. Both situations are described by (10.41). However, it is practical to express the friction force by a dust-neutral collision frequency ν_d for momentum loss, which is defined by $F_\mathrm{n} = \nu_\mathrm{d} m_\mathrm{d}(v_\mathrm{d} - v_\mathrm{n})$ and reads

$$\nu_\mathrm{d} = \delta\frac{4}{3}\pi a^2 \frac{m_\mathrm{n}}{m_\mathrm{d}} n_\mathrm{n} v_{\mathrm{th,n}} = \delta\frac{8}{\pi}\frac{p}{a\rho_\mathrm{d} v_{\mathrm{th,n}}} \,. \tag{10.43}$$

Here, p is the gas pressure and ρ_d the density of the dust material. Note that ν_d is *not* the frequency at which neutral atoms hit the dust particle.

10.2.3 Thermophoretic Force

If there is a gradient of gas temperature in the vicinity of a particle, collisions of neutrals with the dust particles will on average impart more momentum to the particle on the hot side than on the cold side. This transfer of a net momentum from the gas to the particle is called thermophoresis. The thermophoretic force is proportional to the temperature gradient and is directed from the higher gas temperature region to the lower gas temperature region. In dusty plasmas, thermophoretic effects were first observed by Jellum and Graves [270]. An analytical expression for the thermophoretic force is [271, 272]

$$\mathbf{F}_{tp} \approx -8a^2 n_n \lambda_c k_B \nabla T_n \tag{10.44}$$

where λ_c is the average collision length in the gas, and T_n the gas temperature. This expression is valid for gas temperatures, $T_n < 500\,\mathrm{K}$, and when the distance of the dust particles from the walls is much larger than a mean free path.

10.2.4 Ion Wind Forces

The orbital motion of a positive ion with initial velocity v_0 in the field of a negatively charged dust particle can lead to a collision with the dust particle when the impact parameter is smaller than the collection radius (7.54), $b_c = a(1 - 2e\Phi/mv_0^2)^{1/2}$. Larger impact parameters lead to scattering of ions (see Fig. 10.19). In both cases, momentum is transferred to the dust particle. Averaging the transferred momentum over the ion distribution function yields the ion wind force on the dust particle, which consists of a collection force and an orbit force [260]

$$F_i = F_c + F_o \,. \tag{10.45}$$

10.2.4.1 The Collection Force

For monoenergetic ions, the collection force is given by the momentum flux entering the collection cross section $\sigma_c = \pi b_c^2$

$$F_c = \underbrace{n_i\,(m_i\,v_0)\,v_0}_{\text{momentum flux density}}\;\sigma_c = n_i\,m_i v_0^2 \pi a^2 \left(1 - \frac{2e\Phi_f}{m_i v_0^2}\right) . \tag{10.46}$$

One could ask, why the incoming momentum flux is calculated from the original velocity v_0 of an ion far away from the dust particle. In fact, the ion is accelerated by

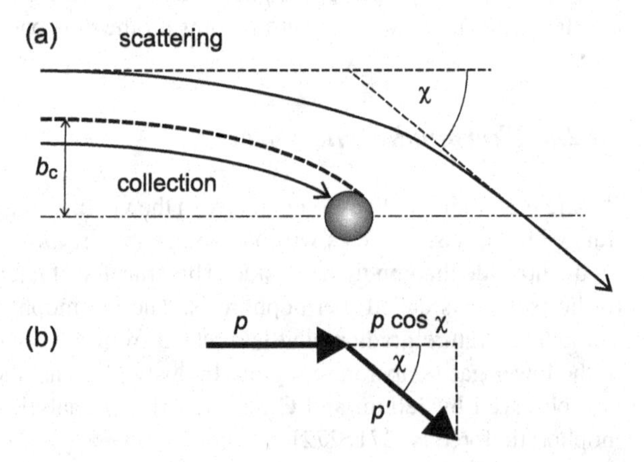

Fig. 10.19 (**a**) Ion collection and ion scattering by a dust particle. (**b**) In a scattering event, the momentum $\Delta p = p[1 - \cos(\chi)]$ is transfered to the dust particle

the attractive force F exerted by the dust particle. Before reaching the surface of the dust particle, the ion has gained an extra momentum $\delta p = \int F\mathrm{d}t$. However, the dust grain has also gained a momentum, $-\delta p$, because of Newton's *actio* = *reactio* law. When the ion hits the dust particle, the ion transfers the gained extra momentum to the dust particle and the original situation is restored. Hence, our calculation based on the unpertubed ion velocity is correct.

For a shifted Maxwellian ion-distribution with a drift velocity $\mathbf{v}_\mathrm{d}$,

$$f_\mathrm{M}(\mathbf{v}) = n_\mathrm{i}\left(\frac{m_\mathrm{i}}{2\pi k_\mathrm{B}T_\mathrm{i}}\right)^{3/2} \exp\left[-\frac{m_\mathrm{i}(\mathbf{v}-\mathbf{v}_\mathrm{d})^2}{2k_\mathrm{B}T_\mathrm{i}}\right] , \tag{10.47}$$

the average collection force becomes [260]

$$\langle F_\mathrm{c}\rangle = n_\mathrm{i}m_\mathrm{i}v_\mathrm{d}v_\mathrm{s}\pi a^2\left(1-\frac{2e\Phi_\mathrm{f}}{m_\mathrm{i}v_\mathrm{s}^2}\right) \tag{10.48}$$

with

$$v_\mathrm{s}^2 = v_\mathrm{d}^2 + \frac{8k_\mathrm{B}T_\mathrm{i}}{\pi m_\mathrm{i}} . \tag{10.49}$$

Again, the average momentum density is $n_\mathrm{i}m_\mathrm{i}v_\mathrm{d}$, but the number of collision processes per second is determined by v_s, which becomes the mean thermal speed for small drift velocities. The collection force can be described by the limiting cases of Stokes friction, $F_\mathrm{c} \propto v_\mathrm{d}$, at low drift velocity, and of aerodynamic ram pressure, $F_\mathrm{c} \propto v_\mathrm{d}^2$, at high drift velocity.

10.2.4.2 The Orbit Force

When an ion is deflected in the field of a heavy dust particle, the ion momentum changes direction, but the magnitude is conserved, $|p| = |p'|$, see Fig. 10.19b. For a scattering angle χ, the transferred momentum is

$$\Delta p = p[1-\cos(\chi)] . \tag{10.50}$$

A general statement about the ion orbit in the shielded potential $\Phi(r)$ of a dust grain can be made from the conservation of energy W and angular momentum L,

$$W = \frac{m_\mathrm{i}}{2}v_0^2 = \frac{m_\mathrm{i}}{2}(\dot{r}^2 + r^2\dot{\theta}^2) + e\Phi(r) \tag{10.51}$$

$$L = m_\mathrm{i}v_0 b = m_\mathrm{i}r^2\dot{\theta}^2 . \tag{10.52}$$

Spherical coordinates r, θ are used in the following, as defined in Fig. 10.20.

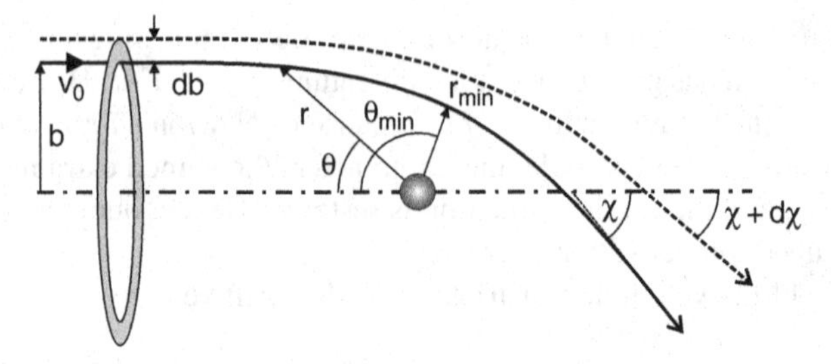

Fig. 10.20 Spherical coordinates r, θ are used to describe the ion trajectory. An increase of the impact parameter $b \to b + \mathrm{d}b$ leads to a smaller scattering angle $\chi + \mathrm{d}\chi$

Replacing $\dot{\theta} = v_0 b / r^2$ in (10.51), the energy equation becomes

$$W = \frac{m_\mathrm{i}}{2}\left[\dot{r}^2 + \frac{L^2}{m_\mathrm{i}^2 r^2}\right] + e\Phi(r)\,. \tag{10.53}$$

The minimum distance r_min of the orbit can be found by setting $\dot{r} = 0$ and solving the resulting quadratic equation with the result (see Problem 10.4)

$$r_\mathrm{min} = -r_\mathrm{C} + \sqrt{r_\mathrm{C}^2 + b^2}\,, \tag{10.54}$$

where we have introduced the Coulomb radius

$$r_\mathrm{C} = \frac{-q_\mathrm{d} e}{4\pi\varepsilon_0 m_\mathrm{i} v_0^2}\,. \tag{10.55}$$

For the further calculations, we must now specify the interaction potential $\Phi(r)$. The particle orbit during this scattering process becomes a hyperbola when the interaction between ion and dust grain can be described by a Coulomb force. This assumption is valid when the impact parameter b of the incoming ion is smaller than the shielding length λ_s of the dust grain. For small ion velocities, $v_0 \ll v_B$, the shielding length is given by the linearized Debye length, $\lambda_s = \lambda_\mathrm{D}$. However, for larger ion speeds, shielding by streaming ions is better described by a modified Debye shielding length [273]

$$\lambda_\mathrm{s} = \frac{\lambda_\mathrm{De}}{1 + k_\mathrm{B}T_\mathrm{e}/(k_\mathrm{B}T_\mathrm{i} + m_\mathrm{i} v_0^2)}\,. \tag{10.56}$$

For $b < \lambda_\mathrm{s}$, the scattering problem is equivalent to the Kepler orbit of a comet in the central field of the Sun. Therefore, we can use the standard textbook result for the scattering angle, $\chi = 2\theta_\mathrm{min} - \pi$, and obtain

$$\tan\left(\frac{\chi}{2}\right) = \frac{r_\mathrm{C}}{b}\,. \tag{10.57}$$

When the impact parameter equals the Coulomb radius, $b = r_C$, the scattering angle becomes $\chi = \pi/2$. Therefore, the Coulomb radius is the impact parameter for 90° scattering, $r_C = b_{\pi/2}$, which was already defined in (4.20). To calculate the orbit force, we must introduce a proper average of the momentum transfer (10.50) over all impact parameters

$$F_o = \left\langle \frac{dp}{dt} \right\rangle_b = \left\langle \underbrace{n_i v_0 \frac{d\sigma}{d\Omega}}_{\text{events per second}} \times \underbrace{m_i v_0 (1 - \cos\chi)}_{\text{momentum transfer}} \right\rangle_b . \tag{10.58}$$

Here, $d\sigma/d\Omega$ is the differential cross section for a scattering event with a scattering angle χ into a fraction $d\Omega$ of solid angle. The concept of a differential cross section assigns to the scattering process a ring-shaped area $d\sigma$ (see Fig. 10.20) that defines the number of scattering events per second, as given by the first factor in (10.58). The calculation of $d\sigma/d\Omega$ involves the relation (10.57) between scattering angle χ and impact parameter b. We have $d\sigma = 2\pi b db$ and $d\Omega = |2\pi \sin\chi d\chi|$. In the latter expression the magnitude has to be taken because $d\chi/db < 0$. Then we obtain for the differential cross section

$$\frac{d\sigma}{d\Omega} = \frac{b}{\sin\chi}\left|\frac{db}{d\chi}\right| = \frac{r_C^2}{\sin^4(\chi/2)} , \tag{10.59}$$

which is the well-known *Rutherford cross-section* that was originally derived for (repulsive) scattering of α-particles on gold atoms. Using this expression, the orbit force becomes

$$F_o = n_i m_i v_0^2 \, 2\pi \int_{\chi_{min}}^{\chi_{max}} (1 - \cos\chi) \frac{d\sigma}{d\Omega} \sin\chi \, d\chi = n_i m_i v_0^2 \, 4\pi r_C^2 \ln\Lambda . \tag{10.60}$$

In this calculation, we have to cut off the integration at a smallest scattering angle χ_{min} that corresponds to the maximum allowed impact parameter, $b = \lambda_s$, and a maximum angle χ_{max} determined by the collection radius, $b = b_c$. The quantity $\ln\Lambda = \ln(\sin\chi_{max}/\sin\chi_{min})$ is called the Coulomb logarithm, which under the present assumptions takes the form

$$\ln\Lambda = \frac{1}{2}\ln\left(\frac{\lambda_s^2 + r_C^2}{b_c^2 + r_C^2}\right) . \tag{10.61}$$

Again, when we average over a shifted Maxwellian, the momentum flux can be approximately replaced by $n_i m_i v_0 v_s$. At last, we obtain the total ion drag force as [260]

$$\langle F_d \rangle = n_i m_i v_0 v_s \left(\pi b_c^2 + 4\pi b_{\pi/2}^2 \ln\Lambda\right) . \tag{10.62}$$

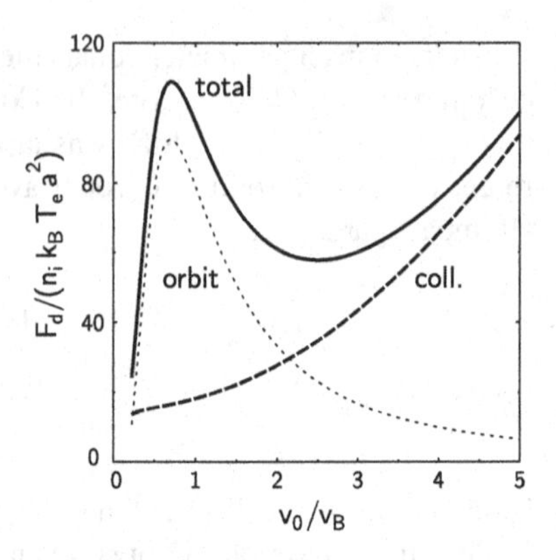

Fig. 10.21 The total ion drag force (*solid line*) and its constituents, the orbit force (*dotted line*) and collection force (*dashed line*) as a function of the normalized ion streaming velocity

The dependence of the ion drag force on the ion drift velocity is shown in Fig. 10.21 for argon ions, $\lambda_s/a = 10$, and $T_e/T_i = 100$. For small drift velocities of the order of the Bohm velocity v_B, the orbit force is dominant. The collection force becomes important for $v_0 \gg v_B$. Since the maximum of the orbit force occurs near $v_0 \approx v_B$, the proper choice of the shielding length is $\lambda_s \approx \frac{2}{3}\lambda_{De}$.

For completeness, it should be mentioned that more precise expressions for the Coulomb logarithm have been introduced that allow for 90° scattering of slow ions that have impact parameters greater than λ_D [274–276]. Approximations to the full non-Keplerian scattering process were obtained from computer simulations [273, 277]. A short discussion of these aspects can be found in [235].

10.2.4.3 Experiments on the Ion Drag Force

The ion drag force has been investigated systematically using the deflection of falling dust particles [278, 279]. There, the influence of ion drag, electric and thermophoretic force was analyzed after the dust particles had reached the terminal velocity that is given by the balance of weight force and neutral drag. The deflection method was also applied in the collisionless plasma of a double-plasma device [280]. There, the deflection angle is given by $\tan\alpha = F_d/F_g$. The experimental arrangement is sketched in Fig. 10.22a.

The experiments were performed in two different modes. A high-velocity ion beam could be generated by applying a bias voltage between the source chamber (S) and target chamber (T). In this limit, the ion drag is determined by the collection force, which is identical in all ion drag models. Figure 10.22b shows how the collection radius derived from the measured ion drag force approaches the limit given by the particle radius.

The more interesting situation is found at low ion drift velocities, where significant differences are found between the models [260] and [275]. This drift is generated by a negative bias voltage on the separation grid and operating source and target chamber at the same plasma potential. In Fig. 10.22c the experimental points indeed come closer to the refined model by Khrapak et al. [275]. However, the model by Barnes et al. [260] with a proper shielding length is not far off. This

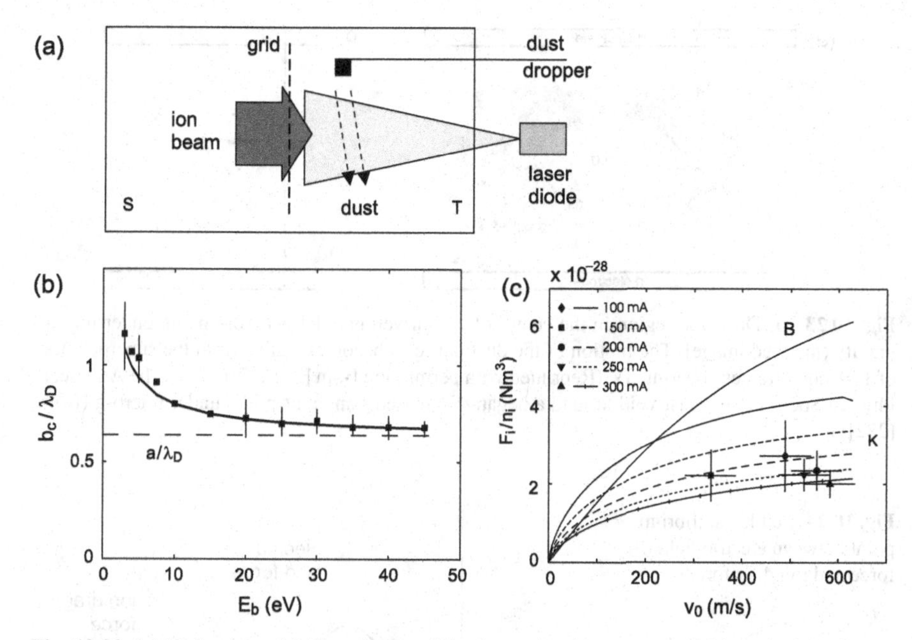

Fig. 10.22 (**a**) Deflection of falling dust particles by an ion beam generated in a double-plasma device. (**b**) Resulting collection radius b_c as a function of beam energy. (**c**) Deflection of falling dust grains by a slow ion drift towards the grid in comparison with the models of Barnes et al. [260] and Khrapak et al. [275] (from [280])

confirms that the simplified treatment in Sect. 10.2.4.2 is able to catch the essential features of ion drag.

10.2.4.4 Dust-free Regions (Voids)

Equilibria of dust clouds involving ion drag and electric forces become important when gravity plays a minor role. This can occur in the laboratory for fine particles of less than 1 μm diameter or for larger particles in experiments under microgravity. (A state similar to microgravity can be established by compensating the weight force by a thermophoretic force.) These equilibria require only weak electric fields such as those associated with ambipolar diffusion, because the orbit force becomes large in the regime $0.3\, v_B < v_0 < v_B$, which is typical of the presheath region.

A characteristic effect of this kind is the formation of dust-free regions (*voids*), which were found in discharges with fine particles [283] and in a dusty plasma on a sounding rocket [281]. An example for the void phenomenon is shown in Fig. 10.23a.

The highest ion density is found in the plasma center and a density profile $n_i(r, z)$ is formed by ambipolar diffusion. The density gradient is steepest near the electrodes and zero in the plasma center. As a consequence, the ambipolar electric field (4.37)

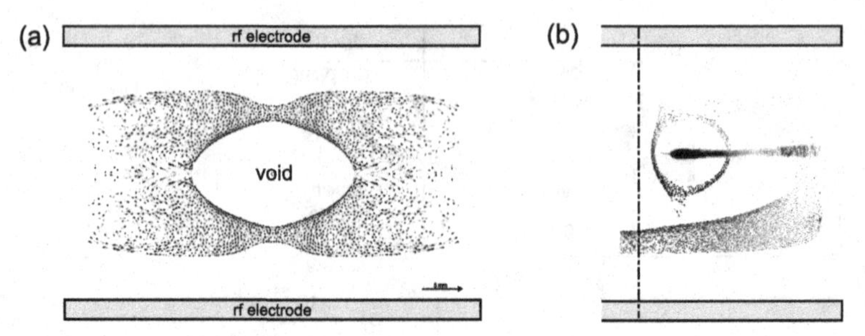

Fig. 10.23 (**a**) Dust free region in the center of a rf-driven parallel plate discharge under microgravity (inverted image). The motion of the dust particles becomes visible from the superposition of 150 video frames covering 3 s. (Reprinted with permission from [281]. ©1999 by the American Physica Society.) (**b**) Dust void around a negatively biased Langmuir probe under micro-g (from [282])

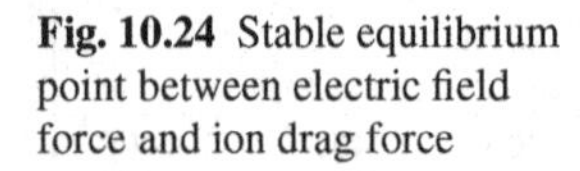

Fig. 10.24 Stable equilibrium point between electric field force and ion drag force

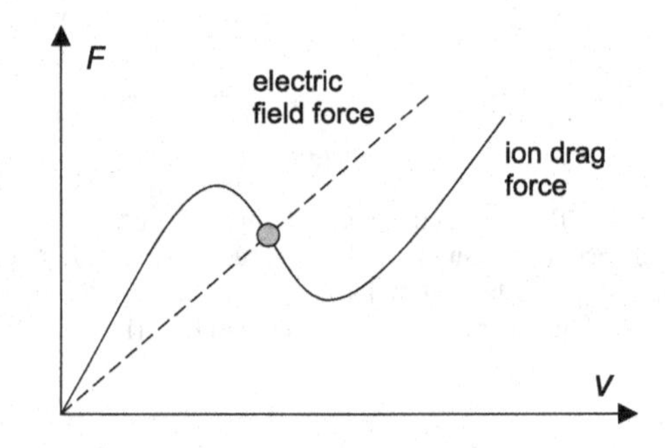

and the ion drift velocity $v_\mathrm{i} = \mu_\mathrm{i} E$ increase from the center to the electrodes. The general dependence of the ion drag force and the electric field force on the local ion velocity is shown in Fig. 10.24. Here, the electric field force $F_E = q_\mathrm{d} v_\mathrm{i}/\mu_\mathrm{i}$ is related to the ion velocity through the ion mobility. The ion drag force points radially outward whereas the electric field force pushes the dust particles towards the center.

The equilibrium position for a single particle is found at the intersection of the two curves in Fig. 10.24. This equilibrium is stable because an outward directed perturbation leads into a region of higher ion velocity where the dominating electric field force pushes the particle back. Similarly, an inward directed perturbation is corrected by the net excess force from ion drag.

A reversed situation for void formation is found near a negatively biased probe [282, 284], as shown in Fig. 10.23b. There, the ion velocity takes its highest values near the probe and the electric field force becomes dominant. Hence, the negative dust is effectively repelled by the negative probe bias creating a dust-free region near the probe. On the other hand, at larger distances, the ion velocity is lower than the critical value and the ion-drag force pushes dust towards the probe forming a stable dust ring around the probe.

10.2.5 Interparticle Forces

The repulsion between two highly-charged dust grains at a distance r_{12} can be described by a shielded interaction energy corresponding to a Yukawa or Debye-Hückel potential (2.27) with a shielding length λ_s,

$$W_Y(r_{12}) = \frac{q_d^2}{4\pi\varepsilon_0 r_{12}} \exp\left(-\frac{r_{12}}{\lambda_s}\right) . \tag{10.63}$$

The shielding length λ_s can have different values depending on the environment of the dust grain. For dust monolayers suspended in the sheath, λ_s describes the interaction in the horizontal direction. Because of the ion streaming velocity, which exceeds the Bohm velocity in the sheath, the shielding length is given by (10.56) and is of the order of the electron Debye length. The directed ion flow in the sheath, however, makes shielding highly anisotropic. Therefore, this value of λ_s does not describe the interaction in the vertical direction, which we will discuss separately in Sect. 10.3.2.

The situation is different for dust grains in the bulk plasma. Here, shielding can be considered as isotropic and the shielding length is close to the linearized Debye length, $\lambda_s \approx \lambda_D$. Small deviations from this limit are due to the nonlinearity of the shielding process [285], which violates the assumption of the classical Debye-Hückel model (2.18) that $|e\Phi| \ll k_B T_i$.

The corresponding repulsive force between identical dust particles with Yukawa interaction potential is

$$F_Y(r_{12}) = -\frac{q_d^2}{4\pi\varepsilon_0 r_{12}^2}\left(1+\frac{r_{12}}{\lambda_s}\right) \exp\left(-\frac{r_{12}}{\lambda_s}\right) . \tag{10.64}$$

10.2.5.1 Particle Pairs in a Horizontal Parabolic Potential Well

The interaction force can be measured by studying pairs of dust particles confined in a parabolic potential trap with an electric potential energy

$$W_{\mathrm{pot}}(r) = \frac{1}{2} m_d \omega_0^2 r^2 , \tag{10.65}$$

as shown in Fig. 10.25. The total force equilibrium is characterized by $\mathbf{F}_E = -(\mathbf{F}_g + \mathbf{F}_{12})$, in which the electric field force from the curved equipotential is balanced by the sum of the weight force and the repulsive force between the particles.

Looking only at the horizontal components of the forces, the restoring force exerted by the potential trap becomes the horizontal component of $\mathbf{F}_E$, which results in

$$F_{\mathrm{trap}}(r) = -m_d \omega_0^2 r . \tag{10.66}$$

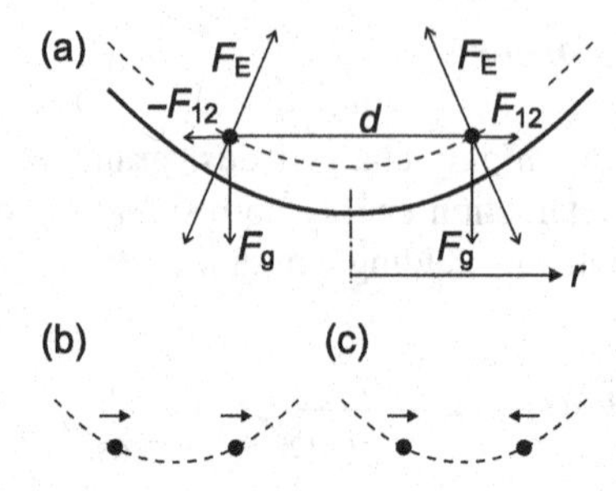

Fig. 10.25 (**a**) Equilibrium positions of two identical dust particles in a parabolic trap. The full line represents the shape of the electrode, the dashed line is the equipotential on which the particles are confined. (**b**) Eigenmodes of the two-particle system: sloshing mode, (**c**) breathing mode

The eigenfrequency of the potential well is ω_0. Then the equilibrium distance d is defined by the force balance

$$\frac{q_d^2}{4\pi\varepsilon_0 d^2}\left(1+\frac{d}{\lambda_s}\right)\exp\left(-\frac{d}{\lambda_s}\right) = m_d\omega_0^2\frac{d}{2}\,. \tag{10.67}$$

This equation contains only two unknowns, the dust charge q_d and the shielding length λ_s! The mass of the particles can be chosen at will by using monodisperse plastic spheres. The eigenfrequency ω_0 can be determined from the (damped) oscillation of a single particle after an initial perturbation. There are now several ways to extract the dust charge and shielding length from experiments with pairs of particles.

1. A first estimate of the dust charge can be obtained by inserting the expected shielding length from (10.56) and solving (10.64) for q_d.
2. In [286] head-on collisions of two identical particles were used to excite coupled oscillations of the two-particle system. The two natural frequencies are the *sloshing mode* at ω_0, which is independent of λ_s, and the symmetric *breathing mode*, which depends on λ_s.
3. Normal modes can also be extracted from a Fourier analysis of the Brownian motion of the trapped particles [287].

From (10.64) the equilibrium distance for an unshielded particle ($\lambda_s \to \infty$) is

$$d_0 = \left(\frac{q_d^2}{2\pi\varepsilon_0 m\omega_0^2}\right)^{1/3}\,. \tag{10.68}$$

It can easily be shown (problem 10.5) that the breathing mode for an unshielded particle pair in a parabolic potential well is given by

$$\omega_{br} = \sqrt{3}\omega_0\,. \tag{10.69}$$

For shielded interaction, the equilibrium distance decreases with increasing shielding factor $\kappa = d_0/\lambda_s$, i.e., shrinking shielding length (see Fig. 10.26). The

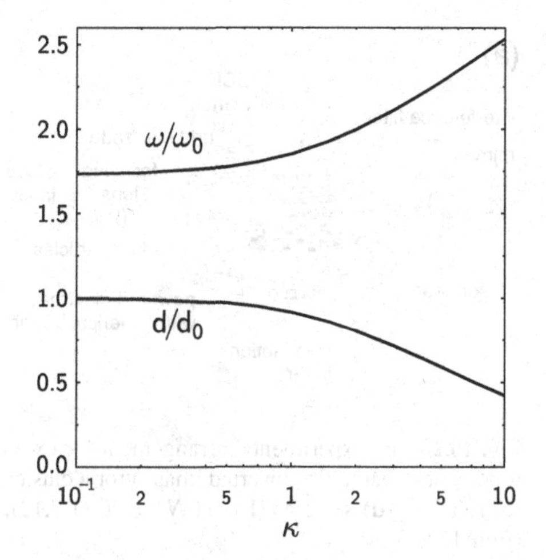

Fig. 10.26 Equilibrium positions d/d_0 of small dust clusters in a parabolic trap potential and breathing frequency ω/ω_0 for each equlilibrium position

frequency of the breathing mode at each of these equilibrium positions, however, increases with κ. This reflects the stiffening of the shielded confinement potential.

10.2.5.2 Small Two-Dimensional Clusters

When a small number of dust particles is levitated in a monolayer with radial confinement by a parabolic potential trap, the particles arrange themselves in regular clusters. The configurations for $N = 1$ to $N = 10$ are shown in Fig. 10.27. The structure of the arrangements is characterized by the formation of distinct shells. The transition from $N = 5$ to $N = 6$ introduces a new inner shell with one particle in the center, whereas the former shell of five particles in the $N = 5$ cluster becomes now the outer shell. The shell occupation numbers are denoted as (1,5). The second particle in the inner shell appears for $N = 9$, which has the configuration (2,7).

The ground states of these 2-D Coulomb clusters have been calculated with the aid of Monte Carlo simulations for pure Coulomb interaction in [288] and for shielded Yukawa interaction in [289]. The shell formation suggests the analogy to a Mendeleev table of atomic shell structure. The influence of shielding is found at

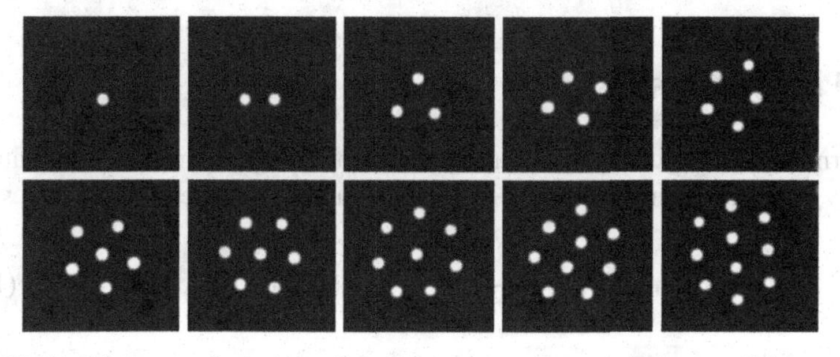

Fig. 10.27 Equilibrium configurations of small dust clusters in a parabolic potential trap

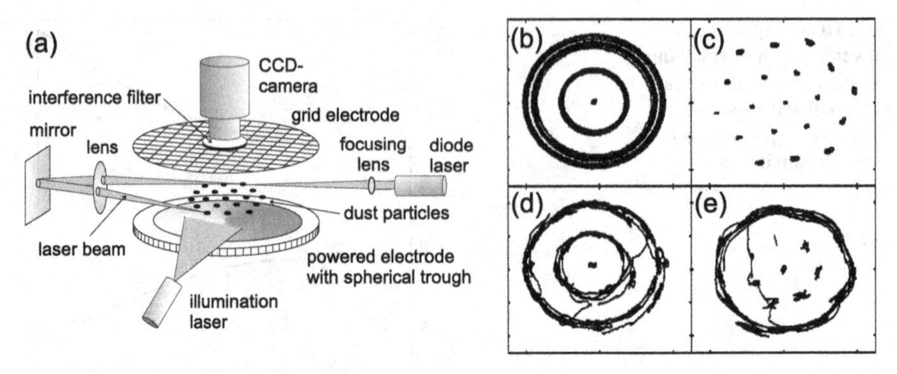

Fig. 10.28 (**a**) Experimental arrangement for excitation of intershell rotation. (**b**) Time-averaged trajectories over 220 s (inverted image) for a cluster with $N = 19$ (1,6,12), (**c**) same, but corrected for rotation. (**d**) same as (**b**) but $N = 20$ (1,7,12), (**e**) corrected for rotation of the middle shell (from [291])

$\kappa > 2$, when the $N = 10$ cluster changes from the (2,8) to the (3,7) configuration or the $N = 19$ cluster, which changes from (1,6,12) to (1,7,11). Apparently, increased shielding makes it energetically favourable to move particles from outer shells to inner shells.

The lowest excited state of small clusters is the intershell rotation mode [290]. An experimental test of the cluster stability was made by Klindworth et al. [291]. The cluster with configuration $N = 19$ (1,6,12) is highly symmetric and has a high energy barrier for intershell rotation. The next larger cluster $N = 20$ (1,7,12) has an incommensurate number of particles in the two outer shells. Therefore, the energy barrier for intershell rotation is low. The intershell rotation is excited by a pair of laser beams, which exert a torque on the particles in the outer shell, see Fig. 10.28a. The resulting cluster rotation is observed from top through an interference filter that discriminates the scattered light of the illumination laser (632 nm) from that of the manipulation laser (690 nm), and from the plasma glow.

Figure 10.28c shows that the $N = 19$ cluster rotates as a solid object. This means that the torque on the outer shell is efficiently communicated to the middle shell. Note that all particles move against friction with the neutral gas. For the $N = 20$ cluster, the outer shell develops a differential rotation w.r.t. the middle shell, which is slowed down by friction, as seen in Fig. 10.28e.

10.3 Plasma Crystals

The formation of regular particle arragements of charged particles is a feature of strongly coupled systems, when the Coulomb coupling parameter of the dust system

$$\Gamma = \frac{q_\mathrm{d}^2}{4\pi\varepsilon_0 a_\mathrm{WS} k_\mathrm{B} T_\mathrm{d}} > 200 \tag{10.70}$$

(see Sect. 2.1.3 for a discussion of Γ). Such conditions are found in laser-cooled ion-crystals (with $q_d = e$) [292–295] or in colloidal suspensions [296–299]. Crystallization of the dust sub-system in a dusty plasma had been predicted by Ikezi in 1986 [300] and experimentally verified, in 1994, by various groups [32–34].

10.3.1 Experimental Observations

Monolayer and mulitilayer plasma crystals are formed in the sheath of capacitively coupled rf-discharges (Fig. 10.29). The plasma crystals are observed with video cameras from top and from the side. Monolayer and bilayer crystals form hexagonal patterns in the plane. Hexagonal order was already found in small 2-D clusters as the energetically favored structure.

A surprising result was the observation in bilayers that the particles in the two layers are vertically aligned rather than being stacked like oranges in bcc or fcc patterns [301]. This observation was the first hint at additional attractive forces between like-charged dust grains.

The formation of vertically aligned particle chains is typical of dust confinement in the sheath [32, 264, 302]. Extremley long chains of dust particles, containing up to 25 particles, were reported from an electrodeless rf discharge [303]. At higher pressures, other authors found typical bulk order with fcc, bcc or hcp structure [35, 304] (see Fig. 10.30). Such 3-D reconstructions of the crystal structure are obtained from a vertical scan. For this purpose the laser generating the horizontal laser sheet and the top-view camera were mounted on a vertical translation stage.

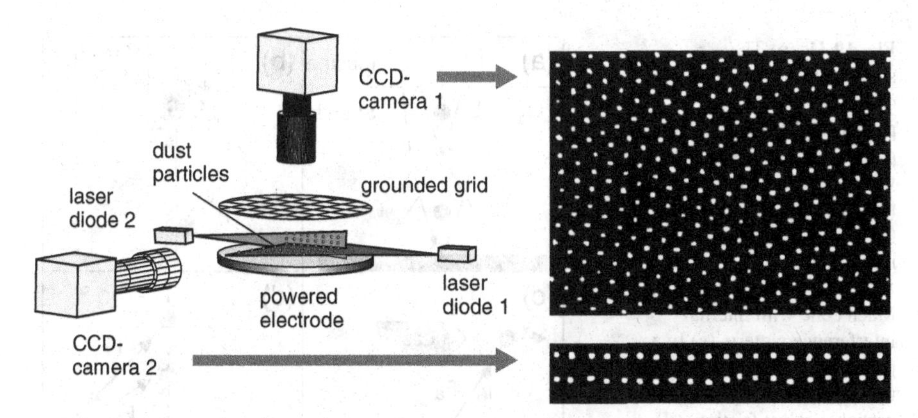

Fig. 10.29 Observation of a two-layer dust cloud from top and from the side. The dust particles are illuminated by a thin layer of laser light. The vertically aligned order is observed for $p < 100\,\mathrm{Pa}$

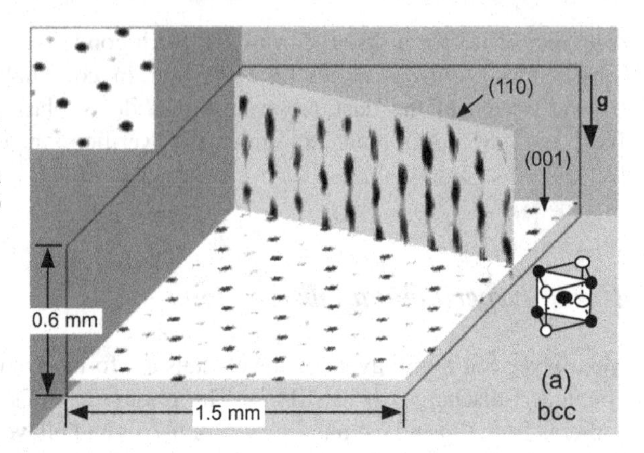

Fig. 10.30 Bulk order (bcc) in multilayer plasma crystal at a pressure of 186 Pa (krypton). The inset at the top left corner shows a top view of the crystal. (Reprinted with permission from [35]. ©1996 by the American Physical Society)

10.3.2 The Role of Ion Wakes

Attraction between like-charged (negative) dust particles was attributed to the scattering of the supersonic ion flow in the sheath region by a dust particle that leads to the formation of positive charges in the wake behind the particle. A cartoon of ion focusing is shown in Fig. 10.31a. This effect can be alternatively described in a particle picture or in a wave picture.

A first theoretical model was based on collective attraction by the exchange of ion-acoustic waves in a stationary plasma [305], the mechanism being similar to Cooper pairing in superconductivity. Alternatively, the focusing of the ion flow by Coulomb scattering on the dust particles was studied by collisionless fluid simulations [306]. Here, the upper particle was considered to act like an electrostatic lens

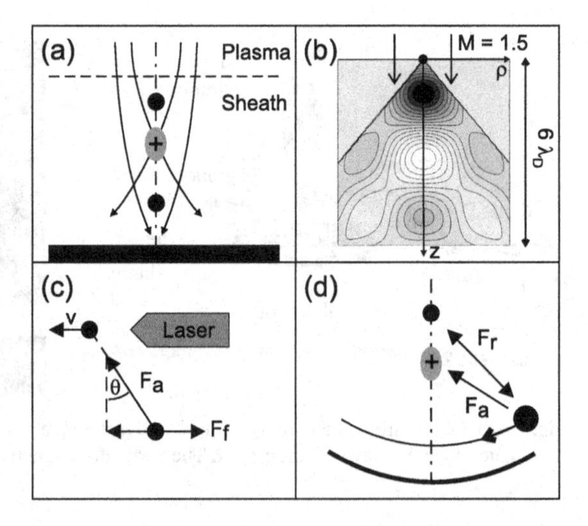

Fig. 10.31 (**a**) The supersonic ion beam is deflected by the upper particle and generates a net positive charge in the wake, which exerts an attractive force on the lower particle. (**b**) An obstacle in a flow with mach number $M = v/v_B = 1.5$ creates a Mach cone with internal interference pattern. (**c**) In a pair of dust particles, the upper particle drags the lower particle against friction. (**d**) Exploration of the attractive force by a test particle moving below the ion focus

that forms a positively charged ion focus in the wake of the particle. Molecular-dynamics simulations [301, 307] demonstrated that this ion focus is fairly robust against ion-neutral collisions, and that the vertically aligned structure is energetically favoured over densely packed structures.

Furthermore, it was concluded that the attractive interaction must be non-reciprocal. This can be understood in terms of the sound barrier: information about the presence of an obstacle in the flow can only be communicated downstream. In the sheath region of the plasma, where the microparticles are suspended, the ion flow is supersonic, i.e., the speed exceeds the ion acoustic or Bohm velocity, v_B. A (negatively charged) particle upstream can exert an attractive force on a like-charged particle in the downstream direction by forming a positive net charge in the ion focus. The particle located downstream, however, cannot modify the supersonic flow on the upstream side. Hence, the ion flow mediates no attractive force on the upper particle. Rather, the upper particle is only pushed upwards by the repulsive force from the lower particle.

The potential structure within the particle wake was studied by several authors using linear response theory, e.g., [305, 308, 309] or by simulations [310–312]. It was conjectured [308] that the regular pattern of potential minima in the wake—see Fig. 10.31b—could be responsible for the transverse structure of plasma crystals. A proper kinetic treatment of the ion-acoustic waves, including Landau damping and ion-neutral collisions, was given in [309].

Those authors found that, in the presence of collisions, the oscillatory wave pattern is preserved on the axis of the wake ($\rho = 0$ in Fig. 10.31b) while the transverse interference maxima and minima disappear. The position of the first minimum in the wake is located at $z = 1.2\lambda_{De}$ (dark region) and is nearly unchanged from its position in the collisionless model [285]. In this way, the wave picture confirms the conclusion from calculations of the ion trajectory that the ion focus is fairly robust against collisions.

There was general agreement that the accumulated positive charge is the reason for the vertical alignment of microparticles. However, it was pointed out [313] that a second force contributes to alignment, which results from the deflected ion flow pattern in the wake. The convergent ion flow pattern in the wake pushes the lower particle from any off-axis position back towards the symmetry axis of the system.

Experimental evidence for wakefield attraction was found by means of laser manipulation [302]. When, in a vertical chain of particles, an upper particle is pushed sideways by the radiation pressure of a focused laser beam, the entire chain follows the motion of the upper particle. In a similar manner, particle pairs of slightly different mass were studied [314], which could move freely within their respective levitation plane and formed a "dust molecule", see Fig. 10.31c. By comparing the response to the laser force on the upper and lower particle, it could be proved that the attractive force is indeed non-reciprocal. Moreover, a quantitative analysis of the particle motion yielded the net attractive force between the particles [315]. In addition, the formation of an ion focus could be suppressed by increasing the ion-neutral collision rate. The destruction of wakefield attraction by ion-neutral collisions explains the existence of densely packed crystal configurations [35, 304] at higher gas pressures.

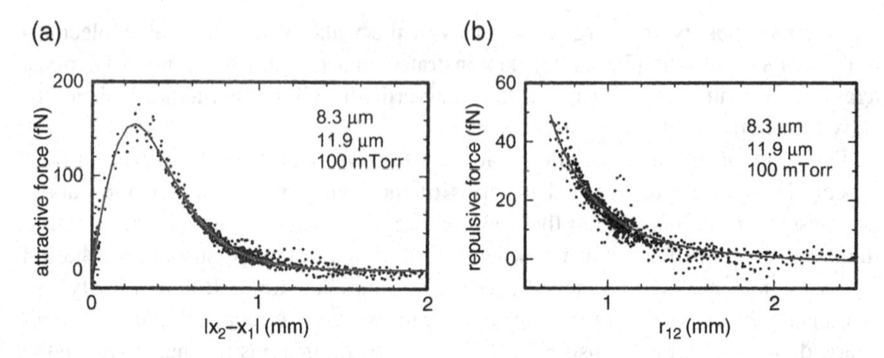

Fig. 10.32 (**a**) Attractive force on a test particle by the wake charge. (**b**). Repulsive force by upper particle. (Reprinted with permission from [316]. ©2003 by the American Physical Society)

A second quantitative experiment to measure the attractive and repulsive forces involved the collision dynamics between suspended particles [316, 317]. In these experiments, a curved electrode provided a radial restoring force, see Fig. 10.31d. While the upper particle was initially at rest, a second particle of higher mass was released from an off-axis position and moved in its own curved levitation plane under the combined repulsion from the upper particle and attraction from the wake charge.

Typical results for the attractive and repulsive force are shown in Fig. 10.32. Here, $|x_2 - x_1|$ is the horizontal distance of the particles, and r_{12} the magnitude of the interparticle distance. The maximum attractive force was found about two times higher than the repulsive force, which is sufficient for the formation of a dust molecule.

10.3.3 Coulomb and Yukawa Balls

Spherical plasma crystals, named *Yukawa balls* were discovered in 2004 [36] when a cloud of dust particles was levitated by the thermophoretic force from a vertical temperature gradient in the gas, and was confined by nearby glass walls. By balancing a large part of the weight force by the thermophoretic force, the dust cloud is embedded in the quasi-neutral bulk plasma, where effects from streaming ions, like wake charging, are minimal. Therefore, the dust confinement is nearly isotropic.

Figure 10.33a,b shows the experimental arrangement and the camera system for scanning video microscopy. For different temperatures of the heated lower electrode, prolate, spherical or oblate dust clouds can be generated.

Different from multilayer crystals suspended in the sheath region of radio-frequency discharges [35, 304], which show bulk order with fcc, bcc or hcp structure, Yukawa balls possess a nested shell structure with mostly hexagonal order on the shells, see Fig. 10.34a,b. The shell structure becomes obvious by plotting all particle positions in a ρ–z plane, thereby ignoring the angular position φ. A detailed analysis of the force field [242] showed that Yukawa balls are confined in a spherical harmonic trap.

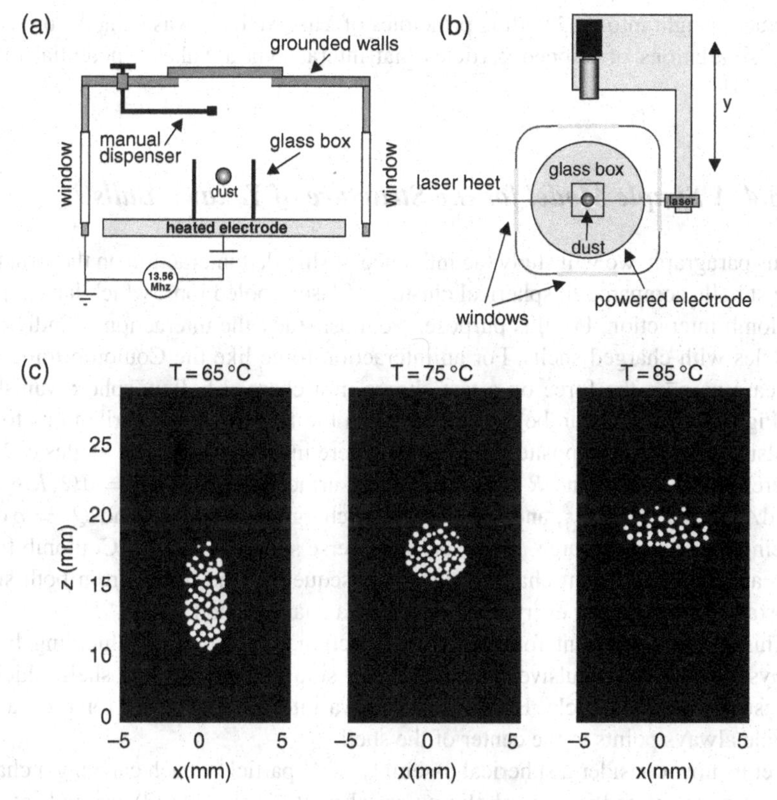

Fig. 10.33 (**a**) Experimental arrangement for generating Yukawa balls by thermophoretic levitation. (**b**) Camera and laser for scanning video microscopy. (**c**) Yukawa balls for different temperature gradients (from [36, 242])

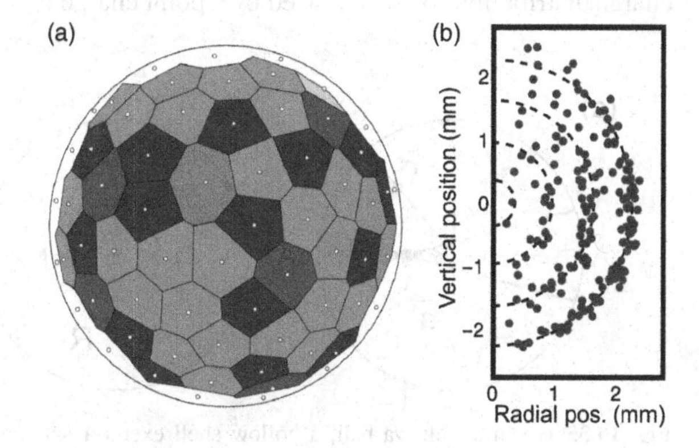

Fig. 10.34 (**a**) Voronoi cells on the surface of a Yukawa ball. (**b**) Shell structure (from [36])

Much insight into the building principles of Yukawa balls was gained from computer simulations of trapped particles that interact via a Yukawa potential [318–322, 234].

10.3.4 A Simple Model for the Structure of Yukawa Balls

In this paragraph, we will study the influence of shielded interaction on the structure of dust balls compared to spherical clusters of laser-cooled ions, which have a pure Coulomb interaction. For this purpose, we must study the interaction of individual particles with charged shells. For an interaction force like the Coulomb force that decrease as r^{-2}, the force on a test charge in a charged hollow sphere vanishes, see Fig. 10.35a. This can be understood from the fact that the contributions to the repulsive force from opposite sides of the sphere involve equal solid angles $\mathrm{d}\Omega$ but different radii, $R+a$ and $R-a$, that define surface elements, $\mathrm{d}A = \mathrm{d}\Omega(R+a)^2$ and $\mathrm{d}A' = \mathrm{d}\Omega(R-a)^2$, and corresponding charges, $Q = \sigma \mathrm{d}A$ and $Q' = \sigma \mathrm{d}A'$, σ being the charge per area. However, the inverse square law of the Coulomb force just cancels the different charge values. Consequently, the forces from both sides are exactly the same for every position inside a charged shell.

This is quite different for shielded interaction, for which the shielding factor always favours the repulsive force from that side of the spherical shell which is nearest to the test particle. Hence, for Yukawa interaction, the net force on a test particle always points to the center of the shell.

Let us now consider a spherical assembly of N particles, each carrying a charge q_d, that are confined in a parabolic potential well $V_\mathrm{t}(r) = (1/2)\alpha r^2$ and interact pairwise, either by a repulsive Coulomb force $F_\mathrm{C}(r_{ij}) = Q^2/(4\pi\varepsilon_0 r_{ij}^2)$ or by a shielded Yukawa force (10.64). For pure Coulomb interaction, the force on a test particle at radius r is only determined by those particles, which have positions $r_1 \leq r$, as shown by the shaded area in Fig. 10.35b. Again, for an inverse square law, this charge distribution can be replaced by a point charge in the center of the sphere. The

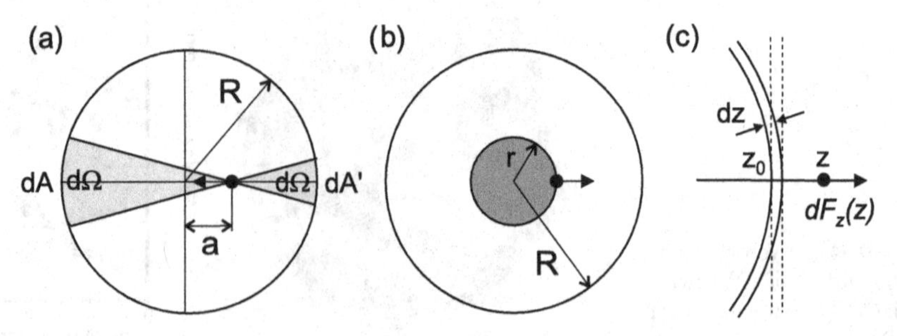

Fig. 10.35 (**a**) In a Yukawa ball, a hollow shell exerts a net force on a particle that pushes it towards the center. (**b**) In a Coulomb ball, a particle experiences only a net force from shells with $r_1 < r$ while outer (*hollow*) shells give no net force. (**c**) The interaction of a particle with a shell of particles is approximated by the interaction with a charged plane (from [323])

net force on the test particle is therefore the repulsion from this equivalent charge and the restoring force $F_t(r) = -\alpha r$ by the trap. When we assume that there are $N(r)$ particles inside the radius r, the force balance requires

$$\frac{[N(r)-1]q_d^2}{4\pi\varepsilon_0 r^2} = \alpha r \, . \tag{10.71}$$

Hence $[N(r) - 1] \propto r^3$. On the other hand, when N is large, we can represent the discrete particle distribution by a continuous density distribution $n(r)$ and obtain

$$N(r) \approx 4\pi \int_0^r n(r_1) r_1^2 \, dr_1 \, . \tag{10.72}$$

Hence, $dN(r)/dr = 4\pi n(r) r^2 \propto r^2$ can only be fulfilled for a constant density

$$n(r) = \frac{3\alpha\varepsilon_0}{q_d^2} =: n_C \, . \tag{10.73}$$

Therefore, a Coulomb ball in a parabolic trap is necessarily homogeneous.

For shielded interaction, these principles do not hold any longer. Consider a point charge q_d at a distance $z - z_0$ from a homogeneously charged (infinitely large) plane sheet of thickness dz that contains a charge density $n(z_0)q_d$ [see Fig. 10.35c]. The test charge interacts with each volume element in this sheet by the shielded force (10.64), and the resulting repulsive force becomes

$$dF_z(z) = n(z_0)\frac{q_d^2}{2\varepsilon_0} \exp\left(-\frac{z - z_0}{\lambda_s}\right) dz \, . \tag{10.74}$$

This force now depends on the distance from the plane whereas the force would be constant for Coulomb interaction. This simple model can be used to approximate the interaction between a point charge and a spherical shell as long as $r \gg \lambda$.

A more quantitative description can be obtained for large Yukawa balls that have $R/\lambda \gg 1$. The force equilibrium for a test particle of charge q_d inside a Yukawa ball is defined by the balance of a net force from a gradient in the density $n(r)$ with the confining force F_t from the trap. For simplicity of calculation, we assume that the test particle is located between an inner and outer half space with a plane interface and a stratified set of density layers parallel to the interface, which have a density distribution $n(r_1) = n(r) + (r_1 - r)n'(r)$. Then, the force from the inner (outer) half space becomes

$$\begin{aligned} F_< &= \frac{q_d^2}{2\varepsilon_0}[\lambda_s n(r) - \lambda_s^2 n'(r)] \\ F_> &= -\frac{q_d^2}{2\varepsilon_0}[\lambda_s n(r) + \lambda_s^2 n'(r)] \, , \end{aligned} \tag{10.75}$$

which defines the force balance $F_< + F_> = -(q_\mathrm{d}^2/\varepsilon_0)\lambda_\mathrm{s}^2 n'(r) = \alpha r$. Hence, for a parabolic confinement, the curvature of the density profile must be a constant,

$$n'' = -\frac{\alpha\varepsilon_0}{q_\mathrm{d}^2\lambda_\mathrm{s}^2} = -\frac{n_\mathrm{C}}{3\lambda_\mathrm{s}^2}\,. \tag{10.76}$$

The same result is obtained for a spherical geometry and for arbitrary R/λ_s [319]. The density profile therefore has the shape of an inverted parabola,

$$n(r) = n(0) - \frac{1}{2}\frac{n_\mathrm{C}}{3\lambda_\mathrm{s}^2}r^2\,. \tag{10.77}$$

The central density $n(0)$, however, still has to be determined.

The force balance at the surface of the Yukawa ball is determined by the force balance of the particles in the inner half space with the trap, $F_< + F_\mathrm{t} = 0$, which yields

$$\lambda_\mathrm{s} n(R) - \lambda_\mathrm{s}^2 n'(R) = \frac{2}{3}n_\mathrm{C} R \tag{10.78}$$

and, using $n'(R) = -(1/3)n_\mathrm{C}(R/\lambda_\mathrm{s}^2)$, we obtain

$$n(R) = \frac{1}{3}n_\mathrm{C}\frac{R}{\lambda_\mathrm{s}}\,. \tag{10.79}$$

Therefore, besides the radial decay of the density, a Yukawa ball also has a finite value of the particle density $n(R)$ at the surface. Finally, the density in the center of the Yukawa ball is obtained as

$$n(0) = \frac{n_\mathrm{C}}{3}\left[\frac{R}{\lambda_\mathrm{s}} + \frac{1}{2}\left(\frac{R}{\lambda_\mathrm{s}}\right)^2\right], \tag{10.80}$$

which gives the asymptotic form of the model in [319]. Note that the density at the surface scales $\propto R/\lambda_\mathrm{s}$, but the central density increases more rapidly $\propto (R/\lambda_\mathrm{s})^2$. Hence, the larger a Yukawa ball becomes, by adding more and more particles, the sharper peaked is the density profile in the center.

The total number of particles in a large Yukawa ball is given by

$$N \approx \frac{4\pi}{6}\int_0^R n_\mathrm{C}\frac{R^2 - r^2}{\lambda_\mathrm{s}^2}r^2\mathrm{d}r = \underbrace{\left(\frac{4\pi}{3}R^3 n_\mathrm{C}\right)}_{N_\mathrm{C}}\frac{1}{15}\left(\frac{R}{\lambda_\mathrm{s}}\right)^2\,. \tag{10.81}$$

The corresponding number N_C of particles in a Coulomb ball is the marked expression in the parentheses. This leads to the useful relation between Coulomb and Yukawa balls with the same number of particles,

$$\frac{R}{\lambda_s} \approx 15^{1/5} \left(\frac{R_C}{\lambda_s} \right)^{3/5} . \tag{10.82}$$

The size of a Yukawa ball grows more slowly than a Coulomb ball for the same number of particles.

The steepening of the profile shape can be seen in Fig. 10.36a. There, the profile function from [319] is used with the asymptotic form $R/\lambda_s \approx 15^{1/5}(R_C/\lambda_s)^{3/5}$ and rescaled to an abscissa r/R_C. The shielding factor is here given as $R_C/\lambda_s = (N/2)^{1/3} d_0/\lambda_s$, d_0 being the equilibrium distance in the parabolic trap of two particles interacting by a Coulomb force given in (10.68). For $d_0/\lambda_s = 1$, the curves represent $N = 2000$, 16000, and 27000 particles.

This continuum model, however, cannot predict the arrangement of the particles on individual shells, which is a result of strong coupling that emphasizes the nearest-neighbor interaction [325, 318].

The increase of the central density in a Yukawa ball by adding more and more particles to the system was studied experimentally and by computer simulation [324]. The comparison is shown in Fig. 10.36b. Here, the number of particles in the outermost shell becomes smaller than the prediction for a Coulomb ball (dashed line) whereas, in the innermost shell, the population is larger than that of a Coulomb ball (solid line). The population densities agree fairly well with simulations for $d_0/\lambda_s = 0.6$.

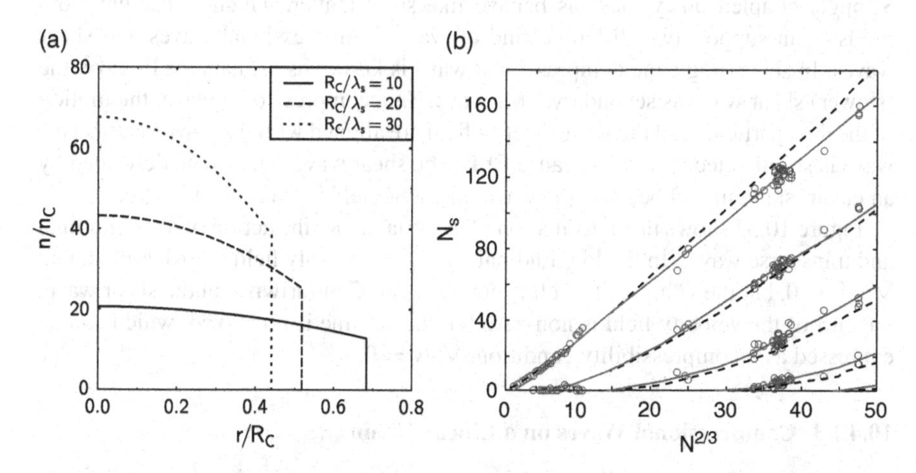

Fig. 10.36 **(a)** Density profiles of Yukawa balls for different values of R_C/λ_s. **(b)** Measured shell populations N_s as a function of the total number N in comparison with the prediction for Coulomb balls (*dashed line*) and Yukawa balls for $d_0/\lambda_s = 0.6$ (*solid line*) (from [322, 324])

10.4 Waves in Dusty Plasmas

Wave processes in which the motion of the dust particles can be observed with video cameras, are interesting phenomena that reveal the wave dynamics with a resolution at the level of kinetic description. Such wave experiments have turned out to be highly reproducible. In the following, we will discuss three characteristic scenarios, which explore typical features of dusty plasmas:

- Laser excited waves in linear chains and monolayers,
- The spectral energy density of a wave from Brownian motion,
- Self-excited density waves.

A weakly-coupled dusty plasma has two kinds of natural waves, the dust-acoustic mode and the dust ion-acoustic mode. In the dust-acoustic wave (DAW), the dust particles move in the wave field while electrons and ions are screening the dust particles. In laboratory plasmas, this wave has typical frequencies between 10 and 100 Hz. In the dust ion-acoustic wave (DIAW), which has typical frequencies between 50 and 500 kHz, the dust is practically immobile and the negative charge bound to the dust particles modifies the ordinary ion-acoustic mode by having $n_\mathrm{i} \neq n_\mathrm{e}$. This special case was already discussed in Sect. 6.5.3. The reduction in free electron density leads to an increase of the phase velocity. This effect could be attributed to the reduced shielding by electrons.

In the following a detailed discussion of lattice waves in strongly coupled systems is given. These wave types are less familiar to plasma physicists.

10.4.1 Compressional and Shear Waves in Monolayers

Strongly-coupled dusty plasmas behave like solid matter, which—different from fluids—can support two different kind of waves, compressional waves and shear waves. In seismology, the compressional wave is known as primary, or P-wave, the (slower) shear wave as secondary, or S-wave. In a compressional wave, the motion of the dust particles and the wave electric field are aligned with the wave vector. This wave is strictly electrostatic, because $\mathbf{E}||\mathbf{k}$. The shear wave can also be described by an electrostatic model because the wave magnetic field is small for $v_\varphi \ll c$.

Figure 10.37 shows the deformation of a crystal under the action of a longitudinal and transverse wave. In the longitudinal wave, the velocity field is irrotational, i.e., $\nabla \times \mathbf{v} = 0$, but the volume of a cell is not constant. Contrariwise, in the shear wave, the curl of the velocity field is non-zero but the volume is preserved, which can be expressed as incompressibility condition, $\nabla \cdot \mathbf{v} = 0$.

10.4.1.1 Compressional Waves on a Linear Chain

Let us start with analyzing the dispersion properties of compressional waves in terms of the interparticle forces. The simplest model for understanding the principles of lattice waves is the linear chain of particles shown in Fig. 10.38. The equilibrium

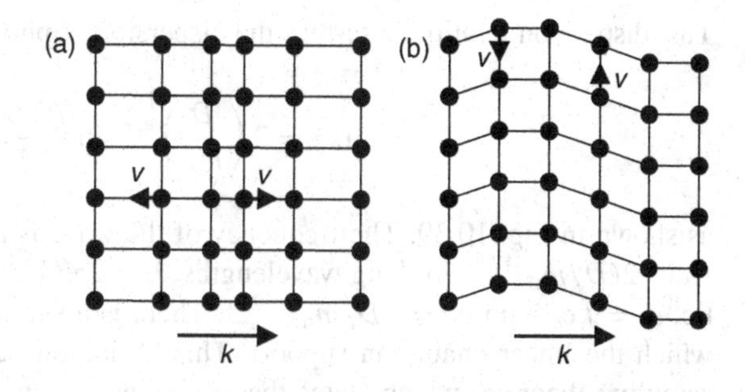

Fig. 10.37 (**a**) Compressional or P-wave. (**b**) Shear or S-wave. The arrows labeled v give the local particle oscillation velocity

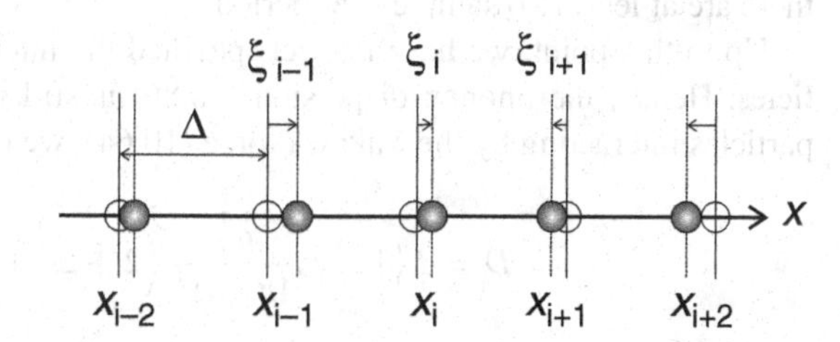

Fig. 10.38 One-dimensional model for a compressional wave on a linear chain of dust particles. The open circles give the equilibrium positions

positions of the particles are $x_i = i\,\Delta + x_0$, where Δ is the interparticle distance, and the displacements from the equilibrium position are designated by ξ_i. The particles interact by a repulsive Yukawa force $F_Y(r_{ij})$ (10.64). We further simplify the analysis by taking only the interaction of a particle i with its neighbors $i-1$ and $i+1$ into account.

This system of mutually repulsive particles can be identified as an arrangement of masses and springs. By expanding the Yukawa force about the equilibrium position

$$F_Y(\Delta + \xi) \approx F_Y(\Delta) - D\xi \tag{10.83}$$

we find Hooke's law, namely that the restoring force is proportional to the elongation ξ of the spring. D can be identified as a spring constant. The minus sign is typical for a preloaded spring, where the force becomes weaker when the spring is expanding. Adding the force contributions from the left and right neighbor, the equation of motion for particle i reads

$$m_d\ddot{\xi} = D(\xi_{i+1} - 2\xi_i + \xi_{i-1})\,. \tag{10.84}$$

Further, we have neglected damping of the particle motion by dust-neutral collisions. We search for longitudinal waves of the type $\xi(x) = \hat{\xi}\exp[\mathrm{i}(kx - \omega t)]$. Inserting this expression into (10.84), we obtain the characteristic equation

$$\begin{aligned} -\omega^2 m_d &= D\left(\mathrm{e}^{\mathrm{i}k\Delta} - 2 + \mathrm{e}^{-\mathrm{i}k\Delta}\right) \\ &= 2D[\cos(k\Delta) - 1] = -4D\sin^2\left(\frac{k\Delta}{2}\right)\,. \end{aligned} \tag{10.85}$$

This dispersion relation describes the dispersion of phonons on the linear chain

$$\omega(k) = 2\left(\frac{D}{m_{\mathrm{d}}}\right)^{1/2} \sin\left(\frac{k\Delta}{2}\right) \tag{10.86}$$

as shown in Fig. 10.39. The frequency of the wave is normalized by its maximum value $2(D/m_{\mathrm{d}})^{1/2}$. For long wavelengths, $\lambda = 2\pi/k \gg \Delta$, the dispersion is acoustic, $\omega = kc_{\mathrm{s}}$ with $c_s = (D/m_{\mathrm{d}})^{1/2}\Delta$. There is a shortest wavelength $\lambda_{\mathrm{min}} = 2\Delta$, which the linear chain can support. This limitation occurs because of Shannon's sampling theorem, which states that a sine wave can only be reconstructed when there are at least two samples per period.

Up to this point, we have not yet specified the interaction law between the particles. Hence, the phonon dispersion (10.86) is still universal. For our system of particles interacting by the Yukawa force (10.64), we obtain

$$D = F'_{\mathrm{Y}}|_{\Delta} = \frac{q_{\mathrm{d}}^2}{4\pi\varepsilon_0\Delta^3}\left(2 + 2\kappa + \kappa^2\right)\mathrm{e}^{-\kappa} \tag{10.87}$$

with $\kappa = \Delta/\lambda_{\mathrm{s}}$. In other words, the spring constant is given by the second derivative of the interaction potential at the equilibrium distance, $D = -W''_{\mathrm{Y}}|_{\Delta}$. This gives the dispersion relation as

$$\omega = \left(\frac{2}{\pi}\right)^{1/2} \omega_{\mathrm{d0}} \left(1 + \kappa + \frac{1}{2}\kappa^2\right)^{1/2} \mathrm{e}^{-\kappa/2} \sin\left(\frac{k\Delta}{2}\right), \tag{10.88}$$

where we have introduced the *dust plasma frequency for strongly coupled systems*

$$\omega_{\mathrm{d0}} = \frac{q_{\mathrm{d}}}{(\varepsilon_0\Delta^3 m_{\mathrm{d}})^{1/2}} \tag{10.89}$$

because of its similarity to the electron plasma frequency (2.32) when we identify Δ^{-3} as the particle density.

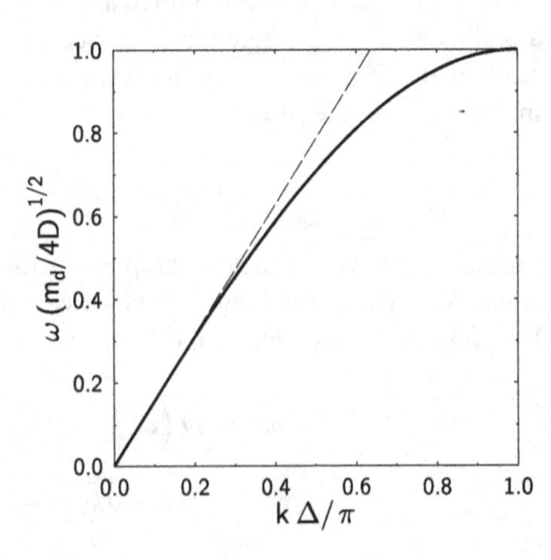

Fig. 10.39 Phonon dispersion on the linear chain. The maximum allowed value of the wavenumber is $k\Delta = \pi$. For small wavenumber (long wavelength), the dispersion is acoustic (*dashed line*)

For comparison with experiments, we must extend this simple model by two aspects, frictional damping of the dust particle motion by dust-neutral collisions and force contributions from particles at positions $i \pm 2$, $i \pm 3 \ldots$ on the linear chain (see problem 10.6). When the wave is excited by a periodic force acting on the particles, frictional damping will make the wavenumber complex, $k = k_{\mathrm{R}} + \mathrm{i}k_{\mathrm{I}}$. The real part describes the wave propagation and the imaginary part the damping.

Waves on a linear chain were studied in [326]. The radiation pressure of a chopped diode laser was applied to exert a periodic force on the first particle in a linear chain. The particles were confined in an oblong potential well formed by barriers on the powered electrode of a rf-discharge, see Fig. 10.40a.

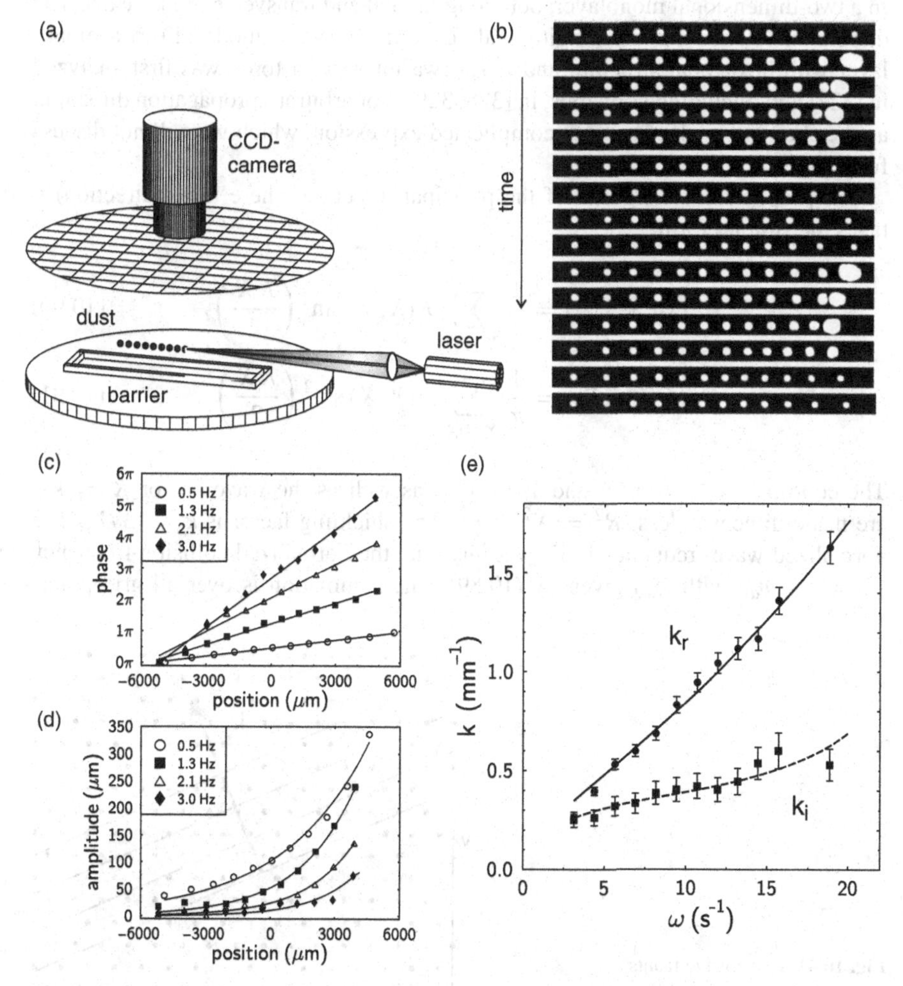

Fig. 10.40 Laser-excited phonon on linear chain. (**a**) Experimental set-up. (**b**) Response of particles on chopped laser radiation pressure (inverted image). (**c**) Phase, (**d**) amplitude response of the linear chain. (**e**) Best fit to dispersion and damping by varying κ (from [326])

The *on* phase of the laser is visible by the enhanced light scattering from the right-most particle. The response of the particles in the linear chain, Fig. 10.40b, is sinusoidal with a phase shift between the individual particles and a decreasing amplitude, see Fig. 10.40c,d. The real and imaginary part of the wave number k are well described by the model (10.87). The dust charge $q_{\mathrm{d}} = -(14000 \pm 4000)e$ was determined independently by the resonance method (Sect. 10.2.1.2) and the gas friction from Epstein's formula (10.41) for diffuse reflection. Then, the shielding factor κ is the only unknown in the wave dispersion. From the best fit shown in Fig. 10.40e, $\kappa = (1.6 \pm 0.6)$ was obtained.

10.4.1.2 Plane Compressional and Shear Waves in Monolayers

In a two-dimensional monolayer, both longitudinal and transverse modes exist. The dispersion relation for the longitudinal (L) and transverse mode (T) in a monolayer with hexagonal structure and a Yukawa interaction force was first analyzed in [327] and, including collisions, in [328, 329]. For arbitrary propagation direction, as sketched in Fig. 10.41, it is a complicated expression, which we will not discuss further.

For propagation along one of the principal directions (here the x direction) it takes the simpler form

$$\Omega_{\mathrm{L}}(\Omega_{\mathrm{L}} + \mathrm{i}\Xi_{\mathrm{d}}) = \frac{1}{\pi} \sum_{X>0,Y} F(X,Y) \sin^2\left(\frac{KX}{2}\right) \tag{10.90}$$

$$\Omega_{\mathrm{T}}(\Omega_{\mathrm{T}} + \mathrm{i}\Xi_{\mathrm{d}}) = \frac{1}{\pi} \sum_{X>0,Y} F(Y,X) \sin^2\left(\frac{KX}{2}\right) . \tag{10.91}$$

The coordinates $X = x/\Delta$ and $Y = y/\Delta$ as well as the wave vector $K = k\Delta$ are made dimensionless. $R^2 = X^2 + Y^2$. The shielding factor is $\kappa = \Delta/\lambda_{\mathrm{s}}$. The normalized wave frequency is $\Omega = \omega/\omega_{d0}$ and the normalized collision frequency $\Xi_{\mathrm{d}} = \nu_{\mathrm{d}}/\omega_{d0}$ with ω_{d0} given by (10.89). The summation is over all grid points

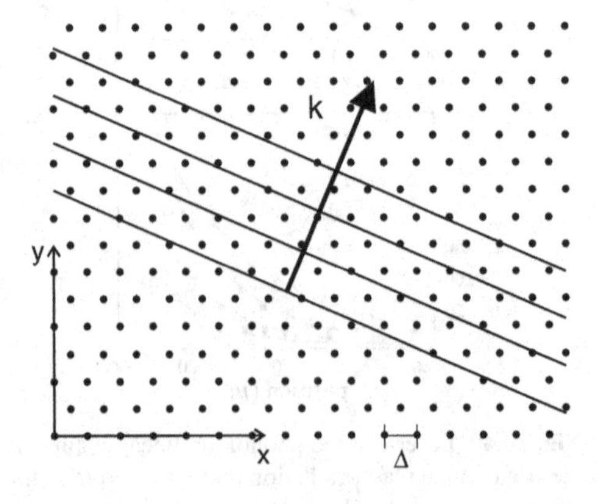

Fig. 10.41 Plane wave fronts in a monolayer of particles with hexagonal structure

(X, Y) of the lattice with $X > 0$. Note that $\sin^2(KX/2)$ terms indicate that particle pairs at $\pm X$ have already been introduced, which yields the restriction $X > 0$. The function $F(X, Y)$ is the (normalized) spring constant for the interaction with a neighboring particle at relative position (X, Y),

$$F(X, Y) = R^{-5}\mathrm{e}^{-\kappa R}\left[X^2\left(3 + 3\kappa R + \kappa^2 R^2\right) - R^2(1 + \kappa R)\right] . \qquad (10.92)$$

The resulting dispersion relations for the compressional and shear wave are shown in Fig. 10.42 for the directions 0° and 90°. The compressional wave has a similar shape as the result for the linear chain in Fig. 10.39. For small k-values, the dispersion is acoustic with a slope $\omega/k = C_{\mathrm{L}}$, the sound speed of the longitudinal wave. For greater k, the compressional wave becomes dispersive and the group velocity $\mathrm{d}\omega/\mathrm{d}k$ vanishes near $k\Delta = \pi$. For large k, the dispersion also depends on the propagation direction in the crystal whereas, for small k, the dispersion is independent of propagation angle. This is due to reaching the continuum limit where the exact position of the particles becomes unimportant and only the area density of dust particles determines the sound speed.

The shear wave shows acoustic behavior over a wider range of k-values. The dispersion in the 0° direction becomes superlinear, i.e., bending upwards. The transverse wave has a smaller sound speed C_T than the longitudinal wave C_L. This is a general observation in solid matter and is the reason for the terminology in seismology of *primary* for the longitudinal wave, and *secondary* for the later arriving transverse wave.

From (10.90) to (10.91) one can see that the normalized frequencies $\Omega_{\mathrm{L,T}}$ are independent of the dust charge q_{d}, which only appears in the frequency ω_{d0} that is used for normalization. Peeters et al. [327] had noticed that the sound speeds $C_{\mathrm{L,T}} = \lim_{k\to 0}\omega_{\mathrm{L,T}}/k$ depend in a different manner on the shielding factor κ. In particular, the ratio of the sound velocities depends only on κ because ω_{d0} cancels in

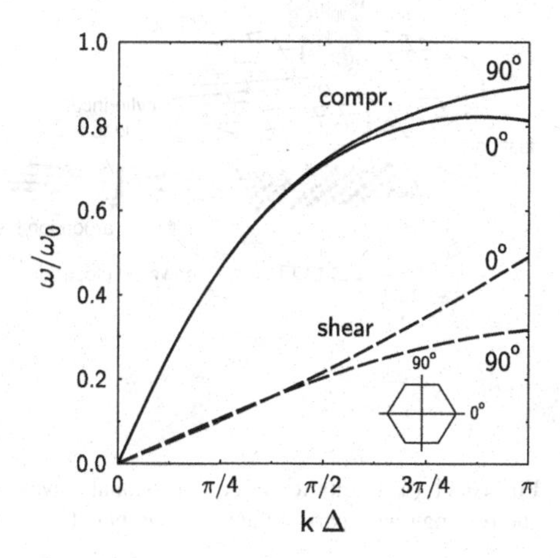

Fig. 10.42 Dispersion of compressional (*solid lines*) and shear wave (*dashed lines*) in a monolayer with hexagonal order for wave propagation in 0° and 90° direction. Dust-neutral collisions are neglected

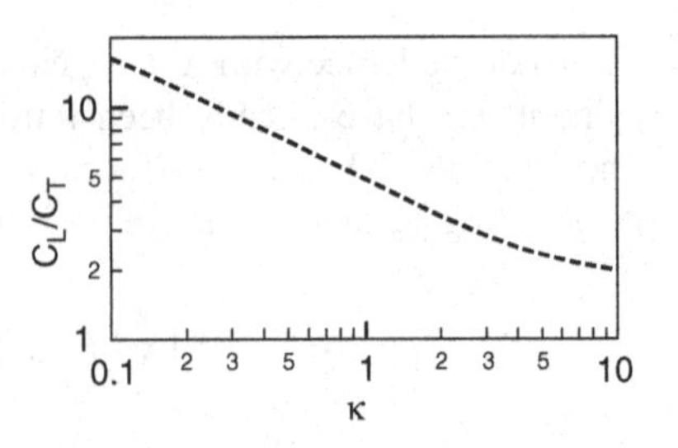

Fig. 10.43 Ratio of sound velocities C_L/C_T vs. shielding factor κ

the ratio. Therefore, the ratio $C_{\mathrm{L}}/C_{\mathrm{T}}$ shown in Fig. 10.43 can be used as a diagnostic method for determining κ [330], as will be discussed below in paragraph 10.4.1.5.

10.4.1.3 Experiments with Plane Compressional Waves

Plane compressional waves were studied in [328] to determine the shielding factor κ. The experiments, shown in Fig. 10.44a, used a monolayer of dust particles trapped in the sheath of a rf discharge at 13.65 MHz and 27 Pa argon pressure. Lateral confinement is provided by a rectangular barrier. The interparticle distance was 740 μm. The first row of particles is periodically pushed by the radiation pressure from an argon-ion laser that is expanded into a line and is modulated by a mechanical chopper.

The wave dispersion and damping are shown as experimental points in Fig. 10.44b,c in comparison with theoretical curves for various values of κ. The dust charge, $q_{\mathrm{d}} = 14500\,\mathrm{e}$, was determined independently by the resonance method

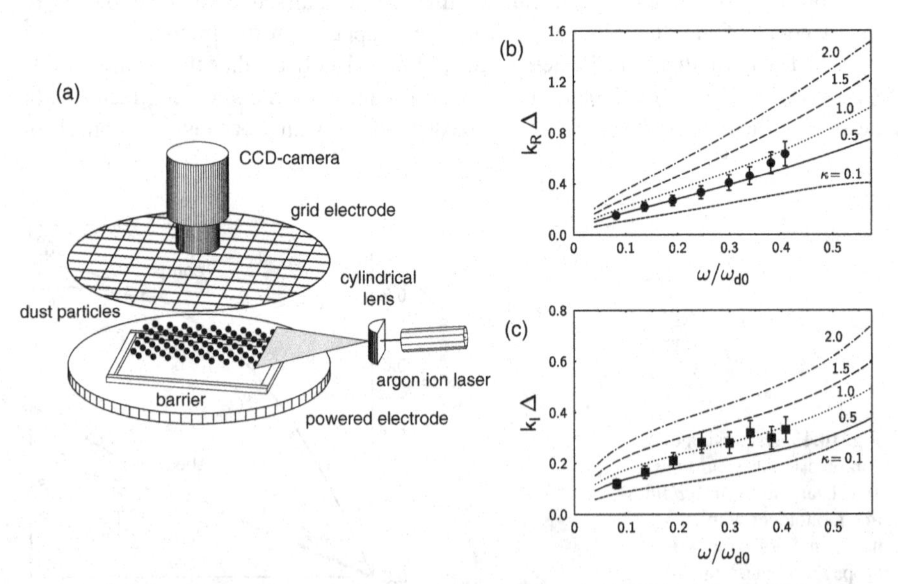

Fig. 10.44 (**a**) Excitation of compressional waves by applying a periodic laser force. (**b**) Real part and (**c**) imaginary part of the wave number. The best fit yields $\kappa = 1 \pm 0.5$ (from [328])

(Sect. 10.2.1.2) and defines the frequency ω_{d0}. For these experimental conditions, the best fit to the experimental data yields $\kappa = 1 \pm 0.5$.

10.4.1.4 Excitation of Plane Shear Waves

Laser-excited shear waves were first demonstrated in the arrangement of Fig. 10.45a [331]. A laser spot hits the monolayer under a small angle and exerts a force in x-direction. Rapidly (200 Hz) moving the laser spot back and forth in x-direction, by means of a scanning mirror, all particles in a line are displaced. A mechanical chopper is used to generate short pulses of applied shear force. In Fig. 10.45b a so-called *velocity map* is shown, in which the arrows represent the instantaneous velocity field. The particle velocities were obtained from the change of the particle positions in subsequent video frames. After the end of the pulsed shear force, the perturbation splits in two waves, which propagate in $\pm y$ direction. Averaging all velocities over the x-direction, see Fig. 10.45c, it becomes evident that a shear wave is formed which propagates at constant speed C_T.

10.4.1.5 Diagnostic Application of Compressional and Shear Waves

The method of determining the dust charge and the shielding factor from the measured sound speeds C_L and C_T was introduced in [330]. The value of κ is determined

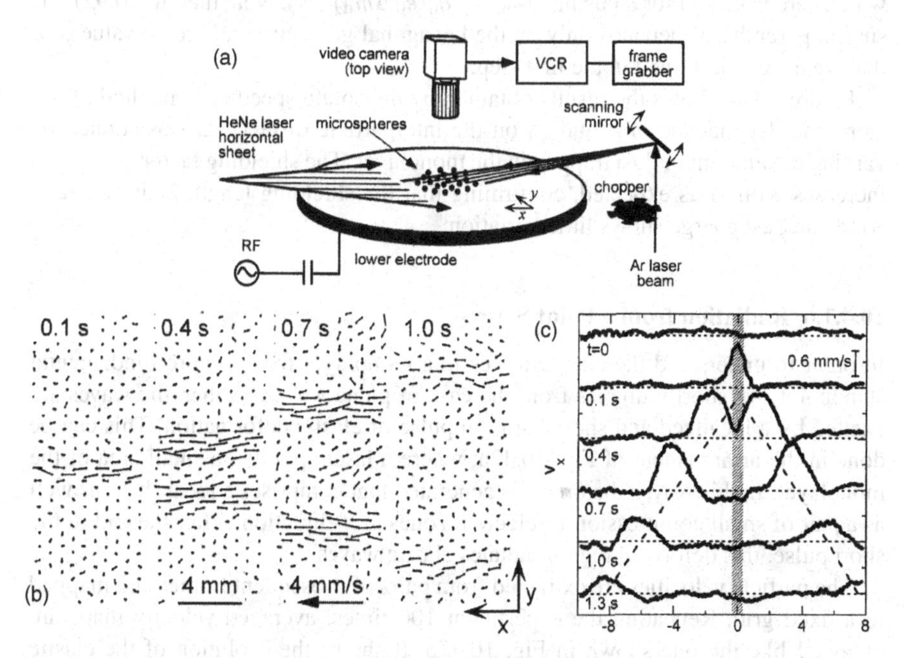

Fig. 10.45 (**a**) Excitation of shear waves by applying a shear pulse with a laser. (**b**) Propagation of the shear pulse perpendicular to the applied shear. (**c**) The shear pulse propagates at a constant sound velocity. (Reprinted with permission from [331]. ©2000 by the American Physical Society)

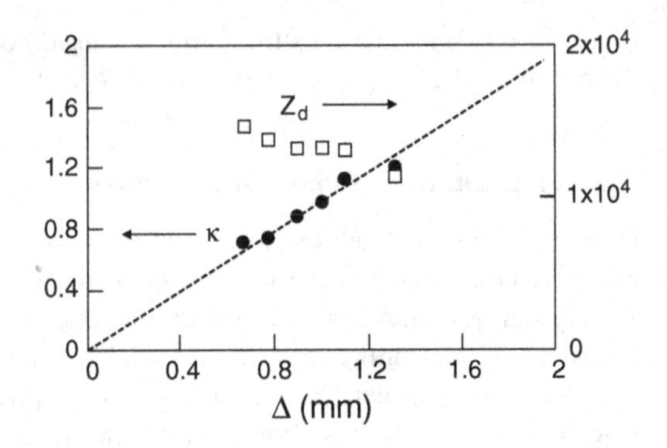

Fig. 10.46 Determination of dust charge number Z_d *(open squares)* and shielding factor κ *(circles)* for different interparticle distances Δ from the ratio of sound velocities C_L/C_T (from [330])

from $C_{\mathrm{L}}/C_{\mathrm{T}}$ (Fig. 10.43). For determining q_{d}, we need the measured sound speed C_{L} and the interparticle distance Δ. In the limit $K \ll 1$ we easily obtain from (10.90)

$$C_{\mathrm{L}} = \lim_{k\to 0}\frac{\omega}{k} = \frac{\Omega}{K}\omega_{d0}\Delta = \left(\frac{1}{\pi}\sum_{X,Y} F(X,Y)\frac{X^2}{4}\right)^{1/2}\omega_{d0}\Delta\,, \tag{10.93}$$

which can be solved for q_{d} using $\omega_{d0}\Delta = q_{\mathrm{d}}(\varepsilon_0\Delta m_{\mathrm{d}})^{-1/2}$. Note that in (10.92), the sum in parentheses depends only on the hexagonal geometry and on the value of κ that we have determined in the first step.

Figure 10.46 shows the results obtained by the sound speed ratio method [330]. Here, the dependence of κ and q_{d} on the interparticle distance Δ was studied by varying the amount of dust trapped in the monolayer. The shielding factor $\kappa = \Delta/\lambda_{\mathrm{s}}$ increases with Δ as expected, confirming that the shielding length λ_{s} is constant. Also, the dust charge shows little variation.

10.4.1.6 Radiation from a Point Source

In order to understand the elastic waves in a monolayer from a more fundamental standpoint, we shortly digress from the case of plane waves. Rather, the waves are excited by a localized and short-duration pulse of elastic deformation. This can be done in the arrangement of Fig. 10.47a, where a laser spot hits a small area of the monolayer. In this way, a distortion is generated that comprises a local shear as well as a pair of small compression-rarefaction zones in x-direction. After the end of the short pulse, this deformed region radiates elastic waves.

The particle velocities are extracted from pairs of subsequent frames and mapped to a fixed grid. Repeating the experiment 100 times, averaged velocity maps are obtained like the one shown in Fig. 10.47b. It shows the evolution of the elastic deformation and should not be confused with any hydrodynamic flow of the dust particles. On average, the particles remain at their lattice sites. The velocity map

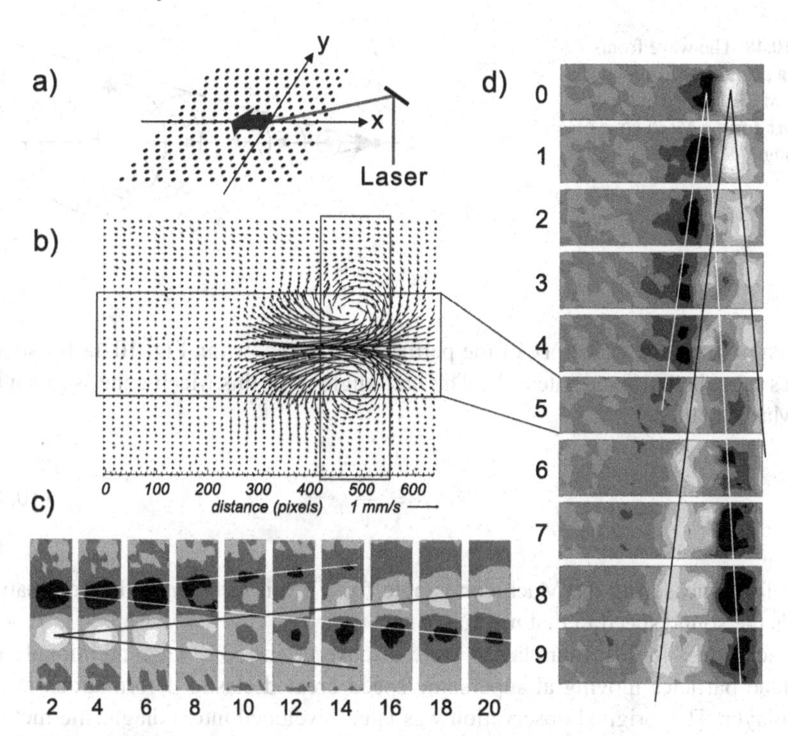

Fig. 10.47 Radiation from a short localized elastic dipole deformation. (**a**) Experimental arrangement. (**b**) Velocity field $\mathbf{v}(x, y)$ at $t = 5$. (**c**) Splitting and propagation of the shear component $\nabla \times \mathbf{v}$. (**d**) Propagation of the density pertubation $\nabla \cdot \mathbf{v}$ (from [332])

shows a large double vortex, which expands in x-direction at the speed of a compressional wave and in y-direction at the speed of the shear wave.

From such velocity maps, we can extract the shear motion by taking the curl of the velocity field, $\nabla \times \mathbf{v}$, shown in Fig. 10.47c. The graph is composed of narrow stripes around the double vortex, which appears as a pair of black and white dot. The numbers below the stripes are the frame number in the video, which was recorded at 30 fps. One observes that each of the vortices splits into a pair (compare with the splitting of the initial shear pulse in Fig. 10.45c), which subsequently propagate in $\pm y$ direction at the sound speed C_T.

The compressional wave is visualized by the divergence of the velocity field, noting that from the continuity equation (5.8) we have $\nabla \cdot \mathbf{v} = -(1/n)\mathrm{d}n/\mathrm{d}t$. The compression (rarefaction) zone appears dark (bright). Again, these perturbations split into pairs that propagate in $\pm x$ direction at the higher sound speed of the compressional wave.

10.4.1.7 Mach Cones

A perturbation in an elastic medium moving at supersonic velocity, $v > c_s$, generates a Mach cone. The Mach cone is the envelope of the elementary waves excited

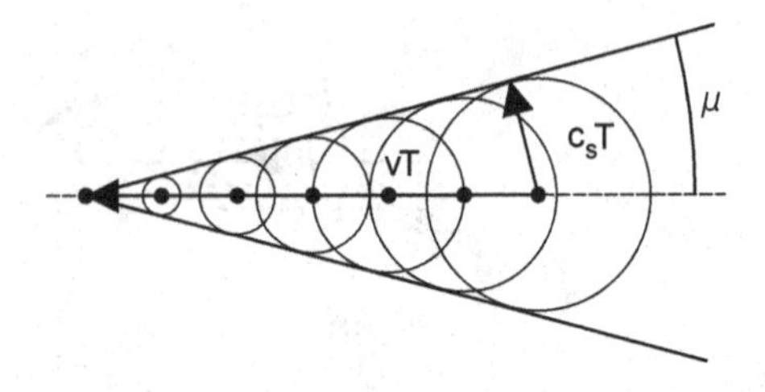

Fig. 10.48 The wave fronts from a sound source moving at speed v launched at times $t_n = n\tau$ form a Mach cone of half angle $\mu = \arcsin(c_s/v)$

at each point of the trajectory of the perturbation, as shown in Fig. 10.48 for sound pulses launched at fixed intervals. The half angle μ of this Mach cone is given by the Mach relation

$$\sin\mu = \frac{c_s}{v} \, . \tag{10.94}$$

Therefore, measuring the Mach cone angle for a given velocity of the pertubation yields the sound speed in that medium.

Mach cones in dust monolayers were discovered in [333] when additional out-of-plane particles moving at supersonic speed created elastic deformations in the monolayer. This original observation was later developed into a diagnostic method by using well-defined perturbations exerted by a moving laser spot [334]. Here we discuss shortly experiments with Mach cones that simultaneously excite compressional and shear waves [335]. The experimental arrangement is the same as in Fig. 10.47a except for using a moving laser footprint instead of applying a short pulse.

The velocity map in Fig. 10.49a shows that there are two distinct Mach cones of different half angle. In the wider cone, the particle vibration is mostly across the cone, which gives a hint at the compressive waves that establish this cone. In the narrower cone, the particle velocities are aligned with the cone indicating the shear waves. This topology is summarized in the cartoon in the right half of the Figure.

Again, the compressional and shear waves can be separated by calculating the density fluctuations $\partial n/\partial t$ and the curl of the velocity field, as shown in Fig. 10.49b,c. The internal structure of the compressional cone, which forms a set of nested *lateral wakes*, is a feature that results from interference effects between different wavelets on the dispersion branch which have different wavelength and propagation velocity [336]. The wake behind a ship has a similar structure. The Mach relation for compressional and shear Mach cones is shown in Fig. 10.49d for various speeds of the laser spot. The resulting sound speeds are $C_L = 23.0\,\mathrm{mm\,s^{-1}}$ and $C_T = 5.7\,\mathrm{mm\,s^{-1}}$.

Mach cones, which may be excited by boulders on Keplerian orbits, were suggested as a possible diagnostic method for the dusty plasma in the ring system of Saturn [337, 338].

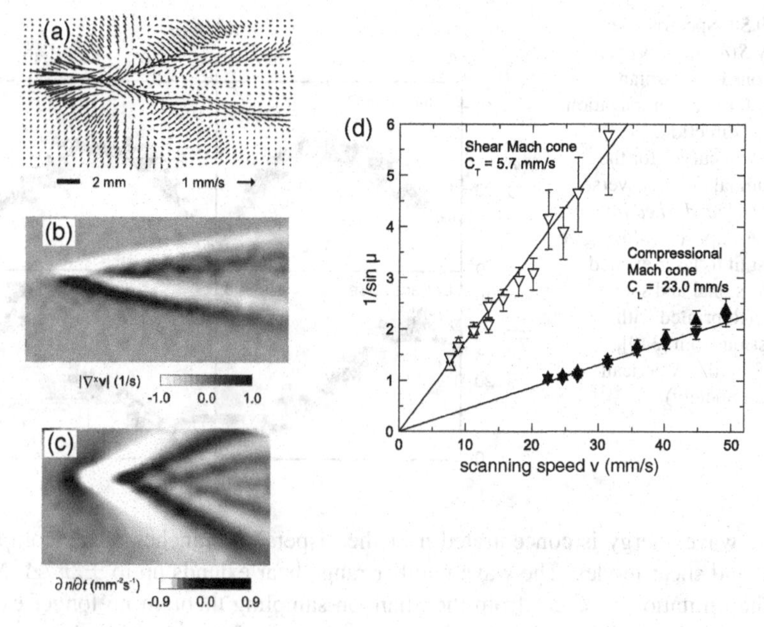

Fig. 10.49 (**a**) Velocity map of the compressional and shear Mach cones in a monolayer excited by a moving laser spot. (**b**) Shear Mach cone, $\nabla \times \mathbf{v}$. (**c**) Compressional Mach cone, $\partial n/\partial t$, with wake structure. (**d**) Mach relation (from [335])

10.4.2 Spectral Energy Density of Waves

The Brownian motion of the dust particles about their equilibrium position in the crystal lattice can be considered as a superposition of thermally excited sound waves. The reconstruction of particle positions with sub-pixel resolution [339] enables the experimenter to investigate these small-amplitude vibrations. The wave spectrum is reconstructed from the Fourier transform (in space and time) of the particle velocities $\mathbf{v}(\mathbf{r}, t)$,

$$\mathbf{v}(\mathbf{k}, \omega) = \frac{2}{TL} \int_0^T \int_0^L \mathbf{v}(\mathbf{r}, t) \mathrm{e}^{-\mathrm{i}(\mathbf{k}\cdot\mathbf{r} - \omega t)} \mathbf{dr}\, \mathbf{dt}\,. \tag{10.95}$$

The spectral power density

$$S(\mathbf{k}, \omega) = \frac{1}{2} m_{\mathrm{d}} |\mathbf{v}(\mathbf{k}, \omega)|^2 \tag{10.96}$$

characterizes the energy distribution over the various modes. Such wave energy spectra [340] are shown in Fig. 10.50.

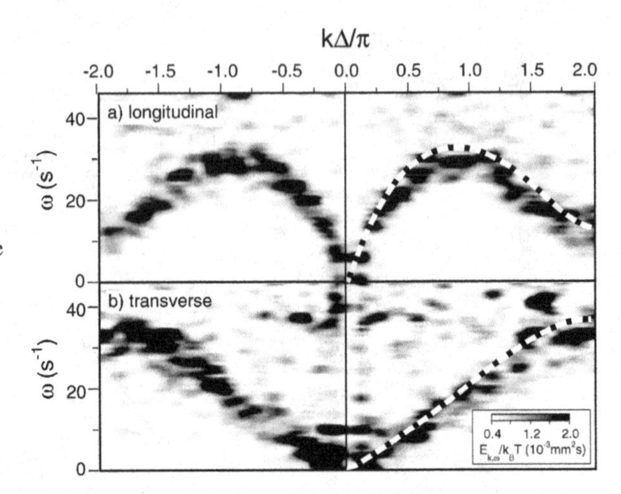

Fig. 10.50 Spectral energy density $S(\omega, k)$ of particle vibrations in Brownian motion for wave propagation in the 0° direction. Dispersion curves for the longitudinal and transverse modes (*white dashed lines*) are superimposed, which give the best fit to laser-excited compressional and shear waves. (Reprinted with permission from [340], ©2002 by the American Physical Society)

The wave energy is concentrated near the dispersion branches of the compressional and shear modes. The wave number range hear extends up to $\pm 2\pi/\Delta$. Note that the limitation $\lambda < 2\Delta$ from the Shannon sampling theorem no longer exists, because there are additional particles between the wave fronts that sample the wave, as shown in Fig. 10.41. The white, dashed curves are theoretical dispersion curves, which give the best fit to small amplitude waves launched by the laser method. The natural phonon spectra are in close agreement with theory.

10.4.3 Dust Density Waves

Historically, the dust acoustic wave (DAW) [341, 342] was the first observation of a collective phenomenon in a dusty plasma that involved the dynamics of dust particles. The DAW is a close relative of the ion-acoustic waves discussed in Sect. 6.5.3. There, we had learned that the wave dynamics is determined by the ion inertia and the shielding by electrons. This led to the ion sound speed $C_s = \omega_{\rm pi}\lambda_{\rm De}$. The DAW involves dust inertia and shielding by electrons and ions. Therefore, the frequencies of DAWs are much lower and are typically found in the range (10–100) Hz.

There is an overwhelming literature on DAWs and the interested reader should refer to reviews [226, 227, 232, 236] or specialized textbooks [238, 239]. At our introductory level, we will here only discuss the linear dispersion properties of the DAW.

We can construct the dielectric function for the dust plasma system in analogy to the ion-acoustic wave (6.75), where the susceptibilities of ions and electrons were given by

$$\chi_{\rm i} = -\frac{\omega_{\rm pi}^2}{\omega^2 - k^2\gamma_{\rm i}k_{\rm B}T_{\rm i}/m_{\rm i}}\,, \quad \chi_{\rm e} = +\frac{1}{k^2\lambda_{\rm De}^2}\,. \tag{10.97}$$

Now, for the DAW, the role of the ions is taken over by the dust, and both electrons and ions contribute to shielding of the dust grains. We describe the dust by a cold-fluid model ($T_d = 0$) with a dust plasma frequency

$$\omega_{pd} = \left(\frac{n_d q_d^2}{\varepsilon_0 m_d}\right)^{1/2} . \tag{10.98}$$

Adopting such a fluid description further implies that the dust system must be in a weakly coupled state with a coupling factor $\Gamma \ll 1$. This becomes evident from the fact that the only information about the dust particle positions is given by their number density n_{d0}. There are no conditions imposed for the mutual arrangement of dust particles, e.g., correlations between particles, which are typical of a fluid phase, or any lattice structure in the solid phase. In this spirit, we can immediately write down the susceptibilities as

$$\chi_d = -\frac{\omega_{pd}^2}{\omega^2} \quad ; \quad \chi_i = \frac{1}{k^2\lambda_{Di}^2} \quad ; \quad \chi_e = \frac{1}{k^2\lambda_{De}^2} . \tag{10.99}$$

Then the dispersion relation is given by the zeroes of the dielectric function

$$0 = \varepsilon(k,\omega) = 1 + \chi_d + \chi_i + \chi_e , \tag{10.100}$$

and the dispersion relation takes the explicit form

$$\omega^2 = k^2 \frac{\omega_{pd}^2 \lambda_D^2}{1 + k^2\lambda_D^2} , \tag{10.101}$$

which is displayed in Fig. 10.51.

In the long-wavelength limit, $k\lambda_D \ll 1$ the DAW has an acoustic dispersion, $\omega = k\, C_{DAW}$, with the dust-acoustic wave speed

$$C_{DAW} = \omega_{pd}\lambda_D = \left[\frac{Z_d^2 n_{d0}/n_{i0}}{1 + \left(\frac{n_{e0}T_i}{n_{i0}T_e}\right)} \left(\frac{k_B T_i}{m_d}\right)\right]^{1/2} \approx Z_d \left(\frac{n_{d0}}{n_{i0}} \frac{k_B T_i}{m_d}\right)^{1/2} . \tag{10.102}$$

The simplification in the last step is valid for $T_e \gg T_i$ as found in many dc or rf discharges. At short wavelength, $k\lambda_D > 1$, the wave frequency approaches the dust plasma frequency.

The first observation of DAWs was made in a Q-machine, where a secondary "firerod" plasma was produced by means of a positively biased (+200 V) small disk electrode, in which a dust cloud is trapped [342]. DAWs of typically 12 Hz frequency and 9 cm/s propagation speed were spontaneously excited by the ion flow

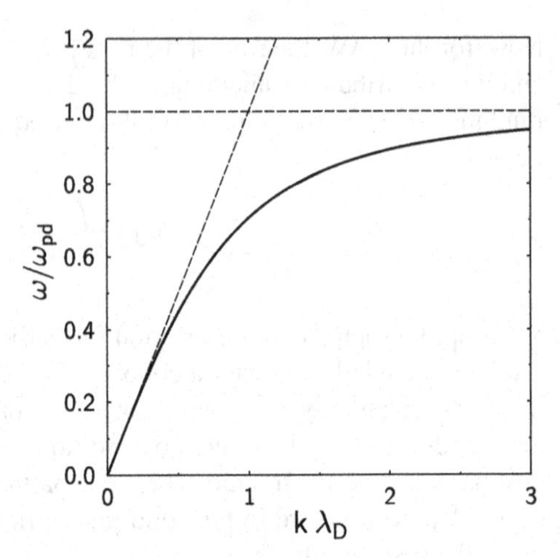

Fig. 10.51 Dispersion of the DAW for $T_\mathrm{d} = 0$. The dashed lines indicate the asymptotic behavior

in the secondary plasma. The dispersion relation of the DAW was investigated in a similar arrangement with a small disk anode in a glow discharge [343].

The situation of a dust cloud trapped in an anodic plasma is sketched in Fig. 10.52a. The dust particles are confined by a combination of ion drag force and electric field force [344]. A snapshot of the DAW is shown in Fig. 10.52b. The frequency of the spontaneously excited DAWs was varied by modulating the anode bias voltage in a frequency range of (6–30) Hz, to which the waves become entrained. The resulting dispersion relation [Fig. 10.52c] shows the acoustic dispersion $\omega/k = C_\mathrm{DAW}$ expected for the long-wavelength limit.

Spontaneously excited DAWs were found in many plasma situations, e.g., in the positive column of a gas discharge [345], in rf-parallel plate discharges [346, 347],

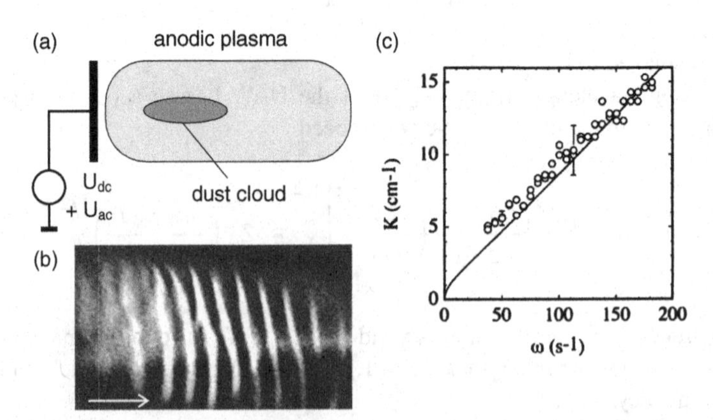

Fig. 10.52 (**a**) Sketch of the experimental arrangement for DAWs with a dust cloud trapped in an anodic plasma. (**b**) Scattered light from the dust cloud. The bright regions have the highest dust density. The arrow indicates the wave propagation direction. (**c**) Wave dispersion observed. (Reprinted with permission from [343]. ©1997, American Institute of Physics)

in fireballs [348] or in magnetized anodic plasmas [344, 349, 350]. When the dust density is estimated from a snapshot that resolves the indivual positions of dust grains, and the ion Debye length is known from the plasma parameters, the dust acoustic speed C_{DAW} can be used as a diagnostic method for the determining the dust charge q_d.

10.4.4 Concluding Remarks

The distinguishing features, by which dusty plasmas introduce new physics into the realm of plasma phenomena, can be summarized as follows:

- Charge variability is a feature that is unknown in classical plasmas. It leads to intrinsic delays, which can trigger instabilities, or imposes the constraint of charge sharing in dense dust clouds.
- Plasma crystallization opens the field of "model solids" for studying crystals with unusual interaction forces of long range (Coulomb) or short range (Yukawa) with respect to their thermodynamic properties (phase transitions) or dynamic properties (phonons).
- Orbital motion of ions in the electric field of highly charged dust particles plays an important role in dusty plasmas and determines charge collection currents or ion drag forces. As a principal feature of dust-ion interaction, it is of similar importance as the gyroorbits in classical magnetized plasmas.

The Basics in a Nutshell

- Dust charging is determined by charge collection (electrons and ions) and electron emission (photoeffect, secondary emission).
- Orbital motion of ions on open trajectories about negative dust particles determines the charge collection and the ion drag force.
- The floating potential Φ_f of a dust particle depends on the electron temperature and the ion mass. The dust charge can be described by the capacitance model as $q_d = 4\pi\varepsilon_0 a \Phi_f$. It depends on the particle radius a and the floating potential.
- Different from the fixed electron and ion charge, the dust charge is a dynamic variable. This leads to charge fluctuations by the discreteness of the absorbed or emitted charges. In dynamic processes, the dust charging process is delayed by a time constant $\tau = RC$.
- Dust particles are subject to the electric force $q_d E$, the gravitational force $m_d g$, drag forces from neutral or ion winds, thermophoretic forces and radiation pressure.
- In laboratory plasmas, dust sediments to the sheath region and forms flat pancake-like clouds. Three-dimensional dust arrangements require microgravity or levitation by thermophoretic forces.

- The coupling factor $\Gamma = q_d^2(4\pi\varepsilon_0 a_{WS} k_B T_d)^{-1}$ determines the thermodynamic phase of a dust cloud.
- Plasma crystals are formed for $\Gamma > 200$. Depending on the geometry of the electrostatic trap that confines a plasma crystal, there are two-dimensional monolayers with hexagonal order, multilayer crystals with fcc, bcc and hcp order, or spherical dust clouds with shell structure.
- Crystalline states of dust clouds support two types of phonons: compressional and shear. In two dimensions, many details of elastic waves could be studied by applying laser forces: dispersion of plane waves, radiation from point sources, or Mach cones. These techniques can be used as diagnostic methods for determining q_d and κ.

Problems

10.1 Consider the lunar-dust fountain effect, in which a dust particle originally sticks to the dust layer covering the surface. Assume that the surface has a potential of $\Phi = +4\,\mathrm{V}$ and the dust grain of 10 nm radius has initially the same potential. The surface potential is shielded within $\lambda_D \approx 1\,\mathrm{m}$. What is the maximum height this dust particle can reach after its bond with the surface is broken? ($\rho_d \approx 3000\,\mathrm{kg\,m^{-3}}$, $g_L = 1.6\,\mathrm{m\,s^{-2}}$)

10.2 Discuss the charge reduction for the case of an isothermal argon plasma with Boltzmann ions and $T_e/T_i = 100$. Use (10.26) in the limit $P \to \infty$, where $\eta_f = \eta_c$. Show that in this limit $\eta_f = 0.078$. What is the normalized ion density n_i/n_∞ in this limit? What does the result mean for the behavior of ions?

10.3 A dust particle of $a = 5\,\mu\mathrm{m}$ radius is embedded in a typical laboratory plasma in argon gas with $T_e = 3\,\mathrm{eV}$, $T_i = 0.03\,\mathrm{eV}$ and $n = 10^{15}\,\mathrm{m^{-3}}$. What is the collection radius b_c and Coulomb radius r_C of this particle for an ion at Bohm velocity? Compare the result with the electron Debye length. What does this mean for the role of collection force and orbit force?

10.4 Perform the missing steps leading to (10.54).

10.5 Two particles in a parabolic potential well, which interact by a pure Coulomb force, find their equilibrium positions at $r = \pm d/2$. Allow for small-amplitude vibrations about this equilibrium. Show that the frequency of this *breathing mode* is given by (10.69).

10.6 Start from the equation of motion for particles in a linear chain and consider interactions of particle i with all other particles. Show that the dispersion relation for longitudinal modes is given by

$$\omega^2 = \frac{2}{\pi}\,\omega_{d0}^2 \sum_{n=1}^{\infty} \frac{1}{n^3}\left(1 + n\kappa + \frac{n^2}{2}\kappa^2\right) e^{-n\kappa} \sin^2\left(\frac{nk\Delta}{2}\right) .$$

10.7 A linear chain of particles (mass m) has an equilibrium spacing Δ and the repulsive force between neighboring particles at the equilibrium position is F_Δ. The individual masses at location x_i are displaced in transverse direction by $\eta_i \ll \Delta$. Discuss, why the repulsive force can be considered as constant in this analysis. Analyse the equation of motion, in analogy to longitudinal modes, in the nearest-neighbor-approximation and show that the wave frequencies for all wavenumbers k are imaginary. Discuss this type of instability and explain why the zig-zag pattern at $k = \pi/\Delta$ has the highest growth rate.

10.8 What are the differences in the interaction forces between a one-dimensional system of charge sheets as discussed in Sect. 9.4.1 and the linear chain model discussed in Sect. 10.4.1.1?

Chapter 11
Plasma Generation

The clip, the clop! All cla. Glass crash. The (klikkaklakkaklaskaklopatzklatschabattacreppycrotty-graddaghsemmihsammihnouithappladdyappladdypkonpkot!).
James Joyce, Finnegans Wake

Lightnings and technical plasmas are generated by an electric breakdown in a gas. The ignition process leads to a subsequent current flow that generates an electrical discharge. Depending on the power source that feeds the plasma, we distinguish direct current (dc), low-frequency alternating current (ac), and radio-frequency (rf) discharges. This chapter gives a brief introduction into the most common types of discharges and the associated plasma processes with emphasis on the *how*-questions rather than giving answers to all *why*-questions.

11.1 DC-Discharges

This Section is focused on low-pressure dc discharges with cold cathodes or thermionic emitters. First a word of caution: The following description of the discharge types is not a lab manual. Rather, operating dc discharges in the lab requires observing the established lab safety standards. These include proper grounding as well as protective insulation to prevent touching any parts that carry voltages in excess of 60 V that may be lethal.

11.1.1 Types of Low Pressure Discharges

The different types of low-pressure dc discharges are compiled in Fig. 11.1 as a function of the discharge current. Assume that a dc voltage is applied to an electrode system consisting of parallel plates of a few square-centimeters area and a few centimeters distance, see Fig. 11.1a. The gap is filled with a gas at low pressure (say, $p = 400\,\mathrm{Pa}$ or 3 torr). To limit the discharge current, a resistor is connected in series, which is initially set at a large value of $1\,\mathrm{M\Omega}$. The voltage U_0 of the power supply is chosen just high enough that electric breakdown occurs. This may happen at $U_0 \approx 600\,\mathrm{V}$ in this example. At discharge currents of up to $1\,\mu\mathrm{A}$ the

A. Piel, *Plasma Physics*, DOI 10.1007/978-3-642-10491-6_11,

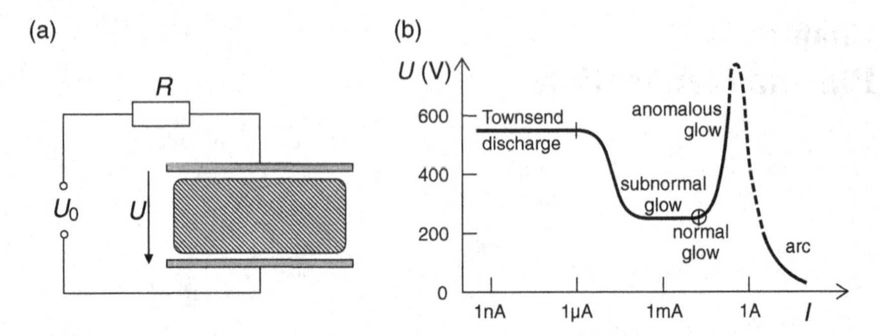

Fig. 11.1 The self-sustaining discharge. (**a**) A gas discharge between parallel plates is operated with a current-limiting resistor R in series and a discharge voltage U develops across the discharge gap. (**b**) By gradually reducing the resistance, four distinct discharge modes are established. Note that the arc discharge has a negative differential resistance

discharge voltage U is independent of the flowing current and remains close to the breakdown voltage. There is no luminosity in the discharge gap. This regime is called the *Townsend dark discharge*, named after John S. Townsend (1868–1957).

When the series resistor is lowered, the discharge current increases and a *subnormal glow* develops that covers only part of the cathode surface. This discharge mode has a lower discharge voltage that is nearly constant over a current regime of $I \approx (0.1\text{–}10)$ mA, as shown in Fig. 11.1b, until the cathode is completely covered with the glow. This endpoint is the normal glow discharge. The discharge now shows various luminous regions which fill the gap. The discharge voltage rises again for even higher discharge currents, forming an *anomalous glow discharge*. Raising the discharge current further, this discharge type becomes unstable (dashed part of the curve) and a different discharge, the *arc discharge* is formed with discharge currents of 1 A and more. In the arc discharge, the discharge channel has contracted to a small part of the surface of the negative electrode. Note that the arc discharge has a characteristic with a negative differential resistance, i.e., the discharge voltage decreases for higher currents.

11.1.2 Regions in a Glow Discharge

The normal glow discharge in a long glass tube is a good example to dissect a gas discharge into its functional blocks. Various regions in the discharge can be distinguished by their brightness or darkness, as shown in Fig. 11.2a. The negative (positive) electrode of a glow discharge is called cathode (anode). When the glow discharge is operated with a plane cathode, a complex pattern of dark spaces and luminous layers becomes visible on the cathode side, as sketched in Fig. 11.2a. Towards the anode, an extended volume of luminosity appears, the positive column. When the length of the discharge tube is increased while the discharge current and gas pressure are kept constant, the cathodic features remain unchanged, but the

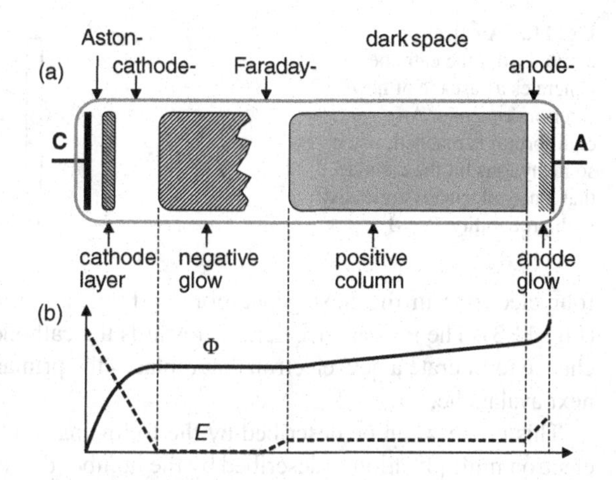

Fig. 11.2 Anatomy of a glow discharge in a long cylindrical tube with plane cathode (C) and anode (A). (**a**) Various dark spaces and luminous regions are indicated. (**b**) Sketch of the electric potential Φ and axial electric field E in the normal glow discharge

positive column becomes longer and fills the additional volume. It is found that when the discharge gap is made shorter, the discharge is maintained even as the positive column and the Faraday dark space have been "consumed", leaving only the negative glow and the dark space adjacent to the electrodes.

This behavior shows that the glow discharge can be split into two functional units: the cathodic part, which is responsible for liberating a sufficient number of electrons from the cold cathode and ends in the negative glow, and the anodic part consisting of positive column and a transition layer to the anode.

The positive column is the active part in fluorescent tubes, in which a small admixture of mercury to a low-pressure argon discharge generates UV light that is converted to "white" visible light in a fluorescent coating of the inner tube wall. The typical red colour of the neon spectrum is found in neon displays.

The potential distribution in a glow discharge and the corresponding axial electric field are sketched in Fig. 11.2b. The largest voltage drop occurs in the cathode region. The negative glow is mostly field-free, $E \approx 0$. In the positive column, the axial electric field is constant. There is a small voltage drop across the anode layer.

11.1.3 Processes in the Cathode Region

The key to understanding electrical breakdown and the steady-state of the cathode region is the interplay of electron multiplication in the gas and release of electrons from the cathode by ion bombardment.

11.1.3.1 Gas Breakdown

When a primary electron is released from the cathode and is accelerated in the electric field, which we assume to be homogeneous before breakdown, it can ionize gas atoms after a typical mean free path for ionization, λ_i. In this ionization process an electron-ion pair is created. Now two electrons are accelerated that generate

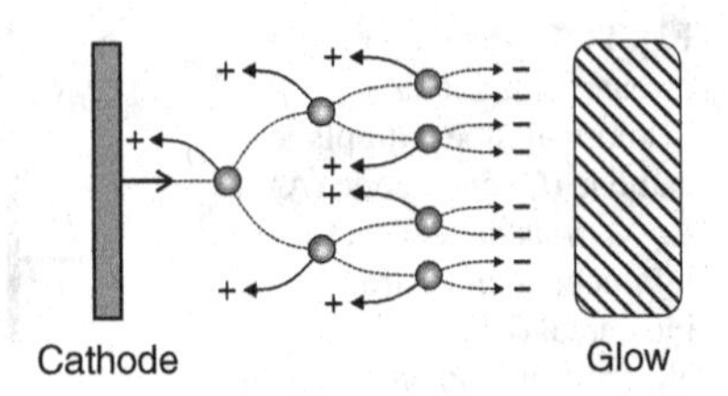

Fig. 11.3 An electron emitted from the cathode generates a cascade of new electron-ion pairs. An equilibrium is reached, when so many ions hit the cathode that a new electron is released with probability $P = 1$

four electrons in the next generation, and so on, creating an electron avalanche (Fig. 11.3). The ions are accelerated towards the cathode, where they have a certain chance to liberate a new electron that replaces the primary electron and launches the next avalanche.

This process can be described by the following model. In a statistical sense, the electron multiplication is described by the number dn_e of newly created electrons in an interval dz

$$dn_e = \alpha\, n_e\, dz\,, \tag{11.1}$$

which leads to exponential growth $n_e(x) = n_e(0)e^{\alpha z}$. Here, $\alpha = \lambda_i^{-1}$ is the gas multiplication factor introduced by Townsend. At the positive electrode, an electron flux

$$\Gamma_e(d) = n_e(d)\mu_e E \tag{11.2}$$

leaves the gap at $z = d$. The fact that equal amounts of electrons and ions were produced in the avalanche process, requires that the ion flux at the cathode is given by $\Gamma_i(0) = -\Gamma_e(d) + \Gamma_e(0)$. Hence,

$$\Gamma_i(0) = -\Gamma_e(0)\left(e^{\alpha d} - 1\right). \tag{11.3}$$

The probability to release an electron from the cathode by ion impact is described by a coefficient γ,[1] which is defined as the ratio of emitted electron flux to incoming ion flux, $\gamma = |\Gamma_e(0)/\Gamma_i(0)|$. Breakdown occurs when the avalanche process can maintain itself. This can be described by the balance equation

$$\gamma\left(e^{\alpha d} - 1\right) = 1\,. \tag{11.4}$$

Townsend used an empirical law for the dependence of α on the electric field,

$$\frac{\alpha}{p} = A\,\exp\left(-\frac{B}{E/p}\right) \tag{11.5}$$

[1] In practice, this coefficient describes the sum of secondary emission by ions and metastable atoms and the photoemission by UV radiation in the cathode fall.

with constants A and B that are characteristic for each gas. Here, α/p represents the electron multiplication per mean free path and E/p the energy gain of an electron per mean free path, $\propto E\lambda_{\mathrm{mfp}}$, which makes (11.5) independent of gas pressure. Using Townsend's stationarity condition (11.4), and defining the breakdown voltage $U_{\mathrm{bd}} = E_{\mathrm{bd}}\, d$, we find Paschen's law, named after Friedrich Paschen (1865–1947)

$$U_{\mathrm{bd}} = \frac{Bpd}{C + \ln(pd)} \quad \text{with} \quad C = \ln\left[\frac{A}{\ln(1 + 1/\gamma)}\right]. \tag{11.6}$$

The number of ionizing events in the discharge gap is proportional to the neutral gas density and the width d of the gap. When we plot the breakdown voltage U_{bd} vs. pd, a so-called *Paschen curve* is obtained (Fig. 11.4), which has a pronounced minimum. Left of the minimum, there are too few atoms for ionization, right of the minimum the energy gain per mean free path becomes insufficient to maintain the electron energy needed for ionization. Both effects have to be compensated by a higher electric field leading to a higher breakdown voltage. The minimum voltage depends on the combination of gas and cathode material. Typical experimental values are compiled in Table 11.1 [351].

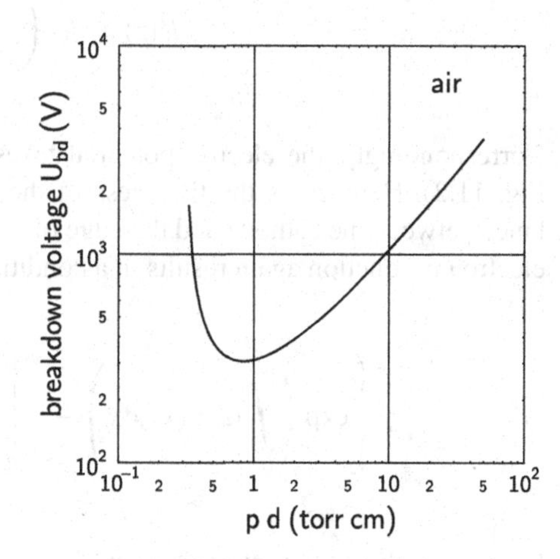

Fig. 11.4 Breakdown voltage U_{bd} in air for $\gamma = 10^{-2}$ as a function of the product pd (1 torr cm = 1.33 Pa m)

Table 11.1 Minimum breakdown voltage

Gas	Cathode	V_{min} (V)	$(pd)_{\mathrm{min}}$ (torr cm)
He	Fe	150	2.5
Ne	Fe	244	3
Ar	Fe	265	1.5
Air	Fe	330	0.57
Hg	W	425	1.8

11.1.3.2 Steady-State Operation

Now let us consider the steady-state operation of the cathode region. In the Aston dark space, primary electrons from the cathode have insufficient energy to excite neutral atoms. Excitation of various levels becomes possible in the cathode layer. In the cathode dark space, the energy of most electrons lies far beyond the maximum of the excitation function and little visible light is emitted. At the boundary of the negative glow, there are many slow electrons from the last generation of the avalanche process, which are again able to excite the atoms. This sequence of events is confirmed by spectroscopic observations: In the cathode layer the spectral lines appear in the order of increasing excitation energy; at the edge of the negative glow, the spectral lines appear in reversed order. This is a hint that the electron velocity decreases in this region. Historically, this was an important finding made by Ernst Gehrcke (1878–1960) and Rudolf Seeliger (1886–1965) that eventually led to the famous experiments of Franck and Hertz, which were fundamental for the understanding of the internal structure of the atom.

After ignition, the electric field near the cathode is no longer homogeneous. Rather, the cathode dark space is predominantly filled with ions giving rise to positive space charge. It is an empirical observation that the electric field in the cathode region varies as

$$E(z) = E_{\mathrm{c}}\left(1 - \frac{z}{d_{\mathrm{n}}}\right) . \tag{11.7}$$

Correspondingly, the electric potential takes a parabolic shape (see the sketch in Fig. 11.2). Here, d_{n} is the thickness of the *normal cathode fall*, roughly the distance between the cathode and the edge of the negative glow. The stationarity of the electron production again results in a condition of the type

$$\gamma\left[\exp\left(\int_0^{d_{\mathrm{n}}} \alpha[E(x)]\mathrm{d}x\right) - 1\right] = \gamma\left[\exp(\bar{\alpha} d_{\mathrm{n}}) - 1\right] = 1 , \tag{11.8}$$

but now with a different electron multiplication coefficient $\bar{\alpha}$ that is based on the electric field profile (11.7) because the electric field determines the electron energy. The voltage drop of the normal cathode fall, $U_{\mathrm{n}} = -\int E(x)\mathrm{d}x$, turns out smaller than the breakdown voltage U_{bd}. The thickness of the normal cathode fall can be obtained from (11.8) as

$$d_{\mathrm{n}} = \frac{1}{\bar{\alpha}} \ln\left(1 + \frac{1}{\gamma}\right) . \tag{11.9}$$

11.1.4 The Hollow Cathode Effect

When the plane cathode of a glow discharge is replaced by a cylinder that is longer than its diameter, the cathode yield can be enhanced by one or two orders of magnitude. A photo of a neon discharge with a hollow cathode is shown in Fig. 11.5.

The radius of the hollow cathode is chosen smaller than the thickness d_n of the normal cathode fall of a plane cathode, which again gives a scaling of the dimensions of the hollow cathode with the product pd. Then, the cathode layers on opposite sides of the inner surface of the hollow cathode tend to overlap and electrons leaving the surface will be decelerated when they enter the opposite cathode layer. Moreover, most of the UV photons, which were free to escape in the plane geometry, are now contributing to the cathode yield. This may explain the increased efficiency of a hollow cathode. Outside the hollow cathode, only a negative glow and the Faraday dark space are found.

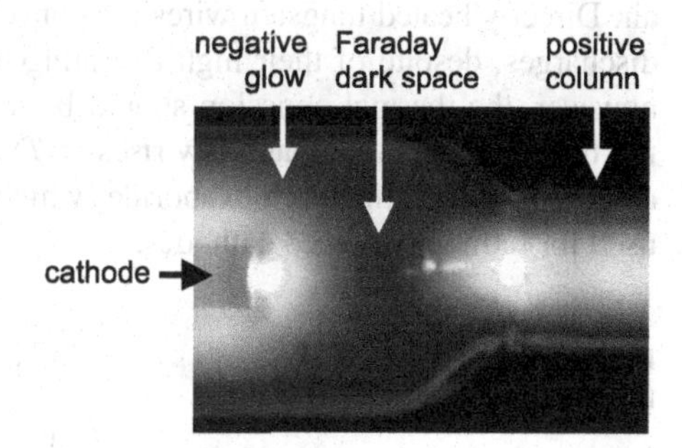

Fig. 11.5 Neon discharge with a hollow cylindrical cathode

11.1.5 Thermionic Emitters

The disadvantage of glow discharges with cold cathodes are the high operating voltages, lying between 300 and 3000 V, typically. When electrons are liberated by a different means, the overall voltage drop can be kept small and the discharge can be operated at a much lower input power. One of these processes is thermionic emission.

The release of electrons from a hot metal (or oxide) surface is described by Richardson's law, named after the British physicist Owen Willans Richardson (1879–1959),

$$j(T) = A_R T^2 \exp\left(-\frac{W_R}{k_B T}\right) . \tag{11.10}$$

A_R and W_R (an effective work function) are empirical constants to describe the emission characteristic of a thermionic emitter. Typical values for selected cathode materials are compiled in Table 11.2.

Table 11.2 Coefficients in the Richardson formula (11.10) for selected cathode materials (from [352])

Material	A_R ($\mathrm{A\,cm^{-2}\,K^{-2}}$)	W_R (eV)
W (polycrystalline)	60	4.51
W (thoriated)	3	2.63
LaB_6	29	2.66
BaSrO	0.5	1.0

The typical ranges of operating temperature for polycrystalline tungsten wires, thoriated tungsten wires, oxide coated nickel cathodes, and lanthanum hexaboride are shown in Fig. 11.6.

Oxide cathodes are commonly used in fluorescent tubes. Their advantage is the high electron emission at convenient low operating temperatures. For plasma experiments, oxide cathodes are not suitable when they are repeatedly exposed to air. Directly heated tungsten wires are convenient electron emitters in low-pressure discharges, despite of their high operating temperature. In planning such devices, however, the thermal emission should be taken into consideration, which according to the Stefan-Boltzmann law rises $\propto T^4$. Lower operating temperatures can be achieved with lanthanum hexaboride, which is available as plates, rods and tubes used for indirectly heated cathodes.

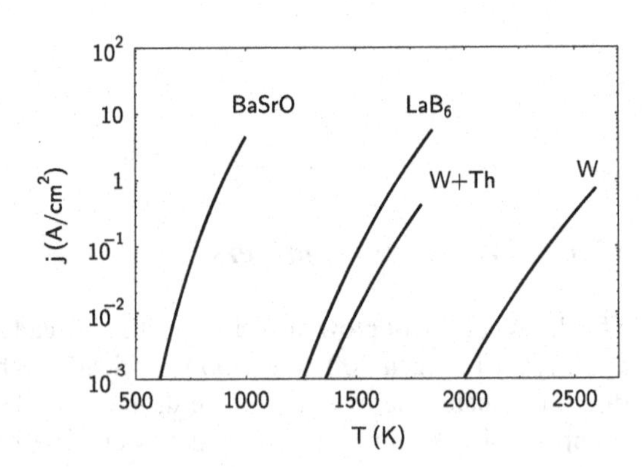

Fig. 11.6 Operating regimes for various cathode materials

11.1.6 The Negative Glow

At the edge of the negative glow, two populations of electrons are found: thermal electrons from the last generation of the avalanche process and energetic electrons originating from the cathode that have gained nearly the entire energy $-e\Delta\Phi$ of the cathode fall. Therefore, the negative glow is produced by a non-local process. Like a heir to a fortune that was amassed by hardworking ancestors, the negative glow has no need to work for sustaining its glittery life. This is the reason why the electric field (and the associated ohmic power deposition, jE) can nearly vanish in the negative glow. With increasing distance from the cathode, the negative glow gets ever fainter and ends up in the Faraday dark space. Remember that this vanishing of the luminosity can be attributed to exhausting the energy of the thermal electrons. The primary electrons with energies of 500 eV and more will only produce some weak ionization and will finally crash into the anode. The Faraday dark space is terminated by the increase of the axial electric field that again leads to electron heating and the establishment of the positive column (which in our analogy has to care for its own living).

By shortening the distance between anode and cathode, the positive column can be completely suppressed. Then a glow discharge is formed, which consists only of the cathodic parts and the negative glow. Glow-covered cathodes are used in indicator lights or in *nixie tubes*—an early type of single-digit display tube (Fig. 11.7). The individual cathodes have the shape of the digits 0–9. The active cathode is covered with a negative glow while the others remain dark. The square grid visible in the front serves as anode.

Fig. 11.7 Glow discharge covering the number-shaped cathodes in *nixie tubes* that were used in early frequency counters, digital voltmeters or computing devices

11.1.7 The Positive Column

The positive column can have different appearances: an axially homogeneous state, a striated appearance of standing bright structures (Fig. 11.8), or an apparently homogeneous state, which at close inspection consists of striations that move rapidly from anode to cathode. For simplicity, we will only discuss the homogeneous state, and otherwise refer the reader to more specialized literature [68].

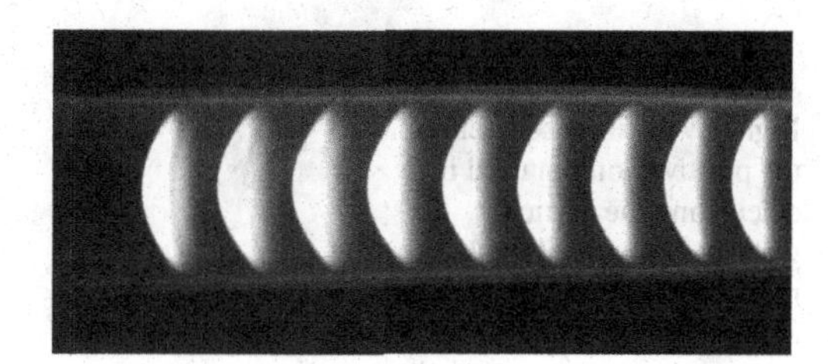

Fig. 11.8 Selforganisation into standing striations of the positive column of a low-pressure glow discharge in hydrogen gas. The cathode is located on the left side

The homogeneous positive column of a low-pressure glow discharge is a quasineutral plasma region with $T_e \gg T_i$. The equilibrium state of the positive column can be described by a sequence of four steps:

1. The energy gain of the electron gas in the axial electric field establishes an equilibrium temperature T_e.
2. The ionization rate of gas atoms, i.e., the number of electron-ion pairs produced per volume and time, is an increasing function of electron temperature.
3. The production rate in a segment Δz of the positive column is balanced by an equal amount of losses by ambipolar diffusion within this segment.
4. For fixed discharge current, this equilibrium is stable: When the electron density falls under the equilibrium value, the conductivity of the segment becomes lower and the electric field rises and increases the electron temperature and production rate.

We had already discussed ambipolar diffusion in Sect. 4.3.3.1. The ambipolar flux is given by (4.35) and (4.36). Solutions for the equilibrium density profile of the positive column in a cylindrical discharge tube were discussed in Sect. 4.3.3.2. The geometry for the balance of plasma production and (radial) diffusion losses is sketched in Fig. 11.9.

For the (grey-shaded) cylindrical shell between r and $r + \mathrm{d}r$ we obtain the production rate in this volume $\mathrm{d}V$ as

$$\left.\frac{\mathrm{d}n_e}{\mathrm{d}t}\right|_{\mathrm{gain}} \mathrm{d}V = \nu_{\mathrm{ion}} n_e 2\pi r \mathrm{d}r \mathrm{d}z \,. \tag{11.11}$$

The loss rate is the difference of the ambipolar fluxes entering at r and leaving at $r + \mathrm{d}r$:

$$\begin{aligned}\left.\frac{\mathrm{d}n_e}{\mathrm{d}t}\right|_{\mathrm{loss}} \mathrm{d}V &= 2\pi D_a \mathrm{d}z \left[r \left.\frac{\mathrm{d}n}{\mathrm{d}r}\right|_r - (r + \mathrm{d}r) \left.\frac{\mathrm{d}n}{\mathrm{d}r}\right|_{r+\mathrm{d}r} \right] \\ &= -2\pi D_a \mathrm{d}z \frac{\mathrm{d}}{\mathrm{d}r}\left(r \frac{\mathrm{d}n}{\mathrm{d}r}\right) \mathrm{d}r \,. \end{aligned} \tag{11.12}$$

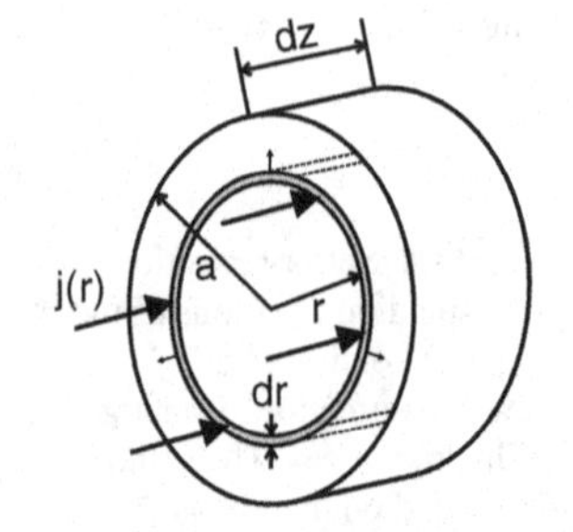

Fig. 11.9 A short segment of the positive column used in calculating the particle gain-loss balance. The short radial arrows indicate the ambipolar flow

Combining (11.11) and (11.12), we recover the Bessel-type equation (4.41), which has the solution $n_e(r) = n_e(0)J_0(2.405r/a)$ shown in Fig. 4.14. This example shows how a short segment of the positive column maintains its existence. Further, we can conclude that the positive column can be extended to an arbitrary length,[2] provided that the discharge current is kept constant. In other words, the positive column requires a constant voltage drop per unit length (= axial electric field), as can be seen in Fig. 11.2b.

11.1.8 Similarity Laws

Two gas discharges in cylindrical tubes have the same plasma parameters n_e and T_e, when the following similarity conditions are fulfilled.

The mean energy gain of an electron in the dc electric field is proportional to the product of the electric field and the mean free path, $E\lambda_{mfp}$. Since the mean free path is inversely proportional to the gas density, the energetics in the different regions of a glow discharge is determined by the quantity E/p, where the gas pressure p represents gas density. Therefore, the electron temperature in the positive column is only a function of E/p and the kind of gas used.

In particular, Townsend's electron multiplication coefficient can be written as $\alpha = pf(E/p)$. It is then straightforward to conclude that the thickness of the normal cathode fall (11.8) must be a fixed number of mean free paths for ionization, hence $d_n \propto 1/p$. Likewise, the voltage drop U_n of the normal cathode fall is a fixed multiple of the energy gain per mean free path and therefore independent of gas pressure. This allows the following conclusions about the similarity parameter for the discharge current. The total discharge current is the same at each axial position of the discharge tube. Hence, we can calculate it at the cathode. There, we have a mobility-limited ion current, $j_i = en_i\mu_i E_0$, with the electric field at the cathode $E_0 = 2U_n/d_n$. The ion density is obtained from the ion space charge in the cathode region $n_i e = \varepsilon_0 E_0/d_n$. Then, with $j_e = \gamma j_i$, the total current becomes

$$j = \varepsilon_0 \mu_i \frac{E_0^2}{d_n}(1+\gamma) \propto p^2 . \tag{11.13}$$

The proportionality to p^2 results from $E_0 \propto p$, $d_n \propto p^{-1}$, and $\mu_i \propto p^{-1}$. Hence, j/p^2 is a similarity parameter.

---

[2] Prof. Sanborn C. Brown told the anecdote of the gas-discharge pioneer Wilhelm Hittorf, who attempted to find the maximum length of the positive column [353]. "Week after week his discharge tube grew as he added meter after meter [...] His tube went all the way across the room, turned and came back, turned again until his laboratory seemed full of thin glass tubing. It was summer [...] and he opened the window to make it bearable. Suddenly from outside came the howl of a pack of dogs in full pursuit and flying through the window came a terrified cat to land [...] in the middle of the weeks and weeks of labor. 'Until an unfortunate accident terminated my experiment', Hittorf wrote, 'the positive column appeared to extend without limit.'"

The influence of the tube radius a on the discharge is given by the production-loss balance (4.43), $\nu_{ion} = D_a (2.405/a)^2$. Noticing, that $\nu_{ion} \propto p$ and $D_a \propto p^{-1}$, we find that the product of gas pressure and tube radius, $p\,a$, must be a similarity parameter. Similar scalings with $p\,d$ were discussed for the breakdown voltage, the thickness of the normal glow, and the dimensions of the hollow cathode.

Last not least, the $p\,d$ scaling explains, why the diameter of fluorescent tubes could be reduced from 1.5″ to 1″, or even to 1/4″ in compact fluorescent tubes, with a corresponding increase in gas pressure, after phosphor coatings were developed that could withstand the increased heat flux. It is also not surprising, that modern gas discharges at atmospheric pressure are tiny objects of sub-millimeter dimensions.

11.1.9 Discharge Modes of Thermionic Discharges

Low-pressure discharges with thermionic emitters are influenced by the formation of electron space-charge in the cathode region. This space-charge cloud can limit the discharge current by forming a virtual cathode (see Sect. 9.2.2), or by a sufficient amount of ion current, can vanish and leave a temperature-limited emission current. Four distinct discharge modes could be identified [354]:

1. the anode-glow mode
2. the ball-of-fire mode
3. the Langmuir mode
4. the temperature-limited mode

Figure 11.10 shows a typical volt-ampere characteristic of a thermionic discharge [355] and identifies the topology of the potential distribution between cathode and anode. The anode glow mode is found in the regime A–B′ on the hysteresis curve. The corresponding potential distribution shows the formation of a potential barrier (virtual cathode) before the cathode. In a layer before the anode, the potential rises

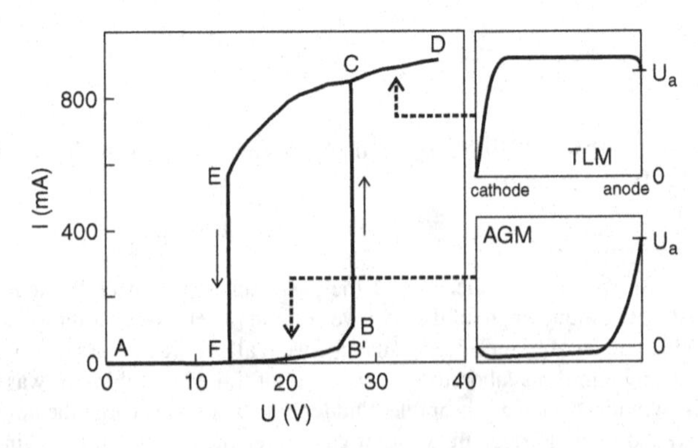

Fig. 11.10 Hysteresis curve for a thermionic discharge (from [355]). The potential distribution for the anode glow mode and the temperature limit mode are sketched on the right

to the anode bias U_a. Only in this layer, the electrons gain sufficient energy to ionize and to excite atoms, which explains the name *anode glow mode* (AGM).

Raising the discharge voltage, an unstable mode is reached between B′–B. This is the ball-of-fire mode, in which the anode layer has expanded towards the cathode and a sudden transition to C can occur. When the discharge is operated with a series resistor that forces the point of operation to lie half-way between B and C, the *Langmuir mode* is established, in which the anodic plasma fills a significant part of the total volume but the current limitation by the virtual cathode is still active.

The discharge topology changes dramatically in the regime C–D, in which most of the plasma volume has reached a potential, which lies slightly above the anode potential because of the ambipolar field between plasma and anode. The voltage drop occurs in the cathode sheath and the virtual cathode has vanished. Hence, the full emission current from the cathode is available for electron injection, which is only limited by the cathode temperature and justifies the name *temperature limited mode* (TLM). Care must be taken that the ion bombardment of the cathode does not lead to a thermal run-away of the cathode. When the discharge voltage is reduced again, the transition to the AGM occurs at a significantly lower voltage (point E) thus creating a large hysteresis in the volt-ampere characteristic.

11.2 Capacitive Radio-Frequency Discharges

The parallel plate discharge operated at 13.56 MHz frequency is the workhorse among the low-pressure rf discharges for plasma etching or plasma-enhanced chemical vapour deposition (PECVD).[3] Under these conditions, the electrons have a thermal energy of a few eV to bring atoms into excited states and dissociate molecules, which facilitates chemical reactions. On the other hand, the heat content of the electron gas is still small because of the low electron density. This allows bringing the plasma into contact with sensitive surfaces. Sometimes the principle, on which these plasma applications are based, is named "cold heat".

Why is it necessary to operate these discharges at radio frequency rather than with dc? When a dielectric substrate is put on one of the electrodes of a parallel plate reactor, the dc current is interrupted and the plasma will seek a connection to the uncovered part of the electrode. This hardly gives the desired homogeneous contact between plasma and substrate. However, when an rf voltage is applied, a displacement current will flow in the substrate that establishes the connection between plasma and electrode, and provides homogeneity.

The parallel plate discharge belongs to the family of capacitive rf discharges, which means that the rf electric field results from surface charges on electrodes or

[3] For operating powerful rf generators, safety issues must be observed according to accepted national lab standards. Operating power generators at other frequencies than 13.56 MHz may be in conflict with the rules set by communications authorities, such as the FCC in the United States.

dielectrica. This distinguishes it from inductive discharges where the electric field is generated by a time-varying magnetic field from an external antenna.

A first impression of the plasma in a parallel-plate discharge can be gained from the photo in Fig. 11.11. The space between the electrodes is filled with plasma of different luminosity. There are dark spaces adjacent to the electrodes and two glow maxima.

Parallel plate discharges fall into different classes of operation:

- The applied rf voltage determines whether the discharge operates in the α-regime, which is governed by ionization from electron avalanches, or in the γ-regime, where electrons are produced at the electrodes by secondary emission from ion bombardment.
- The two rf electrodes of the discharge can have equal or different areas.
- The discharge can be operated through a blocking capacitor, which leads to a dc *self-bias* from rectifying the rf voltage in the sheath region.

Fig. 11.11 The plasma luminosity in a symmetric parallel-plate r.f. discharge. The plasma is separated from the electrodes by two dark spaces. The luminosity of the bulk plasma has two peaks

11.2.1 The Impedance of the Bulk Plasma

The distribution of the luminosity in the plasma suggests that we can split the plasma into a central quasineutral bulk of thickness b and space charge sheaths of width s_1 and s_2 as sketched in Fig. 11.12. Let us first discuss the voltage drop across the bulk plasma and neglect the sheaths. Then the problem is to determine the voltage drop across a plasma-filled capacitor.

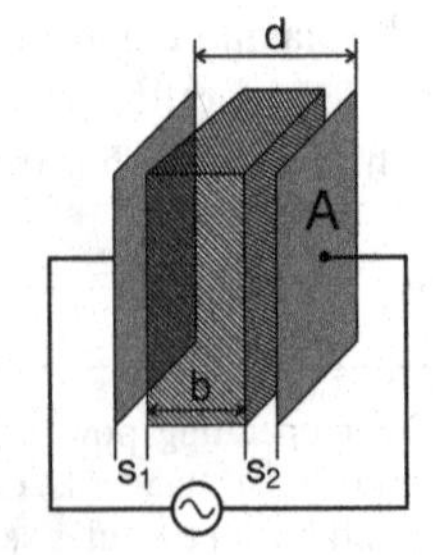

Fig. 11.12 Schematic of a parallel plate discharge. The electrodes form a capacitor of area A and width d. The capacitor is filled with a plasma of thickness b and dielectric constant ε_p, which is separated from electrodes by sheaths of width s_1 and s_2

In electrical engineering, the time evolution of currents and voltages is described by an exponential function that has the opposite sign in the phase factor compared to our convention for plane waves in (6.8). For instance, the voltage drop across the plasma bulk is given as

$$U_b = \hat{U}_b \, e^{+i\omega t} . \tag{11.14}$$

In this notation, the dielectric constant of the (collisional) plasma is

$$\varepsilon_p = 1 - \frac{\omega_{pe}^2}{\omega(\omega - i\nu_m)} . \tag{11.15}$$

Here, ν_m is the electron-neutral collision frequency for momentum loss. The plasma impedance is defined as the (complex) ratio of voltage and current for current flow in a capacitor

$$Z_b = \frac{\hat{U}_b}{\hat{I}_b} = \frac{1}{i\omega C_b} . \tag{11.16}$$

Introducing the capacitance of the empty capacitor, $C_0 = \varepsilon_0 A/b$, and noting that $C_b = \varepsilon_p C_0$, we obtain

$$\frac{1}{Z_b} = i\omega C_0 + \frac{\omega_{pe}^2 C_0}{\nu_m + i\omega} = i\omega C_0 + \frac{1}{R_b + i\omega L_b} . \tag{11.17}$$

Here, we have defined the inductance of the bulk plasma as $L_b = \omega_{pe}^{-2} C_0^{-1}$, and the resistance as $R_b = \nu_m L_b$. The structure of (11.17) shows that the plasma impedance is described by the network shown in Fig. 11.13. The inductor represents the electron inertia and the resistor the electron friction with the neutral gas.

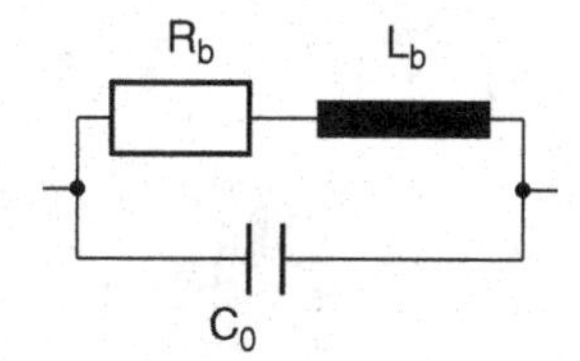

Fig. 11.13 The impedance of the bulk plasma comprises the capacitor of the electrodes, an inductor (electron inertia) and a resistor (electron collisions)

11.2.2 Sheath Expansion

The sheath of a rf discharge has a fast dynamical evolution. At $f = 13.56$ MHz, the response of electrons and ions to the rf electric field is quite different. Remembering that the response time of a particle species is given by the inverse plasma frequency, we have the ordering $f_{pi} \ll f \ll f_{pe}$. This means that the ions cannot follow the changing rf field and are only subject to the time-averaged fields. For the moment we assume that the ions are immobile.

The electrons, however, can follow the instantaneous rf field. We will discuss the electron dynamics in the sheath for the simplified situation of a *matrix sheath*, which has a homogeneous ion density n_{i0}, but is electron-free. A moving sheath-edge separates the sheath from the quasineutral plasma region (Fig. 11.14). This model was proposed by Godyak for the α-regime, e.g., [356], where electron release from the electrode is negligible.

Let us consider the sheath of thickness s_1 on the left side. In the homogeneous ion space charge the electric field increases linearly with $E(s_1) = 0$, and we have $E(0) = E_0 = -en_{i0}s_1/\varepsilon_0$. Now let us assume that the potential on the left electrode becomes more negative and the sheath expands. Then the sheath edge has a velocity $v_1 = \mathrm{d}s_1/\mathrm{d}t$. This means that electrons are removed from the sheath with a total conduction current density

$$j_{\mathrm{c}} = -en_{i0}\frac{\mathrm{d}s_1}{\mathrm{d}t} \,. \tag{11.18}$$

At the same time, the sheath expansion by an amount $\mathrm{d}s_1$ leads to an increase in the electric field

$$\mathrm{d}E = -\frac{e}{\varepsilon_0}n_{i0}\,\mathrm{d}s_1 \,. \tag{11.19}$$

This results in a displacement current inside the sheath,

$$j_{\mathrm{d}} = \varepsilon_0\frac{\mathrm{d}E}{\mathrm{d}t} = -en_{i0}\,\frac{\mathrm{d}s_1}{\mathrm{d}t} \,, \tag{11.20}$$

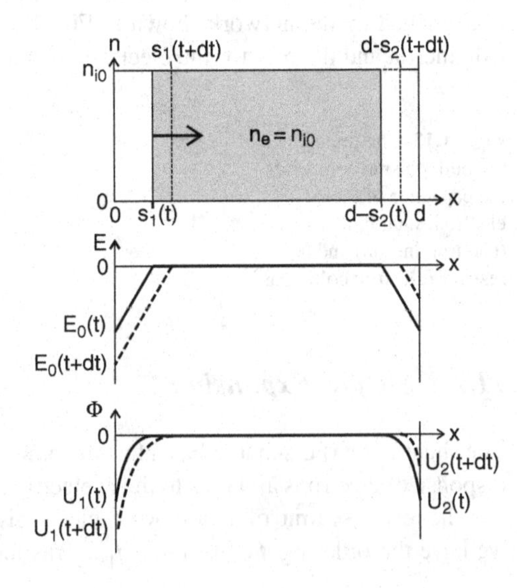

Fig. 11.14 Density, electric field and potential distribution in a parallel plate discharge. The sheath on the left side is assumed to expand. The bulk plasma is shifted to the new position marked by the dashed frame

which equals the conduction current on the plasma side. In this way, displacement current and conduction current establish current continuity across the sheath boundary.

Now, what happens to the surplus electrons, which have been injected into the bulk plasma by the expanding sheath? This question is difficult to answer, and requires kinetic theory or simulations with particle codes. However, the net effect which is established in a few electron plasma periods is evident: Since we have required that the bulk plasma is quasineutral and has the same ion density as the sheath, $n_e = n_{i0}$, the surplus of electrons has to be injected into the sheath on the right side. Hence, the sheath width s_2 must shrink at the same rate as the sheath s_1 expands, which gives $\mathrm{d}s_2/\mathrm{d}t = -\mathrm{d}s_1/\mathrm{d}t$ and the thickness b of the bulk plasma is constant.

When the parallel-plate discharge is operated with a current density $j = \hat{j}\exp(i\omega t)$, it can be seen by integrating (11.18) that the sheath width has a linear response

$$\hat{s}_1 = -\frac{1}{i\omega}\frac{\hat{j}}{en_{i0}}. \tag{11.21}$$

The sheath edge performs a sinusoidal motion about an equilibrium width s_0 with amplitude $\hat{s}_1$. The requirement that both sheaths have to collapse completely during each rf cycle gives $s_0 = \hat{s}_1$.

The voltage drop across the sheath is obtained as

$$U_1 = \int_0^{s_1} E\,\mathrm{d}x = -\frac{en_{i0}}{2\varepsilon_0}s_1^2 = -\frac{en_{i0}}{2\varepsilon_0}\hat{s}_1^2\left(1 + 2\mathrm{e}^{i\omega t} + \mathrm{e}^{2i\omega t}\right) \tag{11.22}$$

and has a non-linear response with contributions from the harmonic at 2ω. When probe measurements are made in parallel plate discharges, filters are necessary to block the first harmonic at 13.56 MHz and the second harmonic at 27.12 MHz.

In the same manner, the voltage drop across the r.h.s. sheath can be calculated as

$$U_2 = \int_{d-s_2}^{d} E\,\mathrm{d}x = -\frac{en_{i0}}{2\varepsilon_0}s_2^2 = -\frac{en_{i0}}{2\varepsilon_0}\hat{s}_2^2\left(1 - 2\mathrm{e}^{i\omega t} + \mathrm{e}^{2i\omega t}\right) \tag{11.23}$$

For a negligible contribution from the bulk plasma, and using $\hat{s}_2^2 = \hat{s}_1^2$, the entire voltage drop across the discharge is then

$$U = U_1 - U_2 = -\frac{2en_{i0}}{\varepsilon_0}\hat{s}_1^2\,\mathrm{e}^{i\omega t}\,. \tag{11.24}$$

Because of the symmetry of the system, the constant and the second harmonic contributions have canceled in the total discharge voltage U. This result is also true when a non-negligible voltage drop occurs across the bulk plasma.

11.2.3 Electron Energetics

11.2.3.1 Ohmic Heating

The total time-averaged power deposition inside the bulk plasma is determined by the ohmic heat in the resistor R_b

$$\langle P\rangle = \frac{1}{2}\hat{I}^2 R_\mathrm{b}\,. \tag{11.25}$$

The factor 1/2 arises from the average of $\sin^2(\omega t)$ over one period.

11.2.3.2 Wave Rider or α-Regime

Ohmic heating of the plasma bulk is not the dominant process at low rf power and low gas pressure. Rather, electrons gain energy from the expanding sheath. This process is often called wave riding, although a better analogy is the momentum exchange between tennis ball and tennis rack. Because the sheath electric field is built up by all the ions inside the sheath, an electron that enters the sheath from the plasma side with velocity v_0 experiences a collision with a massive object. In the frame of the expanding sheath, which moves at v_1, the plasma electron enters the sheath with $v = v_0 + v_1$ and leaves the sheath with $v' = -v$. Hence, in the rest frame the electron finally has $v' = v_0 + 2v_1$.

The characteristic energy gained in one collision with the expanding sheath can be obtained from (11.24) and using $\hat{v}_1 = \mathrm{i}\omega\,\hat{s}_1$. For $v_1 \gg v_0$, we have

$$W \approx \frac{1}{2}m_\mathrm{e}(2\hat{v}_1)^2 = \frac{\varepsilon_0 m_\mathrm{e}}{e^2 n_{i0}}\omega^2 e\hat{U} = \frac{\omega^2}{\omega_{pe}^2}\,e\hat{U}\,. \tag{11.26}$$

The energy gain is therefore higher for increasing frequency. This is why in some applications frequencies up to 100 MHz are used.

Now, what will the energetic electrons do after being reflected from the sheath? Depending on gas pressure, the electrons can perform elastic and inelastic collisions with gas atoms or, at very low pressure, even reach the sheath on the opposite side. When they are scattered by elastic collisions, only a small fraction of their energy is lost (see Sect. 4.4.1) and, for $\lambda_\mathrm{mfp} < s_0$, the electrons can perform multiple collisions with the expanding sheath and gain the energy (11.26) at each collision.

It was shown in [357] that for low applied rf voltage (200 $\mathrm{V_{pp}}$, typically) ionization preferentially occurs near the sheath edge during the expansion phase of the sheath. Since the expanding sheath sweeps up the new electrons from the ionization process, an avalanche process is established. This justifies the name α-regime because of its resemblance to the avalanches in the cathode region of a dc glow discharge.

When the electron mean free path is comparable with the width of the bulk plasma, *stochastic heating* sets in. Then, the electron is bouncing back and forth between the two sheaths gaining energy in a random way. This process is similar to the Fermi acceleration mechanism for cosmic particles. There, the particles bounce between moving magnetic mirrors that approach each other [358].

11.2.3.3 Secondary Emission or γ-Regime

When the applied rf voltage of a low-pressure rf discharge in helium is increased above 400 V_{pp}, typically, a transition from the α-regime to a new γ-regime is found [359], where electrons are liberated from the electrode by secondary emission after ion bombardment, and gain their energy from the voltage drop across the sheath rather than from sheath expansion. These electrons from the γ-process penetrate the plasma and form a new intense glow in the plasma center [360] as shown in (Fig. 11.15).

For low discharge power, the intensity distribution is characterized by two humps near the electrodes and a lower intensity in the plasma center. This compares well with the visual impression in Fig. 11.11. For high power, the γ-regime is established with a strongly enhanced luminosity in the plasma center.

At the same time, the electron temperature in this central *secondary glow* drops dramatically, e.g., from $\approx$ 3 to 0.5 eV in argon [361] or from 2.5 to 0.2 eV in helium [361, 362]. This demonstrates that the secondary glow is produced by electrons which have gained their energy in the sheath. The secondary glow is the analogon to the negative glow of a dc discharge, which is maintained by beam electrons from the cathode fall.

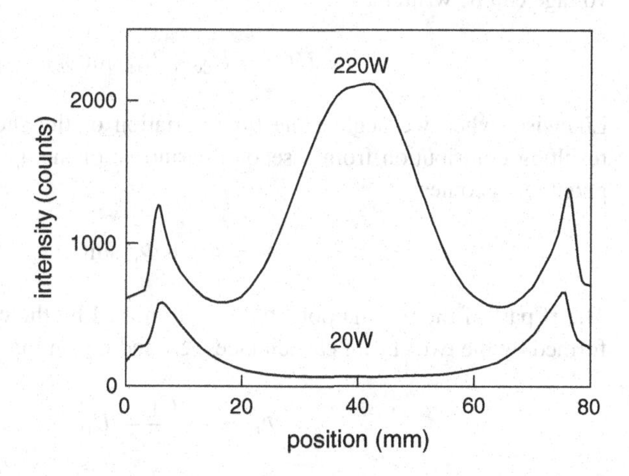

Fig. 11.15 Intensity distribution of the helium line 388.9 nm ($2\,^3S$–$3\,^3P$) for low and high discharge power (from [360])

11.2.4 Self Bias

In this paragraph, we discuss the influence of unequal areas of powered electrode and grounded wall on the plasma potential, and the development of a dc contribution to the sheath voltage (*self bias*), which is an intrinsic feature of capacitively coupled discharges.

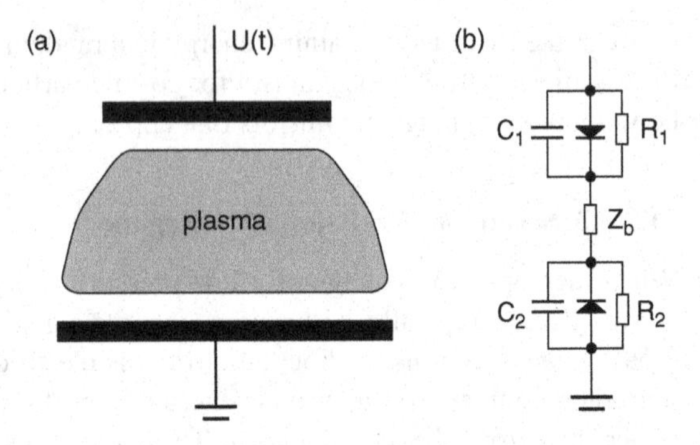

Fig. 11.16 (**a**) Asymmetric parallel plate discharge. (**b**) Equivalent circuit representing the sheath regions and the bulk plasma

In Fig. 11.16 an asymmetric discharge is sketched, in which the upper electrode is powered and the lower grounded. The equivalent circuit for the discharge system consists of two sheaths and the bulk plasma connected in series. The sheath is modeled by a diode representing the electron flow in the sheath, a resistor $R_{1,2}$ mimicing the ion current, and the sheath capacity $C_{1,2}$.

11.2.4.1 High-Frequency Regime

We will first discuss the situation at high frequency where the impedance of the sheath is dominated by the capacitance, i.e., we take the limit $R_{1,2} \rightarrow \infty$, and the impedance of the plasma bulk becomes negligible $|Z_b| \rightarrow 0$ [363]. The discharge voltage can be written as

$$U(t) = U_{dc} + U_{rf} \sin(\omega t) . \tag{11.27}$$

Likewise, when we neglect the time variation of the sheath capacitance and the resulting contribution from a second harmonic of the applied voltage, the plasma potential becomes

$$\Phi_p(t) = \bar{\Phi}_p + \Phi_{rf} \sin(\omega t) . \tag{11.28}$$

The rf part of the plasma potential is determined by the capacitive voltage divider formed by the two sheath capacitances, C_1 and C_2, in Fig. 11.16b

$$\Phi_{rf} = \frac{C_1}{C_1 + C_2} U_{rf} . \tag{11.29}$$

Now, we have to consider the electron dynamics in the sheaths. When the sheath collapses, the high electron mobility leads to an electron flow to the electrode as soon as the plasma potential equals the electrode potential. This is described by the diode symbols that short-circuit the sheath when the plasma potential becomes negative w.r.t. one of the electrodes. Hence, we obtain limiting conditions for the plasma potential

$$\Phi_{p,max} = \bar{\Phi}_p + \Phi_{rf} \geq U_{dc} + U_{rf} , \qquad \Phi_{p,min} = \bar{\Phi}_p - \Phi_{rf} \geq 0 . \tag{11.30}$$

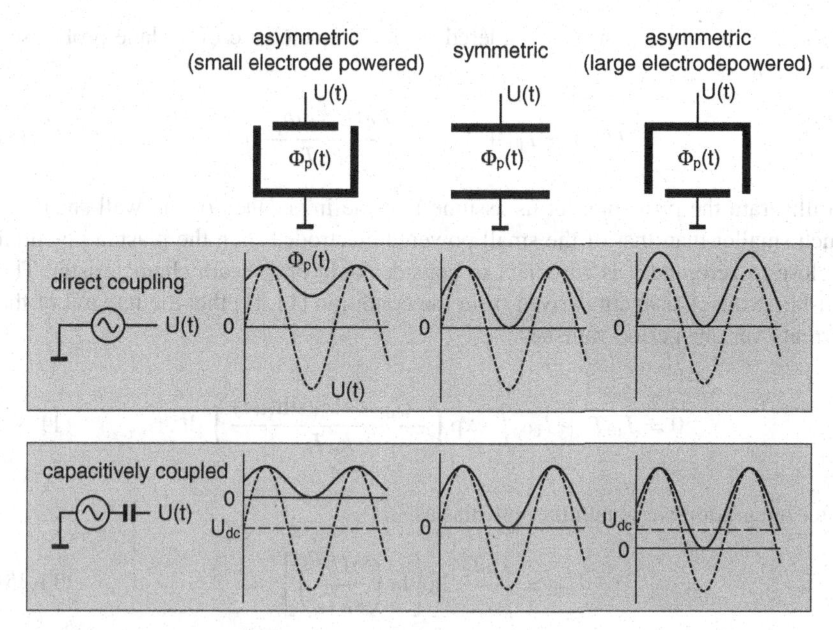

Fig. 11.17 Synopsis of the operation modes of parallel plate discharges with direct coupling and capacitive coupling. The six panels show the applied voltage $U(t)$ (*dashed line*) and plasma potential $\Phi_p(t)$ (*solid line*) vs. time

For direct coupling between rf generator and electrodes, at least one of these inequalities becomes an equality [363] as shown in the conceptual sketch in Fig. 11.17.

When a blocking capacitor is inserted in series with the rf generator, the discharge is called *capacitively coupled*. Then, in steady state, the net electric current over one rf period must be zero

$$\int_0^T I(t)\,\mathrm{d}t = 0\,. \tag{11.31}$$

This can only be established, when an electron flow is flowing for a short time during the rf cycle to *both* electrodes. This has the consequence that both inequalities in (11.30) become equalities and we can solve this set of equations for the mean plasma potential $\bar{\Phi}_p$ and self-bias voltage U_{dc}

$$\bar{\Phi}_p = \frac{1}{2}(U_{dc} + U_{rf})\,, \qquad U_{dc} = \frac{C_1 - C_2}{C_1 + C_2}\,U_{rf}\,. \tag{11.32}$$

11.2.4.2 Low-Frequency Regime

At low frequencies, the conduction current in the sheath becomes larger than the displacement current from sheath expansion. Then, a capacitively-coupled discharge develops a self bias because of the non-linearity of the current-voltage characteristic of the sheath, which can be considered as the characteristic of a plane probe (see Sect. 7.4)

$$I(U) = I_{i0} + I_{e0} \exp\left(\frac{e(U - \Phi_p)}{k_B T_e}\right) . \tag{11.33}$$

To illustrate the principle, let us assume that the impedance of the wall sheath is much smaller than that of the small powered electrode. Then the plasma potential is close to zero and it is sufficient to consider only one sheath characteristic. The self-bias voltage is again derived from the condition (11.31) that the integral of the current over one period vanishes

$$0 = I_{i0}T + I_{e0} \int_0^T \exp\left(\frac{eU_{dc} + U_{rf}\sin(\omega t)}{k_B T_e}\right) dt . \tag{11.34}$$

After integration we obtain the self-bias as

$$U_{dc} = \frac{k_B T_e}{e} \ln\left[\mathrm{I}_0\left(\frac{eU_{rf}}{k_B T_e}\right)\right] , \tag{11.35}$$

where $\mathrm{I}_0(x)$ is the modified Bessel function. In the language of electronics, the sheath acts as a rectifier that charges the blocking capacitor.

11.2.5 Application: Anisotropic Etching of Silicon

The development of a high dc voltage across the sheath adjacent to the smaller powered electrode has a dramatic influence on the energy at which ions impinge on the electrode or on a substrate lying on this electrode [364]. Ions of more than 400 eV energy lead to sputtering of the surfaces. The combination of sputtering and chemical etching, a process known as *reactive ion etching* (RIE) is an efficient means for highly anisotropic etching of silicon, see Fig. 11.18a. The highly directional ion flow preferentially hits the bottom of the etched trench.

Experiments on the action of ion bombardment and the presence of an etching gas (XeF_2) [365] have shown that the combination of sputtering and chemical etching is a cooperative mechanism which leads to an etch rate that is far greater than the sum of the individual processes, see Fig. 11.18b. This cooperative effect makes

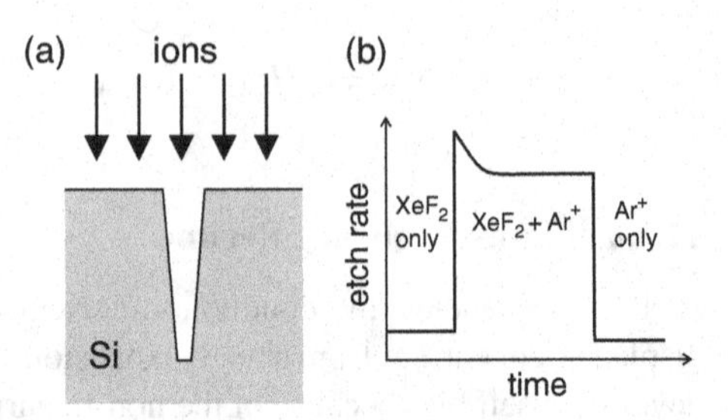

Fig. 11.18 (**a**) Sketch of deep-trench etching in silicon. (**b**) Cartoon of the dependence of the etch rate on the combination of reactive gas (XeF_2) and ion bombardment

anisotropic etching of deep trenches possible because it acts predominantly at the trench bottom while there is apparently no such cooperation at the trench walls. Aspect ratios of up to 40:1 can be achieved. This technology finds applications in dynamic memory chips, where the storage capacitor for a single bit can be efficiently stowed in such a deep trench, which makes economic use of the chip area. Other applications are in etching of micro-mechanical devices from silicon.

When ion bombardment is an unwanted effect, say in film deposition, the substrates are positioned on the larger grounded electrode that develops a much smaller dc voltage across the sheath.

11.3 Inductively Coupled Plasmas

Inductively coupled plasmas (ICPs) became relevant for the semiconductor industry because of the necessity to achieve higher plasma densities of $n_\mathrm{i} = (1\text{–}3)\times 10^{17}\,\mathrm{m}^{-3}$ in low pressure (p=(0.11–2) Pa) discharges. Higher densities lead to faster reaction rates that boost the economy of the process [366–368]. Capacitive discharges were limited to achieve this regime because the plasma density was only increasing with the square root of the applied power. Moreover, the necessity to transport the applied power through the sheath regions generated high voltage drops across the sheath, and ions gained energies of several hundred eV from the sheath potential, which could damage the integrated circuits. Inductive power transport to the plasma keeps the voltage drop across the sheath low and leads to moderate ion energies of (24–40) eV.[4]

An ideal ICP acts like a transformer with the primary being a coil that is positioned close to the plasma, and the plasma forming a single-turn secondary. This type of discharge was also known as *ring discharge*. Typical arangements for helical and flat coil designs are sketched in Fig. 11.19. An essential feature of ICPs is that, because of the skin effect, rf fields can only penetrate into a limited zone of the plasma close to the exciter coil.

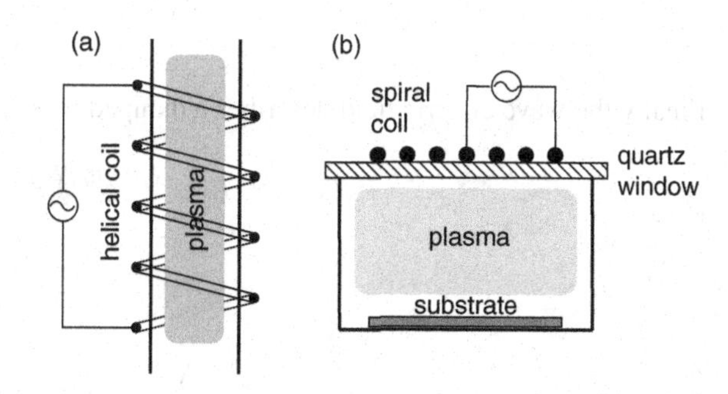

Fig. 11.19 (**a**) Inductively coupled plasma (ICP) with a helical coil wound around discharge tube. (**b**) ICP with a flat spiral coil on a quartz window. Quartz glass has low dielectric losses and can withstand high coil temperatures

[4] ICPs are intrinsically high-power rf discharges. Therefore, the safety issues and radio-interference problems need even more attention.

11.3.1 The Skin Effect

When an electromagnetic wave hits a conducting surface, the wave fields penetrate into the conducting material. There, according to Lentz's rule, electric currents of opposite polarity weaken the wave fields and lead to an exponential decay of the wave amplitude. This is the skin effect, which is well known in radio science where it describes the fact that rf currents flow only in a thin outer layer of conductors.

The skin effect can be easily understood assuming that a plane electromagnetic wave propagates in x-direction and hits a conductor that fills the halfspace $x > 0$ at normal incidence, as shown in Fig. 11.20. The wave electric field is assumed to lie in the y direction and the wave magnetic field in z-direction. The wave fields are described by $E_y(x,t) = \hat{E}_y \exp[\mathrm{i}(kx - \omega t)]$ and $B_z(x,t) = \hat{B} \exp[\mathrm{i}(kx - \omega t)]$. Faraday's induction law (5.2) and Ampere's law (5.4) then read

$$\frac{\partial E_y}{\partial x} = -\frac{\partial B_z}{\partial t}, \qquad -\frac{\partial B_z}{\partial x} = \mu_0 \sigma E_y, \tag{11.36}$$

where we have set $\mathbf{j} = \sigma \mathbf{E}$ and neglected the displacement current $\varepsilon_0 \partial E / \partial t$. We are allowed to do so when $\sigma \gg \omega \varepsilon_0$. Then, the electric field obeys a wave equation of the type

$$\frac{\partial^2 E_y}{\partial x^2} = \mu_0 \sigma \frac{\partial E_y}{\partial t}. \tag{11.37}$$

Inserting the harmonic wave ansatz, we obtain

$$-k^2 \hat{E}_y = -\mathrm{i}\omega \mu_0 \sigma \hat{E}_y. \tag{11.38}$$

which leads to a complex wavenumber

$$k = (\mathrm{i}\omega \mu_0 \sigma)^{1/2} = (\omega \mu_0 \sigma)^{1/2} \frac{1 + \mathrm{i}}{\sqrt{2}}. \tag{11.39}$$

Finally the wave electric field describes a damped wave of the type

$$E_y(x,t) = \hat{E}_y \mathrm{e}^{-x/\delta_0} \mathrm{e}^{\mathrm{i}(k_r x - \omega t)} \tag{11.40}$$

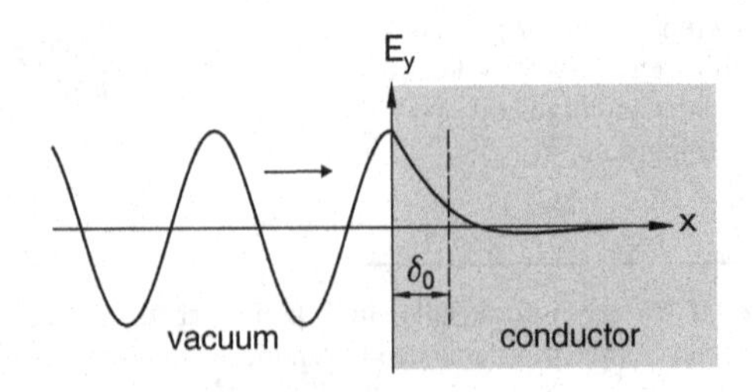

Fig. 11.20 Penetration of an electromagnetic wave into a conducting halfspace $x > 0$. The skin depth is marked with δ

with a characteristic length δ_0, the *classical skin depth*, given by

$$\delta_0 = \left(\frac{2}{\omega\mu_0\sigma}\right)^{1/2} . \tag{11.41}$$

When the electron-neutral collision frequency fulfills $\nu_m > \omega$, we can approximate the conductivity by its dc value (4.30) and obtain

$$\delta_0 = \frac{c}{\omega_{pe}}\left(2\frac{\nu_m}{\omega}\right)^{1/2} . \tag{11.42}$$

For discharges operated at 13.56 MHz, this assumption is valid for $p > 2$ Pa in argon and $T_e \approx 4$ eV. For lower pressures, $p < 0.1$ Pa, the *collisionless skin depth* is approached

$$\delta_{cl} = \frac{c}{\omega_{pe}} . \tag{11.43}$$

This limiting case is obtained from the full expression (6.44) for the complex wavenumber in the limit $\omega^2 \ll \omega_{pe}^2$ and $\nu_m \ll \omega$. A detailed discussion of the skin effect in plasmas can be found in [369].

11.3.2 E and H-Mode

ICPs can be operated in a low-power electrostatic mode, or E-mode, which is mostly found in the plasma density range $10^{14} - 10^{16}\,\mathrm{m}^{-3}$. The real inductive mode, or H-mode,[5] is found for higher plasma densities of $10^{16} - 10^{18}\,\mathrm{m}^{-3}$. The mechanism of E-mode and H-mode is illustrated in Fig. 11.21.

The E-mode is characterized by an rf electric field, that originates from the rf voltage drop across the exciter coil of the ICP discharge. Plasma density in the E-mode is low and leads to a skin depth that is larger than the plasma dimension. The E-mode resembles the parallel plate discharge and leads to energetic ions from

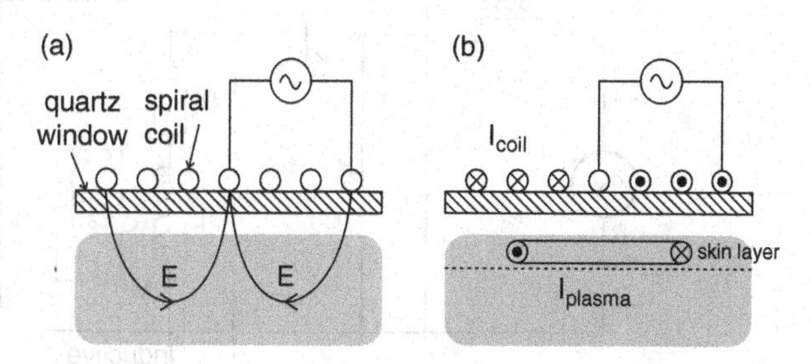

Fig. 11.21 **(a)** The E-mode is governed by the electric field of the spiral coil. **(b)** In the H-mode, a ring-shaped electric field and corresponding ring-current is established in the skin layer, which heats the electrons

[5] This terminology should not be confused with the high-confinement mode in tokamaks.

the sheath region. It is possible to suppress the E-mode by using a so-called *Faraday shield* consisting of a thin grounded copper sheet with slits that are oriented at right angle to the current flow in the antenna.

The H-mode (Fig. 11.21b) deposits the rf energy in the plasma by accelerating electrons by the ring-shaped induction field inside the skin layer. The electron flow represents a ring-current that has the opposite direction as the rf current in the exciter coil. This current system is the single winding of an air-core transformer, which consists of the multiwinding exciter coil and the skin layer of the plasma. The transition between E-mode and H-mode is characterized by a sudden jump of electron density when the discharge power is raised above a critical value (Fig. 11.22).

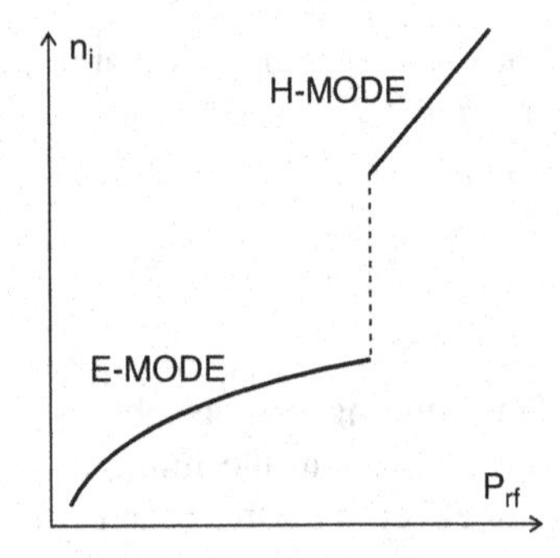

Fig. 11.22 The transition between the E-mode and H-mode is marked by a sudden jump in plasma density

11.3.3 The Equivalent Circuit for an ICP

The ICP plasma can be described by an equivalent circuit that contains the plasma properties as lump circuit elements. Figure. 11.23 shows the rf generator with a typical impedance of 50 Ω, the matching network and the plasma. Basically, the capacitors C_1 and C_2 of the matching network and the inductance L_1 of the exciter coil form a resonant circuit that is tuned to the frequency of the generator. This resonant circuit is damped by the coil resistance R_1 and the plasma load R_{electron}. In the H-mode, the plasma impedance is low, typically 1 Ω. Hence the exciter coil acts

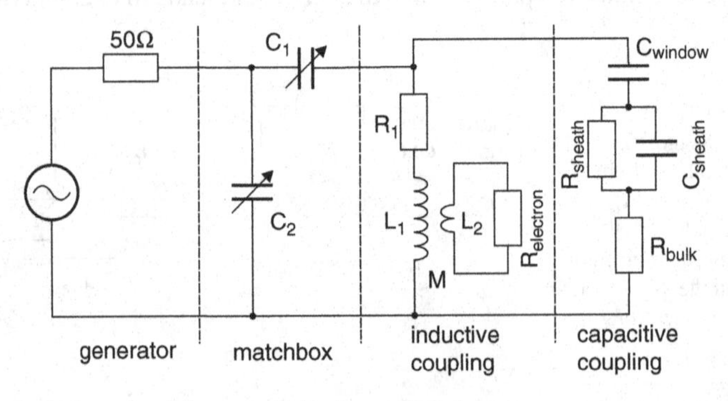

Fig. 11.23 Equivalent circuit for an inductively coupled plasma

as a step-down transformer with about seven primary windings and the plasma as a single winding. The ratio C_1/C_2 is often determined experimentally to fulfill both the resonance condition and the matching to the generator impedance.

The section denoted *capacitive coupling* consists of the capacitance of the quartz window, the RC model for the sheath and the impedance R_{bulk} of the bulk plasma. A detailed discussion of matching an ICP to the generator can be found in [370].

11.4 Concluding Remark

The abridged discussion of plasma generation given so far cannot replace the rich details given in textbooks on plasma discharges, e.g., [68, 370–372] or monographs, e.g., [373]. Nor can it cover the up-to-date understanding of fine details that are reported in a huge number of original papers. The lack of space is the only excuse for the author to omit many important new discoveries in this vivid field of research.

The Basics in a Nutshell

- Electrical breakdown of a gas by an applied dc voltage occurs when the ions produced in an initial ionization avalanche can liberate new electrons from the cathode. The ignition condition for a gap of width d is given by $1 = \gamma(e^{\alpha d} - 1)$ with the secondary yield γ and gas multiplication factor α.
- Paschen's law states that, for any combination of gas and cathode material, there is a minimum breakdown voltage at a specific value of $p\,d$.
- A dc glow discharge is characterized by the cathode fall, the negative glow and, for sufficient length, the positive column. The negative glow is produced by energetic electrons from the cathode fall.
- The particle balance of the positive column is determined by radial losses from ambipolar diffusion and ionization by electrons. The electron temperature is determined by the energy gain of an electron in the axial electric field between two elastic collisions, and is therefore a function of E/p.
- The parallel-plate discharge provides two operation regimes, a low-power regime in which electrons are heated by sheath expansion, and a high-power regime where secondary electrons released from the electrode gain energy from the voltage drop across the sheath.
- Asymmetric parallel-plate discharges with a blocking capacitor develop a large voltage drop across the sheath of the smaller electrode, called self-bias. Therefore, ions can gain considerable energy in the sheath and hit a substrate with a highly directive flow. A synergism of directed ion bombardment and chemical etching leads to highly effective, anisotropic etching of deep trenches in silicon.
- At high frequencies, self bias results from a capacitive voltage divider formed by the two sheath capacitances and the condition of sheath collapse

at both electrodes. For low frequencies, the diode-like characteristic of the sheath rectifies the rf voltage and charges the blocking capacitor.

- The bulk plasma of a parallel-plate discharge can be represented by a combination of an inductor (representing electron inertia) in series with a resistor (representing electron-neutral collisions), and a parallel capacitor (representing the electric field in the bulk plasma).
- Inductively coupled plasmas represent an air-core transformer consisting of a spiral coil and a ring current in the skin-layer of the plasma. This is the operation in the inductive, or H-mode. At low rf power, an electric or E-mode is found, where the electric field of the spiral coil penetrates the plasma and accelerates electrons.

Problems

11.1 In the gap between parallel electrodes, an electron avalanche is triggered by a first electron generated by cosmic rays at the negative electrode. When does the electric current start to flow in the outer circuit? Sketch the expected current vs. time.

11.2 The afterglow of a long cylindrical plasma (of radius a) is described by the time-dependent diffusion equation

$$\frac{\partial n}{\partial t} - D_\mathrm{a}\left(\frac{\partial^2 n}{\partial r^2} + \frac{1}{r}\frac{\partial n}{\partial r}\right) = 0 \, .$$

(a) Separate the variables by the product-ansatz $n(r,t) = R(r)T(t)$ and show that the time dependence can be written as $T(t) = \exp(-t/\tau)$.
(b) Show that the radial profile can be a Bessel profile of the type $R(r) \propto J_0(r/\ell)$, and determine ℓ from the boundary condition $R(a) = 0$.
(c) Show that the decay time is $\tau = D_\mathrm{a}^{-1}(a/2.405)^2$.

11.3 Consider the impedance (11.17) of the bulk plasma in a parallel-plate rf discharge in the limit $\omega^2 \ll \omega_\mathrm{pe}^2$ and $\omega\nu_\mathrm{m} \ll \omega_\mathrm{pe}^2$. Compare the sum of the impedances of the inductor L_b and resistor R_b in the equivalent network with the impedance of the capacitor C_0, and show that the current in the capacitor is negligible.

11.4 What is the collisionless skin depth for a plasma of $n_\mathrm{e} = 10^{17}\,\mathrm{m}^{-3}$ density?

Glossary

Symbol	Definition	Equation	Section
a	minor radius of a torus	(3.53)	3.6.3
a	acceleration	(4.55)	4.4.2.1
a	radius of a dust particle		10.1.3.2
a_{WS}	Wigner-Seitz radius	(2.16)	2.1.3
A_{br}	coefficient for bremsstrahlung	(4.64)	4.4.2.3
A_R	coefficient in Richardson's law	(11.10)	11.1.5
b	impact parameter		4.2.5
b_c	impact parameter for charge collection	(10.5)	10.1.3
b_{90}	impact parameter for 90° deflection	(4.20)	4.2.5
B_{bc}, B_{cb}	Einstein coefficients		8.1.1
B_m	maximum magnetic field in a mirror		3.4.2
B_p	poloidal magnetic field	(3.55)	3.6.1
B_r	radial magnetic field		4.3.5
B_t	toroidal magnetic field	(3.52)	3.6
B_{tot}	total magnetic mirror field		3.4.2
$\mathbf{B}$	magnetic field vector		3.1
$\hat{\mathbf{B}}$	Fourier amplitude of $\mathbf{B}$	(6.8)	6.1.2
$b_{\pi/2}$	impact parameter for 90° deflection		10.2.4
C	capacitance of a dust particle	(10.10)	10.1.3.2
C_0	capacitance of bulk plasma	(11.16)	11.2.1
C_L	sound speed of longitudinal mode	(10.93)	10.4.1.5
c_s	sound velocity	(6.79)	6.5.3
C_s	ion acoustic velocity	(6.78)	6.5.3
C_T	sound speed of transverse mode		10.4.1.5
D	diffusion coefficient	(4.31)	4.3.3
D	Stix parameter	(6.90)	6.6
D	spring constant	(10.87)	10.4.1.1
$D(\omega, \mathbf{k})$	dispersion function	(6.29)	6.2
D_a	ambipolar diffusion coefficient	(4.36)	4.3.3.1

A. Piel, *Plasma Physics*, DOI 10.1007/978-3-642-10491-6

d_0	equilibrium distance of dust particles	(10.68)	10.2.5
$d_{\rm n}$	thickness of the normal cathode fall	(11.8)	11.1.3
$\hat{\mathbf{D}}_\omega$	Fourier coefficient of displacement vector	(6.12)	6.1.3
$E_{\rm a}$	ambipolar electric field	(4.37)	4.3.3.1
E_p	electric field at grid point p	(9.81)	9.4.1
$\mathbf{E}$	electric field vector		3.1
$\hat{\mathbf{E}}$	Fourier amplitude of **E**	(6.8)	6.1.2
$f_{\rm b}$	burn fraction of a pellet	(4.74)	4.4.3.4
$F_{\rm c}$	collection force	(10.46)	10.2.4
$F_{\rm defl}$	deflection force		8
$f_{\rm e1}$	perturbed electron distribution function	(9.35)	9.3.1
F_g	gravitational force on dust grain		10.2.1
F_i	force on i-th particle	(9.82)	9.4.1
f_m	resonance frequencies of a cavity	(6.58)	6.4.4
$F_{\rm i}$	ion wind force	(10.45)	10.2.4
$f_{\rm M}$	Maxwell distribution, general	(2.10)	2.1.1
$f_{\rm M}(v)$	Maxwell distribution of speeds	(4.6)	4.1
$f_{\rm M}(v_x)$	shifted Maxwellian	(5.5)	5.1.2
$f_{\rm M}^{(1)}$	1D Maxwell distribution	(4.1)	4.1
$f_{\rm M}^{(3)}$	3D Maxwell distribution	(4.5)	4.1
$F_{\rm M}(W)$	Maxwell distribution of energies	(4.9)	4.1.4
$\mathbf{F}_{\rm n}$	Epstein drag force	(10.41)	10.2.2
$F_{\rm o}$	orbit force	(10.58)	10.2.4
$F_{\rm p}$	ponderomotive force	(3.63)	3.7.1
$F_{\rm rest}$	restoring force		8
$\mathbf{F}_{\rm tp}$	thermophoretic force	(10.43)	10.2.3
$F_{\rm trap}$	force exerted by potential trap	(10.66)	10.2.5
$F_{\rm Y}$	Yukawa interaction force	(10.64)	10.2.5
$f_+(x, v)$	part of distribution function	(9.25)	9.2.1
$f_-(x, v)$	part of distribution function	(9.27)	9.2.1
$F_<, F_>$	force from inside and outside	(10.75)	10.3.4
g	gravitational acceleration on Earth	(10.32)	10.2.1.1
$g_{\rm L}$	Lunar gravitational acceleration		10.1.2.1
g_i	degeneracy factor of atomic level i		2.1.1
$g(T)$	used in calculation of burn fraction	(4.74)	4.4.3.4
$H_{\rm p}$	poloidal magnetic field	(3.54)	3.6.1
$H_{\rm t}$	toroidal magnetic field	(3.51)	3.6
H_0	scale height	(1.2)	
I	moment of inertia	(9.85)	9.4.2
$I_{\rm e,sat}$	electron saturation current of a probe	(7.29)	7.4.2
$I_{\rm i,sat}$	ion saturation current of a probe	(7.27)	7.4.1
$I_{\rm p}$	probe current		7.4
I_N	particle flux	(5.14)	5.1.4
I_P	momentum flux	(5.16)	5.1.4

I_{ph}	Photoelectron current	(10.3)	10.1.2
j	current density	(4.29)	4.3.2
j_{e}	electron current density	(4.29)	4.3.2
j_{i}	ion current density	(4.29)	4.3.2
$j_{\text{max,Pierce}}$	maximum current in a Pierce diode	(8.39)	
j_{t}	toroidal current density	(3.53)	3.6.3
$\hat{\mathbf{j}}$	Fourier amplitude of $\mathbf{j}$	(6.8)	6.1.2
J	longitudinal invariant		3.4.3
J_0	initial angular momentum		7.5.4
k	wavenumber	(6.8)	6.1.2
K	normalized wavenumber	(10.90)	10.4.1.2
k_{I}	spatial growth rate	(8.20)	8.1.5
k_{B}	Boltzmann's constant		2.1
$\mathbf{kk}$	tensor product of wave vecors	(6.27)	6.2
k_+, k_-	wavenumbers of beam modes	(8.26)	8.3.2
ℓ	length scale		2.2.2
L_{b}	inductance of bulk plasma	(11.17)	11.2.1
m_{a}	mass of an atom	(4.49)	4.4.1
m_{e}	electron mass		
m_{i}	ion mass		
m^*	effective mass describing collisions	(6.43)	6.3.2
m_{d}	dust mass	(10.32)	10.2.1.1
m_{e}^*	effective electron mass		6.3.2
$m_\odot$	solar mass	Table 1.1	1.2
M	Mach number	(7.18)	7.3.2
$\mathbf{M}$	magnetic dipole moment		3.1.3
M_{s}	total mass of spherical pellet	(4.75)	4.4.3.4
n	density		2.1
n_{co}	cut-off density	(6.50)	6.4
n_{e}	electron density		2.1
n_{e0}	unpertubed electron density		2.2.1
n_{i}	ion density		2.1
n_{i0}	unperturbed ion density		2.2.1
n_{A^+}	ion density		2.1.2
n_i	population density of atomic state i	(2.10)	2.1.1
N	number of particles		
$N(r)$	number of particles inside r	(10.72)	10.3.4
N_{a}	number of atomic targets	(4.13)	4.2.2
N_{De}	number of particles in electron Debye sphere	(2.33)	2.3.1
$N^{(\alpha)}$	total number of particles of species α	(9.5)	9.1.1
$\mathcal{N}$	refractive index	(6.24)	6.1.6
$\mathcal{N}_{\text{L}}$	refractive index of L-mode	(6.96)	6.6.2
$\mathcal{N}_{\text{R}}$	refractive index of R-mode	(6.95)	6.6.2

p	pressure		2.1
P	Stix parameter	(6.90)	6.6
P	Havnes parameter	(10.27)	10.1.6
$p(t)$	time-dependent probability	(4.44)	4.3.4
P_{br}	Power in bremsstrahlung	(4.59)	4.4.2.1
P_{DT}	Power from D–T reaction	(4.63)	4.4.2
$P_{\mathrm{e}}, P_{\mathrm{i}}$	probability of capturing an electron (ion)	(10.18)	10.1.5
P_{fus}	Power from fusion reaction		4.4.2.5
P_{ij}	stress tensor	(5.29)	5.1.5
P_{H}	heat loss rate	(4.62)	4.4.2
p_{mag}	magnetic pressure	(5.47)	5.2.2
P_x	x-component of momentum vector	(5.23)	5.1.4
Q	electric charge		2.2.1
Q	quality factor of a resonator		6.4.4
q_{d}	dust charge	(10.11)	10.1.3.2
Q_{DT}	fusion yield of D–T reaction		4.4.2.2
Q_{α}	energy of α-particles		4.4.2.4
r	radius		2.2.1
r_{L}	Larmor radius	(3.5)	3.1.2
R	major radius of a torus		3.6.3
R	equivalent resistance of charging characteristic	(10.15)	10.1.4
R_{b}	resistance of bulk plasma	(11.17)	11.2.1
r_{C}	Coulomb radius	(10.55)	10.2.4
$R(T)$	recombination rate		2.1.2
R_{c}	radius of curvature		5.2.2
R_{C}	radius of a Coulomb ball	(10.81)	10.3.4
R_{hs}	radius of hot spot	(4.70)	4.4.3.3
R_{m}	mirror ratio		3.4.2
R_1, R_2	electrical resistors		2.2.1
R_{total}	resistance of parallel circuit		2.2.1
s	distance		2.2.1
S	entropy	(2.3)	2.1.1.1
S	Stix parameter	(6.90)	6.6
s_1, s_2	sheath thickness	(11.18)	11.2.2
$S(\mathbf{k}, \omega)$	spectral energy density	(10.96)	10.4.2
$S(W_{\lambda})$	energy density of solar spectrum	(10.3)	10.1.2
S_{ion}	ionization rate	(4.18)	4.2.3
$S(T)$	ionization rate		2.1.2
T	temperature		
$\mathscr{T}$	tension		
T_{d}	dust temperature		10.4.3
T_{e}	electron temperature		2.1
T_{i}	ion temperature		2.1
Δt_k	variable time step for charge capture	(10.20)	10.1.5

Symbol	Description	Equation	Section
T_{ph}	effective temperature of photoelectrons	(10.4)	10.1.2
$\mathbf{u}$	drift velocity		5.1.1
U_1, U_2	sheath voltage	(11.22)	11.2.2
u_{i}	ion streaming velocity		7.2
U_{b}	voltage drop in bulk plasma	(11.16)	11.2.1
U_{bd}	breakdown voltage	(11.6)	11.1.3.1
U_{dc}	self bias voltage	(11.27)	11.2.4
U_{ind}	induction voltage		5.1.1
U_{n}	normal cathode fall voltage		11.1.3
U_{p}	probe voltage		7.4
U_{rf}	radio-frequency voltage	(11.27)	11.2.4
v	velocity		
$\mathbf{v}$	velocity vector		3.1
v_x, v_y, v_z	components of the velocity vector		3.1
$\bar{v}_x, \bar{v}_y$	average velocities	(4.46)	4.3.4
v_{A}	Alfvén velocity	(5.83)	5.3.5
v_{b}	velocity of a beam electron	(8.11)	8.1.4
v_{c}	cut-off velocity	(9.24)	9.2.1
v_{B}	Bohm velocity	(7.17)	7.3.2
v_{d}	drift velocity	(4.27)	4.3.1
v_{D}	diamagnetic drift velocity	(5.50)	5.2.3
v_E	E×B drift velocity	(3.12)	3.1.4
v_{exhaust}	exhaust velocity		4.4.3.2
v_{g}	gravitational drift velocity	(3.14)	3.1.5
$\mathbf{v}_{\mathrm{gr}}$	group velocity vector	(6.23)	6.1.5
v_{m}	mass flow velocity	(5.56)	5.3.1
v_{p}	polarisation drift velocity	(3.47)	3.5.1
v_R	curvature drift velocity	(3.23)	3.2.3
v_{shell}	velocity of imploding shell	(4.69)	4.4.3.2
v_{th}	mean thermal speed	(4.7)	4.1.3
v_T	most probable speed	(4.4)	2.3.1
v_0	initial beam velocity	(8.3)	8.1.2
$v_+, v_-, \bar{v}$	contributions to the Pierce mode	(8.33)	8.3.3
$v_\perp$	perpendicular velocity		3.4.2
$v_{\nabla B}$	gradient drift velocity	(3.21)	3.2.2
$\mathbf{v}_\varphi$	phase velocity	(6.18)	6.1.4
W	energy		2.1.1
$W(x)$	charge assignment function	(9.79)	9.4.1
W_{b}	binding energy		
W_{ex}	excess energy		
W_0	initial energy		7.5.4
W_{ion}	ionisation energy		4.2.4
W_{kin}	kinetic energy	(4.8)	4.1.3
W_{m}	maximum energy in secondary emission	(10.2)	10.1.1
$\langle W_{\mathrm{kin}} \rangle$	mean beam kinetic energy	(8.17)	8.1.4

W_{pot}	potential energy	(9.84)	9.4.2
W_R	work function in Richardson's law	(11.10)	11.1.5
W_{tot}	total energy	(8.40), (9.86)	9.4.2
W_Y	energy of Yukawa interaction	(10.63)	10.2.5
$W_\perp$	energy of perpendicular motion	(3.49)	3.5.2
x_{min}	position of potential minimum in diode	(9.25)	9.2.1
Z	partition function		2.1.1
Z_b	impedance of bulk plasma	(11.16)	11.2.1
Z_d	dust charge number = $\lvert q_d \rvert / e$	(10.13)	10.1.3.2
Z_j	charge number of species j		2.2.2
α	index denoting particle species	(6.80)	6.6
α	Townsend's electron multiplication factor	(11.1)	11.1.3.1
α_b	beam fraction in beam-plasma system	(8.3)	8.1.2
α_P	Pierce parameter	(8.37)	8.3.1
$\alpha(L)$	angle of Faraday rotation	(6.100)	6.6.2.1
β	ratio of kinetic and total pressure	(5.51)	5.2.3
γ	growth rate		8.1.3
γ	coefficient for secondary emission by ion impact	(11.4)	11.1.3.1
Γ	coupling parameter	(2.15)	2.1.3
Γ_a	ambipolar flux	(4.34)	4.3.3.1
$\Gamma_{e,i}$	flux of electrons (ions)	(4.31)	4.3.3
δ	accomodation coefficient for Epstein friction	(10.42)	10.2.2
δ_0	skin depth	(11.41)	11.3.1
δ_{cl}	collisionless skin depth	(11.43)	11.3.1
$\delta(W)$	secondary emission coefficient	(10.1)	10.1.1
δ_m	maximum of $\delta(W)$	(10.2)	10.1.1
ε	dielectric function for cold plasma	(6.45)	6.4
ε	...for beam-plasma system	(8.2)	8.1.2
ε	...for ion-acoustic wave	(6.75)	6.5.3
$\boldsymbol{\varepsilon}$	dielectric tensor	(6.89)	6.6
$\varepsilon_{xx}, \varepsilon_{yy}, \varepsilon_{zz}$	elements of the dielectric tensor	(6.34)	6.3.1
η	resistivity	(4.22)	4.2.5
η	efficiency of energy conversion	(4.63)	4.4.2
η_c	normalized cloud potential	(10.25)	10.1.6
η_f	normalized grain potential	(10.25)	10.1.6
η_s	Spitzer resistivity	(4.23)	4.2.5
θ	angle w.r.t. magnetic field direction		3.4.2
θ	normalized frequency for Pierce modes		8.3.3
θ_c	resonance cone angle	(6.105)	6.7
θ_m	loss cone angle	(3.56)	3.6.3
λ	wavelength		6.4
λ	Lagrange multiplier		2.1.1.1
$\ln(\Lambda)$	Coulomb logarithm	(10.61)	10.2.4

λ_B	de Broglie wavelength	(2.36)	2.3.2
λ_D	Debye length		2.2.1
λ_{De}	Electron Debye length	(2.28)	2.2.1
λ_{Di}	Ion Debye length	(2.28)	2.2.1
λ_{max}	maximum of Planck curve	(2.14)	2.1.2
λ_{min}	shortest wavelength in crystal		10.4.1.1
λ_{mfp}	mean free path	(4.14)	4.2.2
λ_s	modified shielding length	(10.56)	10.2.4
μ	Lagrange multiplier		2.1.1.1
μ	magnetic moment	(3.34)	3.3.2
μ	ion-to-electron mass ratio		
μ	Mach cone angle	(10.94)	10.4.1.7
μ_e	electron mobility	(4.28)	4.3.1
μ_i	ion mobility	(4.28)	4.3.1
ν_{coll}	collision frequency	(4.15)	4.2.2
ν_{DT}	deuterium-tritium collision frequency	(4.60)	4.4.2.2
ν_{ei}	electron–ion collision frequency	(4.21)	4.2.5
ν_{ion}	ionization frequency	(4.17)	4.2.3
ν_m	momentum loss frequency	(4.25)	4.3.1
$\xi(W_\lambda)$	photoelectric efficiency	(10.3)	10.1.2
Ξ_d	normalized dust collision frequency	(10.90)	10.4.1.2
ρ	charge density	(5.1)	5.1.1 / 5.1.3
ρ_m	mass density in MHD	(5.38)	5.2
ρ_d	mass density of dust material		10.2.1
ρ_p	assigned charge to gridpoint p	(9.78)	9.4.1
ρ_0	initial mass density		4.4.3.4
ρR	density-radius product for a pellet		4.4.3.4
σ	cross section	(4.13)	4.2.2
σ	conductivity		4.3.2
σ_c	cross section for charge collection	(10.5)	10.1.3
$\sigma_{e,i}$	electron (ion) conductivity	(4.30)	4.3.2
$\sigma_{xx}, \sigma_{yy}, \sigma_{zz}$	elements of conductivity tensor	(6.32)	6.3.1
$\boldsymbol{\sigma}$	conductivity tensor	(4.47)	4.3.4
σ_{ion}	cross-section for ionization	(4.16)	4.2.3
τ	elapsed time	(9.20)	9.2.1
τ	relaxation time of dust charge	(10.16)	10.1.4
τ	temperature ratio T_e/T_i	(10.26)	10.1.6
τ_B	magnetic diffusion time	(5.62)	5.3.2
τ_c	inertial confinement time	(4.68)	4.4.3.1
τ_E	energy confinement time		4.4.2.5
φ	phase angle of a wave	(6.16)	6.1.4
φ_{air}, φ_p	phase angle in air (plasma)		6.4.3
φ_1,φ_2	phase angles of second harmonic		6.4.3
Φ	electric potential		2.2.1 / 7.2
Φ_f	floating potential (probe)	(7.34)	7.4.4

Φ_f	floating potential (dust grain)	(10.8)	10.1.3.1
Φ_p	potential at gridpoint p	(9.80)	9.4.1
Φ_p	plasma potential		7.4
Φ_m	magnetic flux		5.1.1
Φ_min	minimum potential in a diode	()9.28	9.2.2
χ	empirical constant for photoemission	(10.4)	10.1.2
χ	scattering angle	(10.57)	10.2.4
χ_b	permittivity of beam electrons	(8.3)	8.1.2
χ_d	permittivity of dust system	(10.99)	10.4.3
χ_e	permittivity of cold electrons	(8.21)	8.2
χ_i	permittivity of ions	(8.21)	8.2
χ_p	permittivity of plasma electrons	(8.2)	8.1.2
ψ	angle between **k** and **B**	(6.91)	6.6.1
ω	angular frequency		3.1 / 6.6.2
Ω	normalized frequency		10.4.1.2
ω_br	breathing frequency of dust cluster	(10.69)	10.2.5
ω_c	cyclotron frequency	(3.4)	3.1.2
ω_ce	electron cyclotron frequency		3.1.2
ω_ci	ion cyclotron frequency	(6.84)	6.6.1
ω_I	imaginary part = growth rate	(8.9)	8.1.2
Ω_L	normalized frequency of long. mode	(10.90)	10.4.1.2
ω_lh	lower hybrid frequency	(6.104)	6.6.3
ω_pd	dust plasma frequency	(10.98)	10.4.3
ω_pe	electron plasma frequency	(2.32)	2.2.3
ω_pi	ion plasma frequency	(6.75)	6.5.3
ω_R	real part of wave frequency	(8.8)	8.1.2
Ω_T	norm. frequency of transverse mode	(10.91)	10.4.1.2
ω_uh	upper hybrid frequency	(6.103)	6.6.3
ω_0	plasma frequency for strongly coupled system	(10.89)	10.4.1.1

Appendix: Constants and Formulas

"Reeling and Writhing, of course, to begin with," the Mock Turtle replied; "and then the different branches of Arithmetic—Ambition, Distraction, Uglification and Derision."

Lewis Carroll, Alice in Wonderland

1 Physical Constants

The physical constants are given here in SI-units with four digits accuracy. For problem solving in plasma physics, often two digits are sufficient in view of the uncertainty of measured plasma parameters.

Table 1 Physical constants

Electron mass	m_e	$= 9.109 \times 10^{-31}$	kg
Proton mass	m_p	$= 1.673 \times 10^{-27}$	kg
Proton-electron mass ratio	m_p/m_e	$= 1836$	
Elementary charge	e	$= 1.602 \times 10^{-19}$	A s
Specific charge of electron	e/m_e	$= 1.759 \times 10^{11}$	$\mathrm{C\,kg^{-1}}$
Speed of light in vacuum	c	$= 2.998 \times 10^{8}$	$\mathrm{m\,s^{-1}}$
Permittivity of free space	ε_0	$= 8.854 \times 10^{-12}$	$\mathrm{A\,s\,V^{-1}\,m^{-1}}$
	μ_0	$= 1.257 \times 10^{-6}$	$\mathrm{V\,s\,A^{-1}\,m^{-1}}$
Boltzmann constant	k_B	$= 1.381 \times 10^{-23}$	$\mathrm{J\,K^{-1}}$
Temperature associated with 1 eV	e/k_B	$= 11{,}600$	$\mathrm{K\,V^{-1}}$
Stefan-Boltzmann constant	σ	$= 5.670 \times 10^{-8}$	$\mathrm{Wm^{-2}K^{-4}}$
Planck's constant	h	$= 6.626 \times 10^{-34}$	Js
	$\hbar = h/2\pi$	$= 1.055 \times 10^{-34}$	Js
Avogadro's constant	N_A	$= 6.022 \times 10^{23}$	$\mathrm{mol^{-1}}$
Molar gas constant	R	$= 8.314$	$\mathrm{J\,K^{-1}\,mol^{-1}}$
Standard temperature		273.15	K
Earth mass	M_E	$= 5.974 \times 10^{24}$	kg
Mean Earth radius	R_E	$= 6.371 \times 10^{6}$	m
Universal gravitational constant	G	$= 6.673 \times 10^{-11}$	$\mathrm{Nm^2kg^{-2}}$
Gravitational acceleration	$g = GM_E/R_E^2$	$= 9.807$	$\mathrm{ms^{-2}}$

2 List of Useful Formulas

This compilation was inspired by the NRL plasma formulary, which is available free of charge at `http://wwwppd.nrl.navy.mil/nrlformulary/`. However, SI-units are used here for consistency with the rest of the book. Temperatures are given in eV, as indicated.

2.1 Lengths

The mass number of an ion is $\mu = m_\mathrm{i}/m_\mathrm{p}$.

- electron Debye length

$$\lambda_\mathrm{De,Di} = 7.43 \times 10^3\,\mathrm{m}\sqrt{\frac{T_\mathrm{e,i}(\mathrm{eV})}{n_\mathrm{e,i}(\mathrm{m}^{-3})}}$$

- thermal electron gyroradius

$$r_\mathrm{Le} = \frac{v_\mathrm{Te}}{\omega_\mathrm{ce}} = 3.37 \times 10^{-6}\,\mathrm{m}\,\frac{\sqrt{T_\mathrm{e}(\mathrm{eV})}}{B(\mathrm{T})}$$

- thermal ion gyroradius

$$r_\mathrm{Li} = \frac{v_\mathrm{Ti}}{\omega_\mathrm{ci}} = 1.45 \times 10^{-4}\,\mathrm{m}\,\frac{\sqrt{\mu T_\mathrm{i}(\mathrm{eV})}}{B(\mathrm{T})}$$

2.2 Frequencies

- electron plasma frequency

$$\omega_\mathrm{pe} = \sqrt{\frac{n_\mathrm{e}e^2}{\varepsilon_0 m_\mathrm{e}}} = 56.4\,\mathrm{s}^{-1}\sqrt{n_\mathrm{e}(\mathrm{m}^{-3})}$$

- ion plasma frequency

$$\omega_\mathrm{pi} = \sqrt{\frac{n_\mathrm{i}Z^2e^2}{\varepsilon_0 m_\mathrm{i}}} = 1.32\,\mathrm{s}^{-1}\,Z\sqrt{\frac{n_\mathrm{i}(\mathrm{m}^{-3})}{\mu}}$$

- electron gyrofrequency

$$\omega_\mathrm{ce} = \frac{eB}{m_\mathrm{e}} = 1.76 \times 10^{11}\mathrm{s}^{-1}\,B(\mathrm{T})$$

- ion gyrofrequency

$$\omega_{ci} = \frac{ZeB}{m_i} = 9.58 \times 10^7 \mathrm{s}^{-1} \frac{Z}{\mu} B\,(\mathrm{T})$$

2.3 Velocities

- electron thermal speed

$$v_{Te} = \sqrt{\frac{2k_B T_e}{m_e}} = 5.93 \times 10^5\,\mathrm{m\,s^{-1}} \sqrt{T_e(\mathrm{eV})}$$

- ion thermal speed

$$v_{Ti} = \sqrt{\frac{2k_B T_i}{m_i}} = 1.38 \times 10^4\,\mathrm{m\,s^{-1}} \sqrt{\frac{T_i(\mathrm{eV})}{\mu}}$$

- ion sound speed

$$C_s = \sqrt{\frac{k_B(T_e + 3T_i)}{m_i}} = 9.79 \times 10^3\,\mathrm{m\,s^{-1}} \sqrt{\frac{(T_e + 3T_i)(\mathrm{eV})}{\mu}}$$

- Alfvén speed

$$v_A = \frac{B}{\sqrt{\mu_0 n_i m_i}} = 2.18 \times 10^{16}\,\mathrm{m\,s^{-1}} \frac{B(\mathrm{T})}{\sqrt{\mu\, n_i(\mathrm{m^{-3}})}}$$

3 Useful Mathematics

3.1 Vector Relations

The definitions of dot product and vector product of two vectors can be summarized as follows:

$$\mathbf{A}\cdot\mathbf{B} = A_xB_x + A_yB_y + A_zB_z$$

$$\mathbf{A}\times\mathbf{B} = \begin{vmatrix} \mathbf{e}_x & \mathbf{e}_y & \mathbf{e}_z \\ A_x & A_y & A_z \\ B_x & B_y & B_z \end{vmatrix}$$

$$= \mathbf{e}_x(A_yB_z - A_zB_y) + \mathbf{e}_y(A_zB_x - A_xB_z) + \mathbf{e}_z(A_xB_y - A_yB_x)$$

$$\mathbf{A}\cdot\mathbf{B} = \mathbf{B}\cdot\mathbf{A}$$

$$\mathbf{A}\times\mathbf{B} = -\mathbf{B}\times\mathbf{A}$$

The operator $\mathbf{A}\cdot\nabla$ is a scalar operator

$$\mathbf{A}\cdot\nabla = A_x\frac{\partial}{\partial x} + A_y\frac{\partial}{\partial y} + A_z\frac{\partial}{\partial z}\,.$$

The gradient of a vector function is defined as a tensor, which can be written as a matrix (see next paragraph)

$$\nabla\mathbf{A} = \begin{pmatrix} \frac{\partial A_x}{\partial x} & \frac{\partial A_x}{\partial y} & \frac{\partial A_x}{\partial z} \\ \frac{\partial A_y}{\partial x} & \frac{\partial A_y}{\partial y} & \frac{\partial A_y}{\partial z} \\ \frac{\partial A_z}{\partial x} & \frac{\partial A_z}{\partial y} & \frac{\partial A_z}{\partial z} \end{pmatrix}$$

The following list gives some standard rules for operations involving multiple dot products, vector products, and derivatives. Note that differential operators act only on the term on their r.h.s., A_c means that A is not affected by the differential operator. f is a scalar function.

$$\mathbf{A}\cdot(\mathbf{B}\times\mathbf{C}) = \mathbf{B}\cdot(\mathbf{C}\times\mathbf{A}) = \mathbf{C}\cdot(\mathbf{A}\times\mathbf{B})$$

$$\mathbf{A}\times(\mathbf{B}\times\mathbf{C}) = (\mathbf{A}\cdot\mathbf{C})\mathbf{B} - (\mathbf{A}\cdot\mathbf{B})\mathbf{C}$$

$$\nabla(f\,g) = f\nabla g + g\nabla f$$

$$\nabla\cdot(f\,\mathbf{A}) = f\nabla\cdot\mathbf{A} + \mathbf{A}\cdot\nabla f$$

$$\nabla\times(f\,\mathbf{A}) = f\nabla\times\mathbf{A} + \nabla f\times\mathbf{A}_c$$

$$\nabla\cdot(\mathbf{A}\times\mathbf{B}) = \mathbf{B}\cdot(\nabla\times\mathbf{A}) - \mathbf{A}\cdot(\nabla\times\mathbf{B})$$

$$\nabla\times(\mathbf{A}\times\mathbf{B}) = \mathbf{A}(\nabla\cdot\mathbf{B}) - \mathbf{B}(\nabla\cdot\mathbf{A}) + (\mathbf{B}\cdot\nabla)\mathbf{A} - (\mathbf{A}\cdot\nabla)\mathbf{B}$$

$$\nabla(\mathbf{A}\cdot\mathbf{B}) = (\mathbf{A}\cdot\nabla)\,\mathbf{B} + (\mathbf{B}\cdot\nabla)\,\mathbf{A} + \mathbf{A}\times(\nabla\times\mathbf{B}) + \mathbf{B}\times(\nabla\times\mathbf{A})$$

$$\mathbf{A}\times(\nabla\times\mathbf{B}) = (\nabla\mathbf{B})\cdot\mathbf{A}_\mathrm{c} - (\mathbf{A}\cdot\nabla)\mathbf{B}$$

$$\Delta\mathbf{A} = \nabla(\nabla\cdot\mathbf{A}) - \nabla\times(\nabla\times\mathbf{A})$$

$$\nabla\cdot(\nabla\times\mathbf{A}) = 0$$

$$\nabla\times(\nabla f) = 0$$

3.2 Matrices and Tensors

Matrices describe the mapping of a vector into a new vector, usually of different length and direction. A matrix is a special tensor of rank two. There are also tensors of higher rank, which we, however, omit from the discussion.

The product of a 3×3 matrix M and a vector $\mathbf{A}$ can be written as

$$\mathsf{M} \cdot \mathbf{A} = \begin{pmatrix} M_{xx} & M_{xy} & M_{xz} \\ M_{yx} & M_{yy} & M_{yz} \\ M_{zx} & M_{zy} & M_{zz} \end{pmatrix} \cdot \begin{pmatrix} A_x \\ A_y \\ A_z \end{pmatrix} = \begin{pmatrix} M_{xx}A_x + M_{xy}A_y + M_{xz}A_z \\ M_{yx}A_x + M_{yy}A_y + M_{yz}A_z \\ M_{zx}A_x + M_{zy}A_y + M_{zz}A_z \end{pmatrix} .$$

The elements of the resulting vector are the dot product (scalar product) of a row of the matrix M with the vector $\mathbf{A}$.

Matrices can also be used to represent systems of linear equations. The following homogeneous system of equations

$$\begin{aligned} 3x + 5y - 2z &= 0 \\ 2x - 3y + z &= 0 \\ 5x + 2y - z &= 0 \end{aligned}$$

can be rewritten in matrix form as

$$\begin{pmatrix} 3 & 5 & -2 \\ 2 & -3 & 1 \\ 5 & 2 & -1 \end{pmatrix} \cdot \begin{pmatrix} x \\ y \\ z \end{pmatrix} = 0 .$$

This system of equations has a non-zero solution, when the determinant of the matrix vanishes

$$\begin{vmatrix} 3 & 5 & -2 \\ 2 & -3 & 1 \\ 5 & 2 & -1 \end{vmatrix} = 3(3 - 2) + 5(5 + 2) - 2(4 + 15) = 0$$

The unit tensor is

$$\mathsf{I}_{\alpha\beta} = \delta_{\alpha\beta} = \begin{pmatrix} 1 & 0 & 0 \\ 0 & 1 & 0 \\ 0 & 0 & 1 \end{pmatrix} .$$

The dyadic product of two vectors $\mathbf{A}$ and $\mathbf{B}$ is a tensor of rank two containing the products of the vector components, $\mathsf{T}_{\alpha\beta} = A_\alpha B_\beta$,

$$\mathbf{A}\mathbf{B} = \mathsf{T} = \begin{pmatrix} A_xB_x & A_xB_y & A_xB_z \\ A_yB_x & A_yB_y & A_yB_z \\ A_zB_x & A_zB_y & A_zB_z \end{pmatrix} .$$

3.3 The Theorems of Gauss and Stokes

Assume that V is a volume bounded by a closed surface S, with $\mathrm{d}S$ positive outward from the enclosed volume, then

$$\oint_S \mathbf{A} \cdot \mathrm{d}S = \int_V (\nabla \cdot \mathbf{A})\,\mathrm{d}V\,.$$

If S is an open surface bounded by the contour C, of which the line element is $\mathrm{d}s$, then

$$\oint_C \mathbf{A} \cdot \mathrm{d}s = \int_S (\nabla \times \mathbf{A}) \cdot \mathrm{d}S\,.$$

Solutions

"*Forty-two!*"
Douglas Adams, The Hitchhikers Guide to the Galaxy

Problems of Chapter 2

2.1

$$\lambda_{De} = \sqrt{\frac{\varepsilon_0 k_B T_e}{n_e e^2}} = \sqrt{\frac{\varepsilon_0 k_B}{e^2}}\sqrt{\frac{T_e}{n_e}} = 69.0\,\mathrm{m}\sqrt{\frac{T_e(\mathrm{K})}{n_e(\mathrm{m}^{-3})}}$$

2.2 (a) $\lambda_{De} = \lambda_{Di} = 69\sqrt{3000 \times 10^{-12}}\,\mathrm{m} = 3.8 \times 10^{-3}\,\mathrm{m}$.
(b) Note that $3\,\mathrm{eV} \mathrel{\widehat{=}} 3 \times 11600\,\mathrm{K}$. $\lambda_{De} = 1.3 \times 10^{-4}\,\mathrm{m}$, $\lambda_{Di} = 1.2 \times 10^{-5}\,\mathrm{m}$.

2.3 Poisson's equation states that the curvature of the potential is proportional to the (negative) space charge. $\Phi'' = -en_i/\varepsilon_0$. (a) The electric field increases linearly in the positive space charge region from $E(-d) = 0$ to $E(0) = E_{max}$ and decreases in the negative space charge region to $E(d) = 0$. Hence, there is no electric field at the edges of the quasineutral plasma. The potential decreases in the positive space charge region as $\Phi(x) = -\frac{1}{2}n_i e(x+d)^2/\varepsilon_0$ and reaches $\Phi(0) = -9.0 \times 10^3$ V. In the negative space charge region, $\Phi(x)$ has the opposite curvature and reaches a final value $\Phi(d) = -18.0 \times 10^3$ V.

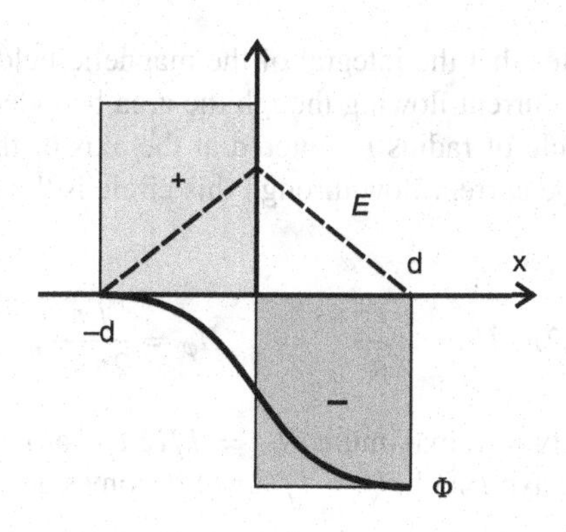

2.4 Calculate the derivatives

$$\Phi(r) = \frac{a}{r} f(r)\,, \quad \Phi' = -\frac{a}{r^2} f(r) + \frac{a}{r} f'\,, \quad \Phi'' = \frac{2a}{r^3} f - \frac{2a}{r^2} f' + \frac{a}{r} f''$$

Inserting into (2.21) gives

$$\frac{a}{r}\left[f'' - \frac{1}{\lambda_{\mathrm{D}}^2} f\right] = 0\,.$$

2.5 Inserting the Wigner-Seitz radius $a_{\mathrm{WS}} = [3/(4\pi n_{\mathrm{i}})]^{1/3}$ into (2.15) gives

$$\Gamma_{\mathrm{i}} = \frac{(4\pi/3)^{1/3}}{4\pi} \frac{e^2 n_{\mathrm{i}}^{1/3}}{\varepsilon_0 k_{\mathrm{B}} T_{\mathrm{i}}} = \frac{1}{3}\left(\frac{4\pi}{3} n_{\mathrm{i}} \lambda_{\mathrm{Di}}^3\right)^{-2/3} = \frac{1}{3} N_{\mathrm{Di}}^{-2/3}$$

2.6 Starting from the definitions $S = -k_{\mathrm{B}} \sum_i n_i \ln n_i$ and $U = \sum_i n_i W_i$ we use the thermodynamic definition of temperature $1/T = \partial S/\partial U$.

$$\frac{1}{T} = \frac{\partial S}{\partial \lambda}\frac{\partial \lambda}{\partial U} = \frac{-k_{\mathrm{B}} \sum\limits_i \left[\frac{\partial n_i}{\partial \lambda}(\lambda W_i - \ln Z) + \frac{\partial n_i}{\partial \lambda}\right]}{\sum\limits_i \frac{\partial n_i}{\partial \lambda} W_i}$$

$$= \frac{-k_{\mathrm{B}}\lambda \sum\limits_i \frac{\partial n_i}{\partial \lambda} W_i + k_{\mathrm{B}}(\ln Z - 1)\sum\limits_i \frac{\partial n_i}{\partial \lambda}}{\sum\limits_i \frac{\partial n_i}{\partial \lambda} W_i} = k_{\mathrm{B}}\lambda\,.$$

Using $\sum_i n_i = 1$, we obtain the result $\lambda = (k_{\mathrm{B}} T)^{-1}$.

Problems of Chapter 3

3.1 Ampere's law states that the integral of the magnetic field strength H along a closed path equals the current flowing though the area bounded by this path, $\oint \mathbf{H} \cdot \mathrm{d}\mathbf{s} = I$. Choose a circle of radius r centered at the axis of the wire for the path. Then, for any $r < a$ the current flow through this circle is the fraction $I\,r^2/a^2$ and we obtain

$$2\pi r H_\varphi = I \frac{r^2}{a^2} \quad \Rightarrow \quad H_\varphi = \frac{I\,r}{2\pi a^2}\,,$$

which increases linearly to a maximum $H_\varphi = I/(2\pi a)$ at $r = a$. For $r > a$ the encircled current is always I and $H_\varphi = I/(2\pi r)$ becomes a decreasing function of radius.

3.2 The current $I(r)$ flowing through a circle of radius $r < a$ determines the magnetic field

$$I(r) = 2\pi \int_0^r r' j(r') \mathrm{d}r' = 2\pi j_0 \left(\frac{r^2}{2} - \frac{r^4}{4a^2} \right) \quad \Rightarrow \quad H_\varphi = j_0 \left(\frac{r}{2} - \frac{r^3}{4a^2} \right).$$

3.3 $B = 49.4\,\mu\mathrm{T}$. (a) $\omega_{\mathrm{ce}} = 8.7 \times 10^6 \mathrm{s}^{-1}$. (b) $r_{\mathrm{ce}} = 0.22\,\mathrm{m}$.

3.4 (a) Use $\mathbf{M} \cdot \mathbf{r} = Mr \cos\theta$ to define the angle θ. The unit vector $\mathbf{e}_r = \mathbf{r}/r$.

$$B_r = \mathbf{B} \cdot \mathbf{e}_r = \frac{\mu_0}{4\pi} \frac{3r^2(\mathbf{r} \cdot \mathbf{M}) - r^2(\mathbf{r} \cdot \mathbf{M})}{r^6} = \frac{\mu_0 M}{4\pi} \frac{2\cos\theta}{r^3}$$

$$B_\theta = |\mathbf{e}_r \times \mathbf{B}| = \frac{\mu_0 M}{4\pi} \frac{\sin\theta}{r^3}.$$

(b) Solve the last line for M, insert $\theta = 90°$, $B_\theta = 30\,\mu\mathrm{T}$ and use the Earth radius plus about (100–200) km altitude for the ionosphere to obtain $M \approx 8 \times 10^{22}\,\mathrm{A\,m^{-2}}$.

3.5 At the magnetic equator, $\theta = 90°$ and the magnetic field has only a θ-component $B_\theta = (\mu_0 M/4\pi) r^{-3}$. (a) Hence, $\mathrm{d}B_\theta/\mathrm{d}r = -3(\mu_0 M/4\pi) r^{-4}$ and $R_\mathrm{c} = r/3 = (R_\mathrm{E} + 500\,\mathrm{km})/3 = 2,290\,\mathrm{km}$. (b) $v_R = 2v_{\nabla B} = 6W/(qrB)$. At $H = 500\,\mathrm{km}$ altitude, the equatorial magnetic field of $30\,\mu\mathrm{T}$ has decreased by a factor $[R_\mathrm{E}/(R_\mathrm{E} + H]^3 = 0.80$ to $24\,\mu\mathrm{T}$. Then $v_R = 0.11\,\mathrm{m\,s^{-1}}$.

3.6 Starting from the equations of motion

$$\dot{v}_x = -\frac{e}{m_\mathrm{e}} E - \omega_{\mathrm{ce}} v_y \quad \text{and} \quad \dot{v}_y = \omega_{\mathrm{ce}} v_x$$

we see that the electron velocity performs harmonic oscillations at the frequency ω_{ce} and $v_x = v_\perp \sin(\omega_{\mathrm{ce}} t)$. There is no cosine term because $v_x(0) = 0$. Then, $v_y = \omega_{\mathrm{ce}} \int v_x \,\mathrm{d}t + C = v_\perp[1 - \cos(\omega_{\mathrm{ce}} t)]$ to fulfill $v_y(0) = 0$. From $\dot{v}_x(0) = 0$ we obtain $v_\perp = -E/B$. Hence,

$$v_x = -\frac{E}{B} \sin(\omega_{\mathrm{ce}} t), \qquad v_y = \frac{E}{B} [1 - \cos(\omega_{\mathrm{ce}} t)]$$

$$x = \frac{E}{B\omega_{\mathrm{ce}}} [\cos(\omega_{\mathrm{ce}} t) - 1], \qquad y = -\frac{E}{B} t + \frac{E}{B\omega_{\mathrm{ce}}} \sin(\omega_{\mathrm{ce}} t)$$

3.7 The differential equation for a field line with the dipole source at the origin reads

$$\frac{\mathrm{d}z}{\mathrm{d}x} = \frac{2}{3}\frac{z}{x} - \frac{1}{3}\frac{x}{z}.$$

Set $z = wx$ to obtain $\mathrm{d}z/\mathrm{d}x = w + x(\mathrm{d}w/\mathrm{d}x)$. Inserting into the differential equation

$$x\frac{dw}{dx} = -\frac{1}{3}w - \frac{1}{3}\frac{1}{w} \quad \Rightarrow \frac{w dw}{w^2+1} = -\frac{1}{3}\frac{dx}{x} \quad \Rightarrow \frac{1}{2}\ln(w^2+1) = -\frac{1}{3}\ln(x)+C\,.$$

This results in $w = (-1+c_1x^{-2/3})^{1/2}$ and $z = x(-1+c_1x^{-2/3})^{1/2}$. The maximum distance $x_{\max}$ of the field line is determined by $z = 0$, which defines $c_1 = x_{\max}^{2/3}$, hence the shape of the field line becomes

$$z = \sqrt{x_{\max}^{2/3}x^{4/3} - x^2}\,.$$

Problems of Chapter 4

4.1 Let A be the normalizing factor of the velocity distribution $f_{\mathrm{M}}(|v|)$. Then

$$0 = A\frac{\mathrm{d}}{\mathrm{d}v}v^2\exp\left(-\frac{mv^2}{2k_{\mathrm{B}}T}\right) = A\left(2v - v^2\frac{mv}{k_{\mathrm{B}}T}\right)\exp\left(-\frac{mv^2}{2k_{\mathrm{B}}T}\right),$$

which requires that the expression in parantheses vanishes. This gives the desired result $v_T = (2k_{\mathrm{B}}T/m)^{1/2}$.

4.2 The mean thermal velocity is the first moment of the distribution of speeds

$$v_{\mathrm{th}} = 4\pi\left(\frac{m}{2\pi k_{\mathrm{B}}T}\right)^{3/2}\int_0^\infty v^3\exp\left(-\frac{mv^2}{2k_{\mathrm{B}}T}\right)\mathrm{d}v = \left(\frac{8k_{\mathrm{B}}T}{\pi m}\right)^{1/2}\underbrace{\int_0^\infty y\mathrm{e}^{-y}\mathrm{d}y}_{=1}$$

4.3 Reduce the degree of the velocity moment by partial integration

$$\begin{aligned}
\int_0^\infty v^4\mathrm{e}^{-av^2}\mathrm{d}v &= -\frac{1}{2a}\int_0^\infty v^3(-2av\mathrm{e}^{-av^2})\mathrm{d}v \\
&= -\frac{1}{2a}\underbrace{\left[v^3\mathrm{e}^{-av^2}\right]_0^\infty}_{=0} + \frac{3}{2a}\int_0^\infty v^2\mathrm{e}^{-av^2}\mathrm{d}v \\
&= -\frac{3}{4a^2}\underbrace{\left[v\mathrm{e}^{-av^2}\right]_0^\infty}_{=0} + \frac{3}{4a^2}\int_0^\infty \mathrm{e}^{-av^2}\mathrm{d}v = \frac{3}{8a^2}\sqrt{\frac{\pi}{a}}\,.
\end{aligned}$$

Set $a = m/2k_{\mathrm{B}}T$, then the mean square velocity becomes

$$\langle v^2 \rangle = 4\pi \left(\frac{m}{2\pi k_B T} \right)^{3/2} \frac{3\sqrt{\pi}}{8} \left(\frac{2k_B T}{m} \right)^{5/2} = 3\frac{k_B T}{m}.$$

Finally: $\frac{m}{2}\langle v^2 \rangle = \frac{3}{2}k_B T$.

4.4 Let the electron momentum before and after the collision be $\mathbf{p}_e$ and $\mathbf{p}'_e$. The scattering angle is θ and the momentum transfered to the atom $\mathbf{p}_a = \mathbf{p}_e - \mathbf{p}'_e$.

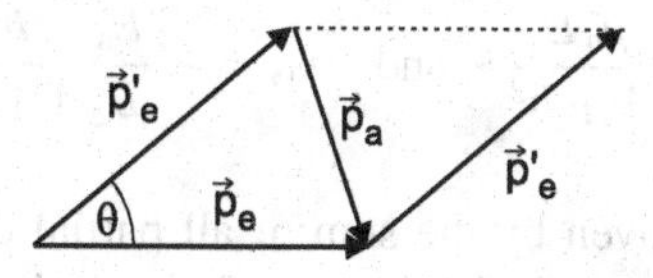

Then the kinetic energy transfered to the atom, which equals the energy loss of the electron, is (for $|\mathbf{p}'_e| \approx |\mathbf{p}_e|$)

$$\Delta W = \frac{\mathbf{p}_e^2 - 2\mathbf{p}_e \cdot \mathbf{p}'_e + \mathbf{p}'^2_e}{2m_a} \approx \frac{\mathbf{p}_e^2}{2m_e} 2\frac{m_e}{m_a} [1 - \cos(\theta)] .$$

4.5 The average fractional energy loss is

$$\left\langle \frac{\Delta W}{W} \right\rangle = 2\frac{m_e}{m_a}\frac{1}{4\pi}\int_0^{2\pi} d\varphi \int_0^{\pi} d\theta \sin\theta[1 - \cos\theta] = 2\frac{m_e}{m_a}\frac{1}{2}\int_{-1}^{1}(1-x)dx = 2\frac{m_e}{m_a} .$$

4.6 The average velocity $\bar{v}_x$ is given by

$$\begin{aligned}
\bar{v}_x &= \frac{E_x \nu_{m,i}^2}{B_z \omega_{ci}} \int_0^\infty [1 - \cos(\omega_{ci} t)] e^{-\nu_{m,i} t} dt = \frac{E_x \nu_{m,i}}{B_z \omega_{ci}} \left[1 - \nu_{m,i} \int_0^\infty \cos(\omega_{ci} t) e^{-\nu_{m,i} t} dt \right] \\
&= \frac{E_x \nu_{m,i}}{B_z \omega_{ci}} \left[1 - \frac{1}{2}\nu_{m,i} \int_0^\infty \left(e^{(i\omega_{ci} - \nu_{m,i})t} + e^{(-i\omega_{ci} - \nu_{m,i})t} \right) dt \right] \\
&= \frac{E_x \nu_{m,i}}{B_z \omega_{ci}} \left[1 + \frac{1}{2}\nu_{m,i} \left(\frac{1}{i\omega_{ci} - \nu_{m,i}} + \frac{1}{-i\omega_{ci} - \nu_{m,i}} \right) \right] \\
&= \frac{E_x \nu_{m,i}}{B_z \omega_{ci}} \frac{\omega_{ci}^2}{\omega_{ci}^2 + \nu_{m,i}^2} = \frac{E_x}{B_z} \frac{\omega_{ci}/\nu_{m,i}}{1 + (\omega_{ci}/\nu_{m,i})^2}
\end{aligned}$$

The solution for $\bar{v}_y$ is analogous, except for $\sin(\alpha) = (1/2i)(e^{i\alpha} - e^{-i\alpha})$.

4.7 The equation of motion for an "average ion" reads

$$m_i \nu_{m,i} \mathbf{v}_i = e(\mathbf{E} + \mathbf{v}_i \times \mathbf{B}) .$$

Introducing the Hall paramater $h = \omega_{ci}/\nu_{m,i}$ and using the usual convention $\mathbf{E} = (E_x, 0, 0)$ and $\mathbf{B} = (0, 0, B_z)$, we have

$$v_{ix} = \mu_i E_x + h v_{iy} \quad \text{and} \quad v_{iy} = -h v_{ix} ,$$

from which we easily obtain

$$v_{ix} = \frac{\mu_i E_x}{1+h^2} \quad \text{and} \quad v_{iy} = -\frac{E_x}{B_z}\frac{h^2}{1+h^2} .$$

4.8 The total current is given by the sum of all partial currents in ring segments $\mathrm{d}I \propto n(r) 2\pi \mathrm{d}r$. Hence, the equivalent density for a parabolic density profile $n(r) = n_0(1 - r^2/a^2)$ becomes

$$\bar{n} = \frac{1}{\pi a^2} 2\pi \int_0^a n(r) r \mathrm{d}r = \frac{2n_0}{a^2} \int_0^a \left(r - \frac{r^3}{a^2}\right) \mathrm{d}r = \frac{1}{2} n_0 .$$

4.9 Ignition occurs, when the α-heating balances the losses by Bremsstrahlung and finite particle confinement time, $P_\alpha = P_{\mathrm{br}} + P_{\mathrm{H}}$. Noting that at each DT-reaction one α-particle is generated, we have

$$P_\alpha = \frac{\eta}{1-\eta} P_{\mathrm{DT}} \quad \Rightarrow \quad \frac{\eta}{1-\eta} = \frac{Q_\alpha}{Q_{\mathrm{DT}}} \quad \Rightarrow \quad \eta = 0.154 .$$

Problems of Chapter 5

5.1 Start from $H_\varphi = I/(2\pi a)$ and $p_{\mathrm{mag}} = B^2/(2\mu_0)$. (a) Then

$$p_{\mathrm{mag}} = \frac{\mu_0 I^2}{8\pi^2 a^2} = 1.6 \times 10^8\,\mathrm{Pa} .$$

Set 1 atm $\approx$ 1 bar, then $p_{\mathrm{mag}} = 1600$ bar. (b) Note that the magnetic pressure has to balance the sum of electron and ion pressure

$$n k_{\mathrm{B}}(T_{\mathrm{e}} + T_{\mathrm{i}}) = 1.6 \times 10^8\,\mathrm{Pa} \quad \rightarrow \quad \frac{k_{\mathrm{B}} T_{\mathrm{e}}}{e} = 500\,\mathrm{V} .$$

The plasma temperature reaches 500 eV in the compressed pinch state.

5.2 $p_{\mathrm{mag}} = B^2/(2\mu_0)$ and $\beta = p_{\mathrm{kin}}(0)/p_{\mathrm{total}}$. Then $B = (2\mu_0 \beta^{-1} 2n k_{\mathrm{B}} T)^{1/2} = 3.6\,\mathrm{T}$.

5.3 The Alfvén velocity becomes

$$v_{\rm A} = 2.18 \times 10^{16}\,{\rm m\,s^{-1}}\,\frac{3}{\sqrt{2 \times 10^{20}}}. = 4.6 \times 10^{6}\,{\rm m\,s^{-1}}\,.$$

5.4 The comparison of Alfvén velocity and sound velocity gives

$$v_{\rm A} = 2.18 \times 10^{16}\,{\rm m\,s^{-1}}\,\frac{3 \times 10^{-5}}{\sqrt{16 \times 10^{12}}} = 1.6 \times 10^{5}\,{\rm m\,s^{-1}}$$

$$C_s = 9.79 \times 10^{3}\sqrt{\frac{0.256 * 4}{16}}\,{\rm m\,s^{-1}} = 2.5 \times 10^{3}\,{\rm m\,s^{-1}}\,.$$

5.5 Use $\omega_\odot = 2.7 \times 10^{-6}\,{\rm s^{-1}}$, $r_\odot = 7 \times 10^{8}$ m and $u_r = 400\,{\rm km\,s^{-1}}$. The figure displays $B(r)/B_0$ (heavy line), B_r/B_0 (dotted line) and B_φ/B_0 (short-dashed line). It is found that the magnetic field in the Parker spiral decays $\propto r^{-2}$ for $r < 1$ AU and $\propto r^{-1}$ for $r > 1$ AU.

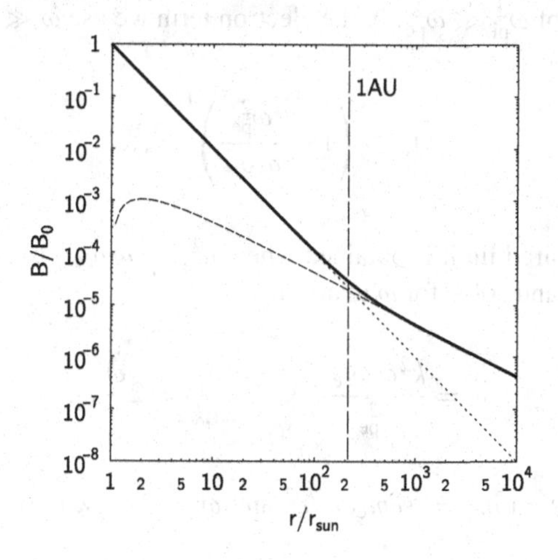

5.6 The total change in magnetic flux is given by the integral

$$\Delta\Phi_{\rm m} = \int_0^\infty [B_0(r) - B_0] 2\pi r {\rm d}r$$

$$= 2\pi\,B_0 \int_0^\infty \left[\left(1 - \frac{2\mu_0 k_{\rm B} T_{\rm e} n_0}{B_0^2}\,{\rm e}^{-(r/a)^2}\right)^{1/2} - 1\right] r\,{\rm d}r$$

$$\approx -B_0 \pi a^2 n_0 \beta \int_0^\infty x{\rm e}^{-x^2}\,{\rm d}x = -\frac{1}{2} B_0 \pi a^2 n_0 \beta\,.$$

Problems of Chapter 6

6.1 The phase and group velocities are

$$(a)\ v_\varphi = \frac{\omega}{k} = \frac{\omega_{\rm pi}\lambda_{\rm De}}{\sqrt{1+k^2\lambda_{\rm De}^2}} \quad \text{and} \quad v_{\rm gr} = \frac{{\rm d}\omega}{{\rm d}k} = v_\varphi \frac{1}{1+k^2\lambda_{\rm De}^2} .$$

(b) For $k^2\lambda_{\rm De}^2 \ll 1$, the phase velocity equals the group velocity, $v_\varphi \approx v_{\rm gr}$, and is independent of k. There is no dispersion of waves of different frequency.

6.2 Start from $v_\varphi v_{\rm gr} = c^2$. Then: $\omega\,{\rm d}\omega = c^2 k\,{\rm d}k$, which after integrating yields $\frac{1}{2}\omega^2 = \frac{1}{2}c^2k^2 + D$. Since ω is real, $\omega^2 > 0$, hence the integration constant must be positive and can be chosen as $D = \frac{1}{2}c^2k_0^2 > 0$. Then, $\omega^2 = c^2(k^2 + k_0^2)$. This means, there can be a cut-off frequency $\omega_{\rm co} = k_0 c$.

6.3 (a) Start from (6.95) for the refractive index of the R-wave. The ion term can be neglected because of $\omega_{\rm pi}^2 \ll \omega_{\rm pe}^2$. In the electron term we use $\omega \ll \omega_{\rm ce}$ to arrive at

$$\mathscr{N}_{\rm R} \approx \left(1 + \frac{\omega_{\rm pe}^2}{\omega\omega_{\rm ce}}\right)^{1/2} ,$$

from which the desired limit is obtained when $\omega_{\rm pe}^2 \gg \omega\omega_{\rm ce}$. (b) Now, use the definition $\mathscr{N} = kc/\omega$ and solve for ω to arrive at

$$\omega = \frac{k^2c^2\omega_{\rm ce}}{\omega_{\rm pe}^2} \quad \text{and} \quad \frac{{\rm d}\omega}{{\rm d}k} = 2\frac{\omega}{k} .$$

6.4 Use the definition $n_{\rm co} = \varepsilon_0 m_{\rm e}\omega^2/e^2$ and $\omega = 2\pi c/\lambda$ to obtain $n_{\rm co} = 2.8 \times 10^{27}\,{\rm m}^{-3}$.

6.5 The cut-off is defined by $\mathscr{N} = 0$. The we have

$$0 = \mathscr{N}^2 = \varepsilon = 1 - \frac{\omega_{\rm pe}^2}{\omega^2} - \frac{\omega_{\rm pp}^2}{\omega^2}$$

resulting in $\omega_{\rm co} = 2^{1/2}\omega_{\rm pe}$.

6.6 $f_{\rm ce} = 1.4\,{\rm MHz}$, $f_{\rm pe} = 12.7\,{\rm MHz}$, $f_{\rm uh} = (f_{\rm ce}^2 + f_{\rm pe}^2)^{1/2} = 12.8\,{\rm MHz}$.

6.7

$$\frac{{\rm d}\omega}{{\rm d}k} = \frac{\omega}{k} \Rightarrow \frac{{\rm d}\omega}{\omega} = \frac{{\rm d}k}{k} \Rightarrow \ln\omega = \ln k + c \Rightarrow \omega = v_\varphi k .$$

Problems of Chapter 7

7.1 Equate the electron retardation current (7.31) with the ion saturation current (7.27) and set $\Phi_p = 0$ for convenience. This balance defines the floating potential Φ_f. Solving for the floating potential and using the quasineutrality of the unperturbed plasma $n_{e0} = n_{i0}$ gives

$$\Phi_f = \frac{k_B T_e}{e} \ln \left[0.61(2\pi)^{1/2} \left(\frac{m_e}{m_i} \right)^{1/2} \right] = \begin{cases} -3.3 k_B T_e/e & (H^+) \\ -5.2 k_B T_e/e & (Ar^+) \end{cases}$$

7.2 The one-dimensional electron current to the probe is

$$j_e = -e \int_{v_{min}}^{\infty} v_z f(v_z) dv_z \quad \text{and} \quad v_{min} = \left(\frac{2eU_p}{m_e} \right)^{1/2} .$$

Then the derivative of the probe characteristic can be written as

$$\frac{dj_e}{dU_p} = \frac{dj_e}{dv_{min}} \frac{dv_{min}}{dU_p} = +ev_{min} f(v_{min}) \frac{e/m_e}{\sqrt{2eU_p/m_e}} = \frac{e^2}{m_e} f[v(U_p)] .$$

7.3 Use $U_2 = U_1 + U_p$ to rewrite (7.46) as

$$-I_p = I_{i0} + I_{e0} \exp\left(\frac{e(U_1 - \phi_p)}{k_B T_e} \right) \exp\left(\frac{eU_p}{k_B T_e} \right)$$

Then eliminate the exponential containing $U_2 - \phi_p$ by means of (7.46) yielding

$$-I_p = I_{i0} + (I_p - I_{i0}) \exp\left(\frac{eU_p}{k_B T_e} \right) ,$$

which by simple rearrangement gives the tanh-shape of the double probe characteristic.

7.4 The slope of the double probe characteristic in the origin is $dI_p/dU_p = eI_{i0}/(2k_B T_e)$. Then a straight line through the origin with this slope intersects the asymptote $I = I_{i0}$ at $U_p = 2k_B T_e/e$.

Problems of Chapter 8

8.1 Set $x = \hat{x} \exp(-i\omega t) = \hat{x} \exp(-i\omega_R t) \exp(\omega_I t)$, which has unstable solutions for $\omega_I > 0$. Overdamped or purely growing modes have $\omega_R = 0$. The characteristic equation for the differential equation then becomes $\omega^2 + i\omega a - b = 0$ with the solution

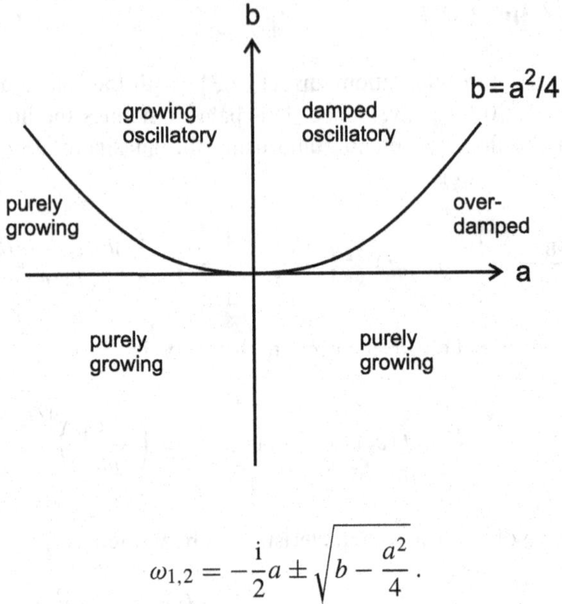

$$\omega_{1,2} = -\frac{\mathrm{i}}{2}a \pm \sqrt{b - \frac{a^2}{4}}\,.$$

8.2 The dielectric function reads

$$\varepsilon = 1 - \frac{\omega_{\mathrm{b}}^2}{(\omega - kv)^2} - \frac{\omega_{\mathrm{b}}^2}{(\omega + kv)^2}\,,$$

which is equivalent to a quartic equation with real or pairs of complex conjugate roots. The roots are given by

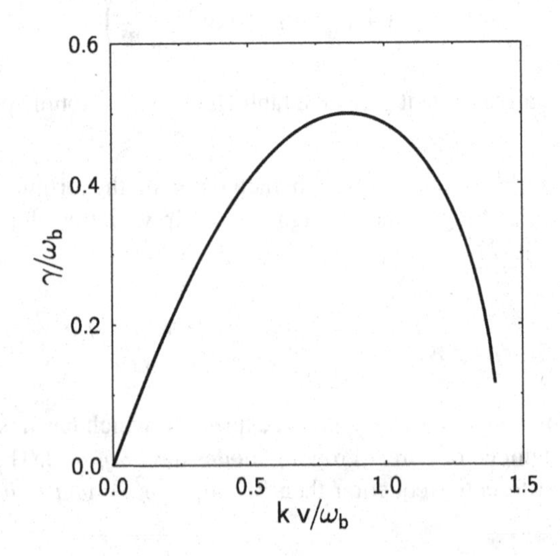

$$\omega = \pm\left[k^2v^2 + \omega_b^2 \pm \left(\omega_b^4 + 4k^2v^2\omega_b^2\right)^{1/2}\right]^{1/2}.$$

Two of these roots are purely imaginary and the one with positive imaginary part is unstable.

8.3 Insert $\omega = \omega_R + i\omega_I$ in (8.22) and require that the imaginary terms cancel:

$$0 = \omega_I - \frac{\omega_R\omega_I}{|\omega|^4}\omega_{pe}\omega_{pi}^2 .$$

Noting that $\omega_R = |\omega|\cos(\theta)$ yields (8.23). Now, set $\omega_I = |\omega|\sin(\theta)$ with $|\omega|$ from (8.23) and calculate $d\omega_I/d\theta = 0$ yielding $\tan(\theta) = \sqrt{3}$ and $\theta = \pi/3$. From $\omega = [\omega_{pi}^2\omega_{pe}\cos(\theta)]^{1/3}\exp(i\pi/3)$ we obtain (8.24).

Problems of Chapter 9

9.1 Note that the total energy W_{tot} is a constant of motion. Let $g(W_{tot}) = g(\frac{1}{2}mv^2 + q\Phi)$ be the distribution function. Then the Vlasov equation can be written as

$$\frac{dg}{dW_{tot}}\left(\underbrace{\frac{\partial W_{tot}}{\partial t}}_{=0} + v\frac{\partial W_{tot}}{\partial x} - \frac{q}{m}\Phi'\frac{\partial W_{tot}}{\partial v}\right) = \frac{dg}{dW_{tot}}\left(+q\Phi' v - mv\frac{q}{m}\Phi'\right) = 0 .$$

9.2

$$\frac{1}{n_{e0}}\int_0^\infty v f_M^{(1)}(v)dv = \left(\frac{m_e}{2\pi k_B T_e}\right)^{1/2}\frac{k_B T_e}{m_e}\int_0^\infty e^{-y}\,dy = \left(\frac{k_B T_e}{2\pi m_e}\right)^{1/2}.$$

9.3 The integral is of the same type as in the previous problem, but now the lower bound of the integral is $v_c = (2e(\Phi - \Phi_{min})/m)^{1/2}$.

9.4 Follow the advice preceding (9.55).

9.5 The unstable mode has a frequency difference from the plasma frequency given by (8.8). Then the difference between the phase velocity and the beam velocity is

$$v_0 - v_\varphi = \frac{\Delta\omega}{k} = \frac{\omega_{pe}}{k}\frac{1}{2}\left(\frac{\alpha_b}{2}\right)^{1/3}.$$

With $\hat{\Phi}_t = m(v_0 - v_\varphi)^2/4$ and $\omega_{pe}/k = v_0$ we obtain the desired result.

9.6 The trapping condition reads $m\Delta v^2 = 4e\hat{\Phi}$ for a difference between phase velocity and beam velocity $\Delta v = (\omega_{pe} - \omega)/k = \Delta\omega/k$. The fastest growing mode has $\Delta\omega = 2^{-4/3}\alpha_{\rm b}^{1/3}\omega_{\rm pe}$. Then the energy density becomes

$$W_E = \frac{\varepsilon_0}{2}\langle E^2\rangle = \frac{\varepsilon_0}{4}k^2\hat{\Phi}^2 = \frac{\varepsilon_0}{4}k^2\left(\frac{m}{4e}\right)^2\Delta v^4 \,.$$

Eliminating $k = \omega_{\rm pe}/v_0$ and using $\alpha_{\rm b} = n_{\rm b}/n_{\rm p}$ gives the desired result.

Problems of Chapter 10

10.1 The dust charge is $q_{\rm d} = 4\pi\varepsilon_0 a\Phi = 4.45\times10^{-18}$ C which is $Z_{\rm d} = 27.8$. This gives an initial (electric) potential energy $W_{\rm pot} = 1.78\times10^{-17}$ J corresponding to 111 eV. From energy conservation, $W_{\rm pot} = mg_L h$, a maximum height of $h = 883$ m is obtained.

10.2 For $P\to\infty$, (10.25) requires $\eta_{\rm f} = \eta_{\rm c}$. Therefore, (10.26) reduces to

$$(\mu\tau)^{-1/2} = {\rm e}^{-(\eta_{\rm f}+\tau\eta_{\rm c})} \quad \text{or} \quad \eta_{\rm f} = \frac{1}{2(1+\tau)}\ln(\mu\tau) = 0.078\,.$$

The normalized ion density becomes $n_{\rm i}/n_\infty = {\rm e}^{\tau\eta_{\rm c}} = {\rm e}^{7.8} = 2440$, a tremendously high value, which demonstrates the inadequacy of assuming a Boltzmann response for the ions.

10.3 The electron Debye length is $\lambda_{\rm De} = 410\,\mu$m. At the Bohm velocity $v_{\rm B} = k_{\rm B}T_{\rm e}/m_{\rm i}$, an ion has the kinetic energy $m_{\rm i}v_{\rm B}^2/2 = k_{\rm B}T_{\rm e}/2$. The normalized floating potential of a sphere is $e\Phi_{\rm f}/k_{\rm B}T_{\rm e} = 2.41$. The collection radius then becomes

$$b_{\rm c} = a\left(1+\frac{2.41\,k_{\rm B}T_{\rm e}}{0.5k_{\rm B}T_{\rm e}}\right)^{1/2} = 2.41\,a = 12.2\,\mu{\rm m}\,.$$

The dust charge is $q_{\rm d} = (1675\times5\times3)e = 4.03\times10^{-15}$ C. This gives a Coulomb radius $r_{\rm C} = 12\,\mu$m. Because the collection radius is much smaller than the electron Debye length, the ion wind force is dominated by the orbit force. Since the Coulomb radius equals the collection radius, the grazing trajectory touches the grain at the "midnight" position, i.e., exactly opposite to the illumination direction.

10.4 At $r_{\rm min}$ and for $\dot{r} = 0$ the energy equation (10.53) reads

$$\frac{m_{\rm i}}{2}v_0^2 = \frac{m_{\rm i}}{2}\frac{(m_{\rm i}v_0 b)^2}{m_{\rm i}^2 r_{\rm min}^2} - \frac{q_{\rm d}e}{4\pi\varepsilon_0 r_{\rm min}}$$
$$0 = r_{\rm min}^2 - b^2 + 2r_{\rm min}r_{\rm C}\,,$$

which gives the desired result.

10.5 Consider a symmetric displacement of both particles by δr from their equilibrium positions. Expand the restoring force into a Taylor series, which gives a first order term

$$\delta F_\mathrm{r} = -m\omega_0^2 \delta r - \frac{2q_\mathrm{d}^2}{4\pi\epsilon_0 d_0^3}(2\delta r)\,.$$

Using the definition (10.68) for d_0, we obtain the equation of motion $\delta\ddot{r} + 3\omega_0^2\delta r = 0$, which yields the frequency of the breathing mode $\omega_\mathrm{br} = \sqrt{3}\omega_0$.

10.6 Define the spring constant D_n for an interaction with a neighbor at distance $n\Delta$. Then,

$$D_n = \frac{q_\mathrm{d}^2}{4\pi\varepsilon_0 n^3 \Delta^3}\left(2 + 2n\kappa + n^2\kappa^2\right)\mathrm{e}^{-n\kappa}\,,$$

which gives the desired r.h.s. of the dispersion relation after summing over all pairs of neighbors at $\pm n\Delta$.

10.7 The deflecting force for the i-th particle is given by

$$F_i = F_\Delta\left(\frac{\eta_i - \eta_{i-1}}{\Delta} + \frac{\eta_i - \eta_{i+1}}{\Delta}\right).$$

The distance between particles i and $i+1$ increases as

$$s = \left[\Delta^2 + (\eta_i - \eta_{i+1})^2\right]^{1/2} \approx \Delta\left[1 + \frac{1}{2}\left(\frac{\eta_i - \eta_{i+1}}{\Delta}\right)^2\right],$$

which contains only a second-order correction to Δ that can be neglected. Assuming a wave-like perturbation $\eta(x,t) = \hat{\eta}\mathrm{e}^{\mathrm{i}(kx-\omega t)}$ we have

$$-\omega^2 m = \frac{F_\Delta}{\Delta}\left(2 - \mathrm{e}^{-\mathrm{i}k\Delta} - \mathrm{e}^{+\mathrm{i}k\Delta}\right) = \frac{4F_\Delta}{\Delta}\sin^2\left(\frac{k\Delta}{2}\right),$$

which gives a purely imaginary ω that attains a maximum at $k\Delta/2 = \pi/2$ or $\Delta = \lambda/2$, i.e., a zig-zag arrangement of nearest neighbors.

10.8 The force between infinitely large charged planes is independent of the distance between the planes. Point-like particles in a linear chain interact via a three-dimensional force law, which may be Coulomb or Yukawa and decays as r^{-2} or faster.

Problems of Chapter 11

11.1 A charge in a parallel-plate capacitor causes image charges on the plates. As the charge moves, the image charges change and a displacement current flows in the outer circuit. The exponentially growing electron avalanche generates a current waveform of exponential shape. When the electrons move at a constant drift velocity $v_\mathrm{d} = -\mu_\mathrm{e} E$, the current rises as $I \propto \mathrm{e}^{\alpha v_\mathrm{d} t}$. The current is zero again when the last electron has reached the electrode.

11.2 (a) Inserting the separation ansatz into the diffusion equation, and dividing by $R(r)T(t)$, we have

$$\frac{1}{T(t)}\frac{\mathrm{d}T(t)}{\mathrm{d}t} - \frac{D_\mathrm{a}}{R(r)}\left(\frac{\mathrm{d}^2R(r)}{\mathrm{d}r^2} + \frac{1}{r}\frac{\mathrm{d}R(r)}{\mathrm{d}r}\right) = 0\,.$$

Because each of these two terms is only dependent on t or on r, respectively, the terms must be a constant (with the dimension of a reciprocal time), say $-1/\tau$. Then $T(t) \propto \exp -t/\tau$.
(b) The constancy of the term containing $R(t)$ can be rewritten as

$$\frac{\mathrm{d}^2R}{\mathrm{d}r^2} + \frac{1}{r}\frac{\mathrm{d}R}{\mathrm{d}r} + \frac{R}{D_\mathrm{a}\tau} = 0$$

$$\frac{\mathrm{d}^2R}{\mathrm{d}x^2} + \frac{1}{x}\frac{\mathrm{d}R}{\mathrm{d}x} + R = 0, \text{ with } x = r(D_\mathrm{a}\tau)^{-1/2}\,.$$

This is Bessels differential equation for $J_0(x)$.
(c) From $R(a) = 0$ we obtain $2.405 = a(D_\mathrm{a}\tau)^{-1/2}$ with the first zero $J_0(2.405) = 0$. and $\tau = D_\mathrm{a}^{-1}(a/2.405)^2$.

11.3 Multiplying (11.17) by the applied voltage $\hat{U}$ gives the total current $\hat{I} = \hat{U}/Z_\mathrm{b}$ in the two branches. Then the ratio of the current in the capacitor to that in the RL-branch is

$$\frac{\hat{I}_C}{\hat{I}_{R+L}} = \frac{\mathrm{i}\omega(\nu_\mathrm{m} + \mathrm{i}\omega)}{\omega_\mathrm{pe}^2}\,,$$

which is a small quantity in the given limit.

11.4 The collisionless skin depth is $\delta_\mathrm{cl} = c/\omega_\mathrm{pe}$. At $n_\mathrm{e} = 10^{17}\,\mathrm{m}^{-3}$ we have $\omega_\mathrm{pe} = 1.78 \times 10^{10}\,\mathrm{s}^{-1}$ and obtain $\delta_\mathrm{cl} = 0.017\,\mathrm{m}$.

References

1. R.N. Franklin, N.S.J. Braithwaite, Plasma Sources Sci. Technol. **18**, 010201 (2009)
2. H.M. Mott-Smith, Nature **233**, 219 (1971)
3. D.A. Frank-Kamenezki, *Plasma – Der Vierte Aggregatzustand* (Verlag Progress, Moskau, 1963)
4. F. Fraunberger, *Illustrierte Geschichte der Elektrizität* (Aulis, Köln, 1985)
5. C.G. Suits, *The collected works of Irving Langmuir* (Macmillan, Pergamon, New York, 1961)
6. W. Schottky, J.V. Issendorff, Z. Phys. **31**, 163 (1925)
7. A.V. Engel, M. Steenbeck, *Elektrische Gasentladungen*, Vol. 1& 2 (Springer, Berlin, 1934)
8. R. Seeliger, *Einführung in die Physik der Gasentladungen* (Barth, Leipzig, 1934)
9. M.J. Druyvesteyn, F.M. Penning, Rev. Mod. Phys. **12**, 87 (1940)
10. E.V. Appleton, J. Inst. Elec. Engrs. **71**, 642 (1932)
11. S. Chapman, *Solar Plasma, Geomagnetism and Aurora* (Blackie, London, 1964)
12. J.A. Ratcliffe, *The Magnetoionic Theory and its Application to the Ionosphere* (Cambridge, New-York, 1959)
13. K.G. Budden, *Radio Waves in the Ionosphere* (Cambridge University Press, New York, 1961)
14. K. Rawer, *Die Ionosphäre* (Noordhoff, Groningen, 1953)
15. A. Unsöld, *Die Physik der Sternatmosphären* (Springer, Berlin, 1968)
16. P. Sweet, *IAU Symposium* **6**, 123 (1958)
17. E.N. Parker, Phys. Rev. **107**, 830 (1957)
18. K. Birkeland, *The Norwegian Aurora Polaris Expedition 1902–1903*, Vol. I (J.A. Barth, Leipzig, 1908), p. 98
19. L. Biermann, Z. Astrophys. **29**, 274 (1951)
20. E.N. Parker, Ap. J. **128**, 664 (1958)
21. K.I. Gringauz, V.V. Bezrukikh, V.D. Ozerov, R.E. Rybchinskii, Planet. Space Sci. **9**, 97 (1962)
22. R.G. Marsden, Space Sci. Rev. **78**, 67 (1996)
23. D.J. McComas, B.L. Barraclough, H.O. Funsten, J.T. Gosling, E. Santiago-Munoz, R.M. Skoug, B.E. Goldstein, M. Neugebauer, P. Riley, A. Balogh, J. Geophys. Res. **105**, 10,419 (2000)
24. W. Feldman, B. Barraclaugh, J. Phillips, Y. Wang, Astronom. Astrophys. **316**, 350 (1996)
25. J. van Allen, L.A. Frank, Nature **183**, 430 (1959)
26. D. Bilitza, B. Reinisch, Adv. Space Res. **42**, 599 (2008)
27. B. Jacob, Lighting Res, Technol. **41**, 219 (2009)
28. Department of the Environment and Heritage, *Final Report on: Public Lighting in Australia - Energy Efficiency Challenges and Opportunities* (Australian Government, 2005)
29. M. Born, Plasma Sources Sci. Technol. **11**, A55 (2002)
30. B.A. Smith, et al., Science **215**, 504 (1982)
31. C.J. Mitchell, M. Horanyi, O. Havnes, C.C. Porco, Science **311**, 1587 (2006)
32. J.H. Chu, L.I, Phys. Rev. Lett. **72**, 4009 (1994)

33. Y. Hayashi, K. Tachibana, Jpn. J. Appl. Phys. **33**, L804 (1994)
34. H. Thomas, G.E. Morfill, V. Demmel, J. Goree, B. Feuerbacher, D. Möhlmann, Phys. Rev. Lett. **73**, 652 (1994)
35. J. Pieper, J. Goree, R. Quinn, Phys. Rev. E **54**, 5636 (1996)
36. O. Arp, D. Block, A. Piel, Phys. Rev. Lett. **93**, 165004 (2004)
37. T.K. Fowler, *The Fusion Quest* (Johns Hopkins University Press, Baltimore MD, 1997)
38. H.S. Bosch, G.M. Hale, Nucl. Fusion **32**, 611 (1992)
39. H.S. Bosch, G.M. Hale, Nucl. Fusion **32**, 1919 (1992)
40. J.H. Nuckolls, L. Wood, A. Thiessen, G.B. Zimmermann, Nature **239**, 139 (1972)
41. R.C. Arnold, J.M. ter Vehn, Rep. Prog. Phys. **50**, 559 (1987)
42. J. Ongena, A.M. Messiaen, Trans. Fusion Sci. Technol. **49**, 425 (2006)
43. F. Wagner, et al., Phys. Rev. Lett. **49**, 1408 (1982)
44. F. Wagner, U. Stroth, Plasma Phys. Control. Fusion **35**, 1321 (1993)
45. F. Wagner, Plasma Phys. Control. Fusion **49**, B1 (2007)
46. P.H. Rebut, The JET-Team, Plasma Phys. Control. Fusion **34**, 1749 (1992)
47. R.J. Hawryluk, Rev. Mod. Phys. **70**, 537 (1998)
48. R.J. Hawryluk, S. Batha, W. Blanchard, M. Beer, et al., Phys. Plasmas **5**, 1577 (1998)
49. P.H. Rebut, Plasma Phys. Control. Fusion **48**, B1 (2006)
50. K. Walter, *LLNL Science & Technology Report*, 4–14 September 2003
51. J. Lindl, Phys. Plasmas **2**, 3933 (1995)
52. J.D. Lindl, P. Amendt, R.L. Berger, S.G. Glendinning, S.H. Glenzer, S.W. Haan, R.L. Kauffman, O.L. Landen, L.J. Suter, Phys. Plasmas **11**, 339 (2004)
53. National Research Council, *Plasma Science—Advancing Knowledge in the National Interest* (NAS, Washington D.C., 2007)
54. M.N. Saha, Phil. Mag. **40**, 472 (1920)
55. P. Debye, E. Hückel, Phys. Z. **24**, 185 (1923)
56. D.H.E. Dubin, T.M. O'Neil, Rev. Mod. Phys. **71**, 87 (1999)
57. J.J. Bollinger, D.J. Wineland, D.H.E. Dubin, Phys. Plasmas **1**, 1403 (1994)
58. G. Baur, G. Boero, S. Brauksiepe, et al., Phys. Lett. B **368**, 251 (1996)
59. G. Blanford, D.C. Christian, K. Gollwitzer, M. Mandelkern, C.T. Munger, J. Schultz, G. Zioulas, Phys. Rev. Lett. **80**, 3037 (1998)
60. M. Amoretti, C. Amsler, G. Bonomi, A. Bouchta, P. Bowe, C. Carraro, C.L. Cesar, et al., Nature **419**, 456 (2002)
61. V. Erckmann, U. Gasparino, Plasma Phys. Control. Fusion **36**, 1869 (1994)
62. R. Prater, Phys. Plasmas **11**, 2349 (2004)
63. H. Wobig, talk on the occasion of the retirement of G. Grieger (2000)
64. R. Jaenicke, E. Ascasibar, P. Grigull, I. Lakicevic, A. Weller, M. Zippe, Nucl. Fusion **33**, 687 (1993)
65. R.B. Brode, Rev. Mod. Phys. **5**, 257 (1933)
66. R. Rejoub, B.G. Lindsay, R.F. Stebbings, Phys. Rev. A **65**, 042713 (2002)
67. L. Spitzer, R. Harm, Phys. Rev. **89**, 977 (1953)
68. Y.P. Raizer, *Gas Discharge Physics* (Springer, Berlin, 1991)
69. A. Bouchoule, J.P. Boeuf, A. heron, O. Duchemin, Plasma Phys. Control. Fusion **46**, B407 (2004)
70. G.D. Racca, A. Marini, L. Stagnaro, J. van Dooren, L. di Napoli, B.H. Foing, R. Lumb, J. Volp, et al., Planet. Space Sci. **50**, 1323 (2002)
71. D.M. Di Cara, D. Estublier, Acta Astronaut. **57**, 250 (2005)
72. A. Lazurenko, V. Krasnoselskikh, A. Bouchoule, IEEE Trans. Plasma Sci. **36**, 1977 (2008)
73. D. Gawron, S. Mazouffre, N. Sadeghi, A. Héron, Plasma Sources Sci. Technol. **17**, 025001 (2008)
74. W. Panofsky, M. Phillips, *Classical Electricity and Magnetism* (Addison-Wesley, Reading MA, 1969)

75. S. Glasstone, R.H. Lovberg, *Controlled Thermonuclear Reactions* (van Nostrand, Princeton NJ, 1960)
76. J.D. Lawson, Proc. Phys. Soc. (Lond.) **B70**, 6 (1957)
77. J. Lister, H. Weisen, Europhys. News **36**, 47 (2005)
78. S. Nakai, H. Takabe, Rep. Prog. Phys. **59**, 1071 (1996)
79. M.D. Rosen, Phys. Plasmas **6**, 1690 (1999)
80. C. Truesdell, Phys. Rev. **78**, 823 (1950)
81. W.H. Bennett, Phys. Rev. **45**, 890 (1934)
82. R.B. Spielman, C. Deeney, G.A. Chandler, M.R. Douglas, et al., Phys. Plasmas **5**, 2105 (1998)
83. M.E. Cuneo, D.B. Sinars, E.M. Waisman, D.E. Bliss, et al., Phys. Plasmas **15**, 056318 (2006)
84. H. Alfvén, Nature **150**, 405 (1942)
85. H. Alfvén, C.G. Fälthammar, *Cosmical Electrodynamics* (Clarendon, Oxford, 1950)
86. S. Lundquist, Nature **164**, 145 (1949)
87. B. Lehnert, Phys. Rev. **94**, 815 (1954)
88. W.H. Bostick, M.A. Levine, Phys. Rev. **87**, 671 (1952)
89. D.F. Jephcott, Nature **183**, 1652 (1959)
90. J.M. Wilcox, F.I. Boley, A.W. De Silva, Phys. Fluids **3**, 15 (1960)
91. H. Rishbeth, J. Atmos. Solar-Terr. Phys. **63**, 1883 (2001)
92. O. Heaviside, *Telegraphy*, vol. 33 (Encyclopedia Britannica, London, 1902)
93. A.E. Kennelly, Elect. World Eng. **6**, 473 (1902)
94. G. Breit, M.A. Tuve, Nature **116**, 357 (1925)
95. G. Breit, M.A. Tuve, Phys. Rev. **28**, 554 (1926)
96. E.V. Appleton, J.A. Ratcliffe, Proc. Roy. Soc. **A117**, 576 (1928)
97. I. Langmuir, L. Tonks, Science **68**, 598 (1928)
98. H.G. Booker, *Cold Plasma Waves* (Martinus Nijhoff Dordrecht, 1984)
99. D.G. Swanson, *Plasma Waves* (Academic New York, 1989)
100. T.H. Stix, *Waves in Plasmas* (AIP New York, 1992)
101. M. Brambilla, *Kinetic Theory of Plasma Waves* (Oxford University Press, Oxford, 1998)
102. H.G. Adler, A. Piel, J. Quant. Spectrosc. Radiat. Transf. **45**, 11 (1991)
103. F.A. Hopf, A. Tomita, G. Al-Jumaily, Opt. Lett. **5**, 386 (1988)
104. F.A. Hopf, A.Q. Tomita, G. Al-Jumaily, M. Cervantes, T. Liepmann, Opti. Comm. **36**, 487 (1981)
105. F. A.Hopf, M. Cervantes, Appl. Opti. **21**, 668 (1982)
106. T.W. Liepmann, F.A. Hopf, Appl. Optics **24**, 1485 (1985)
107. N. Bretz, F. Jobes, J. Irby, Rev. Sci. Instrum. **68**, 713 (1997)
108. F.C. Jobes, N.L. Bretz, Rev. Sci. Instrum. **68**, 709 (1997)
109. E. Stoffels, W.W. Stoffels, D. Vender, M. Kando, G.M.W. Kroesen, F.J. de Hoog, Phys. Rev. E **51**, 2425 (1995)
110. D. Bohm, E.P. Gross, Phys. Rev. **75**, 1851 (1949)
111. D. Bohm, E.P. Gross, Phys. Rev. **75**, 1864 (1949)
112. T.H. Stix, *The Theory of Plasma Waves* (McGraw-Hill, New-York, 1962)
113. J.A. Eilek, F.N. Owen, Ap. J. **567**, 202 (2002)
114. J.V. Hollweg, M.K. Bird, H. Volland, P. Edenhofer, C.T. Stelzried, B.L. Seidel, J. Geophys. Res. **87**, 1 (1982)
115. S. Mancuso, S.R. Spangler, Ap. J. **539**, 480 (2000)
116. S. Ganguly, G.H. Van Bavel, A. Brown, J. Geophys. Res. **105**, 16,063 (2000)
117. D.L. Hysell, J.L. Chau, J. Geophys. Res. **106**, 16,063 (2001)
118. H. Soltwisch, Rev. Sci. Instrum. **57**, 1939 (1986)
119. A.J.H. Donné, T. Edlington, E. Joffrin, H.R. Koslowski, C. Nieswand, S.E. Segre, P.E. Stott, C. Walker, Rev. Sci. Instrum. **70**, 726 (1999)
120. D.L. Brower, Y. Jiang, W.X. Ding, S.D. Terry, N.E. Lanier, J.K. Anderson, C.B. Forest, D. Holly, Rev. Sci. Instrum. **72**, 1077 (2001)

121. D.L. Brower, Y. Yiang, S.D. Terry, et al., Rev. Sci. Instrum. **74**, 1534 (2003)
122. K. Tanaka, K. Kawahata, A. Ejiri, CHS Group, J. Howard, S. Okajima, Rev. Sci. Instrum. **70**, 730 (1999)
123. M.K. Bird, Space Sci. Rev. **33**, 99 (1982)
124. L.R.O. Storey, Phil. Trans. Roy. Soc. (London) Ser. A **246**, 113 (1953)
125. H. Rishbeth, O.K. Garriot, *Introduction to Ionospheric Physics* (Academic, New York, 1969)
126. E.R. Schmerling, J. Atmospheric Terr. Phys. **12**, 8 (1958)
127. J.E. Titheridge, Radio Sci. **23**, 831 (1988)
128. J.E. Titheridge, World Data Center A for STP **UAG -93** (1985)
129. B.W. Reinisch, *Modern Ionosondes* (EGS Katlenburg-Lindau, 1996)
130. P.E. Argo, M.C. Kelley, J. Geophys. Res. **91**, 5539 (1986)
131. G.C. Hussey, K. Schlegel, C. Haldoupis, J. Geophys. Res. **103**, 6991 (1998)
132. M.L. Parkinson, D.P. Monselesan, P.R. Smith, P.L. Dyson, R.J. Morris, J. Geophys. Res. **102**, 24,075 (1997)
133. R.K. Fisher, R.W. Gould, Phys. Rev. Lett. **22**, 1033 (1969)
134. R.K. Fisher, R.W. Gould, Phys. Fluids **14**, 857 (1971)
135. T. Trottenberg, B. Brede, D. Block, A. Piel, Phys. Plasmas **10**, 4627 (2003)
136. E. Michel, C. Béghin, A. Gonfalone, I.F. Ivanov, Ann. Géophys. **31**, 873 (1975)
137. L.R.O. Storey, J. Thiel, Phys. Fluids **21**, 2325 (1978)
138. L.R.O. Storey, J. Thiel, J. Geophys. Res. **89**, 969 (1984)
139. V. Rohde, A. Piel, H. Thiemann, K.I. Oyama, J. Geophys. Res. **98**, 19,163 (1993)
140. A. Piel, G. Oelerich, Phys. Fluids **27**, 273 (1984)
141. A. Piel, G. Oelerich, Phys. Fluids **28**, 1366 (1985)
142. G. Oelerich-Hill, A. Piel, Phys. Fluids **B1**, 275 (1989)
143. C.D. Child, Phys. Rev. **32**, 498 (1911)
144. I. Langmuir, Phys. Rev. **2**, 450 (1913)
145. H.M. Mott-Smith, I. Langmuir, Phys. Rev. **28**, 727 (1925)
146. M.J. Druyvesteyn, Z. Phys. **64**, 787 (1930)
147. E.O. Johnson, L. Malter, Phys. Rev. **80**, 58 (1950)
148. J. Allen, Physica Scr. **45**, 497 (1992)
149. I.H. Hutchinson, *Principles of Plasma Diagnostics* (Cambridge University Press, Cambridge, 1987)
150. F.F. Chen, Plasma Phys. **7**, 47 (1965)
151. N. Hershkowitz, in *Plasma Diagnostics*, ed. by O. Auciello, D. Flamm (Academic, San Diego CA, 1989)
152. J.R. Anderson, K.D. Goodfellow, J.E. Polk, V.K. Rawlin, J.S. Sovey, IEEE Aerospace Conf. Proc. **4**, 99 (2000)
153. J.R. Brophy, Rev. Sci. Instrum. **73**, 1071 (2002)
154. I. Langmuir, Phys. Rev. **33**, 954 (1929)
155. M.A. Raadu, Phys. Rep. **178**, 25 (1989)
156. H. Schamel, Phys. Rep. **140**, 161 (1986)
157. N. Hershkowitz, G.L. Payne, C. Chan, J.R. deKock, Plasma Phys. **23**, 903 (1981)
158. N. Nershkowitz, Space Sci. Rev. **41**, 351 (1985)
159. L. Block, Astrophys. Space Sci. **55**, 59 (1978)
160. P. Coakley, N. Hershkowitz, Phys. Fluids **22**, 1171 (1979)
161. J. Dawson, Phys. Fluids **4**, 869 (1961)
162. P.C. Clemmow, J.P. Dougherty, *Electrodynamics of Particles and Plasmas* (Addison-Wesley, Reading MA, 1969)
163. O. Buneman, Phys. Rev. Lett. **1**, 8 (1958)
164. J.R. Pierce, J. Appl. Phys. **15**, 721 (1944)
165. J.R. Pierce, J. Appl. Phys. **19**, 231 (1948)
166. J.R. Pierce, J. Appl. Phys. **21**, 1063 (1950)
167. J. Frey, C.K. Birdsall, J. Appl. Phys. **36**, 2962 (1965)

168. J.R. Cary, D.S. Lemons, J. Appl. Phys. **53**, 3303 (1982)
169. I.N. Kolyshkin, V. I.Kuznetsov, A.Y. Ender, Sov. Phys. Tech. Phys. **29**, 882 (1984a)
170. S. Kuhn, Phys. Fluids **27**, 1821 (1984)
171. S. Kuhn, Phys. Fluids **27**, 1834 (1984)
172. S. Kuhn, M. Hörhager, J. Appl. Phys. **60**, 1952 (1986)
173. S. Kuhn, Contrib. Plasma Phys. **34**, 495 (1994)
174. A.Y. Ender, S. Kuhn, V.I. Kuznetsov, Phys. Plasmas **13**, 113506 (2006)
175. B.B. Godfrey, Phys. Fluids **30**, 1553 (1987)
176. W.S. Lawson, Phys. Fluids B **1**, 1483 (1989)
177. W.S. Lawson, Phys. Fluids B **1**, 1493 (1989)
178. H. Schamel, V. Maslov, Phys. Rev. Lett. **70**, 1105 (1993)
179. H. Kolinsky, H. Schamel, Phys. Plasmas **1**, 2359 (1994)
180. T.L. Crystal, S. Kuhn, Phys. Fluids **28**, 2116 (1985)
181. H. Matsumoto, H. Yokoyama, D. Summers, Phys. Plasmas **3**, 177 (1996)
182. A.E. Hramov, A.A. Koronovskii, I.S. Rempen, Chaos **16**, 013123 (2006)
183. N. Krahnstöver, T. Klinger, F. Greiner, A. Piel, Phys. Lett. A **239**, 103 (1998)
184. H. Friedel, R. Grauer, K.H. Spatschek, Phys. Plasmas **5**, 3187 (1998)
185. A.A. Koronovskii, I.S. Rempen, A.E. Khramov, Tech. Phys. Lett. **29**, 998 (2003)
186. M. Kočan, S. Kuhn, D. Tskhakaya (jr.), Contrib. Plasma Phys. **46**, 322 (2006)
187. C.K. Birdsall, W.B. Bridges, J. Appl. Phys. **32**, 2611 (1961)
188. C.K. Birdsall, W.B. Bridges, *Electron Dynamics in Diode Regions* (Academic, New York, 1966)
189. D.T. Farley, B.B. Balsley, R.F. Woodman, J.P. McClure, J. Geophys. Res. **75**, 7199 (1970)
190. S.L. Ossakow, J. Atmos. Terr. Phys. **43**, 437 (1981)
191. M.C. Kelley, et al., J. Geophys. Res. **91**, 5487 (1986)
192. R.F. Pfaff, H.A. Sobral, M.A. Abdu, W.E. Swartz, J.W. LaBelle, M.F. Larsen, R.A. Goldberg, F.J. Schmidlin, Geophys. Res. Lett. **24**, 1663 (1997)
193. C.T. Steigies, D. Block, M. Hirt, A. Piel, J. Geophys. Res. **106**, 12765 (2001)
194. L.D. Landau, J. Phys. USSR **10**, 25 (1946)
195. J.H. Malmberg, C.B. Wharton, Phys. Rev. Lett. **17**, 175 (1966)
196. H. Weitzner, Phys. Fluids **6**, 1123 (1963)
197. J. Lacina, Plasma Phys. **14**, 605 (1972)
198. J. Weiland, Eur. J. Phys. **2**, 171 (1981)
199. J. Lacina, Plasma Phys. Control. Fusion **36**, 601 (1884)
200. R.W.B. Best, Physica Scr. **59**, 55 (1999)
201. D.D. Ryutov, Plasma Phys. Control. Fusion **41**, A1 (1999)
202. W.E. Drummond, Phys. Plasmas **11**, 552 (2004)
203. R. Bilato, M. Brambilla, Comm. Nonlin. Sci. Num. Sim. **13**, 18 (2008)
204. I. Alexeff, O. Ishihara, IEEE Trans. Plasma Sci. **6**, 212 (1978)
205. D. Sagan, Am. J. Phys. **62**, 450 (1994)
206. G. Brodin, Am. J. Phys. **65**, 66 (1997)
207. D. Anderson, R. Fedele, M. Lisak, Am. J. Phys. **69**, 1262 (2001)
208. R.W. Gould, T.M. O'Neill, J.H. Malmberg, Phys. Rev. Lett. **19**, 219 (1967)
209. J.H. Malmberg, C.B. Wharton, R.W. Gould, T.M. O'Neill, Phys. Fluids **11**, 1147 (1968)
210. A.Y. Wong, D.R. Baker, Phys. Rev. **188**, 326 (1969)
211. D.R. Nicholson, *Introduction to Plasma Theory* (Wiley, New York, 1983)
212. B.R. Ripin, R.E. Pechacek, Phys. Rev. Lett. **24**, 1330 (1970)
213. D.R. Baker, N.R. Ahern, A.Y. Wong, Phys. Rev. Lett. **20**, 318 (1968)
214. R.W. Hockney, J.W. Eastwood, *Computer Simulation Using Particles* (Adam Hilger, Bristol, 1991)
215. C.K. Birdsall, B. Langdon, *Plasma Physics via Computer Simulation* (Adam Hilger, Bristol, 1991)
216. J.P. Verboncoeur, M.V. Alves, V. Vahedi, C.K. Birdsall, J. Comp. Phys. **104**, 321 (1992)

217. D.J. Sullivan, IEEE Trans. Nucl. Sci. **26**, 4274 (1979)
218. D.J. Sullivan, IEEE Trans. Nucl. Sci. **30**, 3426 (1983)
219. H.A. Davis, R.R. Bartsch, T.J.T. Kwan, E.G. Sherwood, R.M. Stringfield, Phys. Rev. Lett. **59**, 288 (1987)
220. S. Burkhart, J. Appl. Phys. **62**, 75 (1987)
221. W. Jiang, M. Kristiansen, Phys. Plasmas **8**, 3781 (2001)
222. L. Spitzer, *Diffuse Matter in Space* (Wiley, New-York, 1968)
223. K.G. Spears, R.P. Kampf, T.J. Robinson, J. Phys. Chem. **92**, 5297 (1988)
224. A. Bouchoule, A. Plain, L. Boufendi, J.P. Blondeau, C. Laure, J. Appl. Phys. **70**, 1991 (1991)
225. G.S. Selwyn, J. Singh, R.S. Bennett, J. Vac. Sci. Technol. A **7**, 2758 (1989)
226. P.K. Shukla, Phys. Plasmas **8**, 1791 (2001)
227. A. Piel, A. Melzer, Plasma Phys. Control. Fusion **44**, R1 (2002)
228. V.N. Tsytovich, G.E. Morfill, H. Thomas, Plasma Phys. Rep. **28**, 623 (2002)
229. G.E. Morfill, V.N. Tsytovich, H. Thomas, Plasma Phys. Rep. **29**, 1 (2003)
230. H. Thomas, G.E. Morfill, V.N. Tsytovich, Plasma Phys. Rep. **29**, 895 (2003)
231. V.N. Tsytovich, G.E. Morfill, H. Thomas, Plasma Phys. Rep. **30**, 816 (2004)
232. S. Vladimirov, K. Ostrikov, Phys. Rep. **393**, 175 (2004)
233. O. Ishihara, J. Phys. D: Appl. Phys. **40**, R121 (2007)
234. M. Bonitz, P. Ludwig, H. Baumgartner, C. Henning, A. Filinov, D. Block, O. Arp, A. Piel, S. Käding, Y. Ivanov, A. Melzer, H. Fehske, V. Filinov, Phys. Plasmas **15**, 055704 (2008)
235. A. Piel, O. Arp, D. Block, I. Pilch, T. Trottenberg, S. Käding, A. Melzer, H. Baumgartner, C. Henning, M. Bonitz, Plasma Phys. Control. Fusion **50**, 124003 (2008)
236. P.K. Shukla, B. Eliasson, Rev. Mod. Phys. **81**, 25 (2009)
237. A. Bouchoule, *Dusty Plasmas: Physics, Chemistry, and Technological Impacts in Plasma Processing* (Wiley, New York, 1999)
238. F. Verheest, *Waves in Dusty Space Plasmas* (Kluwer, Dordrecht, 2000)
239. P.K. Shukla, M.M. Mamun, *Introduction to Dusty Plasma Physics* (IOP Bristol, 2002)
240. S.V. Vladimirov, K. Ostrikov, A. Samarian, *Physics and Applications of Complex Plasmas* (World Scientific, Singapore, 2005)
241. V. Tsytovich, G.E. Morfill, S.V. Vladimirov, H.M. Thomas, *Elementary Physics of Complex Plasmas* (Springer, Berlin, 2007)
242. O. Arp, D. Block, M. Klindworth, A. Piel, Phys. Plasmas **12**, 122102 (2005)
243. E.C. Whipple, Rep. Prog. Phys. **44**, 1197 (1981)
244. E.J. Sternglass, Westinghouse Res. Lab. **Sci. Pap. 1772** (1954)
245. B.T. Draine, E.E. Salpeter, Ap. J. **231**, 77 (1979)
246. M.K. Wallis, M.H.A. Hassan, Astronom. Astrophys. **121**, 10 (1983)
247. N. Meyer-Vernet, Astronom. Astrophys. **105**, 98 (1982)
248. T.N. Woods, P.C. Chamberlin, J.W. Harder, R.A. Hock, M. Snow, F.G. Eparvier, J. Fontenla, W.E. McClintock, E.C. Richard, Geophys. Res. Lett. **36**, L01101 (2009)
249. D.A. Verner, G.F. Ferland, K.T. Korista, D.G. Yakovlev, Ap. J. **465**, 487 (1996)
250. M.M. Abbas, D. Cantosic, P.D. Craven, R.B. Hoover, L.A. Hoover, L.A. Taylor, J.F. Spann, A. LeClair, E.A. West, ESA **SP-643**, 165 (2007)
251. D.R. Criswell, in *Photon and Particle Interactions with Surfaces in Space*, ed. by R.J.L. Grard (Springer, Heidelberg, 1973), pp. 545–556
252. J.E. McCoy, D.R. Criswell, *Proceedings of 5th Lunar Science Conference* pp. 2991–3005 (1974)
253. C.F. Bohren, D.R. Huffman, *Absorption and Scattering of Light by Small Particles* (Wiley-VCH, Weinheim, 1998)
254. T.J. Stubbs, R.R. Vondrak, W.M. Farrell, Adv. Space Res. **37**, 59 (2006)
255. C. Cui, J. Goree, IEEE Trans. Plasma Sci. **22**, 151 (1994)
256. C. Hollenstein, Plasma Phys. Control. Fusion **42**, R93 (2000)
257. M. Horanyi, C.K. Goertz, Ap. J. **361**, 155 (1990)
258. O. Havnes, Adv. Space Res. **4**, 75 (1984)
259. O. Havnes, G.E. Morfill, C.K. Goertz, J. Geophys. Res. **89**, 10,999 (1984)

260. M.S. Barnes, J.H. Keller, J.C. Forster, J.A. O'Neill, D.K. Coultas, Phys. Rev. Lett. **68**, 313 (1992)
261. E.B. Tomme, D.A. Law, B.M. Annaratone, J.E. Allen, Phys. Rev. Lett. **85**, 2518 (2000)
262. P.S. Epstein, Phys. Rev. **23**, 710 (1924)
263. A. Melzer, T. Trottenberg, A. Piel, Phys. Lett. A **191**, 301 (1994)
264. T. Trottenberg, A. Melzer, A. Piel, Plasma Sources Sci. Technol. **4**, 450 (1995)
265. A. Homann, A. Melzer, A. Piel, Phys. Rev. E **59**, 3835 (1999)
266. T. Nitter, Plasma Sources Sci. Technol. **5**, 93 (1996)
267. A.V. Ivlev, U. Konopka, G.E. Morfill, Phys. Rev. E **62**, 2739 (2000)
268. S. Nunomura, T. Misawa, N. Ohno, S. Takamura, Phys. Rev. Lett. **83**, 1970 (1999)
269. C. Arnas, M. Mikikian, F. Doveil, Phys. Rev. E **60**, 7420 (1999)
270. G.M. Jellum, D.B. Graves, Appl. Phys. Lett. **57**, 2077 (1990)
271. L. Talbot, R.K. Cheng, R.W. Schefer, D.R. Willis, J. Fluid Mech. **101**, 37 (1980)
272. O. Havnes, T. Nitter, V. Tsytovich, G.E. Morfill, T. Hartquist, Plasma Sources Sci. Technol. **3**, 448 (1994)
273. I.H. Hutchinson, Plasma Phys. Control. Fusion **47**, 71 (2005)
274. S.A. Khrapak, A.V. Ivlev, G.E. Morfill, H.M. Thomas, Phys. Rev. E **66**, 046414 (2002)
275. S.A. Khrapak, A. Ivlev, G.E. Morfill, S.K. Zhdanov, Phys. Rev. Lett. **90**, 225002 (2003)
276. S.A. Khrapak, A.V. Ivlev, G.E. Morfill, Phys. Rev. E **70**, 056405 (2004)
277. I.H. Hutchinson, Plasma Phys. Control. Fusion **48**, 185 (2006)
278. C. Zafiu, A. Melzer, A. Piel, Phys. Plasmas **9**, 4794 (2002)
279. C. Zafiu, A. Melzer, A. Piel, Phys. Plasmas **10**, 4582 (2003)
280. M. Hirt, D. Block, A. Piel, Phys. Plasmas **11**, 5690 (2004)
281. G.E. Morfill, H.M. Thomas, U. Konopka, H. Rothermel, M. Zuzic, A. Ivlev, J. Goree, Phys. Rev. Lett. **83**, 1598 (1999)
282. M. Klindworth, A. Piel, A. Melzer, U. Konopka, H. Rothermel, K. Tarantik, G. Morfill, Phys. Rev. Lett. **93**, 195002 (2004)
283. G. Praburam, J. Goree, Phys. Plasmas **3**, 1212 (1996)
284. E. Thomas, K. Avinash, R.L. Merlino, Phys. Plasmas **11**, 1770 (2004)
285. M. Lampe, G. Joyce, G. Ganguli, V. Gavroshchaka, Phys. Plasmas **7**, 3851 (2000)
286. U. Konopka, G.E. Morfill, L. Ratke, Phys. Rev. Lett. **84**, 891 (2000)
287. A. Melzer, Phys. Rev. E **67**, 016411 (2003)
288. V.M. Bedanov, F. Peeters, Phys. Rev. B **49**, 2667 (1994)
289. M. Kong, B. Partoens, F.M. Peeters, New J. Phys. **5**, 32.1 (2003)
290. V.A. Schweigert, F. Peeters, Phys. Rev. B **51**, 7700 (1995)
291. M. Klindworth, A. Melzer, A. Piel, V.A. Schweigert, Phys. Rev. B **61**, 8404 (2000)
292. F. Diedrich, E. Peik, J.M. Chen, W. Quint, H. Walther, Phys. Rev. Lett. **59**, 2931 (1987)
293. D.J. Wineland, J.C. Bergquist, W.M. Itano, J.J. Bollinger, C.H. Manney, Phys. Rev. Lett. **59**, 2935 (1987)
294. S.L. Gilbert, J.J. Bollinger, D.J. Wineland, Phys. Rev. Lett. **60**, 2022 (1988)
295. M. Drewsen, C. Brodersen, L. Hornekaer, J. Hangst, J. Schiffer, Phys. Rev. Lett. **81**, 2878 (1998)
296. W. Luck, M. Klier, H. Wesslau, Ber. Bunsenges. Phys. Chem. **50**, 485 (1963)
297. P.A. Hiltner, I.M. Krieger, J. Phys. Chem. **73**, 2386 (1969)
298. A. Kose, M. Ozaki, K. Takano, Y. Kobayashi, S. Hachiso, J. Coll. Interface Sci. **44**, 330 (1973)
299. R. Williams, R.S. Crandall, Phys. Lett. A **48**, 225 (1974)
300. H. Ikezi, Phys. Fluids **29**, 1764 (1986)
301. A. Melzer, V. Schweigert, I. Schweigert, A. Homann, S. Peters, A. Piel, Phys. Rev. E **54**, R46 (1996)
302. K. Takahashi, T. Oishi, K. Shimomai, Y. Hayashi, S. Nishino, Phys. Rev. E **58**, 7805 (1998)
303. A.V. Zobnin, A.P. Nefedov, V.A. Sinel'shchikov, O.A. Sinkevich, A.D. Usachev, V.S. Filippov, V.E. Fortov, Plasma Phys. Rep. **26**, 415 (2000)

304. M. Zuzic, A.V. Ivlev, J. Goree, G.E. Morfill, H.M. Thomas, H. Rothermel, U. Konopka, R. Sütterlin, D.D. Goldbeck, Phys. Rev. Lett. **85**, 4064 (2000)
305. M. Nambu, S.V. Vladimirov, P.K. Shukla, Phys. Lett. A **203**, 40 (1995)
306. F. Melandsø, J. Goree, Phys. Rev. E **52**, 5312 (1995)
307. V.A. Schweigert, I.V. Schweigert, A. Melzer, A. Homann, A. Piel, Phys. Rev. E **54**, 4155 (1996)
308. O. Ishihara, S.V. Vladimirov, Phys. Rev. E **57**, 3392 (1998)
309. M. Lampe, G. Joyce, G. Ganguli, IEEE Trans. Plasma Sci. **33**, 57 (2005)
310. F. Melandsø, Phys. Rev. E **55**, 7495 (1997)
311. G. Lapenta, Phys. Plasmas **6**, 1442 (1999)
312. D. Winske, W. Daughton, D.S. Lemons, M.S. Murillo, Phys. Plasmas **7**, 2320 (2000)
313. G. Lapenta, Phys. Rev. E **66**, 026409 (2002)
314. A. Melzer, V. Schweigert, A. Piel, Phys. Rev. Lett. **83**, 3194 (1999)
315. A. Melzer, V.A. Schweigert, A. Piel, Physica Scripta **61**, 494 (2000)
316. G.A. Hebner, M.E. Riley, B.M. Marder, Phys. Rev. E **68**, 046401 (2003)
317. G.A. Hebner, M.E. Riley, Phys. Rev. E **69**, 026405 (2004)
318. H. Totsuji, C. Totsuji, T. Ogawa, K. Tsuruta, Phys. Rev. E **71**, 045401 (2005)
319. C. Henning, H. Baumgartner, A. Piel, P. Ludwig, V. Golubnichiy, M. Bonitz, D. Block, Phys. Rev. E **74**, 056403 (2006)
320. C. Henning, P. Ludwig, A. Filinov, A. Piel, M. Bonitz, Phys. Rev. E **76**, 036404 (2007)
321. H. Baumgartner, H. Kählert, V. Golobnychiy, C. Henning, S. Käding, A. Melzer, M. Bonitz, Contrib. Plasma Phys. **47**, 281 (2007)
322. D. Block, M. Kroll, O. Arp, A. Piel, S. Käding, A. Melzer, C. Henning, H. Baumgartner, P. Ludwig, M. Bonitz, Plasma Phys. Control. Fusion **49**, B109 (2007)
323. A. Piel, O. Arp, M. Klindworth, A. Melzer, Phys. Rev. E **77**, 026407 (2008)
324. M. Bonitz, D. Block, O. Arp, V. Golubnychiy, H. Baumgartner, P. Ludwig, A. Piel, A. Filinov, Phys. Rev. Lett. **96**, 075001 (2006)
325. R.W. Hasse, V.V. Avilov, Phys. Rev. A **44**, 4506 (1991)
326. A. Homann, A. Melzer, S. Peters, R. Madani, A. Piel, Phys. Rev. E **56**, 7138 (1997)
327. F.M. Peeters, X. Wu, Phys. Rev. A **35**, 3109 (1987)
328. A. Homann, A. Melzer, S. Peters, R. Madani, A. Piel, Phys. Lett. A **242**, 173 (1998)
329. X. Wang, A. Bhattacharjee, S. Hu, Phys. Rev. Lett. **86**, 2569 (2001)
330. S. Nunomura, J. Goree, S. Hu, X. Wang, A. Bhattacharjee, Phys. Rev. E **65**, 066402 (2002)
331. S. Nunomura, D. Samsonov, J. Goree, Phys. Rev. Lett. **84**, 5141 (2000)
332. A. Piel, V. Nosenko, J. Goree, Phys. Rev. Lett. **89**, 085004 (2002)
333. D. Samsonov, J. Goree, Z. Ma, A. Bhattacharjee, H.M. Thomas, G.E. Morfill, Phys. Rev. Lett. **83**, 3649 (1999)
334. A. Melzer, S. Nunomura, D. Samsonov, J. Goree, Phys. Rev. E **62**, 4162 (2000)
335. V. Nosenko, J. Goree, Z.W. Ma, D.H.E. Dubin, A. Piel, Phys. Rev. E **68**, 056409 (2003)
336. D.H.E. Dubin, Phys. Plasmas **7**, 3895 (2000)
337. O. Havnes, T. Aslaksen, T.W. Hartquist, F. Li, F. Melandsø, G.E. Morfill, T. Nitter, J. Geophys. Res. **100**, 1731 (1995)
338. O. Havnes, F. Li, T.W. Hartquist, T. Aslaksen, A. Brattli, Planet. Space Sci. **49**, 223 (2001)
339. Y. Feng, J. Goree, B. Liu, Rev. Sci. Instrum. **78**, 053704 (2007)
340. S. Nunomura, J. Goree, S. Hu, X. Wang, A. Bhattacharjee, K. Avinash, Phys. Rev. Lett. **89**, 035001 (2002)
341. N.N. Rao, P.K. Shukla, M.Y. Yu, Planet. Space Sci. **38**, 543 (1990)
342. A. Barkan, R.L. Merlino, N. D'Angelo, Phys. Plasmas **2**, 3563 (1995)
343. C. Thompson, A. Barkan, N. D'Angelo, R.L. Merlino, Phys. Plasmas **4**, 2331 (1997)
344. T. Trottenberg, D. Block, A. Piel, Phys. Plasmas **13**, 042105 (2006)
345. S. Ratynskaia, S. Khrapak, A. Zobnin, M.H. Thoma, M. Kretschmer, A. Usachev, V. Yaroshenko, R.A. Quinn, G.E. Morfill, O. Petrov, V. Fortov, Phys. Rev. Lett. **93**, 085001 (2004)

346. A. Piel, M. Klindworth, O. Arp, A. Melzer, M. Wolter, Phys. Rev. Lett. **97**, 205009 (2006)
347. M. Schwabe, M. Rubin-Zuzic, S. Zhdanov, H.M. Thomas, G.E. Morfill, Phys. Rev. Lett. **99**, 095002 (2007)
348. E. Thomas, Jr., Phys. Plasmas **13**, 042107 (2006)
349. I. Pilch, A. Piel, T. Trottenberg, M.E. Koepke, Phys. Plasmas **14**, 123704 (2007)
350. E. Thomas, Jr., R. Fisher, R.L. Merlino, Phys. Plasmas **14**, 123701 (2007)
351. A. von Engel, *Ionized Gases* (Clarendon, Oxford, 1965)
352. W.B. Nottingham, in *Handbuch der Physik*, vol. XXI, ed. by S. Flügge (Springer, Berlin, 1956)
353. J.W. McGowan, P.K. John, *Gaseous Electronics* (North-Holland, Amsterdam, 1974), pp. 9–16
354. L. Malter, E.O. Johnson, W.M. Webster, RCA Rev. **XII**, 415 (1951)
355. R. Timm, A. Piel, Contrib. Plasma Phys. **32**, 599 (1992)
356. V.A. Godyak, N. Sternberg, Phys. Rev. A **42**, 2299 (1990)
357. P. Belenguer, J.P. Boeuf, Phys. Rev. A **41**, 4447 (1990)
358. M.A. Liebermann, V.A. Godyak, IEEE Trans. Plasma Sci. **26**, 955 (1998)
359. V.A. Godyak, A.S. Khanneh, IEEE Trans. Plasma Sci. **41**, 112 (1986)
360. R. Flohr, A. Piel, Contrib. Plasma Phys. **33**, 153 (1993)
361. V.A. Godyak, R.B. Piejak, B.M. Alexandrovich, Plasma Sources Sci. Technol. **1**, 36 (1992)
362. A. Melzer, R. Flohr, A. Piel, Plasma Sources Sci. Technol. **4**, 424 (1995)
363. K. Köhler, J.W. Coburn, D.E. Horne, E. Kay, J. Appl. Phys. **57**, 59 (1985)
364. J.W. Coburn, E. Kay, J. Appl. Phys. **43**, 4965 (1972)
365. J.W. Coburn, H.F. Winters, J. Appl. Phys. **50**, 3189 (1979)
366. J. Hopwood, Plasma Sources Sci. Technol. **1**, 109 (1992)
367. J.H. Keller, J.C. Forster, M.S. Barnes, J. Vac. Sci. Technol. A **11**, 2487 (1993)
368. J.H. Keller, Plasma Sources Sci. Technol. **5**, 166 (1996)
369. V.I. Kolobov, D.J. Economou, Plasma Sources Sci. Technol. **6**, R1 (1997)
370. M.A. Liebermann, A.J. Lichtenberg, *Principles of Plasma Discharges and Material Processing* (Wiley, New York, 1994)
371. B. Chapman, *Glow Discharge Processes* (Wiley, New York, 1980)
372. J.R. Roth, *Industrial Plasma Engineering, Vol. I: Principles* (IOP, Bristol, 1995)
373. R. Hippler, H. Kersten, M. Schmidt, K.H. Schoenbach, *Low Temperature Plasmas*, 2nd edn. (Wiley-VCH, Weinheim, 2008)

Name Index

Subject Index